Lecture Notes in Computer Science

Edited by G. Goos and J. Hartmanis

171

Logic and Machines: Decision Problems and Complexity

Proceedings of the Symposium "Rekursive Kombinatorik" held from May 23 – 28, 1983 at the Institut für Mathematische Logik und Grundlagenforschung der Universität Münster/Westfalen

Edited by E. Börger, G. Hasenjaeger and D. Rödding

Springer-Verlag
Berlin Heidelberg New York Tokyo 1984

Editors

E. Börger
Lehrstuhl Informatik II, Universität Dortmund
Postfach 500 500, D-4600 Dortmund 50

G. Hasenjaeger
Seminar für Logik und Grundlagenforschung, Universität Bonn
Beringstr. 6, D-5300 Bonn 1

D. Rödding
Institut für math. Logik und Grundlagenforschung
Universität Münster
Einsteinstr. 64, D-4400 Münster

ISBN 3-540-13331-3 Springer-Verlag Berlin Heidelberg New York Tokyo
ISBN 0-387-13331-3 Springer-Verlag New York Heidelberg Berlin Tokyo

Library of Congress Cataloging in Publication Data. Symposium „Rekursive Kombinatorik" (1983: Institut für Mathematische Logik und Grundlagenforschung der Universität Münster/Westfalen) Logic and machines. (Lecture notes in computer sciences; 171). 1. Machine theory–Congresses. 2. Computational complexity–Congresses. 3. Decision-making–Congresses. I. Börger, E. (Egon), 1946-. II. Hasenjaeger, G. (Gisbert), 1919-. III. Rödding, D. (Dieter), 1937-. IV. Title. V. Series. QA267.S98 1983 511 84-5546
ISBN 0-387-13331-3 (U.S.)

Printed in Germany

Printing and binding: Beltz Offsetdruck, Hemsbach/Bergstr.
2145/3140-543210

PREFACE

This volume collects written versions of talks presented to the Symposium on "REKURSIVE KOMBINATORIK" which has been held from May 23 to May 28 at the Institut für mathematische Logik und Grundlagenforschung of the University of Münster in Westfalen.

The papers presented by E.Engeler and C.Lautemann have been published in D.Kozen & E.L.Clark: Logic of Programs (Springer Lecture Notes in Computer Science) resp. in Information Processing Letters, and D. Christodoulakis in "Bericht Nr. 127" of the GMD (Gesellschaft für Mathematik und Datenverarbeitung) R. Oldenbourg Verlag, München, Wien 1980.
The editors regret that thwy have received no contribution to these Proceedings from M. Karpinski and A. Selman.
The two papers by D.Schmidt and V.Sperschneider have been presented by title.

The editors of this volume wish to express their thanks to the Stiftung Volkswagenwerk for financial support of the Symposium, and to Springer Verlag for help in publishing these Proceedings.

E.Börger Dortmund

G.Hasenjaeger Bonn

D.Rödding Münster

List of Participants

S.O.Aanderaa, K.Ambos-Spies, A.Blass, E.Börger, H.J.Brämik, A.Brüggemann, B.Buchberger, G.Buntrock, H.-G.Carstens, D.Christodoulakis, V.Claus, D.Cohen, E.Dahlhaus, M.Deutsch, E.Engeler, M.Fürer, E.Grandjean, J.Grollmann, G.Hasenjaeger, U.Hedtstück, G.Henzel, C.Hesse, M.Karpinski, R.Klein, H.Kleine Büning, U.Lattwin, C.Lautemann, J.van Leeuwen, Th.Lettmann, K.Madlener, W.Oberschelp, P.Päppinghaus, L.Priese, P.Pudlák, K.W.Regan, M.M.Richter, D.Rödding, W.Rödding, B.Scarpellini, A.L.Selman, E.Specker, V.Sperschneider, D.Spreen, R.Schätz, B.Schinzel, J.Schulte Mönting, H.Schwichtenberg, J.Tiuryn, H.Volger, I.Wegener, W.Zadrozny.

List of speakers and talks

S.O.Aanderaa: On the solvability of the $\forall\exists \wedge \exists\forall^*$- Ackermann class with identity

K.Ambos-Spies: Splitting recursive sets and their polynomial degrees

A.Blass: Computational complexity and equivalence relations

E.Börger: Spektralproblem and Completeness of Logical Decision Problems

B.Buchberger: The solution of decision problems in polynomial ideal theory based on a critical-pair/completion algorithm

H.G.Carstens/P.Päppinghaus: Extensible algorithms

D.Christodoulakis: Yet another very "small" universal machine

D.E.Cohen: Modular machines as a basis for undecidability and incompleteness

E.Dahlhaus: Reductions to NP-complete problems by interpretations

M.Deutsch: New results in the reduction theory of the Entscheidungsproblem

E.Engeler, Equations in combinatorial algebras

M.Fürer: The computational complexity of the unconstrained limited domino problem (with implications for logical decision problems)

E.Grandjean: Universal quantifiers and time complexity of random access machines

G.Hasenjaeger: Universal Turing machine (UTM) and Jones-Matijasevich masking

U.Hedtstück: Über die Argumentkomplexität mehrfach transitiver boolescher Funktionnen

M.Karpinski: Satisfiability problem in equational theory of the flowchart

H.Kleine Büning: Small universal Turing machines

C.Lautemann: Two theorems on bounded error probability

J.van Leeuwen: Logic circuits, Boolean functions and VLSI

V.Sperschneider: Recursively enumerable extensions of R_1 by finite functions

W.Oberschelp: Fast parallel algorithms for finding all prime implicants for discrete functions

L.Priese: APA-nets

P.Pudlak: New techniques for proving lower bounds of formula size of Boolean functions

Regan,K.W.: Arithmetical degrees of complexity classes

M.M.Richter: On some efficient proof procedures and the permutation of rules

D.Rödding: Modular decomposition of automata

B.Scarpellini: Complete second order spectra

A.Selman: Scholz's Spectrum Problem: after thirty years

E.Specker: What are we looking for in combinatorics?

D.Spreen: On r.e. inseparability of CPO index sets

J.Tiuryn: Implicit definability of terms by (in) finite sets of equations

H.Volger: Rudimentary relations and Turing machines with linear alternation

I.Wegener: Proving lower bounds on the monotone complexity of Boolean functions

TABLE OF CONTENTS

PREFACE

Section I: COMPLEXITY

K. Ambos-Spies
P-mitotic sets 1

A. Blass and Yu. Gurevich
Equivalence relations, invariants, and normal forms 24

Chr. Blatter and E. Specker
Recurrence relations for the number of labeled structures on a finite set 43

W. Menzel and V. Sperschneider
Recursively enumerable extensions of R_1 by finite functions 62

D. Schmidt
On the complement of one complexity class in another 77

V. Sperschneider
The length-problem 88

D. Spreen
On r.e. inseparability of CPO index sets 103

K.W. Regan
Arithmetical degrees of index sets for complexity classes 118

H. Volger
Rudimentary relations and Turing machines with linear alternation 131

Section II: ALGORITHMS

B. Buchberger
A critical-pair/completion algorithm for finitely generated ideals in rings 137

H.G. Carstens and P. Päppinghaus
Extensible algorithms 162

M.M. Richter
Some reordering properties for inequality proof trees 183

Section III: AUTOMATA AND MACHINES

A. Brüggemann, L. Priese, D. Rödding, and R. Schätz
Modular decomposition of automata 198

D.E. Cohen
Modular machines, undecidability and incompleteness 237

G. Hasenjaeger
Universal Turing machines (UTM) and Jones-Matijasevich-masking 248

H. Kleine Büning
Complexity of loop-problems in normed networks 254

Section IV: DECISION PROBLEMS

S.O. Aanderaa
On the solvability of the extended $\forall\exists\wedge\exists\forall^*$-Ackermann class with identity 270

M. Deutsch
Reductions for the satisfiability with a simple interpretation of the predicate variable 285

M. Fürer
The computational complexity of the unconstrained limited domino problem (with implications for logical decision problems) 312

J. Tiuryn
Implicit definability of finite binary trees by sets of equations 320

Section V: SPEKTRALPROBLEM

E. Börger
Spektralproblem and completeness of logical decision problems 333

E. Dahlhaus
Reduction to NP-complete problems by interpretations 357

E. Grandjean
Universal quantifiers and time complexity of random access machines 366

B. Scarpellini
Second order spectra 380

Section VI: COMPLEXITY OF BOOLEAN FUNCTIONS

U. Hedtstück
On the argument complexity of multiply transitive Boolean functions 390

M.R. Kramer and J. van Leeuwen
The VLSI complexity of Boolean functions 397

W. Oberschelp
Fast parallel algorithms for finding all prime implicants for discrete functions 408

P. Pudlák
Bounds for Hodes-Specker theorem 421

I. Wegener
Proving lower bounds of the monotone complexity of Boolean functions 446

P-MITOTIC SETS.

Klaus Ambos-Spies
Lehrstuhl für Informatik II
Universität Dortmund
D-4600 Dortmund 50

There is an analogy between certain areas of recursion theory on one side and complexity theory on the other side. In recursion theory one is interested in a classification of the nonrecursive sets - and in particular the recursively enumerable (r.e.) sets - according to the difficulty of their decision problems. The tool for this classification is Turing reducibility and refinements of it like many-one reducibility. In the subrecursive analogue the role of the computable (i.e. recursive) sets, the nonrecursive sets and the r.e. sets is played by the feasibly computable, i.e. the polynomial time computable (*P*-) sets, the computable but not feasibly computable sets and the *NP*-sets, respectively. The recursive reducibilities are replaced by their polynomial time bounded counterparts. It is not known how deep this analogy is. E.g. while an almost trivial argument shows that the recursive sets are properly contained in the class of r.e. sets, the subrecursive analogue of this question, the *P*=*NP*-problem, is still open. There is some hope, however, that this analogy can be utilized to gain more insight in the structure of the *NP*-sets and possibly to solve the *P*=*NP* problem. In recursion theory some notions played a particular role for the study of the structure of the r.e. sets. One might hope that studying the polynomial analogues of these notions will shed some light on the structure of the class *NP*. This approach was pursued by Homer and Maass [HM] who studied the P-analogues of immune and simple sets. Here we study the counterpart of another recursion theoretic notion, namely that of a *mitotic* set. In recursion theory r.e. mitotic sets were introduced by Lachlan [Lac] and they were exhaustively studied by Ladner [Lad1,Lad2].

A (recursive) set A is *polynomially* (*P*-) *mitotic* if there is a splitting of A into two parts A_0 and A_1 such that A_0 and A_1 both are of the same complexity as A, i.e. A is P-reducible to both A_0 and A_1 and vice versa. Roughly speaking, the encoding of the information in a P-mitotic set is redundant, since the full information can be recovered from two disjoint parts of the set. Since *P*-sets are P-mitotic, the existence of *NP*-sets which are not P-mitotic would imply $P \neq NP$.

In section 1 we formalize the notion of P-mitoticity. Since this notion depends on the chosen P-reducibility and splitting notions, different formalizations are possible. We give four definitions which seem to us to be the most natural ones. Furthermore we state some simple facts on P-mitoticity and its formalizations.

In section 2 we show that non-P-mitotic recursive sets exist and we prove some results on the P-degrees of such sets. Moreover, the answer to the question of whether or not non-P-mitotic *NP*-sets exist is shown to be oracle dependent.

In sections 3 to 5 we study some further properties of P-mitotic sets. In section 3 we discuss the question whether the class of P-mitotic sets is closed under complementation and show that the answer depends on the chosen formalization. In section 4 we generalize the concept of P-mitoticity by introducing multiply P-mitotic sets. A set A is *n-P-mitotic* ($n \geq 2$) if A can be split into n parts, each in the P-degree of A. Intuitively, for increasing n, the encoding of the information in an n-P-mitotic set becomes more and more redundant. This will be made precise by exhibiting a proper hierarchy of multiply P-mitotic sets. Finally, in section 5 we reveal some relations between P-mitotic sets and sets which can be nontrivially P-reduced to themselves.

0. PRELIMINARIES.

The set of nonnegative integers is denoted by $\mathbb{N}$. Lower case letters i,j,k,l, m,n stand for elements of $\mathbb{N}$. For $n \geq 2$, let $< ,\ldots, > : \mathbb{N}^n \to \mathbb{N}$ be a recursive bijection. $\Sigma = \{0,1\}$. x,y,z denote elements of Σ^*, capital letters A,B,C,... denote *recursive* subsets of Σ^*. $|x|$ is the length of x. $<$ denotes the natural ordering of Σ^*, i.e. $x < y$ iff $|x| < |y|$ or $|x| = |y|$ and $\exists\, z,x',y'$ ($x = z0x'$ & $y = z1y'$). We identify a set and its characteristic function, i.e. $x \in A$ iff $A(x) = 1$ and $x \notin A$ iff $A(x) = 0$. $\bar{A}$ is the complement of A, $A \upharpoonright n = \{ x : x \in A \ \& \ |x| < n \}$is is the restriction of A to arguments of length less than n. We write $A = A_0 \sqcup A_1$ if A_0 and A_1 *split* A, i.e. if $A = A_0 \cup A_1$ and $A_0 \cap A_1 = \emptyset$. For $A,B \subseteq \Sigma^*$ we set $AB = \{ xy : x \in A \ \& \ y \in B \}$ and we abbreviate $\{x\}A$ by xA. For $n \in \mathbb{N}$, $\underline{n} = 1^n0$ and $\underline{\mathbb{N}} = \{ \underline{n} : n \in \mathbb{N} \}$. The *P-effective disjoint union* $A \oplus B$ of A and B is defined by $A \oplus B = 0A \cup 1B$. A set A is *tally* if $A \subseteq \{0\}^*$.

P (*NP*) is the class of subsets of Σ^* which are (non)deterministically computable in polynomial time. *PF* is the class of deterministically polynomial time computable functions from Σ^* to Σ^*.$\{ P_n : n \in \mathbb{N} \}$ and $\{ f_n : n \in \mathbb{N} \}$ are effective enumerations of *P* respective *PF*. A bijection $f : \Sigma^* \to \Sigma^*$ is a *P-isomorphism* if f, f^{-1} are elements of *PF*. A set A is a (P-)*cylinder* if there is a set B and a P-isomorphism f such that $A = f(\underline{\mathbb{N}}B)$.$\{ M_n(X) : n \in \mathbb{N} \}$ and $\{ p_n : n \in \mathbb{N} \}$ are a standard enumeration of the deterministic polynomial time oracle machines (with oracle set X) and their respective polynomial bounds. We write $M_n(X)(x) = 1$ (0) iff $M_n(X)$ accepts (refutes) x.

A is *polynomial time many-one* (*P-m*) *reducible* to B, $A \leq^P_m B$, if, for some n, $\forall\, x \in \Sigma^*$ ($A(x) = B(f_n(x))$). A is *polynomial time Turing* (*P-T*) *reducible* to B, $A \leq^P_T B$, if $A = M_n(B)$ for some n. We write $A =^P_{m(T)} B$ iff $A \leq^P_{m(T)} B$ and $B \leq^P_{m(T)} A$. The *P-m(T) degree* of A, $\deg^P_{m(T)} A$, is the set $\{ B : B =^P_{m(T)} A \}$. Whenever we consider P-m-reducibility we systematically ignore the sets $\emptyset$ and Σ^*. Hence *P* consists of a single P-m degree.

P^A is the set of deterministic polynomial time sets relative to A, i.e. $P^A = \{ B : \exists n (B = M_n(A)) \}$. We say B is *relative to A polynomial time Turing* (P^A-T) *reducible* to C, $B \leq_T^{PA} C$, if $B = M_n(A \oplus C)$ for some n.

We conclude this section with some general remarks on the proofs in this paper. The theorems we will prove require us to show that recursive sets A with certain properties $\mathbb{P}$ exist. We then effectively construct such a set A in stages. In general we let $A^{-1} = \emptyset$ and A^s will denote the set of strings enumerated in A by the end of stage s of the construction, $s \in \mathbb{N}$. $A = \cup \{ A^s : s \in \mathbb{N} \}$. To ensure that A is recursive, simultaneously with A we define a *length function* l, $l : \mathbb{N} \cup \{-1\} \to \mathbb{N}$. $l(-1) = 0$ and, for $s \in \mathbb{N}$, $l(s)$ will be defined at stage s. By effectiveness of the construction, l will be a recursive function. Moreover, we will ensure

$$\forall s \in \mathbb{N} \; (l(s) < l(s+1) \;\&\; A^s \subseteq \Sigma^* \upharpoonright l(s) \;\&\; A^s = A \upharpoonright l(s)). \qquad (0.1)$$

Obviously, (0.1) and the effectiveness of the construction imply that A is recursive.

To ensure that A has property $\mathbb{P}$ we will usually break down this condition (or parts of it) into an infinite list of simpler requirements R_n, $n \in \mathbb{N}$. At stage s of the construction we then will guarantee that R_s is met. For this sake it will be crucial that, by (0.1), a computation $M_n(A^s)(x)$ which involves only strings of length less than $l(s)$ (i.e. $p_n(|x|) < l(s)$) is A-correct, i.e. $M_n(A)(x) = M_n(A^s)(x)$. So we can preserve a computation $M_n(A^s)(x)$ by setting $l(s) > p_n(|x|)$.

1. P-MITOTIC SETS.

A recursive set A is *polynomial time* (P-) *mitotic* if it can be split into two sets A_0 and A_1 which both have the same polynomial time degree as A. This informal definition depends on the notions of P-reducibility and splitting.

There is a great variety of P-reducibilities which differ on the recursive sets (cf. [LLS]). Here we restrict ourselves to two P-reducibility notions, namely P-many - one and P-Turing reducibilities, which seem to be the most important P-reducibility notions. While P-Turing reducibility is the most adequate formalization of the intuitive notion of polynomial time bounded reducibility, the more restrictive notion of P-m reducibility is interesting for its conceptual simplicity and the fact that membership in certain complexity classes like *NP* is inherited downward in $\leq_m^P$.

The most general and natural splitting notion is the set theoretical one. A shortcoming of this concept is the fact that the parts of a splitting can be more complex than the whole set - even if we require the parts to be recursive:

1.1. PROPOSITION. Let A be any infinite recursive set. Then there are recursive sets A_o and A_1 such that $A = A_o \sqcup A_1$, $A_o \not\leq^P_T A$ and $A_1 \not\leq^P_T A$.

PROOF. Let $\{x_n : n \in \mathbb{N}\}$ be an enumeration of A in natural ordering. Define A_o and A_1 by

$$A_o(x_{2n}) = 1 - M_n(A)(x_{2n}) \qquad A_o(x_{2n+1}) = M_n(A)(x_{2n+1}) \qquad A_1 = A - A_o.$$

Then A_o and A_1 split A but neither A_o nor A_1 is P-T-reducible to A. □

It is well known that we can avoid this pathology by restricting our splitting notion. We call sets A_o and A_1 a *P-splitting* of A if $A_o = A \cap B$ and $A_1 = A \cap \bar{B}$ for some set $B \in P$. P-splittings have the following wellknown property.

1.2. PROPOSITION. Let A_o, A_1 be a P-splitting of A. Then $A_o \leq^P_m A$ and $A_1 \leq^P_m A$; in fact $A =^P_m A_o \oplus A_1$.

PROOF. Trivial. □

So a P-splitting of a set induces a splitting of the information encoded in it, i.e. a splitting of its degree: If A_o, A_1 P-split A then $\deg^P_m A_o \cup \deg^P_m A_1 = \deg^P_m A$. This property makes P-splittings a very useful tool for investigating P-reducibility and P-degrees (see e.g. [Lad3] and [LLR]).

Taking P-m- and P-T-reducibilities and general splittings as well as P-splittings, we obtain four different formalizations of the notion of a P-mitotic set.

1.3. DEFINITION. (a) A recursive set A is *P-m(T)-mitotic* if A is finite or co-finite or if there is a set $B \in P$ such that

$$A =^P_{m(T)} A \cap B =^P_{m(T)} A \cap \bar{B}. \tag{1.1}$$

Otherwise A is *non-P-m(T)-mitotic*.

(b) A recursive set A is *weakly P-m(T)-mitotic* if A is finite or co-finite or if there are sets A_o and A_1 such that $A = A_o \sqcup A_1$ and

$$A =^P_{m(T)} A_o =^P_{m(T)} A_1.$$

Otherwise A is *strongly non-P-m(T)-mitotic*.

Our convention that finite and co-finite sets are P-mitotic is for convenience. (For P-T-reducibility the convention is superfluous; for P-m-reducibility, however, without this convention Σ^* and one-element sets are strongly non-P-m-mitotic.) This convention has no impact on our investigations since we are mainly interested in recursive non-P sets which automatically are infinite and co-infinite.

Note that, by Proposition 1.2, (1.1) can be replaced by the weaker condition $A \leq^P_{m(T)} A \cap B$ and $A \leq^P_{m(T)} A \cap \bar{B}$.

Results in the following sections will imply that the just introduced P-mitoticity notions are pairwise different on the recursive sets. P-mitoticity differs from weak P-mitoticity since the class of P-m(T)-mitotic sets - but not that of weakly P-m(T)-mitotic sets - is closed under complementation (section 3). Moreover, the complement of a tally language is weakly P-T-mitotic but not necessarily weakly P-m-mitotic (section 2). Finally, that P-T-mitoticity does not imply P-m-mitoticity follows from a stronger result on multiply P-mitotic sets in section 4.

Trivially the following relations hold among our P-mitoticity notions.

1.4. PROPOSITION. (a) If A is (weakly) P-m-mitotic then A is (weakly) P-T-mitotic.

(b) If A is P-m(T)-mitotic then A is weakly P-m(T)-mitotic. □

1.5. PROPOSITION. Each of the formal P-mitoticity notions is polynomially invariant. I.e. if A is P-mitotic and f is a P-isomorphism then $f(A)$ is P-mitotic (in the respective sense). □

P-mitotic sets are abundant. We don't know any example of a "natural" non-P-mitotic set. E.g. all natural NP-complete sets which are known are cylinders ([BH]) and thus P-mitotic (see below). We conclude this section with some simple facts on P-mitotic sets.

1.6. PROPOSITION. (a) If A is a cylinder then A is P-m-mitotic.

(b) For any set A, $A \oplus A$ is P-m-mitotic.

(c) Every P-m-degree contains a P-m-mitotic set.

(d) If $A \in P$ then A is P-m-mitotic.

(e) If A is tally then there is a tally set B such that B is P-m-mitotic. and $B =_m^P A$.

(f) If A is P-immune then there is a P-immune set B such that B is P-m-mitotic and $B =_m^P A$.

(g) For any recursive A and $B \in P$, if $A \cap B$ and $A \cap \bar{B}$ are [weakly] P-m(T)-mitotic then A is [weakly] P-m(T) mitotic.

PROOF. (a),(b) and (d) are immediate. (c) follows from (b). For a tally set A, $B = \{0^{2n+i} : i \leq 1 \ \& \ 0^n \in A\}$ is P-m-mitotic and $A =_m^P B$. This proves (e). In (f) let $B = A \oplus A$. For a proof of (g) fix A recursive and $B \in P$ and assume that A_0, A_1 and A_2, A_3 are splittings which witnessthe [weak] P-m(T)-mitoticity of $A \cap B$ and $A \cap \bar{B}$, respectively. Then the splitting $A_0 \cup A_2$, $A_1 \cup A_3$ is a witness for the [weak] P-m(T)-mitoticity of A. □

2. NON-P-MITOTIC SETS.

In this section we prove some results on non-P-mitotic sets. We start with a theorem which establishes the existence of non-P-mitotic sets in the sense of three of our formalizations.

2.1. THEOREM. Let B and C be disjoint recursive sets such that C is infinite. Then there is a recursive subset A of C such that

(i) $A \cup B$ is non-P-T-mitotic and

(ii) $A \cup B$ is strongly non-P-m-mitotic.

Note that Proposition 1.4 and (i) imply that $A \cup B$ is non-P-m-mitotic.

PROOF. Fix B and C as in the premisse of the theorem and let $\{x_n : n \in \mathbb{N}\}$ be a recursive subset of C such that $|x_n| < |x_{n+1}|$ for every $n \in \mathbb{N}$. We construct a recursive set $A \subseteq \{x_n : n \in \mathbb{N}\}$ such that (i) and (ii) hold. The set A will meet the requirements

$$R_{2<i,j,k>} : A \cup B \neq M_j((A \cup B) \cap P_i) \quad \text{or} \quad A \cup B \neq M_k((A \cup B) \cap \bar{P}_i)$$

and

$$R_{2<m_o,m_1,n_o,n_1>+1} : \text{If } A \cup B = f_{n_o}^{-1}(A \cup B) \sqcup f_{n_1}^{-1}(A \cup B) \text{ then}$$
$$\text{non } A \cup B \leq_m^P f_{n_o}^{-1}(A \cup B) \text{ via } f_{m_o} \text{ or}$$
$$\text{non } A \cup B \leq_m^P f_{n_1}^{-1}(A \cup B) \text{ via } f_{m_1}.$$

for every $i,j,k,m_o,m_1,n_o,n_1 \in \mathbb{N}$.

Obviously, the even index requirements imply (i) while the odd index requirements imply (ii).

We now describe stage s of the construction of A. We ensure at this stage that requirement R_s is met.

<u>Stage s</u>. Let $x = x_n$, where n is chosen minimal such that $|x_n| > l(s-1)$, and distinguish the following cases.

<u>Case 1: s is even</u>, say $s = 2<i,j,k>$. Then let

$$l(s) = \max \{ |x|, p_j(|x|), p_k(|x|) \} + 1 \quad \text{and}$$

$$A^s = \begin{cases} A^{s-1} \cup \{x\} & \text{if } (x \in P_i \ \& \ M_k((A^{s-1} \cup B) \cap \bar{P}_i)(x) = 0) \text{ or} \\ & \quad\ (x \notin P_i \ \& \ M_j((A^{s-1} \cup B) \cap P_i)(x) = 0) \\ A^{s-1} & \text{otherwise.} \end{cases}$$

<u>Case 2: s is odd</u>, say $s = 2<m_o,m_1,n_o,n_1>+1$. Then let

$$l(s) = \max \{ |x|, |f_{m_i}(x)|, |f_{n_i}(f_{m_j}(x))| : i,j \leq 1 \} + 1 \quad \text{and}$$

$$A^s = \begin{cases} A^{s-1} & \text{if } \exists\, i \leq 1(\, f_{n_i}(f_{m_i}(x)) \in A^{s-1} \cup B\,) \text{ or} \\ & \exists\, i \leq 1(\, f_{m_i}(x) \in A^{s-1} \cup B \,\&\, f_{n_{1-i}}(f_{m_i}(x)) \notin A^{s-1} \cup B) \\ A^{s-1} \cup \{x\} & \text{otherwise.} \end{cases}$$

This completes the construction of A. We now show that A has the desired properties. Obviously, $A \subseteq \{x_n : n \in \mathbb{N}\} \subseteq C$ and (0.1) is satisfied, whence A is recursive. It remains to show that, for every s, R_s is met.

First fix s even, say $s = 2\langle i,j,k\rangle$. Then for x as above,
$A^s(x) \neq M_k((A^s \cup B) \cap \bar{P}_i)(x)$ or $A^s(x) \neq M_j((A^s \cup B) \cap P_i)(x)$.
Moreover, by definition of $l(s)$ (and by (0.1)), this disagreement is preserved, whence R_s is met.

Now let s be odd, say $s = 2\langle m_0,m_1,n_0,n_1\rangle+1$. Then to show that R_s is met it suffices to show

$$\exists\, x \in \Sigma^* \,\exists\, i \leq 1\, (\quad \begin{array}{ll} (2.1.1) & A \cup B(x) \neq A \cup B(f_{n_i}(f_{m_i}(x))) \text{ or} \\ (2.1.2) & A \cup B(x) = A \cup B(f_{n_{1-i}}(f_{m_i}(x))) = 1 \text{ or} \\ (2.1.3) & A \cup B(x) > A \cup B(f_{m_i}(x)) \text{ or} \\ (2.1.4) & A \cup B(f_{m_i}(x)) > \max\{\, A \cup B(f_{n_j}(f_{m_i}(x))) : j \leq 1 \,\}). \end{array} \tag{2.1}$$

Namely (2.1.1) implies $A \cup B(x) \neq f_{n_i}^{-1}(A \cup B)(f_{m_i}(x))$, i.e. non $A \cup B \leq_m^P f_{n_i}^{-1}(A \cup B)$ via f_{m_i}. (2.1.2) implies not $A \cup B \leq_m^P f_{n_i}^{-1}(A \cup B)$ via f_{m_i} or $f_{n_0}^{-1}(A \cup B)$ and $f_{n_1}^{-1}(A \cup B)$ are not disjoint. Similarly, (2.1.3) implies not $A \cup B \leq_m^P f_{n_i}^{-1}(A \cup B)$ via f_{m_i} or $f_{n_i}^{-1}(A \cup B) \nsubseteq A \cup B$. Finally,(2.1.4) implies $A \cup B \nsubseteq f_{n_0}^{-1}(A \cup B) \cup f_{n_1}^{-1}(A \cup B)$. So in any case R_s is met.

Now fix x as defined at stage s. Note that, by definition of l and (0.1), $x \notin A^{s-1} \cup B$ and, for any $z \in \{x, f_{m_i}(x), f_{n_i}(f_{m_j}(x)) : i,j \leq 1\}$, $A^s(z) = A(z)$. To verify (2.1) we distinguish the following cases:

Case1: $\exists\, i \leq 1\, (\, f_{n_i}(f_{m_i}(x)) \in A^{s-1} \cup B\,)$. Then $A^s = A^{s-1}$, i.e. $A \cup B(x) = (A^s \cup B)(x) = (A^{s-1} \cup B)(x) = 0$ whereas, for some $i \leq 1$, $A \cup B(f_{n_i}(f_{m_i}(x))) = 1$. So (2.1.1) holds.

Case 2: Case 1 doesn't hold and, for some $i \leq 1$, $f_{m_i}(x) \in A^{s-1} \cup B$ and $f_{n_{1-i}}(f_{m_i}(x)) \notin A^{s-1} \cup B$. Then $A^s = A^{s-1}$. So $f_{m_i}(x) \in A \cup B$ whereas $f_{n_{1-i}}(f_{m_i}(x)) \notin A \cup B$. Moreover, since case 1 does not apply, $f_{n_i}(f_{m_i}(x)) \notin A \cup B$. It follows that (2.1.4) holds.

Case 3: Cases 1 and 2 do not hold and, for some $i \leq 1$, $f_{n_i}(f_{m_i}(x)) \neq x$. Then $A^s = A^{s-1} \cup \{x\}$ and thus $A(x) = 1$ whereas, for some $i \leq 1$,

$A \cup B(f_{n_i}(f_{m_i}(x))) = 0$; i.e. (2.1.1) holds.

Case 4: Cases 1 and 2 do not hold and $f_{n_0}(f_{m_0}(x)) = f_{n_1}(f_{m_1}(x)) = x$. Then $A^s = A^{s-1} \cup \{x\}$. So

$$A \cup B(x) = A \cup B(f_{n_0}(f_{m_0}(x))) = A \cup B(f_{n_1}(f_{m_1}(x))) = 1. \tag{2.2}$$

Now, if $f_{m_0}(x) = f_{m_1}(x)$, then (2.2) implies (2.1.2). So w.l.o.g. $f_{m_0}(x) \neq f_{m_1}(x)$, i.e. we can fix $i \leq 1$ such that $f_{m_i}(x) \neq x$. If $f_{m_i}(x) \notin A \cup B$ then, by (2.2), (2.1.3) holds. So we may assume that $f_{m_i}(x) \in A \cup B$, i.e. $f_{m_i}(x) \in A^s \cup B$ and thus, since $f_{m_i}(x) \neq x$ and $A^s = A^{s-1} \cup \{x\}$, $f_{m_i}(x) \in A^{s-1} \cup B$. Since cases 1 and 2 do not apply, it follows that $f_{n_{1-i}}(f_{m_i}(x)) \in A^{s-1} \cup B$ whence, by (2.2), (2.1.2) holds.

So in any case condition (2.1) is satisfied. This completes the proof. □

By making appropriate choices for B and C in Theorem 2.1 we obtain non-P-T-mitotic and strongly non-P-m-mitotic sets with various properties.

2.2. COROLLARY. Let D be any recursive set. Then there are recursive non-P-T-mitotic and strongly non-P-m-mitotic sets E, F, G, H, I, K such that

(i) E is tally
(ii) $\bar{F}$ is tally
(iii) G is P-immune
(iv) H is not P-immune
(v) $D \leq_m^P I$
(vi) K is *NP*-hard.

PROOF. Apply Theorem 2.1 to the following sets B and C. (i): $B = \emptyset$, $C = \{0\}^*$; (ii): $B = \Sigma^* - \{0\}^*$, $C = \{0\}^*$; (iii) $B = \emptyset$, C any P-immune set; (iv): B,C as in (ii); (v): $B = 0D$, $C = 1\Sigma^*$; (vi): B,C as in (v) with D *NP*-complete. □

Theorem 2.1 fails for strong non-P-T-mitoticity. Namely, in contrast to Corollary 2.2(ii), the complement of any tally set is weakly P-T-mitotic.

2.3. PROPOSITION. Let A be a recursive set which contains a subset B which is P-isomorphic to $1\Sigma^*$. Then A is weakly P-T-mitotic.

PROOF. By Proposition 1.5, w.l.o.g. $1\Sigma^* \subseteq A$. So, for $A_0 = \{ 1x : 0x \in A \}$ and $A_1 = \{ 1x : 0x \notin A \} \cup \{ 0x : 0x \in A \}$, $A = A_0 \sqcup A_1$ and, obviously, $A =_T^P A_0 =_T^P A_1$. □

We now show that - despite this fact - strongly non-P-T-mitotic sets exist.

2.4. THEOREM. There is a recursive tally set A which is strongly non-P-T-mitotic.

PROOF. We construct a set A with the desired properties in stages. For $M_i(A)$ $(M_i(A^s))$ we write A_i (A_i^s). Then to make A strongly non-P-T-mitotic it suffices to meet requirement

$$R_{<i,j,k,l>} : \text{ If } A = A_i \sqcup A_j \text{ then } A \neq M_k(A_i) \text{ or } A \neq M_l(A_j).$$

for every $i,j,k,l \in \mathbb{N}$. The construction of A is as follows.

Stage s, say $s = <i,j,k,l>$. Let

$$m = \max\{\, l(s-1), p_i(l(s-1)), p_j(l(s-1)) \,\} +1,$$

$$n = \max\{\, m, p_k(m), p_l(m) \,\} +1,$$

$$y_k = M_k(A_i^{s-1})(0^m) \quad \text{and} \quad y_l = M_l(A_j^{s-1})(0^m).$$

Then we set

$$l(s) = \max\{\, n, p_i(n), p_j(n) \,\} +1.$$

For the definition of A^s we distinguish two cases.

Case 1: $A^{s-1} \upharpoonright n = A_i^{s-1} \upharpoonright n \sqcup A_j^{s-1} \upharpoonright n$. Then

$$A^s = \begin{cases} A^{s-1} & \text{if } y_k = 1 \text{ or } y_l = 1 \\ A^{s-1} \cup \{0^m\} & \text{otherwise.} \end{cases}$$

Case 2: Otherwise. Then let $A^s = A^{s-1}$.

To prove the construction correct, first by a straightforward induction on s we see that $A \subseteq \{0\}^*$ and (0.1) is satisfied. So, by effectiveness of the construction, the set A is recursive.

It only remains to show that all requirements R_s are met. Fix s, $s = <i,j,k,l>$, and define m,n,y_k,y_l as above. Note that, by definition of $l(s-1)$ and $l(s)$ (and by (0.1)), $A^{s-1}(0^m) = 0$ and

$$A^s \upharpoonright n = A \upharpoonright n, \quad A_i^s \upharpoonright n = A_i \upharpoonright n, \quad A_j^s \upharpoonright n = A_j \upharpoonright n. \tag{2.3}$$

We first assume that case 1 applies to stage s, i.e.

$$A^{s-1} \upharpoonright n = A_i^{s-1} \upharpoonright n \sqcup A_j^{s-1} \upharpoonright n. \tag{2.4}$$

Now, if $y_k = 1$ or $y_l = 1$, then $A^s = A^{s-1}$ and thus, by (2.3),

$$A(0^m) = A^{s-1}(0^m) = 0 \neq 1 = y_k = M_k(A_i^{s-1})(0^m) = M_k(A_i)(0^m)$$

or, similarly, $A(0^m) \neq M_l(A_j)(0^m)$, which implies that R_s is met. So w.l.o.g. we may assume $y_k = y_l = 0$, i.e.

$$M_k(A_i^{s-1})(0^m) = M_l(A_j^{s-1})(0^m) = 0 \tag{2.5}$$

and, by definition of A^s and by (2.3),

$$A \upharpoonright n - A^{s-1} \upharpoonright n = A^s \upharpoonright n - A^{s-1} \upharpoonright n = \{0^m\}. \tag{2.6}$$

Now if $A_i^{s-1} \upharpoonright n = A_i \upharpoonright n$ or $A_j^{s-1} \upharpoonright n = A_j \upharpoonright n$ then, by definition of n, at least one of the computations in (2.5) is preserved, i.e. $M_k(A_i)(0^m) = 0$ or $M_l(A_j)(0^m) = 0$. Since $A(0^m) = 1$, in this case R_s is met. So w.l.o.g. we may assume

$$A_i^{s-1} \upharpoonright n \neq A_i \upharpoonright n \quad \text{and} \quad A_j^{s-1} \upharpoonright n \neq A_j \upharpoonright n. \tag{2.7}$$

Note that, by definition of m and by (2.6),

$$A_i^{s-1} \upharpoonright l(s-1) = A_i \upharpoonright l(s-1) \quad \text{and} \quad A_j^{s-1} \upharpoonright l(s-1) = A_j \upharpoonright l(s-1). \tag{2.8}$$

It follows from (2.7) and (2.8) that there are strings z_i and z_j such that

$$l(s-1) \leq |z_i|,|z_j| < n, \quad A_i^{s-1}(z_i) \neq A_i(z_i) \quad \text{and} \quad A_j^{s-1}(z_j) \neq A_j(z_j). \tag{2.9}$$

Moreover, since, by (2.4) and (0.1), $A_i^{s-1}(z_i) \leq A^{s-1}(z_i) = 0$ and $A_j^{s-1}(z_j) \leq A^{s-1}(z_j) = 0$,

$$A_i(z_i) = A_j(z_j) = 1. \tag{2.10}$$

Since, by (2.6) and (0.1), 0^m is the only string z such that $z \in A$ and $l(s-1) \leq |z| < n$, we can conclude from (2.9) and (2.10) that A_i and A_j don't split A and thus that R_s is met.

Now assume that case 2 applies to stage s. Then $A^s = A^{s-1}$ and thus, by (2.3), $A^{s-1} \upharpoonright n = A \upharpoonright n$, $A_i^{s-1} \upharpoonright n = A_i \upharpoonright n$ and $A_j^{s-1} \upharpoonright n = A_j \upharpoonright n$. So the failure of A_i^{s-1} and A_j^{s-1} to split A^{s-1} is preserved and R_s is met.

This completes the proof of Theorem 2.4. □

Theorem 2.4 can be extended in several directions. E.g. we can obtain a weak analogue of Theorem 2.1 for strong non-P-T-mitoticity.

2.5. THEOREM. Let B and C be recursive sets such that B has infinitely many gaps of exponential length (i.e., for infinitely many n, B doesn't contain any string x such that $n < |x| < 2^n$), C is infinite and B and C are disjoint. Then there is a recursive set $A \subseteq C$ such that $A \cup B$ is strongly non-P-T-mitotic.

We omit the proof which is similar to the one of Theorem 2.4. Theorem 2.5 shows that the sets E, F, G and H in Corollary 2.2 can be chosen to be strongly non-P-T-mitotic.

We now show that the answer to the question whether non-P-mitotic *NP*-sets exist is oracle dependent. Note that Propositions 1.4 and 1.6(d) relativize. So, for any oracle A such that $P^A = NP^A$, there are no (weakly) P^A-m(T)-mitotic sets. The existence of such oracle sets has been shown in [BGS]. So, by the relativization of Proposition 1.4, we only have to give a recursive oracle A such that some NP^A

set is strongly non-P^A-T-mitotic.

2.6. THEOREM. There are recursive sets A and B such that

(i) $B \in NP^A$ and

(ii) B is strongly non-P^A-T-mitotic, i.e. there are no sets C and D such that $B = C \sqcup D$ and $B =_T^{P^A} C =_T^{P^A} D$.

PROOF. The proof is very similar to the one of Theorem 2.4. We construct A in stages and let

$$B = \{ 0^n : \exists\, x \in A \ (|x| = n) \}.$$

Then obviously $B \in NP^A$.

We denote the set $\{0^n : \exists\, x \in A^s \ (|x| = n)\}$ by B^s and abbreviate $M_n(A \oplus B)$ and $M_n(A^s \oplus B^s)$ by B_n and B_n^s, respectively. Then to ensure (ii) it suffices to meet requirement

$$R_{<i,j,k,l>} \ : \ \text{If } B = B_i \sqcup B_j \text{ then } B \neq M_k(A \oplus B_i) \text{ or } B \neq M_l(A \oplus B_j).$$

for every $i,j,k,l \in \mathbb{N}$. The strategy for meeting these requirements is essentially the one used in the proof of Theorem 2.4 to meet the analogous requirements there.

We now describe stage s of the construction, at which we guarantee that R_s is met.

Stage s, $s = <i,j,k,l>$. Let m be the least number such that

$p_k(m) + p_l(m) < 2^m$ and $m > \max\{l(s-1), p_i(l(s-1)), p_j(l(s-1))\}$.

Furthermore, let $n = \max\{m, p_k(m), p_l(m)\} + 1$, $y_k = M_k(A^{s-1} \oplus B_i^{s-1})(0^m)$ and $y_l = M_l(A^{s-1} \oplus B_j^{s-1})(0^m)$.

Define $l(s) = \max\{n, p_i(n), p_j(n)\} + 1$. For the definition of A^s distinguish the following cases.

Case 1: $B^{s-1} \upharpoonright n = B_i^{s-1} \upharpoonright n \sqcup B_j^{s-1} \upharpoonright n$. Then let x be the least string of length m such that $0x$ is not used in the computations $M_k(A^{s-1} \oplus B_i^{s-1})(0^m)$ and $M_l(A^{s-1} \oplus B_j^{s-1})(0^m)$ (i.e. $M_k(A^{s-1} \cup \{x\} \oplus B^{s-1})(0^m) = M_k(A^{s-1} \oplus B_i^{s-1})(0^m)$ and $M_l(A^{s-1} \cup \{x\} \oplus B_j^{s-1})(0^m) = M_l(A^{s-1} \oplus B_j^{s-1})(0^m)$; note that, by choice of m, such an x exists). Define

$$A^s = \begin{cases} A^{s-1} & \text{if } y_k = 1 \text{ or } y_l = 1 \\ A^{s-1} \cup \{x\} & \text{otherwise.} \end{cases}$$

Case 2: Otherwise. Then let $A^s = A^{s-1}$.

This completes the construction. Obviously the construction is effective and (0.1) holds. Moreover, by definition of B and B^s,

$$\forall\, s \in \mathbb{N} \ (B^s \upharpoonright l(s) = B \upharpoonright l(s) \ \& \ B^s \subseteq \{0\}^* \upharpoonright l(s)) \qquad (2.11)$$

To show that the requirements R_s are met, fix s, $s = <i,j,k,l>$, and define m, n, y_k, y_l as in the description of stage s above. Note that $l(s-1) < m < n < l(s)$. Furthermore, the following facts easily follow from the definition of $l(s-1), m, n, l(s)$, and A^s and from (0.1) and (2.11):

$$B^{s-1}(0^m) = 0,$$
$$A^{s-1} \restriction m = A \restriction m \quad \text{and} \quad B^{s-1} \restriction m = B \restriction m,$$
$$B_i^{s-1} \restriction l(s-1) = B_i \restriction l(s-1) \quad \text{and} \quad B_j^{s-1} \restriction l(s-1) = B_j \restriction l(s-1),$$
$$A^s \restriction n = A \restriction n \quad \text{and} \quad B^s \restriction n = B \restriction n,$$
$$B_i^s \restriction n = B_i \restriction n \quad \text{and} \quad B_j^s \restriction n = B_j \restriction n,$$
$$M_k(A^s \oplus B_i^{s-1})(0^m) = M_k(A^{s-1} \oplus B_i^{s-1})(0^m) \quad \text{and} \quad M_l(A^s \oplus B_j^{s-1})(0^m) = M_l(A^{s-1} \oplus B_j^{s-1})(0^m),$$
$$M_k(A^s \oplus B_i^s)(0^m) = M_k(A \oplus B_i)(0^m) \quad \text{and} \quad M_l(A^s \oplus B_j^s)(0^m) = M_l(A \oplus B)(0^m).$$

Now assume that case 1 holds at stage s. Then using the above facts as in the proof of Theorem 2.4 we can argue that R_s is met. Namely, if $y_k = 1$ or $y_l = 1$ then $A^s = A^{s-1}$ and thus $B^s = B^{s-1}$, whence $B(0^m) \neq M_k(A \oplus B_i)(0^m)$ or $B(0^m) \neq M_l(A \oplus B_j)(0^m)$. If $y_k = y_l = 0$ then $A^s = A^{s-1} \cup \{x\}$, $|x| = n$, i.e.

$$B \restriction n - B^{s-1} \restriction n = B^s \restriction n - B^{s-1} \restriction n = \{0^m\}.$$

So if $B_i^{s-1} \restriction n = B_i \restriction n$ or $B_j^{s-1} \restriction n = B_j \restriction n$ then $B(0^m) = 1 \neq 0 = M_k(A \oplus B_i)(0^m)$ or $B(0^m) = 1 \neq 0 = M_l(A \oplus B_j)(0^m)$; otherwise B_i and B_j fail to split B.

If case 2 holds at stage s, then - as in the proof of Theorem 2.4 - the above facts imply that the failure of $B_i^{s-1} \restriction n$ and $B_j^{s-1} \restriction n$ to split $B^{s-1} \restriction n$ is preserved, i.e. not $B = B_i \sqcup B_j$.

This completes the proof of Theorem 2.6. □

We do not know if $P \neq NP$ implies the existence of non-P-mitotic sets in NP. More generally, we do not know if every nonzero P-degree bounds a P-degree which contains a non-P-mitotic set. The following result(which we state without proof), however, shows that not every nonzero P-degree contains a non-P-mitotic set.

2.7. THEOREM. There is a recursive set A such that

(i) $A \notin P$

(ii) $\forall B\ (B =_m^P A \rightarrow B$ is P-m-mitotic).

Moreover A can be chosen so that

(iii) the P-m-degree of A contains infinitely many P-isomorphism types.

Note that Proposition 1.6(g) and Ladner's result in [Lad3] that every recursive non-P set A has a P-splitting A_0, A_1 such that $A \not\leq_T^P A_0, A_1$ imply that the set of P-degrees which contain non-P-mitotic sets has no minimal elements. More results on the P-degrees of non-P-mitotic sets can be obtained using the fact that the class of [weakly] P-m(T)-mitotic sets is locally recursively representable (see [AS]).

3. COMPLEMENTS OF P-MITOTIC SETS.

In this section we distinguish weak P-m/T-mitoticity from P-m/T-mitoticity by showing that in contrast to the P-m/T-mitotic sets the class of weakly P-m/T-mitotic sets fails to be closed under complementation.

3.1. PROPOSITION. The complement $\bar{A}$ of a P-m(T)-mitotic set A is P-m(T)-mitotic.

PROOF. For P-T-reducibility this is immediate, since for any A and any $B \in P$, $A =_T^P \bar{A}$ and $A \cap B =_T^P \bar{A} \cap B$. So let A be P-m-mitotic, say $A \leq_m^P A \cap B$ via g_1 and $A \leq_m^P A \cap \bar{B}$ via g_2, B some P set. W.l.o.g. $A \notin P$. So we can fix $x_1 \in \bar{A} \cap B$, $x_2 \in \bar{A} \cap \bar{B}$. Now $\bar{A} \leq_m^P \bar{A} \cap B$ via g_1' and $\bar{A} \leq_m^P \bar{A} \cap \bar{B}$ via g_2' where

$$g_1'(x) = \begin{cases} g_1(x) & \text{if } g_1(x) \in B \\ x_1 & \text{otherwise} \end{cases} \quad \text{and} \quad g_2'(x) = \begin{cases} x_2 & \text{if } g_2(x) \in B \\ g_2(x) & \text{otherwise.} \end{cases} \quad \square$$

3.2. PROPOSITION. There is a recursive set A such that A is weakly P-T-mitotic whereas its complement $\bar{A}$ is strongly non-P-T-mitotic.

PROOF. By Proposition 2.3 and Theorem 2.4. □

3.3. THEOREM. There is a recursive set $A \subseteq \{0\}^*$ such that A is weakly P-m-mitotic whereas its complement $\bar{A}$ is strongly non-P-m-mitotic.

PROOF. We construct a recursive set $A \subseteq \{0\}^*$ with the desired properties. For every $n \in \mathbb{N}$ we ensure

$$\begin{aligned} &0^{6n}, 0^{6n+1}, 0^{6n+2}, 0^{6n+3}, 0^{6n+4}, 0^{6n+5} \notin A && \text{or} \\ &0^{6n}, 0^{6n+2}, 0^{6n+3}, 0^{6n+4}, 0^{6n+5} \in A \ \& \ 0^{6n+1} \notin A && \text{or} \\ &0^{6n+1}, 0^{6n+2}, 0^{6n+3}, 0^{6n+4}, 0^{6n+5} \in A \ \& \ 0^{6n} \notin A. \end{aligned} \tag{3.1}$$

So, if we let

$$\begin{aligned} A_0 = &\{0^{6n} : 0^{6n} \in A\} \cup \{0^{6n+1} : 0^{6n+1} \in A\} \cup \{0^{6n+2} : 0^{6n} \in A\} \cup \{0^{6n+3} : 0^{6n+1} \in A\} \cup \\ &\cup \{0^{6n+4} : 0^{6n+4} \in A\} \quad \text{and} \\ A_1 = &\{0^{6n+2} : 0^{6n+1} \in A\} \cup \{0^{6n+3} : 0^{6n} \in A\} \cup \{0^{6n+5} : 0^{6n+5} \in A\}, \end{aligned}$$

then $A = A_0 \sqcup A_1$ and - as one can easily verify - $A =_m^P A_0 =_m^P A_1$. Hence A will be weakly P-m-mitotic. To make $\bar{A}$ strongly non-P-m-mitotic we meet requirement

$$R_{\langle m_0, m_1, n_0, n_1 \rangle} : \text{ If } \bar{A} = f_{n_0}^{-1}(\bar{A}) \sqcup f_{n_1}^{-1}(\bar{A}) \text{ then not } \bar{A} \leq_m^P f_{n_0}^{-1}(\bar{A}) \text{ via } f_{m_0} \text{ or not } \bar{A} \leq_m^P f_{n_1}^{-1}(\bar{A}) \text{ via } f_{m_1}.$$

for every $m_0, m_1, n_0, n_1 \in \mathbb{N}$.

We now describe stage s of the construction of A.

Stage s, $s = \langle m_o,m_1,n_o,n_1\rangle$. Given $l(s-1)$ and A^{s-1}, let $k = l(s-1)$, $x = 0^{6k}$,

$C^s = \{x, f_{m_i}(x), f_{n_i}(f_{m_j}(x)) : i,j \le 1\}$, and

$A_i^{s-1} = A^{s-1} \cup \{0^{6k+i}, 0^{6k+2}, 0^{6k+3}, 0^{6k+4}, 0^{6k+5}\}$, $i = 0,1$.

Then define $l(s) = \max\{|z| : z \in C^s\} + 1$ and

$$A^s = \begin{cases} A^{s-1} & \text{if } \exists\, i \le 1\ (\ f_{n_{1-i}}(f_{m_i}(x)) \notin A^{s-1}\ ) \\ A_o^{s-1} & \text{if } \forall\, i \le 1\ (\ f_{n_{1-i}}(f_{m_i}(x)) \in A^{s-1}\ ) \text{ and} \\ & \quad \exists\, i \le 1\ (\ f_{n_i}(f_{m_i}(x)) \notin A_o^{s-1} \text{ or } f_{m_i}(x) \notin A_o^{s-1}\ ) \\ A_1^{s-1} & \text{otherwise.} \end{cases}$$

This completes the construction. It is easy to show that the construction is effective and that (0.1) and (3.1) are satisfied. It remains to show that R_s is met for each s.

Fix s, say $s = \langle m_o,m_1,n_o,n_1\rangle$. As in the proof of Theorem 2.1 we can show that R_s is met if

$$\exists\, x \in \Sigma^*\ \exists\, i \le 1\ (\quad \begin{aligned} &(3.2.1)\quad \bar{A}(x) \ne \bar{A}(f_{n_i}(f_{m_i}(x))) \quad \text{or} \\ &(3.2.2)\quad \bar{A}(x) = \bar{A}(f_{n_{1-i}}(f_{m_i}(x))) = 1 \quad \text{or} \\ &(3.2.3)\quad \bar{A}(x) > \bar{A}(f_{m_i}(x)) \quad \text{or} \\ &(3.2.4)\quad \bar{A}(f_{m_i}(x)) > \max\{\bar{A}(f_{n_j}(f_{m_i}(x))) : j \le 1\}\ ) \end{aligned} \tag{3.2}$$

To verify (3.2) we distinguish the following cases. Note that $x \notin A^{s-1}$ and, for $z \in C^s$, $A^s(z) = A(z)$ (where here and in the following x, C^s, A_o^{s-1} and A_1^{s-1} are defined as in the description of stage s above).

Case 1: $\exists\, i \le 1\ (\ f_{n_{1-i}}(f_{m_i}(x)) \notin A^{s-1}\ )$. Then $A^s = A^{s-1}$ and thus, for some $i \le 1$, $\bar{A}(x) = \bar{A}(f_{n_{1-i}}(f_{m_i}(x))) = 1$; i.e. (3.2.2) holds.

Case 2: $\forall\, i \le 1\ (\ f_{n_{1-i}}(f_{m_i}(x)) \in A^{s-1}\ )$ and $\exists\, i \le 1\ (\ f_{n_i}(f_{m_i}(x)) \notin A_o^{s-1}\ )$. Then $A^s = A_o^{s-1}$. So, for some $i \le 1$, $\bar{A}(x) = 0 \ne 1 = \bar{A}(f_{n_i}(f_{m_i}(x)))$; i.e. (3.2.1) holds.

Case 3: $\forall\, i \le 1\ (\ f_{n_{1-i}}(f_{m_i}(x)) \in A^{s-1}\ \&\ f_{n_i}(f_{m_i}(x)) \in A_o^{s-1}\ )$ and $\exists\, i \le 1\ (\ f_{m_i}(x) \notin A_o^{s-1}\ )$. Then $A^s = A_o^{s-1}$. So, for some $i \le 1$, $\bar{A}(f_{m_i}(x)) = 1$ and $\bar{A}(f_{n_o}(f_{m_o}(x))) = \bar{A}(f_{n_1}(f_{m_1}(x))) = 0$; i.e. (3.2.4) holds.

Case 4: Cases 1 to 3 do not apply and $\exists\, i \le 1\ (\ f_{n_i}(f_{m_i}(x)) \ne x\ )$. Then $A^s = A_1^{s-1}$ and thus $\bar{A}(x) = 1$ whereas, for some $i \le 1$, $\bar{A}(f_{n_i}(f_{m_i}(x))) = 0$, since, by failure of cases 1 and 2, $f_{n_0}(f_{m_0}(x)), f_{n_1}(f_{m_1}(x)) \in A_0^{s-1}$ and thus, for i such that $f_{n_i}(f_{m_i}(x)) \neq x$, $f_{n_i}(f_{m_i}(x)) \in A_1^{s-1}$. It follows that (3.2.1) holds.

Case 5: Otherwise. Then $f_{n_0}(f_{m_0}(x)) = f_{n_1}(f_{m_1}(x)) = x$ and, for $i = 0,1$, $f_{m_i}(x) \in A_0^{s-1}$ and $f_{n_{1-i}}(f_{m_i}(x)) \in A^{s-1}$. Moreover, by construction, $A^s = A_1^{s-1}$. So

$$\bar{A}(x) = \bar{A}(f_{n_0}(f_{m_0}(x))) = \bar{A}(f_{n_1}(f_{m_1}(x))) = 1. \tag{3.3}$$

Now if $f_{m_0}(x) = f_{m_1}(x)$ then (3.3) implies (3.2.2). So we may assume $f_{m_0}(x) \neq f_{m_1}(x)$ and thus we can fix $i \leq 1$ with $f_{m_i}(x) \neq x$. Since $f_{m_i}(x) \in A_0^{s-1}$ this implies $f_{m_i}(x) \in A_1^{s-1}$ and thus $f_{m_i}(x) \in A^s$. Together with (3.3) this implies (3.2.3).

This completes the proof of Theorem 3.3. □

4. MULTIPLY P-MITOTIC SETS.

In this section we generalize the notion of a P-mitotic set by considering splittings into any finite number of parts. A set A is called n-P-mitotic if there is a splitting of A into n parts $A_1, \ldots, A_n$ such that each part has the same complexity as A. If we only consider $\mathcal{P}$-splittings then it is easy to see that any n-P-mitotic set is also (n-1)-P-mitotic. Namely, if $A_1, \ldots, A_n$ is a P-mitotic $\mathcal{P}$-splitting of A, then so is $A_1 \cup A_2, A_3, \ldots, A_n$. We will show that we actually get a hierarchy of multiply P-mitotic sets, i.e. that, for any n, the class of (n+1)-P-mitotic sets is properly contained in the class of n-P-mitotic sets. Moreover, we will show that these classes depend on the chosen P-reducibility notion, namely that the class of n-P-m-mitotic sets is a proper subclass of that of n-P-T-mitotic sets (for any n). Finally we point out that for weak P-mitoticity we don't get an analogous hierarchy, since for set theoretical splittings in general it is impossible to deduce an n-ply P-mitotic splitting of a set from a given (n+1)-ply one.

4.1. DEFINITION. A recursive set A is *n-P-m(T)-mitotic* $(n \geq 2)$ if there are pairwise disjoint sets $Q_1, \ldots, Q_n \in \mathcal{P}$ such that $A \subseteq Q_1 \cup \ldots \cup Q_n$ and $A =^P_{m(T)} A \cap Q_1 =^P_{m(T)} \cdots =^P_{m(T)} A \cap Q_n$. A is *$\omega$-P-m(T)-mitotic* if A is n-P-m(T)-mitotic for every $n \geq 2$.

Note that a set is P-m(T)-mitotic iff it is 2-P-m(T)-mitotic.

4.2. PROPOSITION. (a) A is (n+1)-P-m(T)-mitotic $\rightarrow$ A is n-P-m(T)-mitotic.
(b) A is $n(\omega)$-P-m-mitotic $\rightarrow$ A is $n(\omega)$-P-T-mitotic.

(c) A is a cylinder $\rightarrow$ A is ω-P-m-mitotic.

PROOF. Immediate. □

Since every P-m degree contains a cylinder, part (c) of the proposition shows that every P-m-degree contains an ω-P-m-mitotic set. By (a) and (b) we have the following relations among the multiply P-mitotic sets:

$$\begin{array}{ccc}
\omega\text{-P-m-mitotic} & \rightarrow & \omega\text{-P-T-mitotic} \\
\downarrow & & \downarrow \\
\vdots & & \vdots \\
\downarrow & & \downarrow \\
(n+1)\text{-P-m-mitotic} & \rightarrow & (n+1)\text{-P-T-mitotic} \\
\downarrow & & \downarrow \\
n\text{-P-m-mitotic} & \rightarrow & n\text{-P-T-mitotic} \quad (n \geq 3) \\
\downarrow & & \downarrow \\
\vdots & & \vdots \\
\downarrow & & \downarrow \\
\text{P-m-mitotic} & \rightarrow & \text{P-T-mitotic}
\end{array} \tag{4.1}$$

The following theorems show that in (4.1) only the indicated implications are valid. So, in particular, there is a proper hierarchy of the multiply P-mitotic sets, i.e. for $n \geq 2$, $\{A : A \text{ is } (n+1)\text{-P-m(T)-mitotic}\} \subsetneqq \{A : A \text{ is } n\text{-P-m(T)-mitotic}\}$.

4.3. THEOREM. For any $n \geq 2$, there is a recursive set A such that

(i) A is n-P-m-mitotic

(ii) A is not (n+1)-P-T-mitotic.

PROOF. We construct a recursive set $A \subseteq \{0\}^*$ with the desired properties. To make A n-P-m-mitotic we ensure

$$\forall m \in \mathbb{N} \ \forall i,j < n \ (0^{mn+i} \in A \quad \text{iff} \quad 0^{mn+j} \in A). \tag{4.2}$$

Then, for $Q_i = \{ 0^{mn+i-1} : m \in \mathbb{N} \}$, $i = 1,\dots,n$, $Q_1, \dots, Q_n \in P$, $Q_i \cap Q_j = \emptyset$ for $1 \leq i \neq j \leq n$, $Q_1 \cup \dots \cup Q_n = \{0\}^*$ and $A =_m^P A \cap Q_1 =_m^P \dots =_m^P A \cap Q_n$. So A is n-P-m-mitotic. To make A non-(n+1)-P-T-mitotic it suffices to meet requirements

$R_{\langle s_0,\dots,s_n,t_0,\dots,t_n\rangle}$: If $P_{s_0},\dots,P_{s_n}$ are pairwise disjoint then, for some $i \leq n$, $A \neq M_{t_i}(A \cap P_{s_i})$

for every $s_0,\dots,s_n,t_0,\dots,t_n \in \mathbb{N}$.

Construction of A:

<u>Stage s</u>, $s = \langle s_0,\dots,s_n,t_0,\dots,t_n\rangle$. Let $x = 0^{l(s-1)n}$. Set

$l(s) = \max \{l(s-1), p_{t_0}(|x|),\dots,p_{t_n}(|x|)\} + 1$ and

$$A^s = \begin{cases} A^{s-1} & \text{if } \exists\, i \le n(\ M_{t_i}(A^{s-1} \cap P_{s_i})(x)=1\ ) \\ A^{s-1} \cup \{0^{l(s-1)n},\dots,0^{l(s-1)n+n-1}\} & \text{otherwise.} \end{cases}$$

Obviously the construction is effective and (0.1) and (4.2) are satisfied. To show that the requirements R_s are met, fix $s = <s_o,\dots,s_n,t_o,\dots,t_n>$ and let $x = 0^{l(s-1)n}$. Note that $A^{s-1}(x) = 0$ and, by definition of $l(s)$,

$$A^s(x) = A(x) \text{ and } M_{t_i}(A^s \cap P_{s_i})(x) = M_{t_i}(A \cap P_{s_i})(x),\quad i = 0,\dots,n. \tag{4.3}$$

W.l.o.g. assume that the sets $P_{s_o}, \dots, P_{s_n}$ are pairwise disjoint. Now, if $M_{t_i}(A^{s-1} \cap P_{s_i})(x) = 1$ for some $i \le n$, i.e. $A^{s-1}(x) \ne M_{t_i}(A^{s-1} \cap P_{s_i})(x)$, then $A^s = A^{s-1}$ whence, by (4.3), this disagreement is preserved. Otherwise $A^s(x) = 1$ and $|A^s - A^{s-1}| = n$. So, for some $i \le n$, $A^s \cap P_{s_i} = A^{s-1} \cap P_{s_i}$ and thus

$$A^s(x) = 1 \ne 0 = M_{t_i}(A^{s-1} \cap P_{s_i})(x) = M_{t_i}(A^s \cap P_{s_i})(x).$$

Again by (4.3) this disagreement is preserved. So in any case R_s is met, which completes the proof of the theorem. □

4.4. THEOREM. There is a recursive set A such that

(i) A is ω-P-T-mitotic

(ii) A is non-P-m-mitotic.

PROOF. We construct a recursive set $A \subseteq \underline{\mathbb{N}}\{0\}^*$ with the desired properties. To make A ω-P-T-mitotic we ensure

$$\forall x \in \{0\}^* \ (\ \{n \in \mathbb{N} : \underline{n}x \in A\} = \emptyset \text{ or } \{n \in \mathbb{N} : \underline{n}x \in A\} = \{n : n \text{ even}\} \text{ or } \{n \in \mathbb{N} : \underline{n}x \in A\} = \{0\} \cup \{n : n \text{ odd}\}\ ). \tag{4.4}$$

Then, for any $n \ge 1$, $m \in \mathbb{N}$ and $x \in \Sigma^*$,

$$\underline{m}x \in A \text{ iff } \begin{cases} \underline{2n+1}x \in A & \text{if } m \text{ is odd} \\ \underline{2n}x \in A & \text{if } m \ne 0 \text{ is even} \\ \underline{2n}x \in A \text{ or } \underline{2n+1}x \in A & \text{if } m = 0. \end{cases}$$

Hence, for $n \ge 1$, $A \le_T^P \underline{\{2n,2n+1\}}A$. Since $\underline{\{2n,2n+1\}}\Sigma^* \in P$, it easily follows that A is ω-P-T-mitotic.

To make A non-P-m-mitotic we meet the following requirements:

$R_{<i,j,k>}$: Not $A \le_m^P A \cap P_i$ via f_j or not $A \le_m^P A \cap \bar{P}_i$ via f_k $(i,j,k \in \mathbb{N})$.

As usually we construct a length function l, where $l(-1) = 0$ and, for $s \in \mathbb{N}$, $l(s)$ is defined at stage s of the construction. The function l will be strictly increasing, and - in place of (0.1) - we will have

$$\forall n \in \mathbb{N}\ \forall x \in \{0\}^*\ \forall s \in \mathbb{N}\ ([\ \underline{n}x \in A^s \to |x| < l(s)\ ] \text{ and } [\ |x| < l(s) \to (\ \underline{n}x \in A^s \text{ iff } \underline{n}x \in A\ )\ ]\ ) \tag{4.5}$$

We now describe stage s of the construction of A at which we ensure that R_s is met.

<u>Stage s</u>, $s = \langle i,j,k \rangle$. Let $y = 0^{l(s-1)}$ and $x = \underline{0}y$. Set

$$l(s) = \max\{|x|,|f_j(x)|,|f_k(x)|\} + 1 \qquad \text{and}$$

$$A^s = \begin{cases} A^{s-1} & \text{if } f_j(x) \in A^{s-1} \cap P_i \text{ or } f_k(x) \in A^{s-1} \cap \bar{P}_i \\ A^{s-1} \cup \{x\} \cup \{\underline{n}y : n \text{ odd}\} & \text{if } f_j(x) \notin A^{s-1} \cap P_i,\ f_k(x) \notin A^{s-1} \cap \bar{P}_i \\ & \text{and } f_j(x), f_k(x) \in \underline{\{n : n \text{ even}\}}\Sigma^* \\ A^{s-1} \cup \{\underline{n}y : n \text{ even}\} & \text{otherwise.} \end{cases}$$

Note that the construction is effective, l is strictly increasing, $A \subseteq \underline{\mathbb{N}}\{0\}^*$ and (4.4) and (4.5) are satisfied. It follows that A is recursive and ω-P-T-mitotic. To see that R_s is met for all s, fix $s = \langle i,j,k \rangle$ and let $x = \underline{0}0^{l(s-1)}$. Note that, by (4.5) and definition of l, $x \notin A^{s-1}$ and, for $z \in \{x, f_j(x), f_k(x)\}$, $A^s(z) = A(z)$. To show that R_s is satisfied, according to the definition of A^s, we distinguish the following cases.

Case 1: $f_j(x) \in A^{s-1} \cap P_i$ or $f_k(x) \in A^{s-1} \cap \bar{P}_i$. Then $A^s = A^{s-1}$. Hence $x \notin A$ whereas $f_j(x) \in A \cap P_i$ or $f_k(x) \in A \cap \bar{P}_i$.

Case 2: Otherwise. Then the second or third case in the definition of A^s holds, whence $A(x) = 1$. We distinguish the following subcases.

Case 2.1: $f_j(x), f_k(x) \in \underline{\{n : n \text{ even}\}}\Sigma^*$. Then the second case in the definition of A^s holds. So either $f_j(x) = f_k(x)$ and thus $f_j(x) \notin P_i$ or $f_k(x) \notin \bar{P}_i$, or $f_j(x) \neq x$ or $f_k(x) \neq x$ and thus $f_j(x) \notin A^s \cap P_i$ or $f_k(x) \notin A^s \cap \bar{P}_i$. In any case $A(x) \neq A \cap P_i(f_j(x))$ or $A(x) \neq A \cap \bar{P}_i(f_k(x))$.

Case 2.2: Otherwise. Then $A^s = A^{s-1} \cup \underline{\{n : n \text{ even}\}}\{0^{l(s-1)}\}$ and $f_j(x) \notin (A^{s-1} \cap P_i) \cup \underline{\{n : n \text{ even}\}}\Sigma^*$ or $f_k(x) \notin (A^{s-1} \cap \bar{P}_i) \cup \underline{\{n : n \text{ even}\}}\Sigma^*$. It follows $A(x) \neq A \cap P_i(f_j(x))$ or $A(x) \neq A \cap \bar{P}_i(f_k(x))$.

So in each case R_s is met. This completes the proof. □

Similar to Definition 4.1 the notion of a weak P-mitotic set can be generalized. Here, however, we don't obtain a hierarchy like the one in (4.1) for P-mitotic sets. The reason is that in general Proposition 4.2(a) fails for weak P-mitoticity. This is shown by the following theorem which we state without proof.

4.5. THEOREM. There is a recursive set A such that

(i) A is strongly non-P-m-mitotic

(ii) there are pairwise disjoint sets A_o, A_1, A_2 such that $A = A_o \cup A_1 \cup A_2$ and $A =^P_m A_o =^P_m A_1 =^P_m A_2$ (i.e. A is *weakly 3-P-m-mitotic*).

5. P-AUTOREDUCIBLE SETS.

Every set A can be P-m-reduced to itself via the identity function. Such a reduction is trivial since the question whether or not $x \in A$ for a given x is answered using the information $A(x)$. We call a reduction of A to itself *nontrivial* if, for every x, the reduction doesn't use the information $A(x)$ in order to compute $A(x)$. Sets which can be nontrivially P-reduced to themselves are called *P-autoreducible*. Obviously, for any A, $A \oplus A$ is P-m-autoreducible by the nontrivial reduction f with $f(0x) = 1x$ and $f(1x) = 0x$. So every P-m-degree contains a P-m-autoreducible set.

In the following we show that P-mitoticity and P-autoreducibility are related. P-mitotic sets are P-autoreducible. For tally sets and P-m-reducibility -but not for P-T-reducibility- also the converse is true. Furthermore we show that in general weak P-mitoticity does not imply P-autoreducibility.

5.1. DEFINITION. A recursive set A is P-m-*autoreducible* if A is finite or co-finite or if, for some $f \in PF$, $A \leq_m^P A$ via f and $\forall x \in \Sigma^* \; (f(x) \neq x)$. A is P-T-*autoreducible* if for some $n \in \mathbb{N}$ and every $x \in \Sigma^*$, $A(x) = M_n(A-\{x\})(x)$.

The convention that finite and co-finite sets are P-m-autoreducible is for convenience and -like the similar convention for P-m-mitotic sets- it has no influence on our investigations (Without this convention one-element sets and their complements were not P-m-autoreducible).

The following facts on P-autoreducibility are immediate.

5.2. PROPOSITION. (a) A is P-m-autoreducible $\rightarrow$ A is P-T-autoreducible.
(b) $A \in P$ $\rightarrow$ A is P-m-autoreducible
(c) For any A, $A \oplus A$ is P-m-autoreducible.
(d) A is P-m(T)-autoreducible $\rightarrow$ $\bar{A}$ is P-m(T)-autoreducible.
(e) P-m(T)-autoreducibility is P-invariant. □

We now show that P-mitotic sets are P-autoreducible.

5.3. PROPOSITION. If A is P-m(T)-mitotic then A is P-m(T)-autoreducible.

PROOF. Fix A such that w.l.o.g. $A \notin P$. First assume that A is P-m-mitotic, say $A \leq_m^P A \cap B$ via g_0 and $A \leq_m^P A \cap \bar{B}$ via g_1, $B \in P$. Choose x_0, x_1 such that $x_0 \neq x_1$ and $x_0, x_1 \notin A$. Then A is P-m-reducible to itself via g, where $g(x) = g_1(x)$ if $x \in B$ and $g_1(x) \notin B$, $g(x) = g_0(x)$ if $x \notin B$ and $g_0(x) \in B$, and in all other cases $g(x) = x_1$ if $x = x_0$ and $g(x) = x_0$ if $x \neq x_0$, and obviously $g(x) \neq x$ for all $x \in \Sigma^*$.

Now assume that A is P-T-mitotic, say $A = M_k(A\cap B)$ and $A = M_l(A \cap \bar{B})$, $B\varepsilon P$. Then $A(x) = M_n(A-\{x\})(x)$ for all x, where M_n operates as follows. For $x \varepsilon B$, M_n simulates M_l where any oracle queries about elements of B are suppressed and a negative answer of the oracle is assumed. Similarly, for $x \notin B$, M_n simulates M_k suppressing all queries about strings in $\bar{B}$. □

We now show that in the case of tally sets and P-m-reducibility also the converse of the proposition is true.

5.4. THEOREM. Let A be a tally recursive set. Then A is P-m-mitotic iff A is P-m-autoreducible.

PROOF. By Proposition 5.3 it suffices to show that P-m-autoreducibility implies P-m-mitoticity. So assume that $A \subseteq \{0\}^*$ is P-m-autoreducible via f, i.e.

$$A \leq_m^P A \text{ via } f \quad \text{and} \quad \forall x \varepsilon \Sigma^* \,(\, f(x) \neq x \,). \tag{5.1}$$

Then let

$$B = \{\, 0^n : \forall m < n \,(\, f(0^m) \neq 0^n \,)\,\}.$$

Obviously $B \varepsilon P$. We will define functions g and h such that

$$A \leq_m^P A \cap B \text{ via } g \quad \text{and} \quad A \leq_m^P A \cap \bar{B} \text{ via } h. \tag{5.2}$$

For $x \notin \{0\}^*$, $g(x) = h(x) = x$. For $x = 0^n$, $n \varepsilon \mathbb{N}$, $g(x)$ and $h(x)$ are defined by induction on n:

<u>$n = 0$.</u> $g(0^0) = 0^0$ and $h(0^0) = f(0^0)$.

<u>$n > 0$.</u>

$$g(0^n) = \begin{cases} 0^n & \text{if } 0^n \varepsilon B \\ g(0^i) & \text{if } 0^n \notin B, \end{cases}$$

where i is chosen minimal such that $f(0^i) = 0^n$.

$$h(0^n) = \begin{cases} h(f(0^n)) & \text{if } 0^n \varepsilon B \text{ and } f(0^n) \varepsilon \{0^0,\dots,0^{n-1}\} \\ f(0^n) & \text{if } 0^n \varepsilon B \text{ and } f(0^n) \notin \{0^0,\dots,0^{n-1}\} \\ 0^n & \text{if } 0^n \notin B. \end{cases}$$

Note that for $0^n \notin B$ there is some $i < n$ with $f(0^i) = 0^n$. So g is well defined. Also obviously $g,h \varepsilon PF$. To prove that (5.2) holds we have to show that for all $x \varepsilon \Sigma^*$,

$$A(x) = A \cap B(g(x)) = A \cap \bar{B}(h(x)). \tag{5.3}$$

This is obvious for $x \notin \{0\}^*$. For $x = 0^n$, (5.3) is proved by induction on n using (5.1). □

5.5. COROLLARY. There is a weakly P-m-mitotic set which is not P-m-autoreducible.

PROOF. By Theorem 3.3 and Propositions 1.4 and 3.1 there is a tally recursive set which is weakly-P-m-mitotic but not P-m-mitotic. So the claim follows from Theorem 5.4. □

We don't know if P-m-mitoticity and P-m-autoreducibility are equivalent. For P-T-reducibility, however, P-mitoticity and P-autoreducibility disagree, even on tally sets.

5.6. THEOREM. There is a tally recursive set A which is P-T-autoreducible but not P-T-mitotic.

PROOF. We construct $A \subseteq \{0\}^*$ effectively. We ensure for any $n \in \mathbb{N}$ that

$$\{0^{3n},0^{3n+1},0^{3n+2}\} \subseteq \bar{A} \quad \text{or} \quad |\{0^{3n},0^{3n+1},0^{3n+2}\} \cap A| = 2.$$

So, for any n and $\{i,j,k\} = \{0,1,2\}$,

$$0^{3n+i} \in A \quad \text{iff} \quad (0^{3n+j} \in A \ \& \ 0^{3n+k} \notin A) \ \text{or} \ (0^{3n+j} \notin A \ \& \ 0^{3n+k} \in A),$$

whence A is P-T-autoreducible.

To make A non-P-T-mitotic we have to meet the requirement

$$R_{<i,j,k>} \ : \ A \neq M_j(A \cap B_i) \quad \text{or} \quad A \neq M_k(A \cap \bar{B}_i)$$

for every $i,j,k \in \mathbb{N}$. At stage s, $s = <i,j,k>$, of the construction we ensure that R_s is met by determining which of the strings $x_m = 0^{3\cdot l(s-1)+m}$, $m < 3$, are enumerated in A (By (0.1), which will be satisfied, $x_m \notin A^{s-1}$). If $M_j(A^{s-1} \cap B_i)(x_m) = 1$ or $M_k(A^{s-1} \cap \bar{B}_i)(x_m) = 1$ for some $m < 3$ then we set $A^s = A^{s-1}$. Otherwise we choose $m \neq n < 3$ minimal such that $\{m,n\} \subseteq B_i$ or $\{m,n\} \subseteq \bar{B}_i$ and set $A^s = A^{s-1} \cup \{x_m,x_n\}$. In both cases we set

$$l(s) = \max \{|x_2|,p_j(|x_2|),p_k(|x_2|)\} + 1,$$

whence, for $m < 3$, $A^s(x_m) = A(x_m)$, $M_j(A^s \cap B_i)(x_m) = M_j(A \cap B_i)(x_m)$ and $M_k(A^s \cap \bar{B}_i)(x_m) = M_k(A \cap \bar{B}_i)(x_m)$. So, for some $m < 3$, in the first case

$$A(x_m) = 0 \ \text{whereas} \ \max \{M_j(A \cap B_i)(x_m),M_k(A \cap \bar{B}_i)(x_m)\} = 1$$

and in the second case

$$A(x_m) = 1 \ \text{whereas} \ \min \{M_j(A \cap B_i)(x_m),M_k(A \cap \bar{B}_i)(x_m)\} = 0.$$

It follows that R_s is met. □

5.7. COROLLARY. There is a recursive set which is P-T-autoreducible but not P-m-autoreducible.

PROOF. By Theorems 5.4 and 5.5 and Proposition 1.4. □

Corollary 5.5. implies the existence of sets which are not P-m-autoreducible. We conclude our investigation in the P-autoreducible sets with the proof that also non-P-T-autoreducible sets exist.

5.8. THEOREM. Let B and C be disjoint recursive sets such that C is infinite. Then there is a recursive subset A of C such that $A \cup B$ is not P-T-autoreducible.

PROOF. We construct $A \subseteq \{x_n : n \in \mathbb{N}\}$, where $\{x_n : n \in \mathbb{N}\}$ is a recursive subset of C such that $|x_n| < |x_{n+1}|$ for every $n \in \mathbb{N}$. $A \cup B$ will be non-P-T-autoreducible if for all $n \in \mathbb{N}$,

$$R_n : \quad \exists x \in \Sigma^* \ (A \cup B(x) \neq M_n((A \cup B) - \{x\})(x)).$$

At stage s of the construction we satisfy condition R_s by ensuring

$$\text{for } x = x_{l(s-1)}, \quad A \cup B(x) \neq M_s((A \cup B) - \{x\})(x). \tag{5.4}$$

Namely, if $M_s(A^{s-1} \cup B)(x) = 1$ then $A^s = A^{s-1}$ and $A^s = A^{s-1} \cup \{x\}$ otherwise. Moreover in both cases $l(s) = \max \{|x|, p_s(|x|)\} + 1$. Then (0.1) is satisfied, i.e. $A^{s-1}(x) = 0$ and, by definition of $l(s)$, $A^s(x) = A(x)$ and $M_s((A^s \cup B) - \{x\})(x) = M_s((A \cup B) - \{x\})(x)$. Since, by definition of A^s, $(A^{s-1} \cup B) = (A^s \cup B) - \{x\}$, all this implies (5.4). □

By Theorem 5.8 there are non-P-T-autoreducible sets E, F, G, H, I, K with properties (i) to (vi) of Corollary 2.2. In particular there is a non-P-T-autoreducible set with tally complement. Together with Proposition 2.3 this implies

5.9. COROLLARY. There is a weakly P-T-mitotic recursive set which is not P-T-autoreducible. □

Similarly to the way we modified the proof of Theorem 2.4 to obtain Theorem 2.6, we can modify the proof of Theorem 5.8 to show that the existence of non-P-autoreducible sets in *NP* is oracle dependent.

REFERENCES.

AS K.Ambos-Spies, *Sublattices of the polynomial time degrees*, in preparation.

BGS T.Baker, J.Gill and R.Solovay, *Relativizations of the P=?NP question*, SIAM J. Comput. 4 (1975) 431-442.

BH J.Hartmanis and L.Berman, *On isomorphisms and density of NP and other complete sets*, SIAM J. Comput. 6 (1977) 305 - 322.

HM S.Homer and W.Maass, *Oracle dependent properties of the lattice of NP sets*, to appear.

Lac A.H.Lachlan, *The priority method*.I, Zeitschrift für math. Logik u. Grundlagen d. Mathematik 13 (1967) 1-10.

Lad1 R.E.Ladner, *Mitotic recursively enumerable sets*, J. Symbolic Logic 38 (1973) 199 - 211.

Lad2 R.E.Ladner, A *completely mitotic nonrecursive r.e. degree*,Trans. Amer. Math. Soc. 184 (1973) 479 - 507.

Lad3 R.E.Ladner, *On the structure of polynomial time reducibility*, J. Assoc. Comp. Machinery 22 (1975) 155 - 171.

LLR L.H.Landweber, R.J.Lipton and E.L.Robertson, *On the structure of sets in NP and other complexity classes*, Theor. Comp. Science 15 (1981) 181 - 200.

LLS R.E.Ladner, N.A.Lynch and A.L.Selman, A *comparision of polynomial time reducibilities*, Theor. Comp. Science 1 (1975) 103 - 123.

EQUIVALENCE RELATIONS, INVARIANTS, AND NORMAL FORMS, II

Andreas Blass and Yuri Gurevich
Dept. of Mathematics and Dept. of Computer and Communication Sciences
The University of Michigan
Ann Arbor, MI 48109, U.S.A.

Abstract. For an equivalence relation E on the words over some finite alphabet, we consider the following four problems, listed in order of increasing difficulty. Recognition: Decide whether two words are equivalent. Invariant: Calculate a function constant on precisely the equivalence classes. Normal form: Calculate a particular member of an equivalence class, given an arbitrary member. First member: Calculate the first member of an equivalence class, given an arbitrary member. We consider the questions whether $\mathcal{P}$ solutions for the easier problems yield $\mathcal{NP}$ solutions for the harder ones, or vice versa. We show that affirmative answers to several of these questions are equivalent to natural principles like $\mathcal{NP}$=co-$\mathcal{NP}$, $\mathcal{NP}\cap$co-$\mathcal{NP}=\mathcal{P}$, and the shrinking principle for $\mathcal{NP}$ sets. We supplement known oracles with enough new ones to show that all the questions considered have negative answers relative to some oracles. In other words, these questions cannot be answered affirmatively by means of relativizable polynomial-time Turing reductions. Finally, we show that the analogous questions with "$\mathcal{P}$" replaced by "Borel" have negative answers.

1. Introduction and Review of Part I

Problems of the form "Decide whether two given objects are equivalent" for various equivalence relations occur throughout mathematics. Typical examples include deciding whether two finite presentations of groups define isomorphic groups, whether two (triangulated) topological spaces are homeomorphic, whether two matrices are similar, and whether two Boolean formulas are equivalent. In general, such problems need not admit algorithmic solutions, and even when they do the algorithms may not be practically feasible. When solutions to particular such problems do exist, however, they very often have the form of an algorithm computing a (complete) invariant function, that is, a function F such that any two objects x and y are equivalent if and only if F(x)=F(y).

For example, although the isomorphism problem for finitely presented groups is recursively unsolvable, its restriction to abelian groups is solvable by calculating, from any presentation x, the invariant $F(x)$ consisting of the prime-power (or ∞) orders of cyclic groups whose direct product is the group presented by x. Similarly, although the homeomorphism problem is not recursively solvable for general triangulated spaces (or even for triangulated four-dimensional manifolds), its restriction to two-dimensional manifolds is solvable by calculating, from any triangulation x, the invariant $F(x)$ consisting of the genus of the surface defined by x, the number of boundary components of this surface, and an indicator of whether the surface is orientable. For the other two problems mentioned above, similarity of matrices and equivalence of Boolean formulas, invariants are provided by Jordan canonical forms and truth tables, respectively.

It is natural to ask whether this phenomenon is a general one. If we can decide whether any two given objects are equivalent, can we then always calculate an invariant function or even a normal form function? These questions were considered in [2], of which the present paper is a continuation. The results in [2] concern the questions of reducibility among the following four problems, for an equivalence relation E on the set Σ^* of words over some finite alphabet Σ.

Recognition problem. Given x and y in Σ^*, determine whether xEy.

Invariant problem. Calculate, for x in Σ^*, an invariant $F(x)$ (in Σ_1^* for some finite alphabet Σ_1) such that, for all x and y in Σ^*, xEy if and only if $F(x)=F(y)$.

Normal form problem. Calculate, for x in Σ^*, a normal form $F(x)$ (in Σ^*) such that, for all x and y in Σ^*, $xEF(x)$ and if xEy then $F(x)=F(y)$.

First member problem. Calculate, for x in Σ^*, the first y such that xEy.

In the first member problem, "first" refers to the following standard ordering of Σ^*: y precedes z if either y is shorter than z or they have the same length and y lexicographically precedes z.

The results obtained in [2] are of three general sorts. First, there are some easy positive results. Clearly, a solution to any of the problems yields solutions to all the earlier ones in the list, so we are only interested in implications from earlier to later problems. If we interpret "determine" and "calculate", in the statement of each problem, as referring to recursive computability, then all four problems are equivalent, for if we have a recursive solution of the recognition problem, we can use it to systematically search Σ^*, in standard order, until we find the first y equivalent to a given x, so we can recursively solve the first member problem. Our primary interest, however, is not recursive computability but polynomial time computability, and here this search procedure is useless, for, when the input is x, it may be necessary to search through all the predecessors of x and the number of predecessors increases exponentially in the length $|x|$ of x. Nevertheless, a modification of the search procedure, in which the list of predecessors of x is repeatedly bisected, shows that the first member problem can be solved in polynomial time using an oracle for the set

$\{(x,y) \mid \text{Some member of the equivalence class of } x \text{ precedes } y\}$.

If the recognition problem has a polynomial time solution, then this oracle set is in NP, so the solution of the first member problem is in the class Δ_2^P of the Stockmeyer's polynomial-time hierarchy [9]. In the unlikely event that $P=NP$, it would follow that all four problems are equivalent for polynomial time computability just as for recursive computability. (If we imposed a linear or larger space bound, rather than a time bound, on computations, then the four problems would again be equivalent.)

The second result in [2] is a converse of the preceding. There exists an E for which the recognition problem and even the normal form problem are solvable in polynomial time (with the last member of each equivalence class as the normal form) but the solution of the first member problem is complete for Δ_2^P. In particular, the normal form and first member problems are equivalent, in the context of polynomial time computability, only if $P=NP$.

Finally, it is shown in [2] that no problem on the list is polynomial-time Turing reducible to an earlier one. There is an equivalence relation E such that no solution of its invariant problem belongs to P^E, and there is another equivalence relation having an invariant function F such that no solution of the normal form problem belongs to P^F. (The analogous result for the normal form and first member problems follows, by virtue of oracles in [1], from the Δ_2^P completeness result above.) These results can also be viewed as saying that all four problems are inequivalent in the context of polynomial time computability relative to oracles. In view of the earlier results, we cannot obtain this inequivalence for (unrelativized) polynomial time computability without proving $P \neq NP$.

In the present paper, we extend and improve these non-reducibility results in two ways. First, we consider non-deterministic as well as deterministic polynomial time reducibility. Our results imply, for example, the existence of an equivalence relation with a normal form function F such that the first member function is not in NP^F; similar results hold for the other pairs of consecutive problems in our list. (The notion of a function, rather than a predicate, being in NP will be made precise in a moment.) We also show that this extension of our earlier results, from P to NP, is non-trivial by producing a set A and an equivalence relation E such that none of the four problems for E has a solution in P^A but all four have solutions in NP^A.

These results could be obtained by minor modifications of the methods in [2], but we shall use a different approach that gives considerably more information. In particular, we obtain information about computability without oracles. Propositions like "If the normal form problem for E has a P solution then the first member problem has an NP solution," although not refutable outright without recourse to oracles, can be proved to be equivalent to natural (though sometimes dubious) principles about P and NP. This particular proposition turns out to be equivalent to the principle that NP is closed under complementation, and we obtain similar results for most of the other problems in our list. These results are trivially relativizable to any oracle and therefore imply the results in the preceding paragraph because there are oracles relative to which the natural principles fail. Such oracles are constructed in [1] for two of the three natural principles we consider (including NP=co-NP), and we construct one for the remaining principle.

It seems to be a useful general methodological principle that, when the answer to some question is unobtainable because of the P=NP problem, one should try to show that the question (or perhaps some refinement of it) is equivalent to asking about the truth of some natural principle like the ones considered in this paper. Such a result is preferable to merely constructing oracles showing that either answer is possible in the relativized situation. It implies the existence of such oracles (for oracles for the independence of the relativized natural principles are often already known and generally easy to construct), but it also yields information about the unrelativized case which is the one of primary interest. And, of course, it would be immediately applicable in the event of a breakthrough establishing or refuting the relevant natural principle.

In an appendix to this paper, we pursue the idea [5,8] that the "right" analogs of P and NP in classical recursion theory are not "recursive" and "recursively enumerable" but "Borel" and "analytic". We investigate our list of four problems in the context of Borel relations and functions, showing that all four problems are inequivalent. The proofs use both classical descriptive set theory and some of the techniques of the rest of the paper.

2. Theorems

Since we wish to discuss non-deterministic polynomial time solvability of our four problems, and since three of those problems involve computing functions rather than predicates, we must define what it means for a function $F:\Sigma^* \longrightarrow \Sigma_1^*$ to be in NP. We adopt the convention that this means that there is a non-deterministic Turing machine M, with a polynomial clock bound, that computes F in the following sense. For each x in Σ^*, there is at least one computation of M, with input x, that enters a terminal state - we call such computations successful - and every successful computation has F(x) as output. There are several equivalent ways to formulate this definition. Perhaps the simplest is that F is polynomially bounded (meaning that $|F(x)|$ is bounded by a polynomial function of $|x|$) and its graph is in NP. Another is that F is polynomially bounded and both of the predicates

$$P(x,k,a) \longleftrightarrow \text{a is the } k^{th} \text{ letter in the word } F(x)$$

and

$$Q(x,k) \longleftrightarrow |F(x)|<k$$

are in NP. The condition that Q is in NP could be equivalently replaced by the condition that P is in co-NP.

This last observation and the convention in [2, Section 1] concerning completeness of functions in complexity classes would suggest the name "$NP \cap$co-NP functions" for what we have called NP functions. Further support for this suggestion can be found in the facts that the graphs of these functions belong not only to NP (as already mentioned) but also to co-NP and that the characteristic function of a set is in NP if and only if the set itself is in $NP \cap$co-NP. Nevertheless, we have chosen to use the terminology "NP function", partly for the sake of brevity and partly because of the naturalness of the original definition in terms of machines.

The trivial reducibilities between our four problems and equally trivial facts like $P \subseteq NP \cap$co-NP give us most of the following diagram of implications, in which, for example, P-Inv stands for "the invariant problem for E has a solution P". The only implication that may require comment is the one from NP-Inv to ($NP \cap$co-NP)-

Rec, which is related to the remarks in the preceding paragraph. If the *NP* function F solves the invariant problem for E then, given x and y, we can non-deterministically verify that they are (resp. are not) equivalent by guessing values of F(x) and F(y), guessing witnesses for the correctness of these values, calculating that they really are witnesses, and checking that F(x) and F(y) are (resp. are not) equal. So the recognition problem for E has a solution in *NP*(resp. co-*NP*).

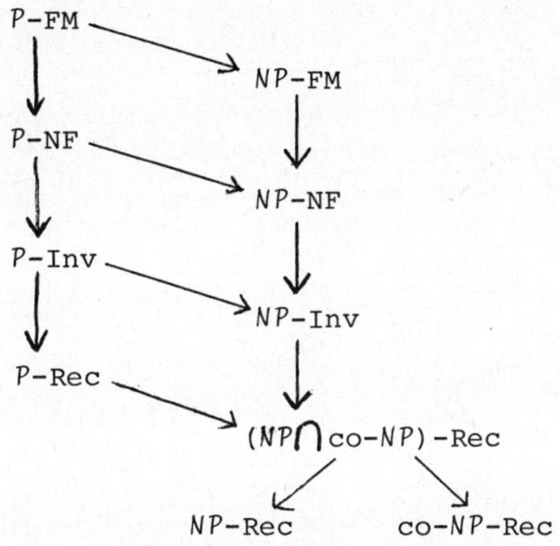

It will follow from our results that this diagram is complete in the sense that no additional arrows can be added if we permit relativization to arbitrary oracles. That is, the only implications, between propositions in the diagram, that hold when relativized to arbitrary oracles are the implications shown in the diagram and those that follow by transitivity of implication (concatenating arrows). To see more clearly what this means, we observe first that, if a new arrow (i.e., one not obtainable by concatenating arrows already shown) were added to the diagram, then its concatenation with arrows already present would produce at least one of the following implications.

(a) *NP*-Rec ⟶ co-*NP*-Rec.

(b) co-*NP*-Rec ⟶ *NP*-Rec.

(c) *P*-Rec ⟶ *NP*-Inv.

(d) *P*-Inv ⟶ *NP*-NF.

(e) *P*-NF ⟶ *NP*-FM.

(f) *NP*-FM ⟶ *P*-Rec.

Thus, the completeness of the diagram amounts to the assertion that each of these six implications fails when relativized to a suitable oracle.

For three of the six implications, (a), (b), and (e), a single, known oracle does the job.

<u>Theorem 1</u>. <u>The following four statements are equivalent</u>.

(1) <u>Every equivalence relation in</u> NP <u>is also in</u> co-NP.

(2) <u>Every equivalence relation in</u> co-NP <u>is also in</u> NP.

(3) <u>If the normal form problem for an equivalence relation has a solution in</u> P, <u>then the first member problem has a solution in</u> NP.

(4) NP=co-NP.

This theorem remains valid, with the same proof, when P and NP are relativized to an arbitrary oracle. Baker, Gill, and Solovay [1] provide an oracle relative to which $NP \neq$co-NP, so we immediately obtain the desired non-implications.

<u>Corollary</u>. <u>There is an oracle relative to which all of</u> (1) <u>through</u> (4) <u>in the theorem fail</u>. □

We have a similar result concerning implication (f).

<u>Theorem 2</u>. <u>The following statements are equivalent</u>.

(1) <u>If the first member problem for an equivalence relation has solution in</u> NP, <u>then the equivalence relation is in</u> P.

(2) $NP \cap$co-NP=P.

Again the proof works relative to any oracle, and again an oracle violating (2) is constructed in [1], so we again have the desired non-implication.

<u>Corollary</u>. <u>There is an oracle relative to which both</u> (1) <u>and</u> (2) <u>in the theorem fail</u>. □

For the implication (d), unlike the four implications just considered, the equivalent natural principles about NP have not, as far as we know, been previously studied, so we introduce them here. They are complexity-theoretic analogs of familiar principles in recursion theory and descriptive set theory (see [4,7]).

The <u>reduction principle</u> or <u>shrinking principle</u> for NP is the assertion that, for any two NP sets A and B (subsets of some Σ^*), there exist disjoint NP subsets $A' \subseteq A$ and $B' \subseteq B$ with $A' \cup B' = A \cup B$. (Terminological remark: "Reduction" is the standard name in recursion theory and descriptive set theory, but, as the same word also has other meanings, e.g. Turing reduction, we shall use "shrinking" instead.)

We call a binary relation R on words <u>polynomially bounded</u> if there is a polynomial p such that, for all words x and y, $R(x,y)$ implies $|y| \leq p(|x|)$. The <u>uniformization principle</u> for NP is the assertion that, for every polynomially bounded binary relation R in NP, there exists R' in NP such that $R' \subseteq R$ and, for all x,

$$\exists y\ R(x,y) \longrightarrow \exists! y\ R'(x,y).$$

In other words, a polynomially bounded NP relation includes a single-valued NP relation with the same domain. In classical recursion theory and descriptive set theory, uniformization easily implies reduction but uniformization is usually significantly harder to prove than reduction. We shall see that in complexity theory the two are equivalent. The uniformization principle is unchanged if R is assumed to be in P, because we can uniformize the NP projection of a P relation by uniformizing the latter and projecting the result.

An equivalence relation on Σ^* is called <u>flat</u> if equivalent words always have equal length. All the equivalence relations constructed in the proofs of the theorems in [2] and in the present paper are flat, and so are many naturally occurring equivalence relations.

<u>Theorem 3. The following three statements are equivalent.</u>

(1) <u>If the invariant problem for a flat equivalence relation has a solution in P, then the normal form problem has a solution in</u> NP.

(2) <u>The shrinking principle holds for</u> NP.

(3) <u>The uniformization principle holds for</u> NP.

We know no similar theorem for arbitrary (non-flat) equivalence relations. Notice that the requirement of flatness makes (1) weaker than (d), so we will have established the desired non-implication if we can find an oracle relative to which (1) fails. As before, the theorem holds relative to any oracle. But the desired oracle can no longer be found in [1]; we construct it.

<u>Theorem 4. There is an oracle relative to which all of</u> (1) <u>through</u> (3) <u>in Theorem</u> 3 <u>fail</u>.

There remains only the implication (c) to consider. Unfortunately, we have not been able to find a natural principle equivalent to it (unless we consider (c) itself a natural principle), so we merely give an oracle refutation for it.

<u>Theorem 5. There is an equivalence relation</u> E <u>such that no solution of its invariant problem is in</u> NP^E.

3. Proofs

Before beginning the proofs of our theorems, we make a few remarks to simplify the exposition. First, we shall make use of the fact that any two alphabets of size at least two are equivalent in the sense that words over any such alphabet can be polynomial-time computably coded by words over any other such alphabet. Thus, we shall often assume that the alphabet under consideration is $\{0,1\}$. Second, we shall sometimes define an equivalence relation E by "requiring xEy" for certain pairs x,y; this means that E is to be the smallest equivalence relation containing these pairs. Third, when we write an implication, like $\mathcal{P}$-Inv $\longrightarrow$ $\mathcal{NP}$-NF, we mean the assertion that every equivalence relation satisfying the hypothesis also satisfies the conclusion.

Proof of Theorem 1. We must show that each of the general implications

(1) $\mathcal{NP}$-Rec $\longrightarrow$ co-$\mathcal{NP}$-Rec

(2) co-$\mathcal{NP}$-Rec $\longrightarrow$ $\mathcal{NP}$-Rec

(3) $\mathcal{P}$-NF $\longrightarrow$ $\mathcal{NP}$-FM

is equivalent to the principle

(4) $\mathcal{NP}$=co-$\mathcal{NP}$.

It is obvious that (4) implies both (1) and (2). To see that it also implies (3), assume (4) and let E be an equivalence relation in $\mathcal{P}$. (Thus, we are actually assuming only $\mathcal{P}$-Rec rather than $\mathcal{P}$-NF for E.) The predicate

$$\exists z(z<y \wedge zEy),$$

saying that y is not the first member of its equivalence class, is then in $\mathcal{NP}$. By our assumption (4), so is the predicate that y is the first member of its class, and therefore so is the conjunction of this predicate with xEy. But this conjunction is precisely the predicate "y is the first member of the class of x", the graph of the first-member function for E. Since this function is polynomially bounded (in fact never increases length), it is in $\mathcal{NP}$ as required.

We turn to the converse implications. First, we assume (1) and deduce (4). Let A be an $\mathcal{NP}$ subset of $\{0,1\}^*$. Define E by requiring $(0x)E(1x)$ for x in A. Then E is in $\mathcal{NP}$ because A is. By (1), the complement of E is also in $\mathcal{NP}$. But the complement of A is trivially reducible to that of E,

$$x \notin A \longleftrightarrow (0x,1x) \notin E.$$

So the complement of A is in $\mathcal{NP}$, as required. The proof that (2) implies (4) is similar.

Finally, we assume (3) and deduce (4). As before, let $A \subseteq \{0,1\}^*$ be in NP. Then A can be represented in the form

$$A(x) \longleftrightarrow \exists y\ B(x,y)$$

for some polynomially bounded relation B in P; fix a polynomial p such that $B(x,y)$ always implies $|y|<p(|x|)$. Define an equivalence relation E on $\{0,1\}^* \times \{0,1\}^*$ by requiring

$$(x,0y)E(x,1^{p(|x|)}) \text{ whenever } B(x,y).$$

A normal form function for E (in fact, the last-member function) can be computed in polynomial time as follows. Given input (x,z) check whether z begins with 0, so $z=0y$, and check whether $B(x,y)$. If so, give output $(x,1^{p(|x|)})$; if not, give output (x,z). By our assumption (3), the first member function F for E is in NP. Therefore, so is the predicate $F(x,1^{p(|x|)})=(x,1^{p(|x|)})$. But this predicate is precisely the complement of A, so A is in co-NP as required. □

An alternate proof of the implication from (3) to (4) can be given by using the equivalence relation E constructed in [2] whose normal form problem has a solution in P while the first member function is Δ_2^P complete. The assumption (3) requires this Δ_2^P complete function, and therefore all Δ_2^P sets, to be in NP. In particular, it requires all co-NP sets to be in NP, which is the desired conclusion (4).

<u>Proof of Theorem 2</u>. We must show that the general implication

(1) NP-FM $\longrightarrow$ P-Rec

is equivalent to

(2) $NP \cap \text{co-}NP = P$.

Since NP-FM implies $(NP \cap \text{co-}NP)$-Rec, the implication from (2) to (1) is trivial. For the converse, assume (1) and let $A \subseteq \{0,1\}^*$ be in $NP \cap \text{co-}NP$. As in the proof of Theorem 1, define an equivalence relation E by requiring

$$(0x)E(1x) \text{ for } x \in A.$$

The first-member function for E is defined by

$$F(z)=\begin{cases} z & \text{if } z \text{ has the form } 0x \\ z & \text{if } z \text{ has the form } 1x \text{ with } x \notin A \\ 0x & \text{if } z \text{ has the form } 1x \text{ with } x \in A. \end{cases}$$

Because both of the predicates $x \in A$ and $x \notin A$ are in NP, so is the function F. By (1), we infer that E is in P. Then so is A, because

$$x \in A \longleftrightarrow (0x,1x) \in E. \quad \square$$

Proof of Theorem 3. We must prove that the implication

(1) $\mathcal{P}$-Inv $\longrightarrow$ $\mathcal{NP}$-NF for all flat E

is equivalent to each of

(2) the shrinking principle for $\mathcal{NP}$

and

(3) the uniformization principle for $\mathcal{NP}$.

We shall establish the implications $(1)\longrightarrow(2)\longrightarrow(3)\longrightarrow(1)$.

$(1)\longrightarrow(2)$. Assume (1), and let A and B be in $\mathcal{NP}$. They can be represented as

$$A(x) \longleftrightarrow \exists y\ P(x,y) \quad \text{and}$$
$$B(x) \longleftrightarrow \exists y\ Q(x,y)$$

where P and Q are polynomially bounded relations in $\mathcal{P}$. By suitably padding the witnesses y, we can arrange that all the y's that satisfy $P(x,y)$ or $Q(x,y)$ for any particular x have the same length; we can also arrange that all the y's satisfying $P(x,y)$ begin with 0 while those satisfying $Q(x,y)$ begin with 1. Define a flat equivalence relation E by requiring

$(x,y)E(x,z)$ if each of the pairs (x,y) and (x,z) satisfies P or Q.

An invariant function for E can be computed in polynomial time as follows. Given input (x,y), check whether it satisfies one of P,Q. If so, give output x; if not, give output (x,y). By our assumption (1), E has a normal form function F in $\mathcal{NP}$; fix such an F. The sets A' and B' defined by

$$A'(x) \longleftrightarrow \exists y(P(x,y) \text{ and } F(x,y) = (x,y)) \text{ and}$$
$$B'(x) \longleftrightarrow \exists y(Q(x,y) \text{ and } F(x,y) = (x,y))$$

are in $\mathcal{NP}$ and are obviously subsets of A and B respectively. Each x_0 in the union of A and B belongs to precisely one of A' and B' because F must fix exactly one member (the normal form) of the equivalence class consisting of all (x_0,y) satisfying P or Q and this member must satisfy exactly one of P and Q. Thus, A' and B' are as required in the shrinking principle.

$(2)\longrightarrow(3)$. It suffices to uniformize a polynomially bounded $\mathcal{P}$ binary relation R on $\{0,1\}^*$. Define $\mathcal{NP}$ predicates

$$A_i(x,u) \longleftrightarrow \exists z\ R(x,uiz) \quad \text{for } i=0,1.$$

Apply the shrinking principle (2) to get disjoint $\mathcal{NP}$ predicates $A_i' \subseteq A_i$ with $A_0' \cup A_1' = A_0 \cup A_1$. Then define R' by

$$R'(x,y) \longleftrightarrow R(x,y) \text{ and, for every prefix } ui \text{ of } y,$$
$$A_i'(x,u) \text{ but not } R(x,u).$$

Obviously $R' \subseteq R$. For each x, there is at most one y satisfying $R'(x,y)$, because suppose y' were another. Clearly, neither of y and y' can be a prefix of the other. Let u be the longest common prefix (possibly empty) of y and y'. Then $u0$ and $u1$ are prefixes of y and y', so, by definition of R', $A_0'(x,u)$ and $A_1'(x,u)$, contradicting disjointness. Finally, given an x such that $\exists y R(x,y)$, we construct a y' satisfying $R'(x,y')$ as follows. We inductively approximate the desired y' by defining progressively longer prefixes w of it. At each stage, we ensure that w is a prefix of some z satisfying $R(x,z)$ and we ensure that, for each prefix of the form ui of w (with $u \varepsilon \{0,1\}^*$ and $i \varepsilon \{0,1\}$), $A_i'(x,u)$. We start the induction with the empty word as w; the two induction hypotheses are trivially satisfied. After an approximation w has been defined, we check whether $R(x,w)$. If so, w can serve as y' and we are done. If not, we lengthen w as follows. Since w is a prefix of some z with $R(x,z)$ and since $z \neq w$ (as $R(x,w)$ fails), either $w0$ or $w1$ is a prefix of such a z. So (x,w) belongs to $A_0 \cup A_1 = A_0' \cup A_1'$. Let i be such that $A_i'(x,w)$. Then we take wi as our new w and observe that the induction hypotheses remain true. The (polynomial) boundedness of R implies that the approximations cannot grow indefinitely long; therefore they eventually terminate with the desired y'.

(3)$\longrightarrow$(1). Assume the uniformization principle (3), and let E be a flat equivalence relation with a polynomial-time computable invariant function F. We may assume that $|F(x)| \geq |x|$ for all x; if necessary replace $F(x)$ by $F(x)01^{|x|}$. (Here we use flatness of E.) Then the relation

$$R(u,x) \longleftrightarrow u=F(x)$$

is polynomially bounded. It is in $\mathcal{P}$ because F is, so let R' be an $\mathcal{NP}$ uniformizing relation, as in (3). Each equivalence class consists of the x's R-related to some specific u, and exactly one of these x's is R'-related to this u. Thus, we have a normal form function G defined by

$$G(x)=y \longleftrightarrow R'(F(x),y).$$

G obviously has $\mathcal{NP}$ graph and is polynomially bounded (as E is flat), so it is in $\mathcal{NP}$, as required. $\square$

<u>Proof of Theorem 4</u>. We shall construct a set $S \subseteq \{0,1\}^*$ such that the $\mathcal{NP}^S$ subsets A_0 and A_1 of $\{0\}^*$ defined by

$$A_i(0^n) \longleftrightarrow \exists x(|x|=n \text{ and } ix \varepsilon S)$$

do not have disjoint $\mathcal{NP}^S$ subsets with the same union. Then (2) of Theorem 3 fails relative to S, and therefore so do (1) and (3).

Enumerate all pairs of non-deterministic query machines (Turing machines equipped to interact with an as yet unspecified oracle) with polynomial clock bounds. Each such pair (M_0, M_1) defines a pair of NP^S sets A_0' and A_1', where A_i' is the set accepted by M_i^S, and our objective is to construct S so that no such pair is a shrinking of A_0 and A_1 as in (2). We construct S in stages. Initially, S is empty. At each stage, one or two words will be added to S and finitely many other words will be "frozen", meaning that they are not to be put in at any future stage. The nth stage will ensure that, regardless of what happens at later stages, the nth pair (M_0, M_1) in our enumeration will not define a shrinking of A_0 and A_1.

At the nth stage, proceed as follows. Let (M_0, M_1) be the nth pair of query machines, and let p be their polynomial clock bound. Fix a natural number d larger than the lengths of all the words added to S or frozen at earlier stages; also take d so large that $2^{d-1} > p(d)$. In what follows, S is the current S, consisting of what was added at previous stages.

<u>Case 1</u>. There exists an $x \varepsilon \{0,1\}^d$ such that $M_0^{S \cup \{0x\}}$ does not accept 0^d. Then we add $0x$ to S, so that 0^d becomes a member of A_0, and we freeze all words about which M_0 queried the oracle $S \cup \{0x\}$ in any of its computations (of length $\leq p(d)$) on input 0^d. We also freeze all words of length $d+1$. The freezing guarantees that, regardless of what happens at future stages, 0^d will belong to A_0 but not to A_1 and not to the set A_0' accepted by M_0 with the ultimate value of S as oracle. Then clearly A_0' and A_1' cannot be a shrinking of A_0 and A_1.

<u>Case 2</u>. There exists a $y \varepsilon \{0,1\}^d$ such that $M_1^{S \cup \{1y\}}$ does not accept 0^d. Proceed analogously to Case 1.

<u>Case 3</u>. Neither of the preceding cases occurs. For each x (resp. y) in $\{0,1\}^d$, fix a computation c_{0x} (resp. c_{1y}) of $M_0^{S \cup \{0x\}}$ (resp. $M_1^{S \cup \{1y\}}$) accepting 0^d. Each of these computations, having length $\leq p(d)$, involves at most $p(d)$ queries. So the number of pairs (x,y) for which c_{0x} queries about $1y$ or c_{1y} queries about $0x$ is at most $2 \cdot 2^d \cdot p(d)$ which is, by choice of d, less than 2^{2d}, the total number of pairs (x,y) from $\{0,1\}^d$. Fix a pair (x,y) from $\{0,1\}^d$ such that c_{0x} (resp. c_{1y}) does not query about $1y$ (resp. $0x$). Add both $0x$ and $1y$ to S; both c_{0x} and c_{1y} remain computations of M_0 and M_1 relative to the new oracle S. Freeze all words about which these computations query the oracle. No matter what happens at future stages, relative to the ultimate S, both M_0

and M_1 will accept 0^d (via computations c_{0x} and c_{1y}), so they will not define a shrinking of A_0 and A_1.$\square$

An alternate proof of Theorem 4 can be obtained by slightly modifying the construction in [2, Thm. 2(b)] of an oracle relative to which P-Inv $\not\rightarrow$ P-NF for flat E. The modification consists of accomodating the non-deterministic nature of the machine M in that construction by choosing specific computations rather than referring to "the computation" on a certain input. The following proof is a similar modification of the proof of Theorem 2(a) in [2].

Proof of Theorem 5. Enumerate all nondeterministic query machines with polynomial clock bounds. We shall construct the desired equivalence relation E on $\{0,1\}^*$ in stages, the nth stage ensuring that, no matter what happens later, the nth query machine with oracle E will not compute an invariant function for E. Initially, E is empty. At each stage, a pair (x,y) along with the symmetric pair (y,x) may be added to E, and finitely many other pairs may be frozen, i.e., made ineligible for future membership in E.

At the nth stage, let E be the equivalence relation constructed in the previous stages, let M be the nth query machine, let p be its clock bound, and let d be a natural number (1) larger than $|x|$ and $|y|$ for all pairs (x,y) previously added to E or frozen and (2) so large that $p(d)<2^{d-1}$.

Case 1. For some $x\varepsilon\{0,1\}^d$, M^E on input x has no successful computation or has successful computations with distinct outputs. Freeze all pairs about which any of these computations (of length$\leq$p(d)) queried E. This ensures that M, with the ultimate E as oracle, fails to compute a function, hence fails to solve the invariant problem for E.

Henceforth, we assume that Case 1 did not occur, so M^E on input x has successful computations, and they all produce the same output which we call x'.

Case 2. There are distinct $x,y\varepsilon\{0,1\}^d$ with x'=y'. Again, we merely freeze all pairs about which any computations of x',y' queried E. This ensures that M, with the ultimate E, produces the same output on the inequivalent inputs x and y and therefore fails to solve the invariant problem for E.

Case 3. $x'\neq y'$ for all $x\neq y$ in $\{0,1\}^d$. For each x in $\{0,1\}^d$, fix one successful computation c_x of M^E on input x. The number of pairs (x,y) of distinct elements of $\{0,1\}^d$ such that c_x or c_y contains a query about (x,y) or (y,x) is, because of the clock bound, at most $2\cdot 2^d\cdot p(d)$. This is less than the total number $2^d(2^d-1)$ of the pairs of distinct elements in $\{0,1\}^d$ because $p(d)\leq 2^{d-1}-1$. So we can find a pair (x,y) such that neither c_x nor c_y queries about (x,y) or (y,x). Add (x,y) and (y,x) to E and freeze all pairs about which c_x and c_y queried E. This ensures that, with the ultimate E as oracle, M will produce, via the computations c_x and c_y, distinct outputs x' and y' for the equivalent inputs x and y. Thus, M^E will fail to solve the invariant problem for E. $\square$

4. Open problems

We list in this section three problems arising from the results presented in the previous sections. The first two problems are to fill obvious gaps in those results. The third concerns the relationship between the shrinking principle (or equivalently the uniformization principle) introduced in Theorem 3 and the familiar principles used in Theorems 1 and 2.

1. Does Theorem 3 remain true if the restriction to flat equivalence relations is removed? If not, find a natural principle equivalent to $\mathcal{P}$-Inv $\longrightarrow \mathcal{NP}$-NF for arbitrary E. If one allows arbitrary equivalence relations then there seems to be no way to make the relation of u and x "x is in the class with invariant u" polynomially bounded. The value of the invariant function on an equivalence class may be much shorter than any element of the class, and there is no quick way to tell whether this occurs if one is given only an element of the class.

2. Find a natural principle (other than a trivial reformulation) equivalent to the general validity of the implication $\mathcal{P}$-Rec $\longrightarrow \mathcal{NP}$-Inv. This implication follows from $\mathcal{NP}$=co-$\mathcal{NP}$, because, in the proof of (4) $\longrightarrow$ (3) in Theorem 1, we used only $\mathcal{P}$-Rec, not $\mathcal{P}$-NF. But there seems to be no reason to expect it to be equivalent to $\mathcal{NP}$=co-$\mathcal{NP}$.

Before stating the third problem, we observe that $\mathcal{NP}$=co-$\mathcal{NP}$ trivially implies the shrinking principle for $\mathcal{NP}$: A and B can be shrunk to A-B and B.

3. Find an oracle relative to which the shrinking principle holds but $\mathcal{NP}\neq$co-$\mathcal{NP}$. Better yet, find an oracle relative to which the shrinking principle holds and $\mathcal{NP}\cap$co-$\mathcal{NP}=\mathcal{P}\neq\mathcal{NP}$.

Appendix. The Borel analog

The recognition, invariant, normal form, and first member problems are all equivalent in the context of recursion theory but inequivalent (under suitable hypotheses) in the context of complexity theory. The reason is basically that in recursion theory, if one is given x, one can search through all predecessors of x, whereas in complexity theory one cannot. In simpler (perhaps oversimplified but suggestive) terms, the power set operation preserves finiteness but does not preserve polynomial size. It also does not preserve countability. This suggests that complexity theory should be somewhat analogous to a theory of computation that deals with countable objects as ordinary recursion theory deals with finite objects.

Such a theory, in which countably long programs operate on countably long inputs (say infinite strings of 0's and 1's) using countably much space and time to produce countably long outputs, was introduced by Kleene [6] as a special case of his theory of computation for objects of higher type. Specifically, Kleene recursion relative to the type two object ^2E (embodying quantification over natural numbers) and arbitrary type 1 objects serves as a computation theory of the sort described. If we consider, for simplicity, functions and predicates defined on the set $\{0,1\}^\infty$ of infinite binary sequences, then, as Kleene essentially showed in [6] (see also [4], Ch. VI), the functions and predicates computable relative to ^2E and type one objects are precisely the Borel functions and predicates with respect to the natural topology on $\{0,1\}^\infty$. This topology has as basic open sets the sets

$$B_w=\{x\varepsilon\{0,1\}^\infty \mid w \text{ is a prefix of } x\}$$

for all $w\varepsilon\{0,1\}^*$; it is determined by the metric that assigns to any two distinct $x,y\varepsilon\{0,1\}^\infty$ the distance 2^{-k} where k is the length of the longest common prefix of x and y.

It is therefore reasonable to expect an analogy between $\mathcal{P}$ and the class $\mathcal{B}$ of Borel relations and functions. The analog of $\mathcal{NP}$ would be the class $\mathcal{A}$ of sets obtainable from Borel sets by existential quantification over $\{0,1\}^\infty$, the analytic sets. Such an analogy has been suggested in [5] and [8].

The Borel theory is somewhat simpler than the polynomial-time theory because of Souslin's classical result [7, Thm. 2E.2] that $\mathcal{A}\cap$co-$\mathcal{A}=\mathcal{B}$. Thus, for example, there is no difference between $\mathcal{A}$ and $\mathcal{B}$ for (total) functions, since the graph of a function, if in $\mathcal{A}$, is automatically also in co-$\mathcal{A}$. (See the analogous discussion for $\mathcal{NP}$ in Section 2.)

The diagram of trivial implications in the Borel situation is thus a bit simpler than the analogous diagram in Section 2.

$$\mathcal{B}\text{-FM} \longrightarrow \mathcal{B}\text{-NF} \longrightarrow \mathcal{B}\text{-Inv} \longrightarrow \mathcal{B}\text{-Rec} \begin{array}{l} \nearrow \mathcal{A}\text{-Rec} \\ \searrow \text{co-}\mathcal{A}\text{-Rec.} \end{array}$$

Here the first member problem refers to the lexicographic ordering of $\{0,1\}^\infty$; this is not a well-ordering, so some equivalence relations have no first member function.

The following theorem shows (without oracles) that no new implications can be added to this diagram.

Theorem 6. (a) There is an equivalence relation on $\{0,1\}^\infty$ such that the normal form problem has a Borel solution and the first member function exists but is not Borel.

(b) There is an equivalence relation on $\{0,1\}^\infty$ whose invariant problem has a Borel solution but whose normal form problem does not.

(c) There is a Borel equivalence relation whose invariant problem has no Borel solution.

(d) There is an analytic but not co-analytic equivalence relation.

(e) There is a co-analytic but not analytic equivalence relation.

Proof. All but (c) are quite easy, using the techniques of Section 3 and results of classical descriptive set theory. For (d) and (e), we observe that, if either of these were false, we could argue just as in the proof of Theorem 1 (from (1) or (2) to (4)) to obtain $\mathcal{A}=\text{co-}\mathcal{A}$ which is known to be false [7,§1E.3]. Similarly, if (b) were false, we could argue as in the proof of Theorem 3 to obtain the shrinking principle for $\mathcal{A}$, which is also known to be false [7, Ex. 4B.10-12]. Part (a) would be just as easy, using the proof of $(3) \longrightarrow (4)$ in Theorem 1, if it were not for the need to ensure that each equivalence class has a first member. Inspection of the proof just cited shows that we need to know that every $\mathcal{A}$ set A can be represented in the form

$$A(x) \longleftrightarrow \exists y\ B(x,y)$$

where B is a Borel relation and for each x in A there is a first $y \in \{0,1\}^\infty$ satisfying B(x,y). This does not seem to be in the literature explicitly but it follows easily from two known facts. First, if we allow y to range over $\{0,1,2,\ldots\}^\infty$ instead of $\{0,1\}^\infty$, then the desired representation is possible; in fact, B can be taken to be a closed set and every closed subset of $\{0,1,2,\ldots\}^\infty$ has a (lexicographically) first element [7,§1E]. Second, there is an order-preserving homeomorphic embedding of $\{0,1,2,\ldots\}^\infty$ into $\{0,1\}^\infty$ with Borel range, namely, $(y_0, y_1, y_2,\ldots)$ is sent to $1^{y_0}01^{y_1}01^{y_2}0\ldots$.

Finally, we prove (c). The proof uses the standard "coin-flipping" measure on $\{0,1\}^\infty$ defined by giving each basic open set B_w measure $2^{-|w|}$. We need two facts about this measure (see [3,§22]). First, every Borel set is measurable. Second, the zero-one law: If $X \subseteq \{0,1\}^\infty$ is measurable and if changing finitely many components of an x in X cannot put it out of X, then the measure of X is 0 or 1.

The desired equivalence relation E contains (x,y) if and only if x and y differ in only finitely many components. This E is Borel; in fact it is the union of countably many closed sets. Suppose a Borel function F solves the invariant problem for E. For each natural number n, the two complementary Borel sets A_0^n and A_1^n given by

$$A_i^n = \{x \mid \text{The } n^{\text{th}} \text{ component of } F(x) \text{ is } i\}$$

satisfying the hypotheses of the zero-one law, so one has measure 0, the other measure 1. Let $i(n)$ be chosen so that $A_{i(n)}^n$ has measure 1. Then, by countable additivity of the measure,

$$\bigcap_n A_{i(n)}^n = \{x \mid F(x)=(i_0,i_1,i_2,\ldots)\}$$

has measure 1. But this set is a single equivalence class of E, because F solves the invariant problem. The equivalence classes of E are countable, hence cannot have measure 1. This contradiction completes the proof.

References

1. T. Baker, J. Gill, and R. Solovay, Relativizations of the P=?NP question, SIAM J. Computing 4(1975) 431-442.
2. A. Blass and Y. Gurevich, Equivalence relations, invariants, and normal forms, SIAM J. Computing, to appear.
3. E. Hewitt and K. Stromberg, Real and Abstract Analysis, Springer, 1965.
4. P.G. Hinman, Recursion-Theoretic Hierarchies, Springer, 1978.
5. T. John, Similarities between NP and analytic sets, preliminary report.
6. S.C. Kleene, Recursive functionals and quantifiers of finite types, I, Trans. Amer. Math Soc., 91(1959) 1-52.
7. Y. Moschovakis, Descriptive Set Theory, North-Holland, 1980.
8. M. Sipser, Borel sets and circuit complexity, extended abstract, 1982.
9. L.J. Stockmeyer, The polynomial-time hierarchy, Theoret. Comp. Sci., 3(1977) 1-22.

Recurrence relations for the number of labeled structures on a finite set

Chr. Blatter and E. Specker (Zürich)

Introduction

Let C be a class of finite binary structures. An element of C is a finite set S endowed with a binary relation R. C is assumed to be closed under isomorphism. With such a class C a function f_C is associated as follows:

$$f_C(n) := \# \text{ structures on } \{1,2,\ldots,n\} \text{ belonging to } C .$$

We give some examples

(1) Class U of all structures

$$f_U(n) = 2^{n^2}$$

(2) Class G of all graphs

$$f_G(n) = 2^{\binom{n}{2}}$$

(3) Class E of all equivalence relations

$$f_E(n) = B_n \quad \text{(Bell numbers)}$$

(4) Class $E2$ of all equivalence relations with two classes of same size

$$f_{E2}(2n) = \frac{1}{2}\binom{2n}{n}$$

(5) Class T of all trees

$$f_T(n) = n^{n-2} \quad \text{(Cayley)}$$

(6) Let A be a nonempty subset of $\mathbb{N}\setminus\{0,1,2\}$. $C(A)$ is the class of all graphs which are n-gons, $n \in A$.

In some of these and many other cases it is well known that f_C satisfies a linear recurrence relation mod m , i.e. there exist integers $k, a_1, \ldots, a_k$ $(1 \leq k)$ such that for all n

$$f_C(n) \equiv \sum_{j=1}^{k} a_j\, f_C(n-j) \bmod m$$

[5,6]. [6] contains an extensive list of references.

In the following we give rather general conditions on C for such relations to hold for all m .

Theorem

If C is the class of R-structures defined by a formula of monadic second order logic, then the function f_C satisfies for every m a linear recurrence relation mod m .

Corollary

If C is as in the theorem, then the function f_C(mod m) is periodic for sufficiently large n .

The proof of the theorem is based on the notion of "class of finite character". A class of R-structures is of finite character iff there exist finitely many R-structures $E_1, \ldots, E_r$ such that the following holds: For every R-structure E there is an i $(1 \leq i \leq r)$ such that for all R-structures S and all x in S :

$$S(E/x) \in C \iff S(E_i/x) \in C ,$$

where S(E/x) is the structure obtained from S by substituting E for x in a natural way. The theorem follows from

Theorem 1

If C is of finite character, then the thesis of the theorem holds.

Theorem 3

If C is as in the hypothesis of the theorem, then C is of finite character.

In the proof of this theorem, Ehrenfeucht-Fraïssé games play an essential role [3,4].

Referring to our examples, the classes introduced in (1), (2), (3) are definable by a formula of first order logic. The class of trees is definable by a formula of monadic second order (but not of first order) logic. In all these cases it is easy to give a direct proof that the character is finite; we shall carry this out for the case of trees. It is equally easy to verify that all the classes $C(A)$ introduced in (6) are of finite character. Since there are uncountably many of these, it follows that not all of them are definable by a formula of monadic second order. The class $E2$ is not of finite character and the function $f_{E2} \pmod 2$ is not periodic for large n, since the values of $f_{E2}(n)$ is odd iff n is an even power of 2. The class $E2$ is definable by a formula of second order logic.

The generalization to structures of several binary relations is immediate. Ternary relations, however, seem to present additional difficulties. The main results of this paper have been announced (and partly proved) in [1,2].

Modular counting

1. A <u>structure</u> for us is a pair $S := (X,R)$ consisting of a nonempty finite set X, the <u>carrier</u> of S, and a binary relation

$$R : X \times X \to \{0,1\}$$

on X. In keeping with common usage we let S denote the carrier of the structure S as well whenever feasible. The one point reflexive structure we denote by 1.

If $S = (S,R)$ is a structure and

$$E. : x \mapsto (E_x, R_x) \quad (x \in S)$$

is a family or list of structures with disjoint carriers E_x, indexed by the elements x of S, we denote by $S(E.)$ the compound structure on the set $\tilde{S} := \bigcup_{x\in S} E_x$ obtained by substituting the structures E_x into the points $x \in S$. The relation $\tilde{R}$ on $\tilde{S}$ is given by

$$\tilde{R}(e_1,e_2) := \begin{cases} R(x_1,x_2) & (e_i \in E_{x_i}\ ,\ x_1 \neq x_2) \\ R_x(e_1,e_2) & (e_1,e_2 \in E_x\ ,\ e_1 \neq e_2) \\ R(x,x) & (e_1 = e_2 \in E_x)\ . \end{cases}$$

This means: elements of different E_{x_i}'s are related by the rules valid in S whereas within an individual E_x we keep the relation R_x defined there but for the case $\tilde{R}(e,e)$, where (somewhat arbitrarily) the relation on S takes over.

If the family $b \mapsto E_b$ of structures to be substituted is restricted to a subset $B \subset S$, in particular to a single point $b \in S$, one may suppose that the list $b \mapsto E_b$ $(b \in B)$ is continued to all of S by the assignment $E_x := 1$ $(x \notin B)$. The resulting compound structure is denoted by $S(E./B)$ resp. by $S(E/b)$.

The following definition will be essential: A class C of structures, closed under isomorphism, is of <u>finite character</u> if there is a finite basis $\mathfrak{B}$ of structures and for any structure E a basis structure $E^{\S} \in \mathfrak{B}$ such that the following holds:

(§) For arbitrary compound structures $S(E.)$ one has

$$S(E.) \in C \quad \Longleftrightarrow \quad S(E.^{\S}) \in C\ .$$

We may assume $1 \in \mathfrak{B}$. The condition (§) is equivalent to the following:

(§') For arbitrary substitutions at a single point one has

$$S(E/b) \in C \quad \Longleftrightarrow \quad S(E^{\S}/b) \in C\ .$$

As a matter of fact a general compound structure founded on $S = \{b_1,\ldots,b_n\}$ is isomorphic to the structure

$$S(E_1/b_1)(E_2/b_2) \; \dots \; (E_n/b_n)$$

obtained by successively substituting at a single point; furthermore the order of the individual substitutions is irrelevant.

In the light of (§') the above definition may be reformulated as follows: Call two structures E , E' C-<u>equivalent</u> and write $E \, \rho^C \, E'$, if for arbitrary substitutions at a single point one has

$$S(E/b) \in C \quad \Leftrightarrow \quad S(E'/b) \in C \; .$$

Then C is of finite character iff there are only finitely many equivalence classes.

<u>Example.</u> Let T denote the class of trees and consider the set

$\mathfrak{B} := \{$  $\}$

of the four structures 1 , S_2 , S_3 , S_4 indicated by our drawing. We will show that $\mathfrak{B}$ is a basis for T .

If $E = 1$ then $E \in \mathfrak{B}$. If an E consisting of $n > 1$ isolated points is substituted at a univalent point b of a tree S , then $S(E/b)$ is a tree; if such an E is substituted at a zero- or multivalent point b of a tree S , then $S(E/b)$ is disconnected or contains a circle, so is not a tree, and finally, if S is not a tree, then $S(E/b)$ is not a tree either. Hence all structures E consisting of more than one isolated points are T-equivalent, in particular T-equivalent to S_2 . Now let E contain an edge but still be no tree. Then $S(E/b)$ will contain a triangle, if b is not isolated and will contain a non-tree as a component, if b is isolated. In other words, $S(E/b)$ will never be a tree, so $E \, \rho^T \, S_3$. Finally, if E is a tree on $n \geq 2$ points, then $S(E/b)$ will be a tree iff $S = 1$, so $E \, \rho^T \, S_4$.

For a given class C of structures we let $f_C(n)$ denote the number of C-structures on $\{1,2,\dots,n\}$. Since C remains fixed in the sequel we may drop the index C and write f instead of f_C , similarly with other objects attached to C .

Theorem 1

If C is a class of finite character, the function f (mod m) satisfies for every module m a linear recurrence relation.

Corollary

If C is a class of finite character, the function f (mod m) is periodic for sufficiently large n .

The proof of Theorem 1 is by induction on the number of prime divisors of m . In the process we shall actually prove a more general theorem (Theorem 2).

2. For a (possibly empty) set A we consider structures S on the carrier $A \cup \{1,2,\ldots,n\}$. A permutation g of A induces a g-image S^g of S through

$$R^g(a,b) := R(g^{-1}(a), g^{-1}(b))$$

where we have tacitly put $g := \mathbb{1}$ on $\{1,2,\ldots,n\}$.

Now let G be a group of permutations of the set A and $\mathbb{A}$ a set of binary relations on A which is invariant under G . We denote by $U(A,G,\mathbb{A};n)$ the set of structures

$$S := (A \cup \{1,2,\ldots,n\} , R)$$

with the following properties:

(2.1) $S \in C$,

(2.2) $R/A^2 \in \mathbb{A}$,

(2.3) $g \in G^* := G \setminus \{\mathbb{1}\} \Rightarrow S^g \neq S$.

Theorem 2

If C is of finite character the function

$$n \mapsto |U(A,G,\mathbb{A}; n)| \mod(|G|\cdot m)$$

satisfies for every $m \geq 1$ a linear recurrence relation.

Theorem 1 is a special case of Theorem 2, for

$$f(n) = |U(\emptyset,\{\mathbb{1}\},\emptyset;\, n)| \;.$$

3. Assume $m = 1$ and consider the equivalence relation

$$S \sim S' \;:\Longleftrightarrow\; \exists g \in G \;:\; S^g = S'$$

on $U(A,G,\mathbb{A};\, n)$. Because of (2.3) each equivalence class consists of $|G|$ elements, hence

$$|U(A,G,\mathbb{A};\, n)| = 0 \qquad \mathrm{mod}\ |G|$$

for all n .

4. For the induction step we count an even more general set of structures. Let $A,G,\mathbb{A}$ be as above and kept fixed. Let

$$(R_i', R_i'') \qquad (1 \leq i \leq r := 4^{|A|})$$

be the pairs of unary relations on A and $\vec{B} := (B_1,\ldots,B_r)$ be an r-tuple of disjoint sets, $B := \bigcup_{i=1}^{r} B_i$. We shall consider structures

$$S := (A \uplus B \uplus \{1,\ldots,n\} \;,\; R)$$

and their compound structures $S(E./B)$ for a given list $E. : B \to \mathbb{B}$ of basis structures. Denote by $V(\vec{B},E.;\, n)$ the set of such structures S with the following properties:

(4.1) $S(E./B) \in C$,

(4.2) $R/A^2 \in \mathbb{A}$,

(4.3) $$\left.\begin{aligned} R(a,b) &= R_i'(a) \\ R(b,a) &= R_i''(a) \end{aligned}\right\} \quad (a \in A \;,\; b \in B_i)$$

(4.4) $g \in G^* \Rightarrow S^g \neq S$.

We will show that the functions

(4.5) $n \mapsto v(\ldots; n) := |V(\ldots; n)| \mod(|G| \cdot m)$

satisfy a linear recurrence relation. In doing so we may assume Theorem 2 for proper divisors of m . Let $\mathfrak{F}$ be the set of all functions (4.5).

5. Let p be a prime divisor of m and let $\mathfrak{F}_o := \{v_j | 1 \leqslant j \leqslant s\}$ be the subset of those $v \in \mathfrak{F}$ for which

$$|B_i| \leqslant (p-1)|\mathfrak{B}| \qquad (1 \leqslant i \leqslant r) .$$

Lemma 1

For each $w \in \mathfrak{F}$ there are numbers $a_j \in \mathbb{Z}$ and a function $q : \mathbb{N} \to \mathbb{N}$ satisfying a linear recurrence relation, such that

$$w = \sum_{j=1}^{s} a_j v_j + q \qquad \mod(|G| \cdot m) .$$

6. We first show that Lemma 1 implies Theorem 2.

Proof

Consider a structure $S \in V(\vec{B}, E.; n)$. The last element n of S is in a certain way related to the individual elements of A , i.e., there is a certain i , $1 \leqslant i \leqslant r$, such that

$$R(a,n) = R_i'(a) , \quad R(n,a) = R_i''(a) \qquad (a \in A) .$$

If one adds the element n to the set B_i and inserts the structure 1 at the proper place into the list $E.$, one obtains objects $\vec{B}^{(i)}$ and $E.^{(i)}$, and S appears as an element of the set $V(\vec{B}^{(i)}, E.^{(i)}; n-1)$.

We thus have the natural identification

(6.1) $V(\vec{B},E.;n) = \dot{\bigcup}_{i=1}^{r} V(\vec{B}^{(i)}, E.^{(i)}; n-1)$.

Consider now a fixed $v_k \in \mathfrak{F}_o$. Because of (6.1) there are functions $w_i \in \mathfrak{F}$ $(1 \leq i \leq r)$ with

$$v_k(n) = \sum_{i=1}^{r} w_i(n-1) ,$$

and by Lemma 1 there are numbers $a_{ij} \in \mathbb{Z}$ and functions p_i satisfying a linear recurrence relation such that

$$w_i = \sum_{j=1}^{s} a_{ij}v_j + p_i \qquad \mathrm{mod}(|G|\cdot m) .$$

Putting $\sum_{i=1}^{r} a_{ij} =: b_{kj}$, $\sum_{i=1}^{r} p_i =: q_k$ we thus have

$$v_k(n) = \sum_{j=1}^{s} b_{kj}v_j(n-1) + q_k(n-1) ,$$

which shows that the $v_k \in \mathfrak{F}_o$ indeed satisfy a linear recurrence relation. Since

$$|U(A,G,\mathfrak{A};\cdot)| = v(\emptyset,\emptyset;\cdot) \in \mathfrak{F}_o ,$$

Theorem 2 follows.

<u>7.</u> Let A , G , $\mathfrak{A}$, $\vec{B}$, $E.$ be as above.

<u>Lemma 2</u>

If $|B_1| > (p-1)|\mathfrak{B}|$, there are an r-tuple of sets $\vec{B}^*$ and t (to be explained later) corresponding lists $E.^{\S k}$ $(1 \leq k \leq t)$ of structures, furthermore a function q satisfying a linear recurrence relation, such that

(7.1) $|B_1^*| = |B_1| - (p-1)$,

(7.2) $B_i^* = B_i \qquad (i \neq 1)$,

(7.3) $v(\vec{B},E.;n) = \sum_{k=1}^{t} v(\vec{B}^*,E.^{\S k};n) + q(n) \qquad \mathrm{mod}(|G|\cdot m)$.

Lemma 1 is a consequence of Lemma 2.

8. Proof of Lemma 2

We may assume

$$B_1 = B_1' \uplus B_1'' , \quad B_1' = \{b_o, b_1, \ldots, b_{p-1}\}$$

with

$$E_{b_\ell} = E_o \qquad (0 \leq \ell \leq p-1) .$$

Consider the cyclic permutation

$$h_o : B_1' \to B_1' , \quad b_\ell \mapsto b_{\ell+1}$$

(the index ℓ taken mod p) which together with G (acting on A) generates a permutation group H of order $|H| = |G| \cdot p$.

The essential step now is to split the set of structures $V := V(\vec{B}, E.; n)$ into two classes according to the effect of h_o :

$$V_1 := \{S \in V | S^{h_o} \neq S\} , \quad V_2 := \{S \in V | S^{h_o} = S\} .$$

9. We first investigate the class V_1 and begin with the following

Lemma 3

If $S \in V_1$, then $S^h \neq S$ for all $h \in H^*$.

Proof

Assume $S^h = S$ for some $h = gh_o^\ell$, $g \in G$. If $\ell = 0$ then $S^g = S$, hence $g = 1\!\!1$ because of (4.4), thus $h = 1\!\!1$. This leaves $\ell \neq 0$, hence $h_1 := h_o^\ell \neq 1\!\!1$. Since p is prime we have $S^{h_1} = S$ as well. From this and $S^{gh_1} = S$ we conclude $g \neq 1\!\!1$. Again by (4.4) we can find elements

$$c_1, c_2 \in A \cup B \cup \{1,2,\ldots,n\}$$

with $R(c_1,c_2) \neq R(g(c_1),g(c_2))$, and since g affects only the points of A we may assume $c_1 \in A$. On the other hand

$$(9.1) \qquad R(c_1,c_2) = R(gh_1(c_1),gh_1(c_2))$$
$$= R(g(c_1),gh_1(c_2)) ,$$

which implies $h_1(c_2) \neq c_2$, hence $c_2 \in B_1'$. This and (4.3) allow us to continue (9.1) in the following way:

$$= R(g(c_1),h_1(c_2)) = R(g(c_1),c_2)$$
$$= R(g(c_1),g(c_2)) .$$

So we have a contradiction to our assumption on c_1, c_2 which proves Lemma 3.

10. Put $A' := A \uplus \biguplus_{b\in B} E_b$. When the $S \in V_1$ are restricted to $A \uplus B$ and the E_b are substituted we obtain on A' a set of structures $\mathbb{A}'$. Because of Lemma 3 we may write

$$V_1(\vec{B},E.;n) = U(A',H,\mathbb{A}';n) .$$

We now apply Theorem 2 with the module $m' := \frac{m}{p}$ and draw the following conclusion: The function

$$n \mapsto |U(A',H,\mathbb{A}';n)| \qquad \mathrm{mod}(|H| \cdot \tfrac{m}{p})$$

satisfies a linear recurrence relation, and because of $|H| = |G| \cdot p$ this in turn implies that the function

$$q := |V_1(\vec{B},E.;\cdot)| \qquad \mathrm{mod}(|G|\cdot m)$$

satisfies a linear recurrence relation.

11. We now turn to the set V_2 which contains the structures from $V(\vec{B},E.;n)$ invariant under h_o. Put

$$B_1^* := \{b_o\} \uplus B_1'' , \quad \vec{B}^* := (B_1^*,B_2,\dots,B_r) ;$$

denote by $W(\vec{B};n)$ the set of structures

$$S = (A \cup B \cup \{1,2,\dots,n\}, R)$$

for which (4.2), (4.3), (4.4) hold (this means that $S(E./B) \in C$ is not required) and by $W_2(\dots)$ the set of those $S \in W(\dots)$ for which $S^{h_o} = S$. Finally let $\mathfrak{V}$ be the set of structures σ_k $(1 \leq k \leq t)$ on B_1' invariant under h_o (so $t := |\mathfrak{V}|$).

12. Lemma 4

The restrictions

$$\alpha_1(S) := S|A \cup B^* \cup \{1,2,\dots,n\}, \quad \alpha_2(S) := S|B_1'$$

define a bijective map

$$\alpha : W_2(\vec{B}; n) \to W(\vec{B}^*; n) \times \mathfrak{V} .$$

Proof

If $g \in G^*$ there is $c_1 \in A$ and c_2 with $R(c_1,c_2) \neq R(g(c_1),g(c_2))$. This already implies $(\alpha_1(S))^g \neq \alpha_1(S)$ in the case $c_2 \notin B_1'$. If $c_2 \in B_1'$ we also have $R(c_1,b_o) \neq R(g(c_1),b_o)$ because of (4.3), hence $(\alpha_1(S))^g \neq \alpha_1(S)$ in this case too. This means that α is a well defined map into the indicated cartesian product.

Assume now $S_1 \neq S_2$ but $\alpha_2(S_1) = \alpha_2(S_2)$. We then can find $c_1 \notin B_1'$ and c_2 with $R_1(c_1,c_2) \neq R_2(c_1,c_2)$. This already implies $\alpha_1(S_1) \neq \alpha_1(S_2)$ in the case $c_2 \notin B_1'$. If $c_2 \in B_1'$ we conclude from $S_i^{h_o} = S_i$ $(i = 1,2)$ that $R_1(c_1,b_o) \neq R_2(c_1,b_o)$, hence $\alpha_1(S_1) \neq \alpha_1(S_2)$ also in this case.

Finally for two given structures

$$\hat{S} \in W(\vec{B}^*; n), \quad \sigma \in \mathfrak{V}$$

consider the structure

$$S = (A \cup B \cup \{1,2,\dots,n\}, R) \in W_2(\vec{B}; n)$$

defined by

$$R(c_1,c_2) := \begin{cases} \hat{R}(c_1,c_2) & (c_1,c_2 \notin B_1') \\ R^{\sigma}(c_1,c_2) & (c_1,c_2 \in B_1') \\ \hat{R}(c_1,b_o) & (c_1 \notin B_1' \, , \, c_2 \in B_1') \, . \end{cases}$$

For this S one has $\alpha(S) = (\hat{S}, \sigma)$, hence Lemma 4 is proven.

13. For each $\sigma_k \in \mathfrak{V}$ we define a list $E_{.}^{k}$ on B^* as follows:

$$E_b^k := \begin{cases} E_b & (b \in B^* \, , \, b \neq b_o) \\ \sigma_k((E_o,\ldots,E_o)) & (b = b_o) \end{cases} .$$

This means that at the point b_o a compound structure has to be substituted, namely the h_o-invariant structure resulting from substituting E_o at every point of σ_k . To this structure there is an equivalent basis structure $E^{\S} \in \mathfrak{B}$ which allows us to form the definitive list

$$E_b^{\S k} := \begin{cases} E_b & (b \in B^* \, , \, b \neq b_o) \\ E^{\S} & (b = b_o) \end{cases} .$$

With these notations we have

$$(13.1) \qquad |V_2(\vec{B},E.;n)| = \sum_{k=1}^{t} |V(\vec{B}^*,E_{.}^{\S k};n)|$$

Proof

A structure S belongs to V_2 iff $S \in W_2(\vec{B};n)$ and $S(E./B) \in C$. Now by Lemma 4 the structures $S \in W_2(\vec{B};n)$ correspond bijectively to the pairs of structures from

$$W(\vec{B}^*;n) \times \{\sigma_1,\ldots,\sigma_t\} .$$

Consider such an S and let $\alpha_2(S) = \sigma_k$. Then one has

$$S(E./B) = \alpha_1(S)(E_{.}^{k}/B^*) ,$$

and because of (§) we may write

$$S(E./B) \in C \iff \alpha_1(S)(E.^k/B^*) \in C$$

$$\iff \alpha_1(S)(E.^{\S k}/B^*) \in C$$

$$\iff \alpha_1(S) \in V(\vec{B}^*, E.^{\S k}; n) .$$

From this we conclude that the structure S is counted on the left of (13.1) iff it is counted on the right.

14. Lemma 2 now follows from the conclusion of section 10 and (13.1).

Logic and classes of finite character

We recall the definition of the Ehrenfeucht-Fraïssé game $G_n(S_1,S_2)$ associated to R-structures S_1, S_2 and a natural number n [3,4]: A play of the game $G_n(S_1,S_2)$ consists of a sequence of n moves of the two players I, II. The ith move $(i = 1,\ldots,n)$ of player I is the choice of either an element or a subset of one of the structures S_1, S_2; the ith move of player II is the choice of an object of the same type as the choice of I in the ith move but in the other structure. So in the ith move two objects s_i^1, s_i^2 belonging to S_1 resp. S_2 are selected; both are either elements or both subsets. After a particular play of the game there have been chosen two sequences $(s_1^1,\ldots,s_n^1)$, $(s_1^2,\ldots,s_n^2)$. Player II has won this play iff these sequences are isomorphic, i.e. iff for all i, j $(1 \leq i, j \leq n)$ the following conditions are satisfied:

(1) If s_i^1, s_j^1 are elements (and therefore also s_i^2, s_j^2):

$$s_i^1 = s_j^1 \text{ iff } s_i^2 = s_j^2 \text{ ; } R(s_i^1,s_j^1) \text{ iff } R(s_i^2,s_j^2) .$$

(2) If s_i^1 is an element (and therefore also s_i^2) and s_j^1 is a

set (and therefore also s_j^2) :

$$s_i^1 \in s_j^1 \text{ iff } s_i^2 \in s_j^2 .$$

The notion of winning strategy for a player is defined in the usual way. Roughly speaking, a strategy tells the player what to do in any situation that might arise during play; the strategy is a winning strategy if the player wins irrespectively of the moves of his opponent. Clearly, if the structures S_1 , S_2 are isomorphic, player II has a winning strategy in the game $G_n(S_1,S_2)$.

The relation defined by "Player II has a winning strategy in the game $G_n(S_1,S_2)$ " will be abbreviated by "$S_1 \ \omega_n \ S_2$ " .

The following two lemmas are known [3,4]:

<u>Lemma 5</u>

For all n , the relation ω_n is an equivalence relation with a finite number of equivalence classes.

<u>Lemma 6</u>

For a formula Φ of monadic second order logic in the predicates R and $=$, there exists a natural number n such that for all R-structures S_1 , S_2 :

If $S_1 \omega_n S_2$ then $S_1 \models \Phi$ iff $S_2 \models \Phi$.

<u>Remark</u>

A formula of monadic second order logic is a formula in individual variables $x_o, x_1, \ldots$ and in set variables $Y_o, Y_1, \ldots$; prime formulas are of the type $x_1 = x_2$, $R(x_1,x_2)$, $x_1 \in Y_1$; formulas are quantified with respect to individual and set variables. The fact that a formula Φ holds in the structure S is abbreviated by $S \models \Phi$.

<u>Lemma 7</u>

If two R-structures E_1 , E_2 satisfy $E_1 \ \omega_n \ E_2$, then for all

R-structures S and all x in S : $S(E_1/x)\ \omega_n\ S(E_2/x)$.

Proof

Let τ be a winning strategy for player II in the game $G_n(E_1,E_2)$. We may (and will) assume that the choice of the empty set by player I is immediately followed by the choice of the empty set by II. Let S be an R-structure and x an element of S ; we put $S_k = S(E_k/x)$, $k = 1,2$. In order to define a winning strategy for player II in the game $G_n(S_1,S_2)$, a partial play of the game $G_n(E_1,E_2)$ is associated to each partial play of the game $G_n(S_1,S_2)$. A partial play of the game $G_n(S_1,S_2)$ defines two sequences $(s_1^1,\dots,s_i^1)$, $(s_1^2,\dots,s_j^2)$, where $i = j$ or $i = j+1$ according to whether it is I's or II's move; the objects s_h^k are elements or subsets of S_k $(k = 1,2)$, s_h^1 and s_h^2 being of the same type. We define

$$t_h^k = t(s_h^k)$$

as the following object of the structure E_k : If s_h^k is an element of the substructure E_k of S_k , then t_h^k is equal to s_h^k ; if s_h^k is another element of S_k , i.e. an element of $S\setminus\{x\}$, then t_h^k is the empty set. If s_h^k is a subset of S_k , then t_h^k is the restriction of this set to E_k . The strategy σ of player II (which we are going to define) will have the following property:
If II plays according to σ and if a partial play results in the sequences $(s_1^1,\dots,s_i^1)$, $(s_1^2,\dots,s_i^2)$ $(i \leqslant n)$, then

(1) The pair of sequences is isomorphic.

(2) If s_h^1 is an element of $S\setminus\{x\}$, then $s_h^2 = s_h^1$.

(3) If s_h^1 is an element of E_1 , then s_h^2 is an element of E_2 .

(4) If s_h^1 is a subset of S_1 , then
$s_h^1 \cap (S\setminus\{x\}) = s_h^2 \cap (S\setminus\{x\})$.

(5) If the pair $(t_1^1,\dots,t_i^1)$, $(t_1^2,\dots,t_i^2)$ is the associated pair, then for all j the choice of t_j^2 corresponds to the strategy τ of II $(1 \leqslant j \leqslant i)$.

In order to define a winning strategy for player II it suffices to extend a pair

$$(s_1^1,\dots,s_i^1)\ ,\ (s_1^2,\dots,s_{i-1}^2)\ —$$

where the pair

$$(s_1^1,\dots,s_{i-1}^1)\ ,\ (s_1^2,\dots,s_{i-1}^2)$$

satisfies conditions (1) - (5) — to a pair $(s_1^1,\dots,s_i^1)\ ,\ (s_1^2,\dots,s_i^2)$ satisfying the conditions (1) - (5). This is done as follows:

If s_i^1 is an element of $S\setminus\{x\}$, then s_i^2 is put equal to s_i^1 ; we then have $t_i^1 = t_i^2 = \emptyset$, satisfying condition (5). Otherwise we consider the associated pair $(t_1^1,\dots,t_i^1)\ ,\ (t_1^2,\dots,t_{i-1}^2)$, which corresponds to a partial play of the game $G_n(E_1,E_2)$, played by player II according strategy τ . Therefore, the strategy τ extends this pair to an isomorphic pair $(t_1^1,\dots,t_i^1)\ ,\ (t_1^2,\dots,t_i^2)$. In this case, the object s_i^2 is defined as follows: If s_i^1 is an element of E_1 , hence equal to t_i^1 , s_i^2 is put equal to t_i^2 , which is then an element of E_2 . If s_i^1 is a subset of S_1 , s_i^2 is defined as the union of $s_i^1 \cap (S\setminus\{x\})$ and t_i^2 ; s_i^1 being a set, also t_i^1 and t_i^2 are sets. From these definitions, it follows immediately that the sequences $(s_1^1,\dots,s_i^1)\ ,\ (s_1^2,\dots,s_i^2)$ satisfy conditions (2), (3), (4), (5). It remains to check the isomorphism of the pair. All relations R , $=$, ε being binary, it suffices to investigate the pairs $(s_h^1,s_j^1)\ ,\ (s_h^2,s_j^2)$, $1 \leq h$, $j \leq i$. Assume that one of the objects s_h^1 , s_j^1 - say s_h^1 - is an element of $S\setminus\{x\}$; then $s_h^2 = s_h^1$. Clearly , if s_j^1 is also an element, then $s_h^1 = s_j^1$ iff $s_h^2 = s_j^2$ and, if s_j^1 is a set $s_h^1 \in s_j^1$ iff $s_h^2 \in s_j^2$. Assume - besides s_h^1 being an element of $S\setminus\{x\}$ - that s_j^1 is an element; we then have to show

$$R(s_h^1,s_j^1) \quad \text{iff} \quad R(s_h^2,s_j^2)\ .$$

If s_j^1 is in $S\setminus\{x\}$, then $s_j^2 = s_j^1$ and the equivalence holds trivially. If s_j^1 is in E_1 , then $R(s_h^1,s_j^1)$ iff $R^S(s_h^1,x)$; s_j^2 being an element of E_2 , we also have $R(s_h^2,s_j^2)$ iff $R^S(s_h^2,x)$ and the equivalence is established because of $s_h^2 = s_h^1$.

Assume now that s_h^1 is an element of E_1 ; then $s_h^1 = t_h^1$ and $s_h^2 = t_h^2$, s_h^2 being an element of E_2 . If s_j^1 is an element of E_1 , then - in the case $s_h^1 \neq s_j^1$ -

$$R(s_h^1,s_j^1) \text{ iff } R^{E_1}(t_h^1,t_j^1) .$$

The isomorphism of the pair (t_h^1,t_j^1) , (t_h^2,t_j^2) implies $R^{E_1}(t_h^1,t_j^1)$ iff $R^{E_2}(t_h^2,t_j^2)$. Together with $R(s_h^2,s_j^2)$ iff $R^{E_2}(t_h^2,t_j^2)$ we obtain $R(s_h^1,s_j^1)$ iff $R(s_h^2,s_j^2)$. In the case $s_h^1 = s_j^1$, we also have $s_h^2 = s_j^2$ and therefore $R(s_h^k,s_j^k)$ iff $R^S(x,x)$, $k = 1,2$, yielding the desired result. The case of the equality relation instead of R is trivial.

Assume finally that s_j^1 is a set; then $s_h^1 \in s_j^1$ iff $t_h^1 \in t_j^1$; furthermore $t_h^1 \in t_j^1$ iff $t_h^2 \in t_j^2$ and $t_h^2 \in t_j^2$ iff $s_h^2 \in s_j^2$.

We thus have considered all cases, h and j being symmetric and no relation being defined if s_h^1 and s_j^1 are sets.

<u>Theorem 3</u>

The class of models of a formula of monadic second order logic in the binary predicates R , $=$ is a class of finite character.

<u>Proof</u>

Let Φ be a formula as in the theorem. According to Lemma 6, there exists an integer n such that for all R-structures S_1 , S_2 : $S_1 \omega_n S_2$ implies $(S_1 \models \Phi \iff S_2 \models \Phi)$. If C is the class of models of Φ , we have

$$(S_1 \models \Phi \iff S_2 \models \Phi) \text{ iff } (S_1 \in C \iff S_2 \in C) ,$$

in other words

$$S_1 \omega_n S_2 \text{ implies } (S_1 \in C \iff S_2 \in C) .$$

Let E_1 , E_2 be R-structures and assume $E_1 \omega_n E_2$. Then, by Lemma 7, for all R-structures S : $S(E_1/x) \omega_n S(E_2/x)$, i.e. $S(E_1/x) \in C \iff S(E_2/x) \in C$.

By the definition of the equivalence relation ρ^C we thus have for all E_1, E_2 :

$$E_1 \; \omega_n \; E_2 \Rightarrow E_1 \; \rho^C \; E_2 \, .$$

ω_n having finitely many classes by Lemma 5, the same holds for ρ^C, proving the theorem.

References

[1] Chr. Blatter et E. Specker, Le nombre de structures finies d'une théorie à caractère fini, Sciences Mathématiques, Fonds National de la Recherche Scientifique, Bruxelles 1981, 41-44

[2] Chr. Blatter and E. Specker, Modular periodicity of combinatorial sequences, Abstracts Am. Math. Soc. 4 (1983), 313

[3] A. Ehrenfeucht, Application of games to some problem of mathematical logic, Bull. Acad. Pol., Cl. III, 5 (1957), 35-37

[4] R. Fraïssé, Sur quelques classifications des systèmes de relations, Publ. Sci. Univ. Alger Ser. A, 1 (1954), 35-182

[5] I.M. Gessel, Combinatorial proofs of congruences, preprint

[6] T.S. Motzkin, Sorting numbers for cylinders and other classification numbers, Proc. of Symposia in pure mathematics 19 (1971), 167-176

Mathematisches Seminar
Eidgenössische Technische Hochschule Zürich
ETH-Zentrum
CH - 8092 Zürich

RECURSIVELY ENUMERABLE EXTENSIONS OF R_1 BY FINITE FUNCTIONS.

W. Menzel & V. Sperschneider
Institut für Informatik I
Universität Karlsruhe

Abstract.

Let R_1 be the total recursive functions from $\mathbb{N}$ to $\mathbb{N}$, and Fin the set of all partial functions from $\mathbb{N}$ to $\mathbb{N}$ having a finite initial segment of $\mathbb{N}$ as domain. Motivated by earlier studies on "simulation-universal automata" (BUCHBERGER&MENZEL 77/ MAIER,MENZEL&SPERSCHNEIDER 82/ MENZEL &SPERSCHNEIDER 82) we ask what it means that $R_1 \cup F$ is recursively enumerable (r.e.), for a subfamily F of Fin. We show that each such family F is, in a certain sense, very rich. A (simple) corollary is that it must be dense in Fin w.r.t. the usual product topology on Fin. As a consequence one obtains simple but useful necessary conditions on F to make $R_1 \cup F$ r.e. . We also consider the class $Ext(R_1) := \{F \subseteq Fin \mid R_1 \cup F \text{ r.e.}\}$ as a whole. It is also quite rich in structure (e.g. if viewed as an upper semi-lattice w.r.t. union of families of functions). A sufficient criterion for F to be in $Ext(R_1)$ provides examples of families "naturally in $Ext(R_1)$", thus demonstrating that richness. On the other hand, there are families which belong to $Ext(R_1)$ in a more nonstandard way.
Main results:

- There is an F in $Ext(R_1)$ which is itself not r.e. .
- For F in $Ext(R_1)$, F is r.e. iff F is "effectively dense in Fin", in some appropriate sense.

We conclude with a first look at $Ext^*(R_1)$, which is the quotient semi-lattice of $Ext(R_1)$ modulo finite families. We prove that $Ext^*(R_1)$ does not contain minimal elements.

Notation.

$\mathbb{N}$	the set of natural numbers, including 0
I	is called an <u>initial segment</u> of $\mathbb{N}$ iff $I=\mathbb{N}$ or $I=\{x \mid x<k\}$, for some $k \in \mathbb{N}$
$f(x)\downarrow$	f is <u>defined</u> at x
$f(x)\uparrow$	f is <u>undefined</u> at x
dom(f)	$\{x \mid f(x)\downarrow\}$, the <u>domain</u> of f
rg(f)	$\{f(x) \mid f(x)\downarrow\}$, the <u>range</u> of f

graph(f)	$\{(x,f(x)) \mid f(x)\downarrow\}$, the <u>graph</u> of f
$f\subseteq g$	$graph(f)\subseteq graph(g)$, g is an <u>extension</u> of f
$f\restriction A$	the <u>restriction</u> of f to the set A
$f\restriction x$	$f\restriction\{0,\ldots,x-1\}$
P_n	the <u>partial recursive</u> functions from $\mathbb{N}^n$ to $\mathbb{N}$
R_n	the <u>total recursive</u> functions from $\mathbb{N}^n$ to $\mathbb{N}$
Fin	$\{\delta\in P_1 \mid dom(\delta)$ is a finite initial segment of $\mathbb{N}\}$
$lg(\delta)=l$	for $\delta\in Fin$, iff $dom(\delta)=\{0,\ldots,l-1\}$, l is the <u>length</u> of δ
$\varphi_i^{(n)}$	is the function in P_n computed by Turing machine i
φ_i	$\varphi_i^{(1)}$
$\varphi_{i,t}(x)=y$	Turing machine i with input x stops within t steps with output y, and $i<t$, $x<t$, $y<t$
W_i	$dom(\varphi_i)$
$W_{i,t}$	$dom(\varphi_{i,t})$
$\lambda x,y.\langle x,y\rangle$	some standard pairing function, the corresponding component functions being $\lambda z.(z)_{.1}$ and $\lambda z.(z)_{.2}$

$F\subseteq P_1$ is <u>r.e.</u> iff there is some $i\in\mathbb{N}$ such that $F=\{\varphi_k \mid k\in W_i\}$; W_i is then called a <u>basis</u> for F

$(D_n)_{n\in\mathbb{N}}$	is the usual <u>canonical numbering</u> (coding) of all finite subsets of $\mathbb{N}$
$(\delta_n)_{n\in\mathbb{N}}$	is some fixed <u>canonical numbering</u> of Fin (obtained for example via our pairing function and the numbering $(D_n)_{n\in\mathbb{N}}$)

For $F\subseteq Fin$, in addition to being r.e., we say that F is <u>canonically enumerable</u> resp. <u>canonically decidable</u> iff $\{n \mid \delta_n\in F\}$ is r.e. resp. decidable.

<u>Introduction</u>.

R_1 is not r.e. , but one immediately obtains an r.e. family by adjoining Fin to R_1. We ask what r.e. families there are between R_1 and $R_1\cup Fin$. Thus define $Ext(R_1):=\{F\subseteq Fin \mid R_1\cup F$ is r.e.$\}$. Our motivation to investigate $Ext(R_1)$ is threefold.

(1) Questions concerning the existence of "simulation-universal automata" (see BUCHBERGER&MENZEL 77, MENZEL&SPERSCHNEIDER 82) or embedding families of total recursive functions into one single such function (see MAIER,MENZEL&SPERSCHNEIDER 82) aroused our interest in the family $R_1\cup\{\delta\in Fin \mid rg(\delta)\subseteq dom(\delta)\}$. (The question was whether this family is "r.e. modulo recursive isomorphism", i.e. whether there exists an r.e. subfamily of the family above which intersects each recursive isomorphism class of the functions in the given family. The answer is "no"

for the family above, hence a fortiory that family is not r.e. .) Proving that the family above is not r.e. involved a method of "projecting function families above R_1 onto subfamilies of R_1", which seemed not to be desribed in the literature so far. This method is presented in a lemma in section 2 below. It has "topological features" somehow, and yields a necessary condition C_1 for some F to be a member of $Ext(R_1)$. Reinforcing C_1, one obtains a sufficient condition C_2 for being an element of $Ext(R_1)$, which in essence is already known from the literature (see MAL'CEV or POUR-EL). C_1 is not sufficient and C_2 is not necessary. Concentrating on the gap between C_1 and C_2, one obtains less standard elements of $Ext(R_1)$, indicating the richness of this class of families of functions. Also, one is motivated to study the structure of $Ext(R_1)$, as partially ordered by inclusion, and of $Ext^*(R_1)$, which is the quotient semi-lattice of $Ext(R_1)$ modulo finite families.

(2) Some algorithms enumerating functions in stages, thereby fulfilling certain requirements, typically involve some method of eliminating mutual dependencies between certain of these requirements. This method, in simpler versions, has also been applied in common diagonal arguments in MAIER,MENZEL&SPERSCHNEIDER 82 and MENZEL&SPERSCHNEIDER 82. Typically, instead of regarding an infinite set of negative requirements simultaneously, it suffices to deal with a single one in a uniform way (this does not remove, of course, eventual dependencies between positive and negative requirements). We regard our results as case studies in order to get further insight into what can be done in such a situation.

(3) Consider the computable real numbers. They can be identified with the total recursive 0,1-functions. But, as these are not r.e., we do not obtain an effectively given datatype. To achieve recursive enumerability we can add the 0,1-functions in Fin (representing certain rational numbers). Of course, one has to pay for that: It is no more possible now to compute uniformly from the index of a function representing some real number x, with arbitrary precision, a sequence of rational intervals defining x. As it is the finite functions which cause this difficulty, one might wish to enumerate the total recursive 0,1-functions together with as few as possible finite ones. With the slight modification (which is indeed unimportant) that functions with arbitrary values are considered, a natural question would be: Are there minimal r.e. extensions of R_1 by functions in Fin? We will give the answer to this in our last theorem.

1. Definitions and elementary facts.

Definition.

For $T \subseteq R_1$ define $Ext(T) := \{F \subseteq Fin \mid T \cup F \text{ is r.e.}\}$.
For $F \subseteq Fin$ and $G \subseteq Fin$ define $F =^* G$ iff $(F \setminus G) \cup (G \setminus F)$ is finite.
For $F \subseteq Fin$ define F^* to be the equivalence class of F in Fin w.r.t. $=^*$.
For $T \subseteq R_1$ define $Ext^*(T) := \{F^* \mid F \in Ext(T)\}$.

Facts.

(1) $\{F^* \mid F \in Ext(T)\}$ defines a partition of Ext(T), i.e. for all $F \subseteq Fin$ and $G \subseteq Fin$ the following holds: $F \in Ext(T) \wedge F =^* G \Rightarrow G \in Ext(T)$

Proof. It suffices to prove that for each $F \in Ext(T)$ and $\delta \in F$ the family $F \setminus \{\delta\}$ is again in Ext(T). Let such an F and δ be given. Let B be an r.e. basis for $T \cup F$. Define

$$A := \{b \in B \mid \exists x < lg(\delta)\ [\ \varphi_b(x){\downarrow} \wedge \varphi_b(x) \neq \delta(x)\ ]\ \vee\ \exists x \geq lg(\delta)\ [\ \varphi_b(x){\downarrow}\ ]\ \}.$$

Then A is an r.e. basis for $T \cup (F \setminus \{\varepsilon \in Fin \mid \varepsilon \subseteq \delta\}$. Hence $T \cup (F \setminus \{\delta\})$ is, as the union of $\{\varphi_a \mid a \in A\}$ and the finite family $\{\varepsilon \in F \mid \varepsilon \subseteq \delta \wedge \varepsilon \neq \delta\}$, r.e. .

(2) Ext(T) (Ext*(T)) is an upper semi-lattice if we define the supremum of two elements F and G (F^* and G^*) to be $F \cup G$ ($(F \cup G)^*$). If Ext(T) is non-empty then Fin (Fin*) is the greatest element in Ext(T) (Ext*(T)). Ext(T) (Ext*(T)) need not be a lattice, in general. Ext(T) (Ext*(T)) has a least element iff T is r.e. .

Proof. Easy exercise which is left to the reader.

(3) For all $T \subseteq R_1$ the following conditions are equivalent:

(a) $Ext(T) \neq \emptyset$

(b) $\{e \mid \varphi_e \in T\}$ is a Σ_3-set in the arithmetical hierarchy

(c) T has a Σ_3-basis

Proof.

(a)⇒(b): Assume that Ext(T) is non-empty. Take some $F \subseteq Fin$ such that $T \cup F$ is r.e. . Let B be an r.e. basis for $T \cup F$. Then for each e, $\varphi_e \in T$ iff $\exists b\ [\ b \in B \wedge \forall x\ [\ \varphi_b(x){\downarrow} \wedge \varphi_e(x){\downarrow} \wedge \varphi_b(x) = \varphi_e(x)\ ]\ ]$. This proves (b).

(b)⇒(c): Trivial.

(c)⇒(a): Let B be a Σ_3-basis for T. Let P be a decidable predicate such that for all $e \in \mathbb{N}$ the following holds: $e \in B \Leftrightarrow \exists x\ \forall y\ \exists z\ Pexyz$.
For each e and x define in stages a function $\psi_{e,x}$ as follows:

$$\psi_{e,x} := \bigcup_{s \in \mathbb{N}} \psi^s_{e,x}\ ,\ \text{with } \psi^s_{e,x} \subseteq \psi^{s+1}_{e,x} \subseteq \varphi_e \text{ and } \psi^s_{e,x} \in Fin, \text{ as follows:}$$

$\psi^0_{e,x} := \emptyset$.

Let $\psi^s_{e,x}$ be already defined. Define $l := lg(\psi^s_{e,x})$. Then define

$$\psi^{s+1}_{e,x} := \begin{cases} \psi^s_{e,x} \cup \{(l,\varphi_e(l))\} & \text{if } \exists z \leq s+1 \; [\; \varphi_{e,s+1}(l)\downarrow \wedge \text{Pexlz} \;] \\ \psi^s_{e,x} & \text{otherwise} \end{cases}$$

Hence, initial segments $\varphi_e \upharpoonright \{0,\ldots,l\}$ are enumerated into $\psi_{e,x}$ provided that $\forall y \leq l \; [\; \varphi_e(y)\downarrow \wedge \exists z \text{ Pexyz} \;]$. Thus, if $\psi_{e,x}$ is total then $e \in B$ ("witnessed by x"). Conversely, if $e \in B$ then there is some x such that $\forall y \; \exists z$ Pexyz, and therefor $\psi_{e,x} = \varphi_e$. It follows that $\{\psi_{e,x} | e,x \in \mathbb{N}\}$ is an r.e. family (since $\psi_{e,x}$ can be computed uniformly in e and x) between T and T$\cup$Fin. Hence, Ext(T) is non-empty.

2. A necessary condition for membership in Ext(R_1).

Definition.

$F \subseteq$ Fin is *dense at* $\delta_0 \in Fin$ iff $\forall \delta \in \text{Fin} \; [\; \delta_0 \subseteq \delta \rightarrow \exists \varepsilon \in F \; [\delta \subseteq \varepsilon] \;]$.
$F \subseteq$ Fin is *dense in Fin* iff F is dense at every $\delta_0 \in$ Fin.

(F being dense in Fin means that it is dense there w.r.t. the usual product topology on Fin defined by the neighbourhoods $\{\varepsilon \in \text{Fin} | \delta \subseteq \varepsilon\}$, for $\delta \in$ Fin.)

Lemma.

Let $F \subseteq Fin$, $E \subseteq Fin$ and $\delta_0 \in Fin$ be given such that

(a) $E \cap F = \emptyset$

(b) E is canonically decidable

(c) E is dense at δ_0

(d) $\forall \varepsilon \in E \; \exists \varepsilon' \in E \; [\; \varepsilon \subseteq \varepsilon' \wedge lg(\varepsilon') = 1 + lg(\varepsilon) \;]$.

If $R_1 \cup F$ is r.e. then $\{f \in R_1 | \delta_0 \subseteq f \wedge \exists m \; \forall x \geq m \; f \upharpoonright x \in E\}$ is r.e., too.

Proof. Let F,E and δ_0 be given as above and assume that $R_1 \cup F$ is r.e. . Then $\{\psi \in R_1 \cup F | \delta_0 \subseteq \psi\}$ is r.e., too. Let B be an r.e. basis for this latter family. For each $\delta \in$ Fin with $\delta_0 \subseteq \delta$ define ext(δ) as follows:

$$\text{ext}(\delta) := \bigcup_{i \in \mathbb{N}} \varepsilon_i \text{ , with}$$

ε_0:=the first element in the canonical numbering of Fin such that $\delta \subseteq \varepsilon_0$ and $\varepsilon_0 \in E$ (ε_0 exists by assumption (c) and can be computed from δ by property (b))

ε_{i+1}:=the first element in the canonical numbering of Fin such that $\varepsilon_i \subseteq \varepsilon_{i+1}$, $\varepsilon_{i+1} \in E$ and $lg(\varepsilon_{i+1}) = 1 + lg(\varepsilon_i)$ (ε_{i+1} exists by assumption (d) and can be computed from ε_i by (b)) .

Then, for each $\delta \in Fin$ with $\delta_0 \subseteq \delta$, the following holds:

$\delta \subseteq ext(\delta)$, $ext(\delta) \in R_1$, $\exists m\ \forall x \geq m\ ext(\delta) \upharpoonright x \in E$, and a φ-index of $ext(\delta)$ can be computed from the canonical index of δ

Now, for each $b \in B$, $m \in \mathbb{N}$ and $t \in \mathbb{N}$, we define in stages a function $\psi_{b,m,t}$ as follows:

$$\psi_{b,m,t} := \bigcup_{s \in \mathbb{N}} \psi^s_{b,m,t} \quad \text{with} \quad [\ \psi^s_{b,m,t} \in Fin \Rightarrow \psi^s_{b,m,t} \subseteq \varphi_b\ ] \quad .$$

$$\psi^0_{b,m,t} := \begin{cases} \varphi_b \upharpoonright m & \text{if } m \geq lg(\delta_0),\ \forall x<m\ \varphi_{b,t}(x)\downarrow \text{ and } \varphi_b \upharpoonright m \in E \\ ext(\delta_0) & \text{otherwise} \end{cases}$$

$$\psi^{s+1}_{b,m,t} := \begin{cases} \varphi_b \upharpoonright (lg(\psi^s_{b,m,t})+1) & \text{if } \psi^s_{b,m,t} \in Fin \text{ and } \psi^s_{b,m,t} \in E \\ ext(\psi^s_{b,m,t}) & \text{if } \psi^s_{b,m,t} \in Fin \text{ and } \psi^s_{b,m,t} \notin E \\ \psi^s_{b,m,t} & \text{if } \psi^s_{b,m,t} \notin Fin \end{cases}$$

Remark. In the first case of the definition of $\psi^{s+1}_{b,m,t}$ we know that $\psi^s_{b,m,t}$ is in E, hence not in F. Since $\psi^s_{b,m,t} \subseteq \varphi_b$ and φ_b is an element of $R_1 \cup F$ we can conclude that φ_b is defined at $lg(\psi^s_{b,m,t})$. Hence, definition of $\psi^{s+1}_{b,m,t}$ makes sense.

The following facts are now immediately verified:

(a) $\psi_{b,m,t}$ is computable, uniformly in b,m and t

(b) $\delta_0 \subseteq \psi_{b,m,t}$ (see choice of B and definition of $\psi^0_{b,m,t}$)

(c) $\exists m_0 \forall x \geq m_0\ \psi_{b,m,t} \upharpoonright x \in E$ (remember the definition of $ext(\delta)$)

(d) for all $f \in R_1$ such that $\delta_0 \subseteq f \wedge \exists m_0\ \forall x \geq m_0\ f \upharpoonright x \in E$, there are b,m and t such that $\psi_{b,m,t} = f$
(as b take some φ-index for f in B, as m take some number $\geq lg(\delta_0)$ such that $\forall x \geq m\ f \upharpoonright x \in E$, as t take some number such that $\forall x<m\ \varphi_{b,t}(x)\downarrow$)

It follows that $\{f \in R_1 \mid \delta_0 \subseteq f \wedge \exists m_0\ \forall x \geq m_0\ f \upharpoonright x \in E\}$ is equal to $\{\psi_{b,m,t} \mid b \in B,\ m \in \mathbb{N} \text{ and } t \in \mathbb{N}\}$, hence it is r.e. . This proves the lemma.

The idea behind this lemma is establishing an appropriate uniform projection of families $R_1 \cup F$ onto corresponding subfamilies of R_1 thereby preserving recursive enumerability. This provides a means for proving $R_1 \cup F$ not to be r.e. .

Theorem.

Let F be a subfamily of Fin. Assume that there are $E \subseteq Fin$ and $\delta_0 \in Fin$ such that the following properties hold:

(a) $E \cap F = \emptyset$

(b) E is canonically decidable

(c) E is dense at δ_0

(d) $\forall \varepsilon \in E\ \exists \varepsilon' \in E\ \exists \varepsilon'' \in E\ [\ \varepsilon' \neq \varepsilon'' \ \wedge\ \varepsilon \subseteq \varepsilon' \ \wedge\ \varepsilon \subseteq \varepsilon'' \ \wedge\ lg(\varepsilon') = lg(\varepsilon'') = lg(\varepsilon)+1\ ]$

Then $R_1 \cup F$ *is not r.e. .*

Proof. Let F,E and δ_0 be such that (a)-(d) hold. Because of the lemma above it suffices to prove that $G := \{f \in R_1 \mid \delta_0 \subseteq f \wedge \exists m\ \forall x \geq m\ f \restriction m \in E\}$ is not r.e. . Assume that G is r.e. . Let B be an r.e. basis for G. By (c),(d) above, G is nonempty. Hence we can take some $g \in R_1$ such that $B = rg(g)$. Fix some $\varepsilon \in E$ with $\delta_0 \subseteq \varepsilon$ (ε exists by property (c)). Now define:

ε_0:=the first element in the canonical numbering of Fin such that $\varepsilon \subseteq \varepsilon_0$, $\varepsilon_0 \in E$, $lg(\varepsilon_0) = 1 + lg(\varepsilon)$, $\varepsilon_0(lg(\varepsilon)) \neq \varphi_{g(0)}(lg(\varepsilon))$
(ε_0 exists by property (d))

ε_{n+1}:=the first element in the canonical numbering of Fin such that $\varepsilon_n \subseteq \varepsilon_{n+1}$, $\varepsilon_{n+1} \in E$, $lg(\varepsilon_{n+1}) = 1 + lg(\varepsilon_n)$, $\varepsilon_{n+1}(lg(\varepsilon_n)) \neq \varphi_{g(n+1)}(lg(\varepsilon_n))$
(ε_{n+1} exists by property (d)).

Define $h := \bigcup_{n \in \mathbb{N}} \varepsilon_n$. Then h is computable (use property (b)), $\delta_0 \subseteq h$ and for all $x \geq lg(\varepsilon)$ we know that $h \restriction x$ is in E. Hence $h \in G$. But h is different from each function $\varphi_{g(n)}$, for $n \in \mathbb{N}$. Contradiction.

Corollary. *Every F in* $Ext(R_1)$ *is dense in Fin.*

Proof. Let F be in $Ext(R_1)$ and assume that, for some $\delta_0 \in Fin$, $F \cap \{\delta \mid \delta_0 \subseteq \delta\} = \emptyset$. Apply the theorem above with $E := \{\delta \mid \delta_0 \subseteq \delta\}$ to conclude that F is not in $Ext(R_1)$.

The last corollary could be proved directly without using the theorem. Indeed, the theorem is much more powerful, as is demonstrated by the following

Example. $\{\delta \in Fin \mid rg(\delta) \subseteq dom(\delta)\}$ is dense in Fin, but not an element of $Ext(R_1)$. (Apply the theorem with $E := \{\delta \in Fin \mid rg(\delta) \not\subseteq dom(\delta)\}$.)

Before giving more applications we prove another corollary which in a more drastic way says that families in $Ext(R_1)$ must be "rich".

Definition. For a function $Op:Fin \to \mathbb{N}$ we say that Op is

- *computable* iff the function $\lambda n.Op(\delta_n)$ is total recursive
- *selectively monotonic* iff for each $\delta \in Fin$ there are $\delta', \delta'' \in Fin$ such that $\delta' \neq \delta''$, $\delta \subseteq \delta'$, $\delta \subseteq \delta''$, $lg(\delta') = lg(\delta'') = 1 + lg(\delta)$, $Op(\delta) \leqq Op(\delta')$, $Op(\delta) \leqq Op(\delta'')$
- *strongly unbounded on F*, for some $F \subseteq Fin$, iff for each $\delta_0 \in Fin$ Op is unbounded on $\{\delta \in F \mid \delta_0 \subseteq \delta\}$.

Examples of functions Op which are computable, selectively monotonic and strongly unbounded on Fin are: $lg(\delta)$, $max(rg(\delta))$, the cardinality of $rg(\delta)$, number of gaps in $rg(\delta)$, number of arguments x such that δ has a "relative maximum" at x, the integral part of $max(rg(\delta))/1+lg(\delta)$, and many more.

Corollary. *If F is in $Ext(R_1)$ then every function $Op:Fin \to \mathbb{N}$ which is computable, selectively monotonic and strongly unbounded on Fin, is strongly unbounded on F, too.*

Proof. Let F in $Ext(R_1)$ be given. Assume that, for some function Op, Op is computable, selectively monotonic, strongly unbounded on Fin, but not strongly unbounded on F. This means that there is some $\delta_0 \in Fin$ such that $\{Op(\delta) \mid \delta_0 \subseteq \delta \wedge \delta \in F\}$ is contained in $\{0,\ldots,n\}$, for some $n \in \mathbb{N}$. Define $E := \{\delta \in Fin \mid \delta_0 \subseteq \delta \wedge Op(\delta) > n\}$. Then it follows that

(a) $E \cap F = \emptyset$

(b) E is canonically decidable (since Op is computable)

(c) E is dense at δ_0 (since Op is strongly unbounded on Fin)

(d) property "(d)" from the last theorem (since Op is selectively monotonic).

By the last theorem we can conclude that F is not in $Ext(R_1)$. Contradiction.

Remarks.

(1) The assumption that Op is selectively monotonic in the last corollary cannot be weakened by deleting the condition that $lg(\delta') = lg(\delta'') = 1 + lg(\delta)$ in the definition of selective monotonicity, as the following example shows:

$$F := \{\delta \in Fin \mid lg(\delta) \text{ is even}\} \text{ and } Op(\delta) := \begin{cases} 0 & \text{if } lg(\delta) \text{ is even} \\ lg(\delta) & \text{otherwise} \end{cases}$$

(2) The conclusion that Op being strongly unbounded on Fin implies Op being strongly unbounded on F cannot be strengthened to: For all δ_0, if Op is unbounded on $\{\delta \in Fin \mid \delta_0 \subseteq \delta\}$ then Op is unbounded on $\{\delta \in F \mid \delta_0 \subseteq \delta\}$. This is shown by the following example. Take $F := \{\delta \mid \exists x < lg(\delta)\ \delta(x) \neq 0\}$, $Op(\delta) := 0$ for $\delta \in F$ and $Op(\delta) := lg(\delta)$ for $\delta \notin F$, and $\delta_0 := \emptyset$. Then F is in $Ext(R_1)$, Op is computable, selectively monotonic, unbounded on Fin, but bounded on F.

(3) In order to show that some given $F \subseteq Fin$ is not in $Ext(R_1)$, one may hence proceed as follows. Look for computable functions Op which have bounded values on F. Among such functions try to find one which is selectively monotonic and strongly unbounded on Fin.
For example, in order to prove that $\{\delta \in Fin \mid rg(\delta) \subseteq dom(\delta)\}$ is not in $Ext(R_1)$ consider $Op(\delta)$:=integral part of $\max(rg(\delta))/1+lg(\delta)$.
Also, the first of our corollaries follows with $Op(\delta):=lg(\delta)$.

3. On the structure of $Ext(R_1)$.

We begin with a simple, but useful sufficient condition for membership in $Ext(R_1)$.

Definition. $F \subseteq Fin$ has the *limit property* iff every total recursive function from $\mathbb{N}$ to $\mathbb{N}$ is the ascending union of a sequence in F.

Theorem. *If $F \subseteq Fin$ is r.e. and has a canonically enumerable subfamily F_0 with the limit property, then F is in $Ext(R_1)$. Moreover, $R_1 \cup F$ is r.e. without repetitions, i.e. there is an r.e. basis B for $R_1 \cup F$ such that different programs in B compute different functions.*

Proof. See MAL'CEV, S.153 ff. From theorem 3 there it follows that $R_1 \cup F$ is r.e. without repetitions. The weaker assertion that $R_1 \cup F$ is r.e. can be proved directly in the following quite simple way. Let F and F_0 be as in the theorem above. Then there is a total recursive function f such that for all i

$$\varphi_{f(i)} = \lambda x. \begin{cases} \varphi_i(x) & \text{if } \exists y \geq x\ [\ \varphi_i(0)\downarrow \wedge \ldots \wedge \varphi_i(y)\downarrow \wedge \varphi_i \upharpoonright y+1 \in F_0] \\ \uparrow & \text{otherwise} \end{cases}$$

Since F_0 has the limit property it follows that $\{\varphi_{f(i)} \mid i \in \mathbb{N}\} = R_1 \cup F_0 \cup \{\emptyset\}$. By eventually deleting the empty function it follows that $R_1 \cup F_0$ is r.e., and so is $R_1 \cup F$ as the union of the two r.e. families $R_1 \cup F_0$ and F.

Applications. For $L \subseteq \mathbb{N}$ define $Fin(L) := \{\delta \in Fin \mid lg(\delta) \in L\}$.
(1) If L is r.e. and infinite then $R_1 \cup Fin(L)$ is r.e. .
(2) If L is r.e., L^c is infinite and L^c contains an infinite r.e. subset L_0 then $R_1 \cup Fin(L^c)$ is r.e. .

Proof.
(1) Apply the theorem above with $F := F_0 := Fin(L)$.
(2) Use the fact that for every r.e. set L the family $Fin(L^c)$ is r.e. (though not always canonically enumerable) and apply the theorem above with $F := Fin(L^c)$ and $F_0 := Fin(L_0)$.

These applications give rise to the following "Length-Problem":

For which sets $L \subseteq \mathbb{N}$ is $R_1 \cup Fin(L)$ r.e. ?

So far this problem has been solved for r.e. sets L by (1), and for the complements of non-simple sets by (2). It is treated in a separate paper by V. SPERSCHNEIDER. The fact used in part (2) of the proof above is shown there. As further results we mention the following theorems.

Theorem. If L is r.e., L^c is infinite and L is not hyper-simple then $R_1 \cup Fin(L^c)$ is r.e. .

Theorem. There is a maximal set M such that $R_1 \cup Fin(M^c)$ is r.e. .

For any simple set L such that $R_1 \cup Fin(L^c)$ is r.e. (for example Post's simple set or the maximal set from the second theorem above), we cannot prove $R_1 \cup Fin(L^c)$ to be r.e. by an application of the sufficient condition in this chapter, since for such an L $Fin(L^c)$ does not contain a canonically enumerable subfamily with the limit property.

We will now concentrate on the other condition in the sufficient criterion of this chapter, namely F being itself r.e. .

Definition. $F \subseteq Fin$ is called *effectively dense in Fin* iff there is some $f \in R_1$ such that, for all $n \in \mathbb{N}$, $\varphi_{f(n)} \in F \wedge \delta_n \subseteq \varphi_{f(n)}$, i.e. proving F to be dense in Fin is possible in a constructive way.

Theorem.

For F in $Ext(R_1)$, F is r.e. iff F is effectively dense in Fin.

Proof. Let F in $Ext(R_1)$ be given. Let $rg(g)$, with $g \in R_1$, be an r.e. basis for $R_1 \cup F$.

(a) Assume that F is r.e., say $F = \{\varphi_{h(i)} \mid i \in \mathbb{N}\}$ with $h \in R_1$. Define $f \in R_1$ by

$$f(n) := h((\mu\langle i,t\rangle . \delta_n \subseteq \varphi_{h(i),t})_1)$$

f is total since F is dense in Fin, and for all n it follows from the definition of f that $\varphi_{f(n)} \in F \wedge \delta_n \subseteq \varphi_{f(n)}$.

(b) Assume that F is effectively dense in Fin, with $f \in R_1$ being such that, for all n, $\varphi_{f(n)} \in F \wedge \delta_n \subseteq \varphi_{f(n)}$. For every i and m define

$$\psi_{i,m} := \begin{cases} \varphi_{g(i)} & \text{if } dom(\varphi_{g(i)}) \subseteq \{0,\dots,m-1\} \\ \varphi_{f(n)}, \text{ with } \delta_n := \varphi_{g(i)} \upharpoonright m & \text{otherwise} \end{cases}$$

$\psi_{i,m}$ guesses that $\varphi_{g(i)}$ is in Fin with length m. As long as this guess is not disproved, $\psi_{i,m}$ is like $\varphi_{g(i)}$. If eventually some $x \geq m$ is found such that $\varphi_{g(i)}(x)\downarrow$, then from now on $\psi_{i,m}$ is $\varphi_{f(n)}$ with n as above. It follows that F is equal to $\{\psi_{i,m} \mid i,m \in \mathbb{N}\}$. Hence F is r.e., since of course $\psi_{i,m}$ can be computed uniformly in i and m.

Corollary. *Being r.e. is hereditary upwards in $Ext(R_1)$, i.e. every family G in $Ext(R_1)$ which contains an r.e. subfamily F in $Ext(R_1)$ is itself r.e.* .

Proof. If F is effectively dense in Fin then G is effectively dense in Fin, too. So the corollary follows immediately from the theorem above.

Theorem. *There is a family F in $Ext(R_1)$ which is not r.e.* .

Proof. We enumerate a family G between R_1 and $R_1 \cup Fin$ such that $G \cap Fin$ is not effectively dense in Fin. Then, by the theorem above, $F := G \cap Fin$ is the desired family. Thus we have to enumerate a family G between R_1 and $R_1 \cup Fin$ such that for each $f \in R_1$

$$\exists n \in \mathbb{N}\ [\ \varphi_{f(n)} \notin G \cap Fin \vee \delta_n \not\subseteq \varphi_{f(n)}\ ]\ .$$

We enumerate G in such a way that for each $e \in \mathbb{N}$, regardless of whether φ_e (= the f from above) is total or not, the following holds:

$$\exists n_e \in \mathbb{N}\ [\ \varphi_{\varphi_e(n_e)} \notin Fin \vee \delta_{n_e} \not\subseteq \varphi_{\varphi_e(n_e)} \vee \varphi_{\varphi_e(n_e)} \notin G\ ]$$

(with the usual understanding that $\varphi_{\varphi_e(n_e)}$ is empty if $\varphi_e(n_e)\uparrow$).

Given e, we will make a special choice for n_e, namely define n_e by

$$\delta_{n_e} := \{(0,e)\}\ .$$

Let $g \in R_1$ be such that, for all e,

$$\varphi_{g(e)} = \varphi_{\varphi_e(n_e)}\ ,\text{ with } n_e \text{ as above.}$$

With these choices and abbreviations our task reads as follows:

Enumerate a family G between R_1 and $R_1 \cup Fin$ such that for each e:
$$[\varphi_{g(e)} \in Fin \wedge \varphi_{g(e)}(0)=e] \Rightarrow \varphi_{g(e)} \notin G\ .$$

Such a family G exists for arbitrary $g \in R_1$:

Lemma. *For each $g \in R_1$ there is an r.e. family between R_1 and $R_1 \cup Fin$ such that for all $e \in \mathbb{N}$: $[\ \varphi_{g(e)} \in Fin \wedge \varphi_{g(e)}(0)=e\ ] \Rightarrow \varphi_{g(e)} \notin G$* .

Proof. Let $g \in R_1$ be given. Uniformly in $i \in \mathbb{N}$ and $l \in \mathbb{N}$ we enumerate functions $\psi_{i,tot}$ and $\psi_{i,l}$. Our family G will then be defined as follows: $G := \{\psi_{i,tot} \mid i \in \mathbb{N}\} \cup \{\psi_{i,l} \mid i,l \in \mathbb{N}\}$. Enumeration of these functions proceeds in stages:

$$\psi_{i,tot} = \bigcup_{s \in \mathbb{N}} \psi^s_{i,tot} \quad \text{and} \quad \psi_{i,l} = \bigcup_{s \in \mathbb{N}} \psi^s_{i,l}\ ,$$ where the following properties hold:

$(*_i)$ For all s, $\psi^s_{i,tot} \subseteq \psi^{s+1}_{i,tot}$, $\psi^s_{i,tot} \in Fin$, $\psi^s_{i,tot} \subseteq \varphi_{i,s}$,

$$\psi^s_{i,l} \subseteq \psi^{s+1}_{i,l} \text{ and } [\ \psi^s_{i,l} \in R_1 \text{ or } [\ \psi^s_{i,l} \in Fin \wedge \psi^s_{i,l} \subseteq \varphi_{i,s}\ ]\ ]\ .$$

$\psi_{i,tot}$ and the $\psi_{i,l}$ are called *followers for* φ_i since we enumerate, as long as possible, only initial segments of φ_i into them. Sometimes, following φ_i via $\psi_{i,l}$, for some l, must be given up; if this happens at stage s then $\psi^s_{i,l}$ is extended to a total recursive function (and there is no longer a chance of $\psi_{i,l}$ becoming equal to φ_i). Following φ_i via $\psi_{i,tot}$ is never given up.

Our lemma is proved if we enumerate $\psi_{i,tot}$ and $\psi_{i,l}$, for i and l, in such a way that for all numbers i and e the following requirements hold:

$(*_i)$ from above

(P_i) $\varphi_i \in R_1 \Rightarrow [\ \psi_{i,tot} = \varphi_i$ or $\exists l\ \psi_{i,l} = \varphi_i\ ]$

$(N_{i,e})$ $[\ \varphi_{g(e)} \in Fin \wedge \varphi_{g(e)}(0) = e] \Rightarrow [\ \psi_{i,tot} \neq \varphi_{g(e)} \wedge \forall l\ \psi_{i,l} \neq \varphi_{g(e)}]$.

For a given i, the condition "$\varphi_{g(e)}(0)=e$" in requirement $(N_{i,e})$, for $e \in \mathbb{N}$, implies that at most one of the infinitely many requirements $(N_{i,e})$, for $e \in \mathbb{N}$, is non-trivial. More precisely:

- If $\varphi_i(0)\uparrow$ then all $(N_{i,e})$, for $e \in \mathbb{N}$, are trivially true.
- If $\varphi_i(0)\downarrow$ then all $(N_{i,e})$, for $e \neq \varphi_i(0)$, are trivially true.

(The proof uses $(*_i)$ and is an easy exercise.)

We summarize what we have to do in order to obtain the assertion of the lemma:

Enumerate (uniformly in i and l) in stages functions

$$\psi_{i,tot} = \bigcup_{s \in \mathbb{N}} \psi^s_{i,tot} \quad \text{and} \quad \psi_{i,l} = \bigcup_{s \in \mathbb{N}} \psi^s_{i,l}$$

such that for all $i \in \mathbb{N}$ the following requirements hold:

$(*_i)$ and (P_i) from above

(N_i) If $\varphi_i(0)\downarrow$ then, for $e := \varphi_i(0)$, the following holds:

$$[\ \varphi_{g(e)} \in Fin \wedge \varphi_{g(e)}(0) = e\ ] \Rightarrow [\ \psi_{i,tot} \neq \varphi_{g(e)} \wedge \forall l\ \psi_{i,l} \neq \varphi_{g(e)}]$$

Remarks. (P_i) and (N_i) now give us some hint how to enumerate $\psi_{i,tot}$ and the functions $\psi_{i,l}$. (P_i) says: "Try to enumerate φ_i via $\psi_{i,tot}$ or one of the $\psi_{i,l}$." This must succeed in case $\varphi_i \in R_1$. (N_i) says: "Wait until $\varphi_i(0)$ is defined and then, for $e := \varphi_i(0)$, try to avoid enumerating $\varphi_{g(e)}$ via $\psi_{i,tot}$ or one of the $\psi_{i,l}$." This must succeed if $\varphi_{g(e)} \in Fin \wedge \varphi_{g(e)}(0)=e$. At first sight, (P_i) and (N_i) seem to be incompatible: Let i be given such that $\varphi_i(0)\downarrow$, say $e := \varphi_i(0)$. We do not know from i whether φ_i is

total, nor do we know whether $\varphi_{g(e)}$ is in Fin or not. It could be the case that $\varphi_i=\varphi_{g(e)}$; even more it could, by accident, be the case that $i=g(e)$, hence we even *know* that $\varphi_i=\varphi_{g(e)}$. Then (P_i) says "Try to enumerate φ_i." , whereas (N_i) says "Try to avoid enumerating φ_i.".
This difficulty is solved by enumerating $\psi_{i,tot}$ and the $\psi_{i,l}$ according to a suitable strategy, which we now describe in an informal way.

Informal description of the algorithm enumerating $\psi_{i,tot}$ and $\psi_{i,l}$.

$\psi_{i,tot}$ and $\psi_{i,l}$ are empty as long as $\varphi_i(0)$ is undefined. If eventually, at some stage s, $\varphi_{i,s}(0)$ is defined then, with $e:=\varphi_i(0)$, enumeration proceeds as follows:

- $\psi_{i,tot}$ "guesses" that $\varphi_{g(e)}$ is total. It follows φ_i in such a way that, at each stage $t\geq s$, if $\varphi_{g(e),t}(0)\downarrow$ then $dom(\psi^t_{i,tot})$ is strictly contained in $dom(\varphi_{g(e),t})$. Thus, regardless of whether the underlying guess is correct or not, if $\varphi_{g(e)}$ is in Fin and $\varphi_{g(e)}(0)=e$ then $\psi_{i,tot}\neq\varphi_{g(e)}$. Furthermore, if the guess is correct and φ_i is total then $\psi_{i,tot}=\varphi_i$.
- $\psi_{i,l}$ "guesses" that $\varphi_{g(e)}$ is not total and l is the least number outside $dom(\varphi_{g(e)})$. As long as indeed $\varphi_{g(e)}(l)$ is undefined, all initial segments of φ_i of length $>l$ are enumerated into $\psi_{i,l}$. If eventually, at some stage t, $\varphi_{g(e),t}(l)$ is defined then we give up following φ_i via $\psi_{i,l}$ and extend the finite function $\psi^t_{i,l}$ to a total recursive function. As above, regardless of whether the underlying guess is correct or not, if $\varphi_{g(e)}$ is in Fin and $\varphi_{g(e)}(0)=e$ then $\psi_{i,l}\neq\varphi_{g(e)}$. Furthermore, if the guess is correct and φ_i is total then $\psi_{i,l}=\varphi_i$.

Finally we translate this informal description into precise algorithms.

Algorithm computing $\psi^s_{i,tot}$:

stage 0: $\psi^0_{i,tot}:=\emptyset$.

stage s+1: Case 1: $\varphi_{i,s+1}(0)\uparrow$. Then define $\psi^{s+1}_{i,tot}:=\emptyset$.

Case 2: $\varphi_{i,s+1}(0)\downarrow$, say $e:=\varphi_{i,s+1}(0)$.

2a: $\varphi_{g(e),s+1}(0)\uparrow$. Then define $\psi^{s+1}_{i,tot}:=\emptyset$.

2b: $\varphi_{g(e),s+1}(0)\downarrow$. Take the maximal $y\leq s+1$ such that $\{0,\dots,y-1\}\subseteq dom(\varphi_{i,s+1}) \wedge \{0,\dots,y\}\subseteq dom(\varphi_{g(e),s+1})$ (by case 2b y exists) and define

$\psi^{s+1}_{i,tot}:=\varphi_i\upharpoonright y$.

Algorithm computing $\psi^s_{i,l}$:

stage 0: $\psi^0_{i,l}:=\emptyset$.

stage s+1: Case 1: $\psi^s_{i,l}$ is total recursive, i.e. following φ_i via $\psi_{i,l}$ has been given up at some stage before via case 2b2. Then define $\psi^{s+1}_{i,l}:=\psi^s_{i,l}$.

Case 2: $\psi^s_{i,l}\in\text{Fin}$.

2a: $\{0,\ldots,l\}\not\subseteq\text{dom}(\varphi_{i,s+1})$. Define $\psi^{s+1}_{i,l}:=\emptyset$.

2b: $\{0,\ldots,l\}\subseteq\text{dom}(\varphi_{i,s+1})$. Define $e:=\varphi_{i,s+1}(0)$.

2b1: $\varphi_{g(e),s+1}(l)\uparrow$. Let $z\leq s+1$ be maximal such that $\varphi_{i,s+1}(0)\downarrow \wedge \ldots \wedge \varphi_{i,s+1}(z)\downarrow$ (note that $z\geq l$). Define $\psi^{s+1}_{i,l}:=\varphi_i\upharpoonright z+1$.

2b2: $\varphi_{g(e),s+1}(l)\downarrow$. Then define $\psi^{s+1}_{i,l}:=\psi^s_{i,l}\cup\{(y,0)\mid y\notin\text{dom}(\psi^s_{i,l})\}$. Say that *following φ_i via $\psi_{i,l}$ has been given up at stage s+1.*

Verification.

Lemma: For all i, requirement $(*_i)$ holds.

Proof. Look at the algorithms.

Lemma: For all i, requirement (P_i) holds.

Proof. Let φ_i be total recursive. Define $e:=\varphi_i(0)$. If $\varphi_{g(e)}$ is total then $\psi_{i,tot}=\varphi_i$ since in the algorithm computing $\psi^s_{i,tot}$, from some stage on, only case 2b occurs and the numbers y tend to infinity. If $\varphi_{g(e)}$ is not total and l is any number outside $\text{dom}(\varphi_{g(e)})$ then $\psi_{i,l}=\varphi_i$ since in the algorithm computing $\psi_{i,l}$, from some stage on, only case 2b1 occurs and the numbers z tend to infinity.

Lemma: For all i, requirement (N_i) holds.

Proof. Assume that $\varphi_i(0)\downarrow$ and, for $e:=\varphi_i(0)$, $\varphi_{g(e)}\in\text{Fin} \wedge \varphi_{g(e)}(0)=e$. Assume that $\psi_{i,tot}=\varphi_{g(e)}$. Take some s such that $\psi_{i,tot}=\psi^{s+1}_{i,tot}$ and $\varphi_{g(e)}=\varphi_{g(e),s+1}$. It follows that $\psi^{s+1}_{i,tot}$ must have been defined via case 2b. But in this case $\text{dom}(\psi^{s+1}_{i,tot})$ is strictly contained in $\text{dom}(\varphi_{g(e),s+1})$. Contradiction. In an analogous way one derives a contradiction from the assumption that $\psi_{i,l}=\varphi_{g(e)}$.

End of the proof of the theorem.

Remark. If $F \in Ext(R_1)$ is not r.e. then, by the corollary above, there is no r.e. subfamily of F in $Ext(R_1)$. Thus such an F is, in some weak sense, minimal in $Ext(R_1)$, where "weak" means that only r.e. families are considered. From chapter 1, fact (1) we know that $Ext(R_1)$ does not contain minimal elements in the usual sense of lattice theory (an element of a partially ordered set is called minimal iff there is no element strictly below it; a least element- if it exists -is the only minimal element). The question arises whether $Ext^*(R_1)$ contains minimal elements.

Theorem. *$Ext^*(R_1)$ does not contain minimal elements.*

Proof. Let $R_1 \cup F$ be r.e. . We enumerate a family $R_1 \cup G$ such that $G \subseteq F$ and $F \setminus G$ is infinite. For this let us define, for each y, $F_y := \{\psi \in R_1 \cup F \mid \psi(0)=y\}$. F_y is r.e. uniformly in y, i.e. there is some $\pi \in R_2$ such that, for all y, $F_y = \{\varphi_{\pi(y,x)} \mid x \in \mathbb{N}\}$. Uniformly in y, we enumerate a family $G_y \subseteq F_y$ such that $F_y \setminus G_y$ contains at least one finite function but no total function. Then, of course, $\bigcup_{y \in \mathbb{N}} G_y$ can be taken as the family $R_1 \cup G$ from above.

G_y is enumerated as follows: Guess some x such that $\varphi_{\pi(y,x)}$ is finite and guess its length l. Begin with guess $\langle x,l\rangle := 0$. As long as the guess "$lg(\varphi_{\pi(y,x)})=l$" is not disproved, generate followers for $\varphi_{\pi(y,z)}$ for all those z such that $lg(\varphi_{\pi(y,z)})>l$. If eventually the actual guess turns out to be wrong, give up all followers generated so far by extending them to total recursive functions, and start again with the new guess $\langle x,l\rangle := \langle x,l\rangle+1$. After a finite number of wrong guesses a correct guess must occur (since F is dense in Fin). From now on, all functions in F_y having a length greater than the guessed l or being total recursive are supplied with a follower which is never given up. Hence, G_y has the desired properties.

Suggestions for further research. (1) Investigate $Ext^*(R_1)$, as an upper semi-lattice, in more detail (density, co-atoms etc.).
(2) Consider $Ext^*(T)$, for arbitrary $T \subseteq R_1$. It is not difficult to write down some $T \subseteq R_1$ such that T is r.e., hence $\emptyset^*$ is the least element of $Ext^*(T)$, and $Ext^*(T)$ has atoms. This gives rise to asking: What semi-lattices appear among the semi-lattices $Ext^*(T)$, for $T \subseteq R_1$?

Literature. BUCHBERGER ,B. & MENZEL,W. (1977), Simulation-universal automata, Interner Bericht Nr. 14/77, Institut für Informatik I, Universität Karlsruhe

MAIER,W.,MENZEL,W. & SPERSCHNEIDER,V. (1982), Embedding properties of total recursive functions, Zeitschrift f. Math. Logik und Grundlagen der Mathematik, Band 28

MAL'CEV,A.I. (1970), Algorithms and recursive functions, Groningen

MENZEL,W. & SPERSCHNEIDER,V. (1982), Universal automata with uniform bound on simulation time, Information and Control, Vol. 52

POUR-EL,M. & HOWARD,W.A. (1964), A structural criterion for recursive enumeration without repetition, Zeitschrift f. Math. Logik u. Grundl. der Mathematik, Band 10

ROGERS,H.JR. (1967), Theory of recursive functions and effective computability,New York

SPERSCHNEIDER,V., The Length-Problem, in this lecture notes volume

On the complement of one complexity class in another.

Diana Schmidt

Institut für Informatik I, Zirkel 2,
75 Karlsruhe 1, Federal Republic of
Germany

Abstract. We investigate the diagonalization theorems in [7] (Theorem 23), [3] (Theorem 6) and [10] (main Theorem) and show that they can be strengthened so as to be applicable to most complexity classes, not just to those closed under polynomial-time reducibility. Thus the applications in [7], [3] and [10] (e.g. $\mathcal{P} \neq \mathcal{NP} \Rightarrow (\mathcal{NP} \setminus \mathcal{P})$ is not recursively presentable) are not peculiar to $\mathcal{P}$, $\mathcal{NP}$, PSPACE etc.; rather, they are examples of properties common to almost all "reasonable" complexity classes.

§1. Introduction.

In the following, we suppose that all sets are languages over some fixed alphabet Γ such that $|\Gamma| \geq 2$. We use ordinary upper-case letters A,B,C,... to denote subsets of Γ^*, and script uppercase letters $\mathcal{A},\mathcal{B},\mathcal{C},\ldots$ to denote classes of such subsets. In [3], Chew and Machtey proved a widely applicable diagonalization theorem:

Let $\mathcal{C}_1$ be a recursively presented list of infinite r.e. sets, $\mathcal{C}_2$ a recursively presented list of recursive sets which is closed under finite variations, and B a recursive set. Then

$$B \notin \mathcal{C}_2 \Rightarrow \{B \cap C \mid C \in \mathcal{P}\} \not\subseteq \mathcal{C}_1 \cup \mathcal{C}_2.$$

And in [10], Schöning proved a similar but different theorem which can be stated as follows:

Let $\mathcal{C}_1$ and $\mathcal{C}_2$ both be recursively presented lists of recursive sets which are closed under finite variations, and let A_2 be a recursive set. Then

$$\mathcal{P} \not\subseteq \mathcal{C}_1 \ \& \ A_2 \notin \mathcal{C}_2 \cup \{\emptyset,\Gamma^*\} \Rightarrow \{A \mid A \leq^p_m A_2\} \not\subseteq \mathcal{C}_1 \cup \mathcal{C}_2.$$

These theorems can be used to deduce easily results like:

$\mathcal{NP} \neq \mathcal{P} \Rightarrow \mathcal{NP} \neq \mathcal{P} \cup \{A \mid A \text{ is } \mathcal{NP}\text{-complete under } \leq^p_T\}$;

$\mathcal{NP} \neq \mathcal{P} \Rightarrow (\mathcal{NP} \setminus \mathcal{P})$ is not recursively presentable (even by r.e. indices);

the class of infinite subsets of Γ^* decidable in polynomial time is not recursively presentable (even by r.e. indices).

But the above theorems do not, as they stand, yield corollaries of the form "$\mathcal{C} \not\subseteq \mathcal{C}_1 \cup \mathcal{C}_2$" or "$(\mathcal{C} \setminus \mathcal{C}_1)$ is not recursively presentable" in which $\mathcal{C}$ is a complexity class properly contained in $\mathcal{P}$. The purpose of this note is to strengthen the thoerems so that they do yield such corollaries: For almost all "reasonable" complexity classes $\mathcal{C}$, $\mathcal{C}_1$ and $\mathcal{C}_2$,

$$\mathcal{C} \not\subseteq \mathcal{C}_1 \ \& \ \mathcal{C} \not\subseteq \mathcal{C}_2 \Rightarrow \mathcal{C} \not\subseteq \mathcal{C}_1 \cup \mathcal{C}_2.$$

In particular, $\mathcal{C} = \mathcal{C}_1 \cup \mathcal{C}_2 \Rightarrow \mathcal{C} = \mathcal{C}_1$ or $\mathcal{C} = \mathcal{C}_2$ ("most reasonable complexity classes cannot be the non-trivial union of two others");
and

$$\mathcal{C} \not\subseteq \mathcal{C}_2 \ \& \ \mathcal{C} \cap \mathcal{C}_2 \neq \emptyset \Rightarrow (\mathcal{C} \setminus \mathcal{C}_2) \text{ is not recursively presentable even by r.e. indices}$$

("the non-trivial complement of one complexity class in another is mostly hard to generate").

We prove the main theorems in §2 and their applications to particular complexity classes in §3. In §4 we use the results of §2 and §3 to prove stronger and more general versions of existing theorems on the structure of polynomial time reducibility.

§2 The main theorem.

We refer the reader to [10] for any unexplained notation. We shall use the following notation for (relativized and unrelativized) Turing machine complexity classes: If $A \subseteq \Gamma^*$ and F is a set of unary functions from $\mathbb{N}$ into $\mathbb{N}$,

$\mathrm{DTIME}^A(F) = \{X \mid$ there is an $f \in F$ and a deterministic f-timebounded Turing machine with oracle A which decides $X\}$;

$\mathrm{NTIME}^A(F) = \{X \mid$ there is an $f \in F$ and a nondeterministic f-timebounded Turing machine with oracle A which recognizes $X\}$;

$\mathrm{DTIME}(F) = \mathrm{DTIME}^{\emptyset}(F)$; $\mathrm{NTIME}(F) = \mathrm{NTIME}^{\emptyset}(F)$;
and for any complexity class $\mathcal{C}$
$\text{co-}\mathcal{C} = \{\Gamma^* \setminus X \mid X \in \mathcal{C}\}$.

The Turing machine space complexity classes $\mathrm{DSPACE}^A(F)$, $\mathrm{NSPACE}^A(F)$ etc. are defined analogously, except that the suffices "on-line" and "off-line" are added to indicate whether the input must be read from left to right only or not (this distinction is only important if the space bounds in F are less than linear) - we refer the reader to [4] for a comparison of the power of on-line and off-line Turing machines.

A class $\mathcal{C}$ of sets is recursively presentable (by recursive indices) if there is an effective enumeration of Turing machines $M_1, M_2, \ldots$, all of which halt on all inputs, such that $\mathcal{C} = \{L(M_i) \mid i \in \mathbb{N}\}$. $\mathcal{C}$ is merely recursively presentable by r.e. indices if the Turing machines are not required to halt on all inputs, i.e. if they merely enumerate the sets of $\mathcal{C}$ rather than deciding them. A class of functions from $\mathbb{N}$ into $\mathbb{N}$ is recursively presentable if there is an effective enumeration of Turing machines which compute the functions.

Definition 1. Let $k \in \mathbb{N}$, $k \geq 1$. $t: \mathbb{N} \to \mathbb{N}$ is time constructible on k tapes iff there is a k-tape Turing machine which on input w always stops after exactly $t(|w|)$ steps.

Lemma 2. For every recursive $t: \mathbb{N} \to \mathbb{N}$ there is a $t': \mathbb{N} \to \mathbb{N}$ which is time constructible on 1 tape and satisfies $t'(n) \geq t(n)$ for all $n \in \mathbb{N}$.

In order to prove a version of Schöning's theorem which is applicable to complexity classes at all levels, we need "gap sets" which are similar to the sets G[r] used in [7] but more easily decidable.

Definition 3. Let $r: \mathbb{N} \to \mathbb{N}$ satisfy $r(n) > 0$ for each $n \in \mathbb{N}$.

$$G^*[r] = \{x \in \Gamma^* \mid r(0) + r^2(0) + \ldots + r^{2n}(0) < |x| \leq r(0) + \ldots + r^{2n+1}(0) \text{ for some } n \in \mathbb{N}, \text{ or } |x| = 0\}.$$

Lemma 4. If $r: \mathbb{N} \to \mathbb{N}$ is time constructible on 1 tape and $r(n) > 0$ for all $n \in \mathbb{N}$, then $G^*[r] \in \mathrm{DTIME}(n)$.

Definition 5. $s: \mathbb{N} \to \mathbb{N}$ is binary space constructible iff there is a deterministic Turing machine which on input bin(n) always produces output bin(s(n)) and is O(log(s(n)))-tape-bounded.

Lemma 6. For every recursive $s: \mathbb{N} \to \mathbb{N}$ there is a binary space constructible $s': \mathbb{N} \to \mathbb{N}$ such that $s'(n) \geq s(n)$ for all $n \in \mathbb{N}$.

Lemma 7. If $r(n) > 0$ for all n and r is binary space constructible, then $G^*[r] \in \mathrm{DSPACE}_{\text{on-line}}(\log n)$.

Definition 8. Let $\mathcal{C}, \mathcal{K}$ be classes of subsets of Γ^*. $\mathcal{C}$ is closed under definition by $\mathcal{K}$-cases if and only if for all $A_1, A_2 \in \mathcal{C}$ and all $G \in \mathcal{K}$, $(G \cap A_1) \cup (\bar{G} \cap A_2) \in \mathcal{C}$; $\mathcal{C}$ is closed under $\mathcal{K}$-intersection if and only if for all $A \in \mathcal{C}$ and all $G \in \mathcal{K}$, $G \cap A \in \mathcal{C}$.

It turns out that almost all "reasonable" time complexity classes are closed under definition by DTIME(n)-cases, and almost all "reasonable" space complexity classes are closed under definition by $\mathrm{DSPACE}_{\text{on-line}}$

(log n)-cases, and that either of these properties is sufficient to make the following theorem hold:

Theorem A. Let $\mathcal{K} \in \{\mathrm{DTIME}(n),\ \mathrm{DSPACE}_{\text{on-line}}(\log n)\}$ and $\mathcal{C}$ be a class of recursive sets which is closed under definition by $\mathcal{K}$-cases. Let $\mathcal{C}_1, \mathcal{C}_2$ be recursively presentable classes of recursive sets such that $\mathcal{C}_1, \mathcal{C}_2$ are closed under finite variations. Then

$$\mathcal{C} \not\subseteq \mathcal{C}_1 \;\; \& \;\; \mathcal{C} \not\subseteq \mathcal{C}_2 \Rightarrow \mathcal{C} \not\subseteq \mathcal{C}_1 \cup \mathcal{C}_2.$$

Proof (essentially the same as in [10]). Suppose that $\mathcal{C}_1, \mathcal{C}_2$ are closed under finite variations and $P_1, P_2, \ldots, Q_1, Q_2, \ldots$ are effective enumerations of Turing machines which present $\mathcal{C}_1$ and $\mathcal{C}_2$ respectively. Given $A_1 \in \mathcal{C} \setminus \mathcal{C}_1$ and $A_2 \in \mathcal{C} \setminus \mathcal{C}_2$, define

$$r_1(n) = \max_{i \le n}\{|(\mu z)(|z|>n \;\&\; z \in L(P_i) \,\Delta\, A_1)|\},$$

$$r_2(n) = \max_{i \le n}\{|(\mu z)(|z|>n \;\&\; z \in L(Q_i) \,\Delta\, A_2)|\}.$$

r_1, r_2 are total (since $A_1 \notin \mathcal{C}_1$, $A_2 \notin \mathcal{C}_2$ and $\mathcal{C}_1, \mathcal{C}_2$ are closed under finite variations) and recursive (since A_1, A_2 are recursive and each P_i, Q_i halts on all inputs). Now let r be a time constructible function if $\mathcal{K} = \mathrm{DTIME}(n)$ and a binary space constructible function if $\mathcal{K} = \mathrm{DSPACE}_{\text{on-line}}(\log n)$, such that $r(n) \ge \max(r_1(n^2),\ r_2(n^2))$ for all $n \in \mathbb{N}$ (by Lemmas 2 and 6). Then $G^*[r] \in \mathcal{K}$ by Lemmas 4 and 7. Define $A = (G^*[r] \cap A_1) \cup (\overline{G^*[r]} \cap A_2)$. $A \in \mathcal{C}$ since $\mathcal{C}$ is closed under definition by $\mathcal{K}$-cases.

$A \notin \mathcal{C}_1$: If $A \in \mathcal{C}_1$ then there is a j such that $A = L(P_j)$.

Let $n \in \mathbb{N}$ be even and $r(0) + \ldots + r^n(0) \ge j$.
Then by the definition of r_1 there is a $z \in \Gamma^*$ such that
$r(0) + \ldots + r^n(0) < |z| \le r_1(r(0) + \ldots + r^n(0))$
and $z \in L(P_j) \,\Delta\, A_1 = A \,\Delta\, A_1$.

$$\begin{aligned} \text{Now } |z| &\le r_1(r(0) + \ldots + r^n(0)) \\ &\le r_1((r^n(0))^2) && \text{since } r_1 \text{ is monotonic} \\ &\le r(r^n(0)) && \text{by the definition of } r \\ &= r^{n+1}(0) \\ &\le r(0) + \ldots + r^{n+1}(0); \end{aligned}$$

hence $z \in G^*[r]$. But then $z \in A \Leftrightarrow z \in A_1$, contradiction.

$A \notin \mathcal{C}_2$: Analogous.

Hence $A \in \mathcal{C} \setminus (\mathcal{C}_1 \cup \mathcal{C}_2)$, and thus $\mathcal{C} \not\subseteq \mathcal{C}_1 \cup \mathcal{C}_2$.

Note that a recursive index for A can be calculated uniformly from recursive indices for A_1 and A_2 and recursive presentations of $\mathcal{C}_1$ and $\mathcal{C}_2$.

Similarly, almost all "reasonable" time complexity classes are closed under DTIME(n)-intersection and almost all "reasonable" space complexity classes are closed unter $DSPACE_{on-line}(\log n)$-intersection, and either of these properties is sufficient for the following theorem:

Theorem B. Let $\mathcal{K} \in \{DTIME(n), DSPACE_{on-line}(\log n)\}$ and $\mathcal{C}$ be a class of recursive sets which is closed under $\mathcal{K}$-intersection. Let $\mathcal{C}_1$ be a recursively presentable class of infinite r.e. sets and $\mathcal{C}_2$ a recursively presentable class of recursive sets which is closed under finite variations. Then

$$\mathcal{C} \not\subseteq \mathcal{C}_2 \Rightarrow \mathcal{C} \not\subseteq \mathcal{C}_1 \cup \mathcal{C}_2.$$

Proof (essentially the same as in [3]). Define r_2 as before, and r_1 by:

$$r_1(n) = \max_{i \leq n}\{|(\mu z)(|z|>n \ \& \ z \in L(P_i))|\}.$$

Then proceed as for Theorem A, but define $A = G^*[r] \cap A_2$.

$A \notin \mathcal{C}_2$ follows as before;

$A \notin \mathcal{C}_1$: If $A \in \mathcal{C}_1$ there is a j such that $A = L(P_j)$. Let $n \in \mathbb{N}$ be odd and $r(0) + \ldots + r^n(0) \geq j$. Then by definition of r_1 there is a $z \in \Gamma^*$ such that $r(0) + \ldots + r^n(0) < |z| \leq r_1(r(0) + \ldots + r^n(0)) \leq r(0) + \ldots + r^{n+1}(0)$ - hence $z \notin G^*[r]$ - and $z \in L(P_j) = A$, contradiction.

$A \in \mathcal{C}$ holds because $\mathcal{C}$ is closed under $\mathcal{K}$-intersection.

§3 Applications to the structure of complexity classes.

In order to apply the main theorem, we shall show that most complexity classes are closed under definition either by DTIME(n)-cases or by $DSPACE_{on-line}(\log n)$-cases (since all "reasonable" complexity classes contain the empty set, it follows that they are also closed under DTIME(n)-intersection resp. $DSPACE_{on-line}(\log n)$-intersection); and that even more classes of sets are recursively presentable and closed under finite variations.

Lemma 9. Let F be a set of unary functions from $\mathbb{N}$ into $\mathbb{N}$ which is closed under maximum (i.e. $f,g \in F \Rightarrow \lambda n.\max(f(n),g(n)) \in F$), A be any oracle set.

a) If we consider only Turing machines which always read the entire input first (which implies $f(n) \geq n$ for all $f \in F$, $n \in \mathbb{N}$), then the following complexity classes are closed under definition by DTIME(n)-cases: DTIME(F), NTIME(F), co-NTIME(F), $DTIME^A(O(F))$,

$\mathrm{NTIME}^A(O(F))$, co-$\mathrm{NTIME}^A(O(F))$.

b) If there is an $f \in F$ such that $f(n) \geq \log n$ for all $n \in \mathbb{N}$, then the following complexity classes are closed under definition by $\mathrm{DSPACE}_{\text{on-line}}(\log n)$-cases:
$\mathrm{DSPACE}^A_{\text{off-line}}(F)$, $\mathrm{DSPACE}^A_{\text{on-line}}(O(F))$, $\mathrm{DSPACE}_{\text{on-line}}(F)$,
$\mathrm{NSPACE}^A_{\text{off-line}}(F)$, co-$\mathrm{NSPACE}^A_{\text{off-line}}(F)$, $\mathrm{NSPACE}^A_{\text{on-line}}(O(F))$,
co-$\mathrm{NSPACE}^A_{\text{on-line}}(O(F))$, $\mathrm{NSPACE}_{\text{on-line}}(F)$, co-$\mathrm{NSPACE}_{\text{on-line}}(F)$.

The assertions for the relativized on-line complexity classes hold only if, in the definition of space complexity for oracle computations, the space used on the oracle tape is not ignored.

Remarks. By Lemma 9, the following complexity classes are closed under definition by either DTIME(n)-cases or $\mathrm{DSPACE}_{\text{on-line}}(\log n)$-cases:
DTIME(n), $\mathrm{DTIME}(n^2)$, $\mathrm{DTIME}(n^3)$,...,
$\mathcal{P}$, $\mathcal{NP}$, co-$\mathcal{NP}$, all classes Σ^p_k, Π^p_k and Δ^p_k in the polynomial hierarchy, $\mathcal{PH}$ (the union of the polynomial hierarchy), $\Sigma^{p,A}_k$, $\Pi^{p,A}_k$, $\Delta^{p,A}_k$ and $\mathcal{PH}^A$ for any oracle set A (see [1]), EXPTIME, NEXPTIME etc.; $\mathrm{DSPACE}_{\text{on-line}}(\log n)$, $\mathrm{DSPACE}_{\text{off-line}}(\log n)$, DSPACE(n), NSPACE(n), $\mathrm{DSPACE}(n^2)$,..., PSPACE, EXPSPACE etc; in fact, all "reasonable" complexity classes which spring to mind except space complexity classes with a space bound smaller than log n.

Thus, in the main theorem, $\mathcal{C}$ can be taken to be almost any "reasonable" complexity class. We shall see that the scope of possibilities for $\mathcal{C}_1$ and $\mathcal{C}_2$ is even wider, though somewhat different (for example, any time- or space-bounding functions which define $\mathcal{C}_1$ and $\mathcal{C}_2$ must be recursive, whereas for $\mathcal{C}$ this is not necessary). We extend Schöning's list ([10]) of recursively presentable classes which are closed under finite variations.

Definition 10. Let $\mathcal{C}$ be any class of sets, $\leq$ any reduction relation. We use

$(\mathcal{C}\leq)$ to denote $\{X \mid (\exists Y \in \mathcal{C})\ (Y \leq X)\}$ and
$(\leq \mathcal{C})$ to denote $\{X \mid (\exists Y \in \mathcal{C})\ (X \leq Y)\}$.

Note that if X is $\leq$-complete for $\mathcal{C}$ then
$\{\leq\text{-complete sets for } \mathcal{C}\} = \mathcal{C} \cap (\{X\} \leq)$.

Lemma 11. The following classes of sets are recursively presentable (by recursive indices) and closed under finite variations:

a) $DTIME^A(F)$, $NTIME^A(F)$, co-$NTIME^A(F)$, $DSPACE^A_{off\text{-}line}(F)$, $NSPACE^A_{off\text{-}line}(F)$, co-$NSPACE^A_{off\text{-}line}(F)$, $DSPACE^A_{on\text{-}line}(F)$, $NSPACE^A_{on\text{-}line}(F)$, co-$NSPACE^A_{on\text{-}line}(F)$,
for any recursively presentable set F of unary recursive functions and any recursive oracle A.

b) $(\mathcal{C}_1 \cap \mathcal{C}_2)$ (provided $\mathcal{C}_1 \cap \mathcal{C}_2 \neq \emptyset$) and $(\mathcal{C}_1 \cup \mathcal{C}_2)$ whenever $\mathcal{C}_1$ and $\mathcal{C}_2$ are themselves recursively presentable and closed under finite variations.

c) $(\leq \mathcal{C})$ whenever $\mathcal{C}$ is recursively presentable, $\mathcal{C} \not\subseteq \{\emptyset, \Gamma^*\}$, F is a recursively presentable set of unary recursive functions and $\leq$ is either the F-time-bounded or the F-space-bounded restriction of $\leq_T$, $\leq_m$, $\leq_{tt}$ or any of the restrictions of truth-table reducibility (see [6]), of the nondeterministic (see [6] §4) or strongly nondeterministic (see [7] §3) variant of any of these reduction relations, or of $\leq_R$ (see [8]).

d) $\mathcal{C}_1 \cap (\mathcal{C}_2 \leq)$ whenever $\leq$ is as in c), $\mathcal{C}_1$ and $\mathcal{C}_2$ are recursively presentable, $\mathcal{C}_1 \cap (\mathcal{C}_2 \leq) \neq \emptyset$ and $\mathcal{C}_1$ is closed under finite variations and $(\forall X \in \mathcal{C}_2)$ $(\forall$ finite $Y \subseteq \Gamma^*)(X \not\leq Y$ & $X \not\leq \Gamma^* \setminus Y)$. (The proofs are easy generalisations of proofs in [10].)

Remarks. By Lemma 11, the following classes are recursively presentable and closed under finite variations:
DTIME(n), DTIME(n^2), DTIME(n^3),..., $\mathcal{P}$, $\mathcal{NP}$, co-$\mathcal{NP}$, all classes Σ^p_k, Π^p_k and Δ^p_k in the polynomial hierarchy, $\mathcal{PH}$, EXPTIME, NEXPTIME etc; $DSPACE_{on\text{-}line}(\log n)$, $DSPACE_{off\text{-}line}(\log \log n)$, $DSPACE_{off\text{-}line}(\log n)$, DSPACE(n), NSPACE(n), DSPACE(n^2),..., PSPACE, EXPSPACE etc; {$\leq$-complete sets for $\mathcal{C}$}
whenever $\leq$ is as in Lemma 11c) and $\mathcal{C}$ is recursively presentable and closed under finite variations and contains a set which is complete under $\leq$ (for if all sets in $\mathcal{C}$ are complete under $\leq$, {$\leq$-complete sets for $\mathcal{C}$} = $\mathcal{C}$; and if $\mathcal{C}$ contains a set which is not complete under $\leq$, then {$\leq$-complete sets for $\mathcal{C}$} = $\mathcal{C} \cap (\{X\}\leq)$, where X is $\leq$-complete for $\mathcal{C}$, fulfil the conditions of Lemma 11d)). Thus, by Lemma 11, all "reasonable" complexity classes and many more classes which arise naturally in complexity theory are recursively presentable and closed under finite variations.

Corollary 12. "Most reasonable complexity classes cannot be the nontrivial union of two others."

Let $\mathcal{C}$ be any of the classes in Lemma 9 and $\mathcal{C}_1, \mathcal{C}_2$ any of the classes in Lemma 11. Then

$$\mathcal{C} = \mathcal{C}_1 \cup \mathcal{C}_2 \Rightarrow \mathcal{C} = \mathcal{C}_1 \text{ or } \mathcal{C} = \mathcal{C}_2.$$

Corollary 13. "The non-trivial complement of one complexity class in another is mostly hard to generate."

Let $\mathcal{C}$ be any of the classes in Lemma 9 and $\mathcal{C}_2$ any class from Lemma 11. Then

$$\mathcal{C} \nsubseteq \mathcal{C}_2 \;\&\; \mathcal{C} \cap \mathcal{C}_2 \neq \emptyset \Rightarrow (\mathcal{C} \setminus \mathcal{C}_2) \text{ is not recursively presentable by recursive indices;}$$

(and, if $\mathcal{C}_2$ contains all finite subsets of Γ^*, $(\mathcal{C} \setminus \mathcal{C}_2)$ is not even recursively presentable by r.e. indices).

Proof. Put $\mathcal{C}_1 = (\mathcal{C} \setminus \mathcal{C}_2)$. Since $\mathcal{C}$ and $\mathcal{C}_2$ are closed under finite variations, so is $\mathcal{C}_1$ and Theorem A can be applied; and if $\mathcal{C}_2$ contains all finite subsets of Γ^*, Theorem B can be applied.

Remarks. Corollaries 12 and 13 between them subsume or immediately imply most of the applications of Schöning's Theorem in [10] and (using [12] for the existence of complete sets at all levels of the polynomial hierarchy) all those in [11], and also Chew and Machtey's result in [3] that, provided $\mathcal{P} \neq \mathcal{NP}$, $(\mathcal{NP} \setminus \mathcal{P})$ is not recursively presentable by r.e. indices. The new applications are to classes $\mathcal{C}$ which are not closed under $\leq^p_m$, and in particular to complexity classes which are properly contained in $\mathcal{P}$. Corollary 13 may be regarded as saying, roughly speaking, that while there may be general methods which generate large classes of instances of a given lower complexity bound (e.g. the generation of EXPTIME-complete sets as instances of sets not in $\mathcal{P}$), there is no method which can generate all such instances.

§4 Applications to the structure of time and space bounded reducibilities.

In this section we shall use the results so far to generalize existing theorems on the structure of polynomial time reducibility. In order to do so, we shall consider classes $\mathcal{C}$ which fulfil the conditions of both Lemma 9 and Lemma 11.

Lemma 14. The following classes are recursively presentable, closed under finite variations and under definition by $\mathcal{K}$-cases for some $\mathcal{K} \in$ {DTIME(n), $\text{DSPACE}_{\text{on-line}}(\log n)$}:

a) DTIME(F), NTIME(F), co-NTIME(F), $DTIME^A(O(F))$, $NTIME^A(O(F))$, co-$NTIME^A(O(F))$,
for any set F of unary functions from $\mathbb{N}$ into $\mathbb{N}$ which is recursively presentable and closed under maximum and any recursive oracle set A, if we consider only Turing machines which read the entire input first;

b) $DSPACE^A_{off\text{-}line}(F)$, $DSPACE^A_{on\text{-}line}(O(F))$, $DSPACE_{on\text{-}line}(F)$, $NSPACE^A_{off\text{-}line}(F)$, co-$NSPACE^A_{off\text{-}line}(F)$, $NSPACE^A_{on\text{-}line}(O(F))$, co-$NSPACE^A_{on\text{-}line}(O(F))$, $NSPACE_{on\text{-}line}(F)$, co-$NSPACE_{on\text{-}line}(F)$,
for any set F of unary functions from $\mathbb{N}$ into $\mathbb{N}$ which is recursively presentable and closed under maximum and contains a function f such that $f(n) \geq \log n$ for all $n \in \mathbb{N}$, and any recursive oracle A.

c) $(\mathcal{C}_1 \cap \mathcal{C}_2)$ (provided $\mathcal{C}_1 \cap \mathcal{C}_2 \neq \emptyset$) whenever $\mathcal{C}_1$ and $\mathcal{C}_2$ themselves have these properties.

d) $(\leq X)$ whenever $X \notin \{\emptyset, \Gamma^*\}$, F is a recursively presentable set of unary recursive functions which is closed under maximum and addition of $\lambda n.n$ [resp. $\lambda n.\log n$] i.e. $f \in F \Rightarrow \lambda n.(f(n)+n) \in F$ [resp. $f \in F \Rightarrow \lambda n.(f(n) + \log n) \in F$]), and $\leq$ is the F-time-bounded [resp. F-space-bounded] restriction of $\leq_T$, $\leq_m$, $\leq_{tt}$ or any of the restrictions of truth-table reducibility or of the nondeterministic or strongly nondeterministic variant of any of these reduction relations.

Proof. a) and b) follow immediately from Lemmas 9 and 11, c) and d) easily from Lemma 11.

Corollary 15. Let $\mathcal{C}$ be any of the classes in Lemma 14 and $\leq$ and of the reduction relations in Lemma 14d). Let $A,B \in \mathcal{C}$ satisfy $B<A$ (i.e. $B\leq A$ and $A \not\leq B$) and $B \notin \{\emptyset,\Gamma^*\}$. Then there is an infinite class $\mathcal{J} \subseteq \mathcal{C}$ of mutually $\leq$-incomparable sets such that, for each set $X \in \mathcal{J}$, $B<X<A$.

Proof. Put $\mathcal{J} = \{B + X_i \mid i \in \mathbb{N}\}$, where X_i is defined by recursion on i as follows:

Put $\mathcal{C}' = \mathcal{C} \cap (\leq A)$, $\mathcal{C}_1 = (\leq B)$,

$$\mathcal{C}_2^i = \{X \in \mathcal{C}' \mid Y \leq B \oplus X \text{ for some } Y \in \{A\} \cup \{B \oplus X_j \mid j<i\}\},$$

$$\mathcal{C}_3^i = \{X \in \mathcal{C}' \mid B \oplus X \leq B \oplus X_j \text{ for some } j<i\},$$

where $B \oplus X = \{0w \mid w \in B\} \cup \{1w \mid w \in X\}$.

Then apply Theorem A (twice) to show that there is an

$$X_i \in \mathcal{C}' \setminus (\mathcal{C}_1 \cup \mathcal{C}_2^i \cup \mathcal{C}_3^i).$$

In order to apply Theorem A at each recursion step it is sufficient to prove that $\mathcal{C}'$ satisfies Lemma 14, $\mathcal{C}_1$, $\mathcal{C}_2^i$ and $\mathcal{C}_3^i$ are recursively presentable and closed under finite variations and $\mathcal{C}' \not\subseteq \mathcal{C}_1$, $\mathcal{C}' \not\subseteq {}^{i}_{2}$, $\mathcal{C}' \not\subseteq {}^{i}_{3}$. All these assertions can be proved easily. Now, for each $i \in \mathbb{N}$,

$$
\begin{array}{rll}
B & \leq B \oplus X_i & \text{(trivial)} \\
B \oplus X_i & \not\leq B & \text{(because } X_i \notin \mathcal{C}_1\text{),} \\
B \oplus X_i & \leq A & \text{(because } B \leq A \text{ and } X_i \in \mathcal{C}'\text{),} \\
A & \not\leq B \oplus X_i & \text{(because } X_i \notin \mathcal{C}_2^i\text{),} \\
B \oplus X_i & \not\leq B \oplus X_j & \text{for all } j<i \text{ (because } X_i \notin \mathcal{C}_3^i\text{),} \\
B \oplus X_j & \not\leq B \oplus X_i & \text{for all } j<i \text{ (because } X_i \notin \mathcal{C}_2^i\text{).}
\end{array}
$$

Remark. If we put $\mathcal{P} = \mathcal{NP}$, Corollary 15 is a stronger version of Corollary 3.2 of [2], in that the lower bound B does not have to be in $\mathcal{P}$. Corollary 15 also strengthens Ladner's Corollary 2.2 in [5] by showing the existence of infinitely many $\leq$-incomparable sets between B and A rather than just finitely many. But what seems to us more interesting than the slight (and maybe predictable) strengthening of existing theorems is the fact that the result follows (via Theorem A) from a few simple properties of the complexity class and the reduction relation, and can be obtained without a great knowledge of either. The same is true for several other results originally proved by the "looking back" method as exemplified in the proof of Theorem 1 in [5]. Thus Theorems A and B are examples of "structural" theorems (see [3]) which are easy to state and apply and can, in many cases, serve as an easy alternative to the "looking back" method.
Regan's paper [9] has much in common with this work; in particular, it contains a "Uniform Diagonalization Theorem" which is reminiscent of Theorems A and B and is probably stronger but is less simple to state and, like Schöning's Theorem, is not applicable to complexity classes $\mathcal{C}$ properly contained in $\mathcal{P}$.

Note that, by [4], any on-line space complexity class whose space bound does not grow as fast as log n is identical to the class of regular languages, and Theorems A and B do not hold for $\mathcal{C}$ = class of regular languages, since finiteness is decidable for this class. The off-line space complexity classes defined by a space bound between log log n and log n properly contain the class of regular languages; I do not know whether Theorems A and B hold for these classes.

REFERENCES

[1] T.P. Baker and A.L. Selman, A second step toward the polynomial hierarchy, Theor. Comput. Sci. 8 (1979) 177-187.

[2] J.L. Balcazar and J. Diaz, A note on a theorem by Ladner, Information Processing Lett. 15 (1982) 84-86.

[3] P.Chew and M. Machtey, A note on structure and looking back applied to the relative complexity of computable functions, J. Comput. System Sci. 22 (1981) 53-59.

[4] J.E. Hopcroft and J.D. Ullman, Some results on tape-bounded Turing machines, J. ACM 16 (1969) 168-177.

[5] R.E. Ladner, On the structure of polynomial time reducibility, J. ACM 22 (1975) 155-171.

[6] R.E. Ladner, N.A. Lynch and A.L. Selman, A comparison of polynomial time reducibilities, Theor. Comput. Sci 1 (1975) 103-123.

[7] L.H. Landweber, R.J. Lipton & E.L. Robertson, On the structure of sets in NP and other complexity classes, Theoret. Comput. Sci. 15 (1981) 181-200.

[8] T.J. Long, Strong nondeterministic polynomial-time reducibilities, Theoret. Comput. Sci. 21 (1982) 1-25.

[9] K. Regan, On diagonalization methods and the structure of language classes, Proceedings of the 1983 FCT-Conference, Springer 1983, 368-380.

[10] U. Schöning, A uniform approach to obtain diagonal sets in complexity classes, Theoret. Comput. Sci. 18 (1982), 95-103.

[11] U. Schöning, On the structure of Δ_2^p, Information Processing Lett. 16(1983), 209-211.

[12] C. Wrathall, Complete sets and the polynomial-time hierarchy, Theoret. Comput. Sci. 3 (1977) 23-33.

THE LENGTH-PROBLEM

V. Sperschneider
Institut für Informatik I
Universität Karlsruhe

Abstract.

In MENZEL & SPERSCHNEIDER "R.e. extensions of R_1 by finite functions" families of partial recursive functions of the sort $R_1 \cup F$ were investigated, where R_1 is the set of all total recursive functions from $\mathbb{N}$ to $\mathbb{N}$ and F is a subfamily of Fin which is the set of all partial functions from $\mathbb{N}$ to $\mathbb{N}$ having a finite initial segment of $\mathbb{N}$ as domain. Here we treat families F which are defined by some "length-set" in the following sense: for a function $\delta \in \mathrm{Fin}$ define $\lg(\delta)$ to be the cardinality of the domain of δ (read "$\lg(\delta)$" as "the length of δ"), and for $L \subseteq \mathbb{N}$ define $\mathrm{Fin}(L) := \{\delta \in \mathrm{Fin} \mid \lg(\delta) \in L\}$. The length-problem is then stated as the following problem:

(LP) Characterize the sets $L \subseteq \mathbb{N}$ such that $R_1 \cup \mathrm{Fin}(L)$ is r.e. .

If $R_1 \cup \mathrm{Fin}(L)$ is r.e., then L is in the level Σ_2 of the arithmetical hierarchy. Conversely, for a set L in Σ_1, (LP) is trivially solved in that $R_1 \cup \mathrm{Fin}(L)$ is r.e. iff L is infinite. Sets L in Π_1 are more difficult to treat. Let L be r.e. with infinite complement. If L is not hyper-simple then $R_1 \cup \mathrm{Fin}(L^c)$ is r.e. . This is shown by providing for each partial recursive function φ_e a "follower" ψ_e which is equal to φ_e in case $\varphi_e \in R_1$ and is an element of $\mathrm{Fin}(L^c)$ otherwise. Providing such a "follower" ψ_e for φ_e is constructive (we need only one follower for each φ_e which is never "given up") and extensional (ψ_e depends only on φ_e and not on e). Enumeration of ψ_e is uniform in e, hence $\{\psi_e \mid e \in \mathbb{N}\}$ is a r.e. family between R_1 and $R_1 \cup \mathrm{Fin}(L^c)$. Since for arbitrary r.e. set L we show that $\mathrm{Fin}(L^c)$ is r.e. it follows that $R_1 \cup \mathrm{Fin}(L^c)$ is r.e. as the union of the two r.e. families $\{\psi_e \mid e \in \mathbb{N}\}$ and $\mathrm{Fin}(L^c)$.

We prove that such a constructive and extensional way of providing a follower for each partial recursive function under the condition that the resulting finite functions have a length in L^c can be applied only to non-hyper-simple sets L. By more sophisticated methods we show that there is even a maximal set M such that $R_1 \cup \mathrm{Fin}(M^c)$ is r.e. .

Notation.

$\mathbb{N}$	set of all natural numbers, including 0
$f(x)\downarrow$	f is defined at x
$f(x)\uparrow$	f is undefined at x
dom(f)	$\{x \mid f(x)\downarrow\}$, the domain of f
rg(f)	$\{f(x) \mid x\in dom(f)\}$, the range of f
$f\subseteq g$	g is an extension of f
$f\upharpoonright A$	the restriction of f to A
P_n	the partial recursive functions from $\mathbb{N}^n$ to $\mathbb{N}$
R_n	the total recursive functions from $\mathbb{N}^n$ to $\mathbb{N}$
Fin	$\{\delta\in P_1 \mid dom(\delta)$ is a finite initial segment of $\mathbb{N}\}$
$lg(\delta)=l$	iff $dom(\delta)=\{0,\ldots,l-1\}$ (the length of δ)
$\varphi_e^{(n)}$	the function in P_n computed by Turing-machine e
φ_e	$\varphi_e^{(1)}$
$\varphi_{e,s}(x)=y$	iff Turing-machine e with input x stops within s steps with output y and $e<s$, $x<s$, $y<s$.
W_e	$dom(\varphi_e)$
$W_{e,s}$	$dom(\varphi_{e,s})$
$(D_n)_{n\in\mathbb{N}}$	is the usual canonical numbering (coding) of all finite subsets of $\mathbb{N}$
$(\delta_n)_{n\in\mathbb{N}}$	is some fixed canonical numbering of Fin

A subfamily F of P_n is called recursively enumerable (r.e.) iff there is a r.e. subset B of $\mathbb{N}$ such that $F=\{\varphi_b^{(n)} \mid b\in B\}$; B is called a basis for F.

For a subfamily F of Fin we additionally call F canonically enumerable (or canonically decidable) iff $\{n \mid \delta_n\in F\}$ is r.e. (or decidable).

The Length-Problem.

For a subset L of $\mathbb{N}$ define $Fin(L):=\{\delta\in Fin \mid lg(\delta)\in L\}$.
Then the Length-Problem is stated as the following question:

(LP) For which sets L is $R_1\cup Fin(L)$ r.e. ?

Some simple facts. *For all $L\subseteq\mathbb{N}$ the following holds:*

(1) If $R_1\cup Fin(L)$ is r.e. then L is in the level Σ_2 of the arithmetical hierarchy.

(2) If L is r.e. then Fin(L) is r.e. .

(3) If L is r.e.. then $R_1\cup Fin(L)$ is r.e. iff L is infinite.

(4) If L is r.e. then $Fin(L^c)$ is r.e. .

(5) If L is r.e., has infinite complement and is not simple then $R_1 \cup Fin(L^c)$ is r.e. .

Proof.

(1) Let B be a r.e. basis for $R_1 \cup Fin(L)$. Then for all $l \in \mathbb{N}$, $l \in B$ iff $\exists b[\, b \in B \wedge \varphi_b(l)\uparrow \wedge \forall x<l\ \varphi_b(x)\downarrow\,]$. Hence L is a Σ_2-set.

(2) Trivial. Fin(L) is even canonically enumerable.

(3) Let L be r.e. . If $R_1 \cup Fin(L)$ is r.e. then of course L must be infinite. Conversely, assume L is infinite. For each e define a "follower" ψ_e for φ_e as follows:

$$\psi_e := \bigcup \{\varphi_e \restriction \{0,\dots,l-1\} \mid l \in L \wedge \forall x<l\ \varphi_e(x)\downarrow\}.$$

ψ_e is computable uniformly in e, hence $\{\psi_e \mid e \in \mathbb{N}\}$ is r.e. . Since L is infinite it follows that $\{\psi_e \mid e \in \mathbb{N}\}$ is equal to $R_1 \cup Fin(L \cup \{0\})$. Hence $R_1 \cup Fin(L)$ is r.e. (Observe that removing the empty function from a r.e. family of functions gives again a r.e. family.).

(4) If L^c is finite, nothing remains to be proved. So assume that L^c is infinite. Take $f \in R_1$ such that $L = rg(f)$. For each n define the graph of a function ψ_n as follows:

$$\psi_n := \bigcup_{s \in \mathbb{N}} \psi_n^s \quad \text{with } \psi_n^s \in Fin,\ \psi_n^s \subseteq \psi_n^{s+1} \quad \text{defined as follows:}$$

$\psi_n^0 := \delta_n$. If ψ_n^s is already defined then define

$$\psi_n^{s+1} := \begin{cases} \psi_n^s & \text{if } lg(\psi_n^s) \notin \{f(0),\dots,f(s)\} \\ \psi_n^s \cup \{(lg(\psi_n^s),0)\} & \text{otherwise} \end{cases}$$

It follows that in case $lg(\delta_n) \in L^c$ ψ_n is equal to δ_n, and in case $lg(\delta_n) \in L$ ψ_n is a finite extension of δ_n with length in L^c (since L^c is infinite the second case in the definition of ψ_n^{s+1} occurs only finitely often). Since $Fin(L^c)$ equals $\{\psi_n \mid n \in \mathbb{N}\}$ and ψ_n can be computed uniformly in n it follows that $Fin(L^c)$ is r.e. .

(5) L^c contains an infinite r.e. subset L_0. Now simply write $R_1 \cup Fin(L^c)$ as the union of $R_1 \cup Fin(L_0)$ and $Fin(L^c)$. Both parts are r.e. by (3) and (4) above, hence $R_1 \cup Fin(L^c)$ is r.e., too.

So far, the Length-Problem has been solved for r.e. sets L as well as for the complements of r.e., non-simple sets. We soon solve it for some simple sets, too.

Informally stated, (LP) is the question of how "thin" the complement of a r.e. set L may be if nevertheless $R_1 \cup Fin(L^c)$ is r.e. .

There are several well known notions in recursive function theory saying that a r.e. set L has a "thin" complement. We repete some of them.

Definitions. For a r.e. set L with infinite complement we say that

- L is *simple* iff L^c does not contain an infinite r.e. subset .
- L is *hyper-simple* iff there is no total recursive function f such that

$$\forall i\ [\ D_{f(i)} \cap L^c \neq \emptyset \ \wedge\ \forall j \neq i\ D_{f(i)} \cap D_{f(j)} = \emptyset\ ]\ .$$

- L is *hyper-hyper-simple* iff there is no total recursive function f such that

$$\forall i\ [\ W_{f(i)} \text{ finite } \wedge\ W_{f(i)} \cap L^c \neq \emptyset \ \wedge\ \forall j \neq i\ W_{f(i)} \cap W_{f(j)} = \emptyset\ ]\ .$$

- L is *maximal* iff for every r.e. set B containing L either $B \setminus L$ or $\mathbb{N} \setminus B$ is finite.

Remark. L maximal $\Rightarrow$ L hyper-hyper-simple $\Rightarrow$ L hyper-simple $\Rightarrow$ L simple. L is maximal iff L is r.e., has infinite complement and for every e either $L^c \subseteq^* W_e$ or $L^c \subseteq^* (\mathbb{N} \setminus W_e)$ (where $A \subseteq^* B$ means that $A \setminus B$ is finite), in pictures, either or

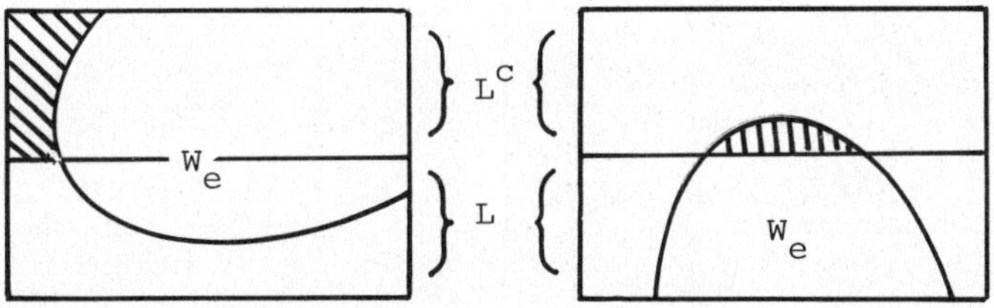

where the shaded areas denote finite sets.

Theorem. *If L is r.e., has infinite complement and is not hyper-simple then $R_1 \cup Fin(L^c)$ is r.e.* .

Proof. Take $g \in R_1$ such that $rg(g) = L$, and $f \in R_1$ such that

$$\forall i\ [\ D_{f(i)} \cap L^c \neq \emptyset \ \wedge\ \forall j \neq i\ D_{f(i)} \cap D_{f(j)} = \emptyset\ ]\ .$$

For each e define a "follower" ψ_e for φ_e as follows:

$\psi_e := \bigcup_{s \in \mathbb{N}} \psi_e^s$, with $\psi_e^s \subseteq \psi_e^{s+1} \subseteq \varphi_e$ and $\psi_e^s \in Fin$ such that the following property (*) holds:

(*) If $\psi_e^s \neq \emptyset$ then between $lg(\psi_e^s)$ and the maximal number $l \leq s$ such that $\forall x < l\ \varphi_{e,s}(x)\downarrow$ there is a number in L^c, in particular $l \geq lg(\psi_e^s)$.

<u>Algorithm defining ψ_e^s.</u>

<u>stage s=0</u>: $\psi_e^0 := \emptyset$.

<u>stage s+1</u>: Let ψ_e^s be already defined such that (*) holds for e and s. Then ψ_e^{s+1} is defined in two steps.

<u>Step 1</u>: Compute the maximal number $l \leq s+1$ such that $\forall x<l\ \varphi_{e,s+1}(x)\downarrow$.

<u>Case a</u>: There is a number $i \leq s+1$ such that $lg(\psi_e^s) < \min(D_{f(i)})$ and $\max(D_{f(i)}) \leq l$.

Then take the maximal such i and define

$$\tilde{\psi}_e^s := \varphi_e \upharpoonright \{0,\ldots,\min(D_{f(i)})-1\}.$$

(Note that between $lg(\tilde{\psi}_e^s)$, which is $\min(D_{f(i)})$, and l there is a number in L^c, by choice of f.)

<u>Case b</u>: There is no such number i as in case a above. Then define

$$\tilde{\psi}_e^s := \psi_e^s \ .$$

<u>Step 2</u>: Define $\tilde{l} := lg(\tilde{\psi}_e^s)$.

<u>Case a</u>: $\tilde{l}=0$ or $\tilde{l} \notin \{g(0),\ldots,g(s+1)\}$. Then define

$$\psi_e^{s+1} := \tilde{\psi}_e^s \ .$$

<u>Case b</u>: $\tilde{l}>0$ and $\tilde{l} \in \{g(0),\ldots,g(s+1)\}$. Then define

$$\psi_e^{s+1} := \tilde{\psi}_e^s \cup \{(\tilde{l},\varphi_e(\tilde{l}))\} \ .$$

(Note that $\varphi_e(\tilde{l})$ is indeed defined because of assumption (*) for e and s and the fact that $\tilde{l}>0$.)

For e and s+1 again (*) holds since in step 2, case b the number $\tilde{l}$ is not an element of L^c.

We can conclude now:

(0) There is some $\pi \in R_2$ such that, for all e and s, $\psi_e^s = \delta_{\pi(e,s)}$.

(1) (Uniform followers) For all e, $\psi_e^s \subseteq \psi_e^{s+1} \subseteq \varphi_e$, and if $\varphi_e \in R_1$ then $\varphi_e = \psi_e$.

Proof. In case $\varphi_e \in R_1$ there are infinitely many s such that in stage s of the algorithm step 1,case a occurs, thereby extending ψ_e^s further.

(2) For all e, $\varphi_e \notin R_1$ implies $\psi_e \in Fin$ with $lg(\psi_e) \in L^c \cup \{0\}$.

Proof. If $\varphi_e \notin R_1$ then there is a stage s such that, from stage s on, case a of step 1 does not occur. If $lg(\psi_e^s)=0$ then $\psi_e=\emptyset$ since from stage s on only case a of step 2 occurs. If $lg(\psi_e^s)>0$ then case b of step 2 occurs only finitely often after stage s, until ψ_e has reached a length in L^c. Hence in this case $lg(\psi_e^s) \in L^c$.

(3) (Extensional followers) For all e,e',s and s' with $\varphi_{e,s}\subseteq\varphi_{e',s'}$ there is a stage s'' such that $\psi_e^{s}\subseteq\psi_{e'}^{s''}$.
In particular, if $\varphi_e=\varphi_{e'}$ then $\psi_e=\psi_{e'}$.

Proof. Simple exercise. As s'' take the maximum of s and s'.
From (0),(1) and (2) it follows that $\{\psi_e | e\in\mathbb{N}\}$ is a r.e. family of functions between R_1 and $R_1\cup Fin(L^c\cup\{0\})$. Hence $R_1\cup Fin(L^c\cup\{0\})$, as the union of $\{\psi_e | e\in\mathbb{N}\}$ and $Fin(L^c\cup\{0\})$, is r.e. too. By eventually deleting the empty function (if $0\in L$) it follows that $R_1\cup Fin(L^c)$ is r.e. .

By the method used in the proof above, namely providing in a uniform and extensional way followers for the partial recursive functions and thereby taking care that the resulting finite functions are empty or have a length in L^c, we cannot solve the Length-Problem for more than the non-hyper-simple sets. This is shown by the following theorem.

__Theorem.__ *Let L be r.e. with infinite complement. Assume $\pi\in R_2$ is given such that for $\psi_e^s:=\delta_{\pi(e,s)}$ and $\psi_e:=\bigcup_{s\in\mathbb{N}}\psi_e^s$ properties (1),(2) and (3) from the proof above hold. Then L is not hyper-simple.*

Proof. Take a fixed e such that $\varphi_e(x)=0$, for all x. Define $f\in R_1$ by recursion as follows:

__f(0)__: Take the first s such that $\psi_e^s\neq\emptyset$ (such a s exists since by property (1) we know that $\psi_e=\varphi_e$). Let l be the maximal number such that $\varphi_{e,s}(l)\downarrow$. Then define f(0) by

$$D_{f(0)}:=\{lg(\psi_e^s),\ldots,\max\{lg(\psi_e^s),l+1\}\} \ .$$

__f(n+1)__: Take the first s such that $lg(\psi_e^s)>\max(D_{f(n)})$ (s exists because of property (1)). Let l be the maximal number such that $\varphi_{e,s}(l)\downarrow$. Then define f(n+1) by

$$D_{f(n+1)}:=\{lg(\psi_e^s),\ldots,\max\{lg(\psi_e^s),l+1\}\} \ .$$

Then of course the sets $D_{f(n)}$, for $n\in\mathbb{N}$, are pairwise disjoint. It remains to prove that $D_{f(n)}$ intersects L^c. We prove this for $D_{f(n+1)}$ ($D_{f(0)}$ is treated in just the same way). Let s and l be the numbers chosen in the definition of f(n+1). Let e' be such that

$$\varphi_{e'}=\{(x,0) | x<\max(D_{f(n+1)})\} \ .$$

Let s' be such that $\varphi_{e',s'}=\varphi_{e'}$.

Since $dom(\varphi_{e,s})\subseteq\{0,\ldots,l\}$, $l\leq\max(D_{f(n+1)})-1$ and $\{0,\ldots,\max(D_{f(n+1)})-1\}$ is the domain of $\varphi_{e',s'}$ it follows that

$\varphi_{e,s} \subseteq \varphi_{e',s'}$.

By property (3) we can conclude that there exists s'' with

$\psi_e^s \subseteq \psi_{e'}^{s''}$.

Since $\psi_{e'}$ is not the empty function it follows from property (2) that $\psi_{e'}$ is in Fin with length in L^c. From (1) we conclude that $\psi_{e'} \subseteq \varphi_{e'}$. It follows that

$$lg(\psi_e^s) \leq lg(\psi_{e'}^{s''}) \leq lg(\psi_{e'}) \leq lg(\varphi_{e'}) = \max(D_{f(n+1)}) .$$

Hence $D_{f(n+1)}$ contains an element of L^c, namely $lg(\psi_{e'})$.

By more sophisticated methods we finally show that there is even a maximal set M such that $R_1 \cup Fin(M^c)$ is r.e. . The method by which we enumerate $R_1 \cup Fin(M^c)$ goes beyond the one described in the theorem above in that

- there are infinitely many followers for each partial recursive function φ_e and we can (non-constructively) prove that in case $\varphi_e \in R_1$ one of these followers will be equal to φ_e
- following φ_e depends on e and not only on φ_e .

<u>Theorem.</u> *There is a maximal set M such that* $R_1 \cup Fin(M^c)$ *is r.e.* .

Proof. We refine the standard maximal set construction using moveable markers (see ROGERS). The algorithm enumerating M is described using a table which, at the beginning of our algorithm (stage 0) looks as follows:

stage 0

marker	position	W_0	W_1	W_2	W_3	....
Λ	0	0	0	0	0	....
Λ_0	1	0	0	0	0	....
Λ_1	2	0	0	0	0	....
Λ_2	3	0	0	0	0	....
Λ_3	4	0	0	0	0	....
⋮	⋮	⋮	⋮	⋮	⋮	

Markers $\Lambda, \Lambda_0, \Lambda_1, \Lambda_2, \Lambda_3, \ldots$, sitting at stage 0 at positions 0,1,2,3,4,..., move down during execution of the following algorithm. Moves are controlled by the entries in the right half of the table. At stage s, under position x and column W_e the entry is 0 if $x \notin W_{e,s}$, and 1 if

$x \in W_{e,s}$. At stage 0, all entries are 0. Hence, with increasing s, some zeroes may change into ones, but not vice versa.

Marker moves will look as follows: Some Λ_e generates the move, thereby moving down to the position of Λ_{e+1} and forcing every marker Λ_{e+1+k}, for $k \in \mathbb{N}$, to move down to the position of Λ_{e+2+k}. Λ will never move. Each marker will move only finitely often. Call m_e the final position of Λ_e. Hence $0<m_0<m_1<m_2<\ldots$. Define $M^c := \{0\} \cup \{m_e \mid e \in \mathbb{N}\}$. Then M^c is infinite and M is r.e. since for all x: $x \in M$ iff there is a stage s such that x, at stage s, is not the position of a marker.

Simultaneous to moving the markers we enumerate functions $\psi_e(z,l)$, for $e \in \mathbb{N}$, $z \in \{0,1\}^{e+1}$ and $l \in \mathbb{N}$, in stages as follows:

$$\psi_e(z,l) := \bigcup_{s \in \mathbb{N}} \psi_e^s(z,l) \quad , \text{ with } \psi_e^s(z,l) \subseteq \psi_e^{s+1}(z,l).$$

For each e,z and l, $\psi_e(z,l)$ is called a "follower of φ_e" what indicates that we try to enumerate, as long as possible, only initial segments of φ_e into $\psi_e(z,l)$. Sometimes following φ_e via $\psi_e(z,l)$ must be given up, but we will be able to prove that in case $\varphi_e \in R_1$ there is some $z \in \{0,1\}^{e+1}$ and some $l \in \mathbb{N}$ such that following φ_e via $\psi_e(z,l)$ is never given up and therefor $\psi_e(z,l) = \varphi_e$. Those among the functions $\psi_e(z,l)$, which are in Fin, will have a length in M^c.

The algorithm.

stage s=0: The marker positions are as indicated in the table above.

$$\psi_e^0(z,l) := \emptyset \text{ , for all } e \in \mathbb{N},\ z \in \{o,1\}^{e+1} \text{ and } l \in \mathbb{N} .$$

stage s+1:

Definition. For $e,x \in \mathbb{N}$, the *e-state of x at stage s+1* is the binary word $b_0 b_1 \ldots b_e \in \{0,1\}^{e+1}$ with $b_i := 0$ if $x \notin W_{i,s+1}$ and $b_i := 1$ if $x \in W_{i,s+1}$, for $i=0,1,\ldots,e$.

Definition. For $0 \leq i \leq e$ say that Λ_e *wants to make an i-move at stage s+1* iff for the position x of Λ_e and the position y of Λ_{e+1} the following conditions hold:

(1) the e-state of x resp. y at stage s+1 is of the form $\underline{b}0\underline{b}_1$ resp. $\underline{b}1\underline{b}_2$ with some $\underline{b} \in \{0,1\}^i$ and $\underline{b}_1, \underline{b}_2 \in \{0,1\}^{e-i}$

(2) for all $e' \leq i$, $z' \in \{0,1\}^{e'+1}$, $l' \in \mathbb{N}$ with $\psi_{e'}^s(z',l') \in$ Fin, the length of $\psi_{e'}^s(z',l')$ is different from x
(this condition will be decidable since by the way the algorithm is defined only finitely many l´ must be checked)

i.e. at the beginning of stage s+1 our table looks as follows:

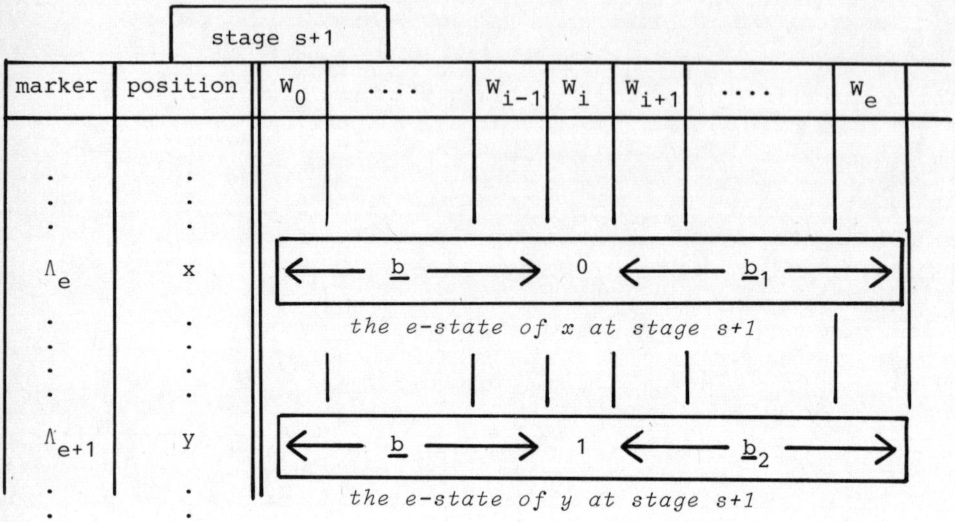

and no one of the followers of $\varphi_0,\ldots,\varphi_i$ has, after stage s, the length x.

Stage s+1 consists now of the following three steps (I),(II),(III):

(I) (Moving the markers) If there is an $e\leq s+1$ such that Λ_e wants to make an i-move at stage s+1, for some (uniquely determined) $i\leq e$, then take the least such e and move each marker Λ_{e+k}, for $k\in\mathbb{N}$, down to the position of Λ_{e+k+1} after stage s. Say that Λ_e *generates this move*. If there is no such e, do nothing.

(II) (Giving up some follower)

(a) (By the marker move from (I) above)

For all $e\in\mathbb{N}$, $z\in\{0,1\}^{e+1}$ and $l\in\mathbb{N}$ define $\tilde{\psi}_e^s(z,l)$ as follows:

- If $\psi_e^s(z,l)\in \mathrm{Fin}$, a move was generated in step (I) above, and $\lg(\psi_e^s(z,l))$ was the position of the marker generating this move before moving, then define $\tilde{\psi}_e^s(z,l):=\psi_e^s(z,l)\cup\{(y,0)\,|\,y\notin \mathrm{dom}(\psi_e^s(z,l))\}$. Say that *following* φ_e *via* $\psi_e(z,l)$ *has been given up at stage s+1.*
- In any other case define $\tilde{\psi}_e^s(z,l):=\psi_e^s(z,l)$.

(Note that in the first case the number $\lg(\psi_e^s(z,l))$ was enumerated into M by removing some marker from this position, hence this length is no longer admissible and we must give up all followers having, after stage s, just this length.)

(b) (By wrong guess (z,l))

For all $e\in\mathbb{N}$, $z\in\{0,1\}^{e+1}$ and $l\in\mathbb{N}$ define $\overline{\psi}_e^s(z,l)$ as follows:

- If $\tilde{\psi}_e^s(z,l)\in Fin$ and there is some x' with $l<x'\leq s+1$ such that, after step (I) of stage s+1, some marker is sitting at position x' and the e-state of x' at stage s+1 is greater (as binary number) than z, then define $\overline{\psi}_e^s(z,l):=\tilde{\psi}_e^s(z,l)\cup\{(y,0)\mid y\notin dom(\tilde{\psi}_e^s(z,l))\}$.

  Say that *following φ_e via $\psi_e(z,l)$ has been given up at stage s+1.*
- In any other case define $\overline{\psi}_e^s(z,l):=\tilde{\psi}_e^s(z,l)$.

(Why we give up the follower in the first case is explained in the following remark (3).)

(III) (Following φ_e)

For all $e\in\mathbb{N}$, $z\in\{0,1\}^{e+1}$ and $l\in\mathbb{N}$ define $\psi_e^{s+1}(z,l)$ as follows:

- If $\overline{\psi}_e^s(z,l)\in Fin$ and there is some $x\leq s+1$ such that $x>l$, $x>lg(\overline{\psi}_e^s(z,l))$, $\{0,\ldots,x-1\}\subseteq dom(\varphi_{e,s+1})$, some marker is sitting after step (I) at position x, and the e-state of x at stage s+1 is z, then take the maximal such x and define $\psi_e^{s+1}(z,l):=\varphi_e\restriction\{0,\ldots,x-1\}$.
- In any other case define $\psi_e^{s+1}(z,l):=\overline{\psi}_e^s(z,l)$.

End of stage s+1.

Remarks.

(1) Though we simultaneously enumerate infinitely many followers $\psi_e(z,l)$ for each φ_e, all decisions in the algorithm above are effectively possible (for example deciding whether Λ_e wants to make an i-move or whether $\psi_e^s(z,l)$ is in Fin). Note that we assume about our gödel numbering φ_e that if $\varphi_{e,s}(x)$ is defined then $x<s$ and $e<s$. Hence, in stage s+1, at most the functions $\psi_e^s(z,l)$ with $e<s+1$ and $l<s+1$ are non-empty.

(2) Moving the markers is guided by the principle that Λ_e wants to move to a position with maximal e-state by making an i-move (with i being uniquely determined at stage s+1 by e) to the position of Λ_{e+1}. It is allowed to make such an i-move only if it does not injure the length of one of the followers of $\varphi_0,\ldots,\varphi_i$, i.e. no one of these followers must be given up in the following step (II)(a) . If several markers want to make a move the one with least index is allowed to generate the move. Note the main difficulty which comes up when compared with the standard maximal set construction:

- moving Λ_e, for example by an i-move, must respect all the infinitely many followers of one of $\varphi_0,\dots,\varphi_i$ (this difficulty will be solved by a "window lemma"),
- following φ_e may be injured (given up) infinitely often by some i-move for some i<e (though each marker moves only finitely often there may be, for some i, infinitely many i-moves; this difficulty is solved by a "convergence lemma").

The reader familiar with the standard maximal set construction is encouraged to convince himself that an algorithm based on the (finite injury) priority ordering (">" means having higher priority)

1 follower of φ_0 > moving Λ_0 > 2 followers of φ_1 > moving Λ_1 > 7 followers of φ_2 ...

(resulting from the observation that Λ_0 moves at most once, Λ_1 moves at most 6 times,...) would not succeed. Maximality of M could not be proved since moving the markers would be too severely restricted.

(3) In enumerating $\psi_e(z,l)$ we make the "guess" that
- z is maximal in $\{0,1\}^{e+1}$ such that there are infinitely many positions x having the property that, at some stage s, some marker is sitting at x and the e-state of x at stage s is z,
- l is the maximum of all positions x´ having the property that, at some stage s, some marker is sitting at x´ and the e-state of x´ at stage s is greater (as binary number) than z.

Note that for each e there is a unique pair (z,l) such that the two conditions above hold. For later use let us say that

(z,l) is the correct guess for e.

Enumeration of $\psi_e(z,l)$ is done under the hypothesis that (z,l) is the correct guess for e. Making this guess we enumerate into $\psi_e(z,l)$ initial segment $\varphi_e \upharpoonright \{0,\dots,x-1\}$ of φ_e at stage s, provided that x>l, $\varphi_{e,s}$ is defined at 0,...,x-1, some marker is sitting at position x at stage s, and the e-state of x at stage s is z.
If eventually, at some stage s, it turns out that (z,l) is not the correct guess for e (this can turn out only in case that some x´>l is found at some stage s such that some marker is sitting at x´ at stage s and the e-state of x´ at stage s is greater than z) then following φ_e via $\psi_e(z,l)$ is given up.

The verification.

Lemma 1. *Each marker moves only finitely often.*

Proof. By induction on e. Assume that after stage s no one of $\Lambda_0,\dots,\Lambda_{e-1}$ moves. Hence, after stage s, each move of Λ_e is generated by Λ_e, hence Λ_e moves to a position with greater e-state. Since there are only 2^{e+1} many different e-states, Λ_e makes, after stage s, at most 2^{e+1} further moves.

Definition. m_e:=final position of Λ_e, $M^c:=\{0\}\cup\{m_e \mid e\in\mathbb{N}\}$.

Lemma 2. *M^c is infinite and M is r.e.* .

Proof. By the way the markers move we know that $0<m_0<m_1<m_2<\dots$, and for all x, $x\in M$ iff there is some stage s such that no marker is sitting at position x at stage s. Hence M^c is infinite and M is r.e. .

Lemma 3. *For all e, if (z,l) is the correct guess for e, then for all l´ ≥ l and all stages s: $\overline{\psi}^s_e(z,l')=\tilde{\psi}^s_e(z,l')=\psi^s_e(z,l')\subseteq\varphi_{e,s}\subseteq\varphi_e$, i.e. following φ_e via $\psi_e(z,l')$ is never given up at steps (II)(a) or (b).*

Proof. Let (z,l) be the correct guess for e, and $l'\geq l$. Then, of course, following φ_e via $\psi_e(z,l')$ is never given up at some step (II)(b). Assume that following φ_e via $\psi_e(z,l')$ is given up at step (II)(a) of some first stage s+1. This means that, after stage s, $\psi^s_e(z,l')$ had some length x>l, some marker $\Lambda_{e'}$ was sitting at position x after stage s, the e-state of x at stage s+1 was z, and $\Lambda_{e'}$ generated the move at stage s+1. It follows that $\Lambda_{e'}$ moves down to a position with increased e´-state. Since the e-state of the new position of $\Lambda_{e'}$ cannot be greater than z (by assumption that (z,l) is the correct guess for e) this implies that the move $\Lambda_{e'}$ made at stage s+1 was an i-move for some i>e. But making such an i-move must respect in particular the length of $\psi^s_e(z,l')$ (see definition of "$\Lambda_{e'}$ wants to make an i-move at stage s+1"). Hence $\Lambda_{e'}$ cannot have been the marker generating the move at stage s+1. Contradiction.

Remark. If, for some e, (z,l) is the correct guess for e and $l\geq l'$ then by the lemma above following φ_e via $\psi_e(z,l')$ is never given up. We cannot already conclude that, in case φ_e is total, $\psi_e(z,l')=\varphi_e$. Although z appears at infinitely many marker positions as e-state at some stage s, this situation might vanish before φ_e reaches a great enough length. What we need is the following

Lemma 4 (Convergence lemma). *For all e, if (z,l) is the correct guess for e then there are infinitely many positions x such that x is the final position of some marker and, from some stage on, the e-state of x is constantly z.*

Proof. Assume there are only finitely many positions x as described above. Let x_0 be the maximal such position. By assumption on (z,l) there is some $x>x_0$, $x>l$, and some $i,s\in \mathbb{N}$ such that Λ_i is sitting at position x st stage s, the e-state of x at stage s is z, and $i\geq e$.
If, at stage s+1, Λ_i generates the move then the i-state of its new position grows. Since $i\geq e$ it follows that the e-state of its new position is not less than z. By assumption on (z,l), the e-state of its new position is, on the other side, not greater than z. Hence the e-state of the new position of Λ_i is again z.
If some Λ_j with $j<i$ generates the move at stage s+1, then after this move, Λ_{i-1} is sitting at the old position of Λ_i and the e-state of this position is still z (again by assumption on (z,l)).
We can conclude that, for all stages $t\geq s$, one of $\Lambda_0,\ldots,\Lambda_i$ is sitting at stage t at some position $x'\geq x$ and the e-state of x' at stage s is z. Moves generated by $\Lambda_0,\ldots,\Lambda_i$ occur only finitely often. Hence, after a while, some Λ_j with $j\leq i$ is constantly sitting at some position $x'\geq x$ and the e-state of x' is constantly z. This contradicts the choice of x_0.

Lemma 5. *For all e, if (z,l) is the correct guess for e, $l'\geq l$ and φ_e is total, then $\psi_e(z,l')=\varphi_e$.*

Proof. By lemma 3, following φ_e via $\psi_e(z,l')$ is never given up, and by lemma 4 there are infinitely many positions x which are final marker positions and constantly, from some stage on, have e-state z. Hence there is time for φ_e to reach such a length x, and then the corresponding initial segment of φ_e is enumerated into $\psi_e(z,l')$ via step (III) of our algorithm.

Lemma 6. *For all $e\in \mathbb{N}$, $z\in\{0,1\}^{e+1}$ and $l\in \mathbb{N}$, $\psi_e(z,l)$ is either total recursive or in Fin with length in M^c.*

Proof. Observe that at each stage s of our algorithm, if $\psi_e^s(z,l)$ is in Fin then it has a length x such that some marker is sitting at position x at stage s. Now lemma 6 follows.

Lemma 5 and 6 say that $\{\psi_e(z,l)\,|\,e,l\in \mathbb{N} \wedge z\in\{0,1\}^{e+1}\}$ is a r.e. family between R_1 and $R_1\cup Fin(M^c)$. It follows that $R_1\cup Fin(M^c)$ is r.e. .
It remains to show that M is maximal.

Lemma 7 (Window lemma). *For every* $e\in\mathbb{N}$ *and* $z\in\{0,1\}^{e+1}$ *there is some* $l_0\in\mathbb{N}$ *such that*

$$\forall l_1>l_0\ \exists s_0\ \forall s\geq s_0\ \forall l\in\mathbb{N}\ [\ \psi_e^{s+1}(z,l)\in Fin \Rightarrow lg(\psi_e^{s+1}(z,l))<l_0 \text{ or } >l_1\ ],$$

i.e. there is some fixed l_0 *such that for arbitrary large "windows"* $\{l_0,\dots,l_1\}$ *there is a stage* s_0 *such that, from* s_0 *on, all the infinitely many functions* $\psi_e^{s+1}(z,l)$, *for* $l\in\mathbb{N}$, *have lengths outside the window.*

Proof. Let $e\in\mathbb{N}$ and $z\in\{0,1\}^{e+1}$ be given.

Case 1: $\varphi_e\notin R_1$. Define $l_0:=1$+the least number outside the domain of φ_e. Since in case $\psi_e^{s+1}(z,l)\in Fin$ (i.e. following φ_e via $\psi_e(z,l)$ has not been given up until stage s+1) we know that $\psi_e^{s+1}(z,l)$ is an initial segment of φ_e, we can conclude that its length is less than l_0.

Case 2: $\varphi_e\in R_1$. Let $(\bar{z},\bar{l})$ be the correct guess for e.

2a: $z>\bar{z}$. Define $l_0:=\bar{l}+1$. Since z appears as the e-state of some position x at some stage s+1 only for $x\leq l$ (see definition of "correct guess") we conclude that for all s, if $\psi_e^{s+1}(z,l)\in Fin$ then the length of $\psi_e^{s+1}(z,l)$ is less than l_0.

2b: $z=\bar{z}$. Define $l_0:=1$. Now let $l_1>l_0$ be given.
For all $l<\bar{l}$ it turns out after some stage that (z,l) is not the correct guess for e. For all l with $\bar{l}\leq l<l_1$ we know from lemma 5 that $\psi_e(z,l)$ is equal to φ_e. For all $l\geq l_1$ and all stages s we know that $\psi_e^{s+1}(z,l)$ is either total or empty or has length $>l$.
So it suffices to take as s_0 a stage s such that up to s all the pairs (z,l) with $l<\bar{l}$ have turned out not to be the correct guess for e, and for all l with $\bar{l}\leq l<l_1$ the function $\psi_e^{s+1}(z,l)$ has reached a length $>l_1$.

2c: $z<\bar{z}$. Define $l_0:=1$. Now let $l_1>l_0$ be given. As s_0 take some stage s such that up to stage s all the pairs (z,l) with $l<l_1$ have turned out not to be the correct guess for e (this turns out since $z<\bar{z}$). The conclusion of lemma 7 follows now as in case 2b.

Lemma 8. *M is maximal.*

Proof. We have to show that, for each e, $M^c\subseteq^* W_e$ or $M^c\subseteq^*(W_e)^c$, where $A\subseteq^* B$ means that $A\setminus B$ is finite. In other words, we have to show that, for each e, the sequence $(\chi_{W_e}(m_i))_{i\in\mathbb{N}}$ converges (χ_A, for $A\subseteq\mathbb{N}$, is the characteristic function of A with value 1 on A and 0 on A^c).
Assume, by induction, that this has been shown for all $e'<e$.
Assume further that the sequence $(\chi_{W_e}(m_i))_{i\in\mathbb{N}}$ does not converge.
By assumption on $e'<e$ we can choose $i_0\in\mathbb{N}$ and $\underline{b}\in\{0,1\}^e$ such that

(1) $\forall i \geq i_0 \quad \chi_{W_0}(m_i) \ldots \chi_{W_{e-1}}(m_i) = \underline{b}$.

Applying lemma 7 finitely often for all $e' < e$ and $z' \in \{0,1\}^{e'+1}$ we get a number l_0 such that for all $l_1 > l_0$

(2) $\exists s_0 \ \forall s \geq s_0 \ \forall e' < e \ \forall z' \in \{0,1\}^{e'+1} \ \forall l \in \mathbb{N}$

$$[\ \psi_{e'}^{s+1}(z',l) \in Fin \Rightarrow lg(\psi_{e'}^{s+1}(z',l)) < l_0 \text{ or } > l_1\ ] .$$

By assumption on $(\chi_{W_e}(m_i))_{i \in \mathbb{N}}$ we can choose a number i such that

(3) $i \geq e$ (4) $i \geq i_0$ (5) $m_i \geq l_0 + 1$ (6) $\chi_{W_e}(m_i) = 0$, $\chi_{W_e}(m_{i+1}) = 1$.

We now apply (2) with $l_1 := m_i$ (by (5) $l_1 > l_0$). This gives us a stage s such that the following conditions hold (use (1),(4) and (6)):

(7) For all $e' < e$, $z' \in \{0,1\}^{e'+1}$ and $l \in \mathbb{N}$: if $\psi_{e'}^{s+1}(z',l) \in Fin$ then its length is outside $\{l_0, \ldots, l_1\}$, in particular it is different from m_i.

(8) $\chi_{W_{e,s+1}}(m_{i+1}) = 1$ and $\chi_{W_{e,s+1}}(m_i) = 0$

(9) $\chi_{W_{0,s+1}}(m_i) \ldots \chi_{W_{e-1,s+1}}(m_i) = \chi_{W_{0,s+1}}(m_{i+1}) \ldots \chi_{W_{e-1,s+1}}(m_{i+1}) = \underline{b}$

(10) Λ_i and Λ_{i+1} have, up to stage s, reached their final positions.

Hence, at the beginning of stage s+1, the table used in our algorithm looks as follows (use (3),(7),(8),(9) and (10)).

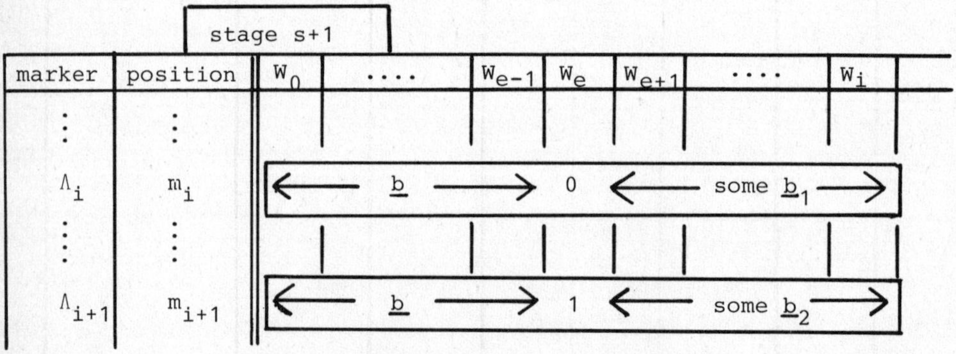

This means that Λ_i wants to make an e-move at stage s+1. Hence Λ_i or some marker with smaller index generates the move at stage s+1. This contradicts the fact that m_i is the final position of Λ_i.

<u>Literature.</u>

MENZEL,W.&SPERSCHNEIDER,V.,Recursively enumerable extensions of R_1 by finite functions, in this lecture notes volume

ROGERS,H. Jr., Theory of recursive functions and effective computability, McGraw Hill, New York

On R.E. Inseparability of CPO Index Sets

Dieter Spreen

Lehrstuhl für Informatik I
Rheinisch-Westfälische Technische Hochschule Aachen
Büchel 29-31, D-5100 Aachen, West Germany

Abstract

In this paper the r.e. inseparability and the effective r.e. inseparability of index sets under certain indexings of the computable elements of an effective cpo are studied. As a consequence of the main result on effective r.e. inseparability we obtain a generalization of a theorem by McNaughton. As a further application we obtain generalizations of results by Myhill/Dekker on the productivity of certain index sets. From this we infer the generalization of theorems by Rice/Shapiro/McNaughton/Myhill and Myhill/Shepherdson. This demonstrates the importance of the r.e. inseparability notion.

1. Introduction

Complete partial orders (cpo) have been introduced by D. Scott [9] as a tool for studies in the semantics of the λ-calculus and programming languages. The basic idea was to study infinitary objects as limits of finite ones. A partial number theoretic function, e.g., can be considered as the union of an increasing sequence of finite subfunctions. In the same way a total number theoretic function is the union of an increasing sequence of finite subfunctions which are defined on initial segments of ω. Moreover, each set of natural numbers in the union of its finite subsets, and each real number is the intersection of a decreasing sequence of closed intervalls with rational endpoints. Computability on cpo-s has been considered by Scott [9], Egli/Constable [2], Plotkin [7], Smyth [11], Sciore/Tang [10], Kanda/Park [5] and Weihrauch/Deil [12]. The aim of these studies was to define exactly what is meant by an effective approximation of an object. Objects that are effectively approximatable are called computable. In the above examples the partial recursive functions, the total recursive functions, the recursively enumerable sets and the computable reals respectively are computable.

In this paper we consider index sets under certain numberings of the computable elements of an effective cpo, and study the r.e. inseparability and the effective r.e. inseparability of such index sets. As a special case we obtain results by J. Case [1] and his school for index sets under certain numberings of the partial recursive functions. The main result on effective r.e. inseparability says that for two disjoint sets S and T of computable cpo elements the index set of S is effectively r.e. inseparable from the index set of T, if there is some algorithmically presentable subset T' of T such that S has a nonempty intersection with the closure of T' in the Scott-topology of the cpo. A consequence of this theorem is that the index set of a class of computable cpo elements which is not open in the relative topology of the Scott-topology on the set of computable cpo elements in productive. This generalizes a result by Myhill/Dekker [6]. As further consequences we obtain generalizations of some well-known theorems by McNaughton, Rice/Shapiro/McNaughton/Myhill and Myhill/Shepherdson (cf.[6,8]). Some of the results also hold for non admissible indexings. In an appendix we show the existence of such indexings.

2. Basic Definitions and Properties

In what follows, let $\langle , \rangle : \omega^2 \to \omega$ be a pairing function, let $R^{(n)}$ denote the set of all n-ary total recursive functions, and let W_i be the domain of the i-th partial recursive function φ_i with respect to some Gödelnumbering φ. $A \subseteq \omega$ is *r.e. inseparable* from $B \subseteq \omega$, if there is no recursively enumerable (r.e.) set C with $A \subseteq C \subseteq \omega \setminus B$. A is *effectively r.e. inseparable* from B, if there is some $h \in R^{(1)}$ such that for all $i \in \omega$, either $h(i) \in A \setminus \omega$ of $h(i) \in B \cap W_i$. h effectively witnesses the r.e. inseparability of A from B.

Let now $(D, \sqsubseteq)$ be a partial order. A set $S \subseteq D$ is *directed*, if $S \neq \emptyset$ and for all $a, b \in S$ there is some $c \in S$ with $a, b \sqsubseteq c$. $\mathcal{D} = (D, \sqsubseteq, \bot)$ is a *complete partial order* (cpo), if $\bot \in D$ is the smallest element of $(D, \sqsubseteq)$ and for directed subsets $S \subseteq D$ there exists $\sqcup S = \sup S$. On D a binary relation $\ll$ is defined as follows: $x \ll y$ iff for directed subsets $S \subseteq D$ the relation $y \subseteq \sqcup S$ always implies the existence of an $u \in S$ with $x \sqsubseteq u$. Note that $\ll$ is transitive.

A subset $B \subseteq D$ is a *basis* of the cpo $\mathcal{D}$, if for any $x \in D$ the set $B_x = \{b \in B \mid b \ll x\}$ is directed and $x = \sqcup B_x$. If the cpo $\mathcal{D}$ has a basis, then it is said to be *continuous*. For such cpo-s it is shown in

[12, Lemma 2,3] that for $x,y,z \in D$

LEMMA 2.1. (i) $\perp \ll x$

(ii) $x \ll y \Rightarrow x \sqsubseteq y$

(iii) $x \ll y \sqsubseteq z \Rightarrow x \ll z$

(iv) $x \ll y \Rightarrow \exists b \in B \quad x \ll b \ll y$

(v) B_x is directed with respect to $\ll$.

Let B be a basis of D. If there exists an indexing $\beta : \omega \to B$ of B such that $\{\langle i,j\rangle \mid \beta(i) \ll \beta(j)\}$ is r.e., then $\mathcal{D} = (D,\sqsubseteq,\perp,B,\beta)$ is called *effective* cpo. Effective cpo-s have been introduced by K. Weihrauch [12]. An element $x \in D$ is said to be *computable*, if $\{i \mid \beta(i) \ll x\}$ $[= \beta^{-1}(B_x)]$ is r.e. Let D_c denote the set of all computable elements of $\mathcal{D}$. An indexing η of the computable elements of effective cpo $\mathcal{D}$ is a map from ω onto D_c. Let $\mathrm{Ind}(D_c)$ denote the set of all indexings of D_c. Moreover, for $\eta \in \mathrm{Ind}(D_c)$ and $S \subseteq D$ let $I_\eta(S) = \eta^{-1}(S)$. $I_\eta(S)$ is the *index set* of S under η. S is η-*recursively enumerable* (η-r.e.), if there is an r.e. set $A \subseteq \omega$ with $S = \eta(A)$. An indexing $\eta \in \mathrm{Ind}(D_c)$ is said to *satisfy the recursion theorem*, if for each two-place partial recursive function k there is a function $g \in R^{(1)}$ such that $\eta k(i,g(i)) = \eta g(i)$, for all i with $(i,g(i)) \in \mathrm{dom}(k)$. Furthermore, η is called *admissible* if it satisfies the following axioms (A1) and (A2):

(A1) $\{\langle i,j\rangle \mid \beta(i) \ll \eta(j)\}$ is r.e.

(A2) There is a function $d \in R^{(1)}$ with $\eta d(i) = \sqcup \beta(W_i)$ for all indices $i \in \omega$ such that $\beta(W_i)$ is directed.

As has been shown in [12, Satz 5] such indexings exist.

LEMMA 2.2. Let $\eta,\eta',\eta'' \in \mathrm{Ind}(D_c)$ such that η is admissible and η' satisfies (A2). Then the following statements hold.

(i) $\beta \leq_m \eta'$, i.e., $\beta = \eta' \circ f$, for some $f \in R^{(1)}$.

(ii) $\eta \leq_m \eta'' \Leftrightarrow \eta''$ satisfies (A2).

(iii) η satisfies the recursion theorem.

PROOF. (i) and (ii) are proved in [12, pp. 43-44]. Moreover, in [12, Lemma 14] it is shown that η is precomplete, and by [3, p. 329] we know that precomplete indexings satisfy the recursion theorem. Thus (iii) is also true.

LEMMA 2.3. Let $\eta \in \mathrm{Ind}(D_c)$ satisfy (A1). Then there is a function

$r \in R^{(2)}$ such that for all $i,j \in \omega$ with $\beta(i) \ll \eta(j)$ the following statements hold

(a) $\varphi_{r(i,j)} \in R^{(1)}$

(b) $\beta(i) \ll \beta\varphi_{r(i,j)}(0)$

(c) $\beta\varphi_{r(i,j)}(n) \ll \beta\varphi_{r(i,j)}(n+1) \quad (n \in \omega)$

(d) $\eta(j) = \bigsqcup_n \beta\varphi_{r(i,j)}(n)$.

PROOF. Let $M_{ij} = \{n \mid \beta(i) \ll \beta(n) \ll \eta(j)\}$. Then $M_{ij} = \{n \mid \beta(i) \ll \beta(n)\} \cap \{n \mid \beta(n) \ll \eta(j)\}$. Hence M_{ij} is r.e. Moreover, there is a function $t \in R^{(2)}$ with $M_{ij} = W_{t(i,j)}$. Since $B_{\eta(j)}$ is directed with respect to $\ll$, it follows that $\beta(M_{ij})$ is also directed with respect to $\ll$. Now, let i,j be such that $\beta(i) \ll \eta(j)$. Then $M_{ij} \neq \emptyset$, by Lemma 2.1(iv). Thus $\bigsqcup\beta(M_{ij})$ exists and we have that $\bigsqcup\beta(M_{ij}) \sqsubseteq \eta(j)$. Since $B_{\eta(j)}$ is directed with respect to $\ll$, for each $b \in B_{\eta(j)}$ there is some $b' \in \beta(M_{ij})$ with $b \ll b'$. With this it follows that $\eta(j) \sqsubseteq \bigsqcup\beta(M_{ij})$. Hence $\eta(j) = \bigsqcup\beta(M_{ij})$.
In the remainder of this proof let $\sigma \in R^{(1)}$ and $k \in R^{(2)}$ such that $\varphi_{\sigma(n)}$ is a total function which enumerates W_n and $W_{k(n,m)} = W_n \cap W_m$. Define $g \in R^{(3)}$ by

$$g(i,j,0) = \varphi_{\sigma t(i,j)}(0)$$

$$g(i,j,n+1) = \varphi_{\sigma k(t(g(i,j,n),j),t(\varphi_{\sigma(i,j)}(n+1),j))}(0).$$

Then $g(i,j,n+1) \in M_{g(i,j,n),j} \cap M_{\bar{\imath}j}$ where $\bar{\imath}$ is the $(n+1)$-st element of M_{ij} in the enumeration $\varphi_{\sigma t(i,j)}$. Observe that $M_{g(k,j,m),j} \cap M_{\bar{\imath}j} \neq \emptyset$, since M_{ij} is directed with respect to $\ll$. Thus for all $n \in \omega$ we have that $\beta(i) \ll \beta g(i,j,n) \ll \beta g(i,j,n+1)$ and $\beta\varphi_{\sigma t(i,j)}(n+1) \ll \beta g(i,j,n+1)$. This implies that $\bigsqcup\beta(M_{ij}) \sqsubseteq \bigsqcup_n \beta g(i,j,n)$. Since $g(i,j,n) \in M_{ij}$, for all $n \in \omega$, we also have that $\bigsqcup_n \beta g(i,j,n) \sqsubseteq \bigsqcup\beta(M_{ij})$. Hence $\bigsqcup_n \beta g(i,j,n) = \bigsqcup\beta(M_{ij}) = \eta(j)$. Now, let $r \in R^{(2)}$ such that $\varphi_{r(i,j)}(n) = g(i,j,n)$. Then r has the desired properties.

COROLLARY 2.4. Let $\eta \in \mathrm{Ind}(D_c)$ satisfy (A1) and let $x \in D_c$. Then there is a function $p \in R^{(1)}$ such that $\beta p(n) \ll \beta p(n+1)$, for all $n \in \omega$, and $x = \bigsqcup_n \beta p(n)$.

PROOF. By Lemma 2.1 we have that $\perp \ll x$ and $\perp \in B$. Let $i,j \in \omega$ be such that $\beta(i) = \perp$ and $\eta(j) = x$. Then set $p = \varphi_{r(i,j)}$.

For continuous cpo-s $D = (D, \sqsubseteq, \bot)$ a canonical topology can be defined (cf.[9]). A subset $X \subseteq D$ is open, if (O1) and (O2) hold:

(O1) $\forall x,y \in D\ (x \in X \wedge x \sqsubseteq y \Rightarrow y \in X)$.

(O2) $\forall x \in X\ \exists y \in X\quad y \ll x$.

The topology thus defined is called the *Scott-topology* of D. If B is a basis of D, then $\{O_b | b \in B\}$ with $O_b = \{x \in D | b \ll x\}$ is a base for this topology. A set $E \subseteq D$ is said to be *inseparable from* $F \subseteq D$ *by an open set*, if there is no open set X with $E \subseteq X \subseteq D \setminus F$. For $H \subseteq D$, let $\overline{H}$ denote the topological closure of H with respect to this topology. By [4, 1.4(i)] we obtain

LEMMA 2.5. Let $V \subseteq D$ and $x \in D$. Then the following statements hold.

(i) If $Y \subseteq \overline{V}$ is directed, then $\sqcup Y \in \overline{V}$.

(ii) If $x \sqsubseteq z$, for some $z \in V$, then $x \in \overline{V}$.

Now, let $D = (D,\sqsubseteq,\bot,B,\beta)$ be an effective cpo. A set $G \subseteq D_c$ is a *Lacombe set*, if $G = \bigcup \{O_{\beta(i)} \cap D_c | i \in A\}$, for some r.e. set A. Hence a Lacombe set is open in the relative topology on D_c. Let $L_j = \bigcup \{O_{\beta(i)} \cap D_c | i \in W_j\}$. Then L is an indexing of the Lacombe sets of D. From [12, Satz 9] it follows

LEMMA 2.6. Let $\eta \in Ind(D_c)$ satisfy (A1). Then there is a function $t \in R^{(1)}$ with $W_{t(j)} = I_\eta(L_j)$.

This lemma generalizes a result by Rice (cf.[8]). Since ω with the discrete topology is a topological space, this lemma implies that η is effectively continuous. Let $S,T \subseteq D_c$. Then S is *Lacombe inseparable from* T, if there is no Lacombe set G with $S \subseteq G \subseteq D_c \setminus T$. Let in addition $\eta \in Ind(D_c)$. Then S is *effectively Lacombe inseparable from* T *with respect to* η, if there is a function $f \in R^{(1)}$ such that for all $j \in \omega$, either $\eta f(j) \in S \setminus L_j$ or $\eta f(j) \in T \cap L_j$. Now, let $D' = (D', \sqsubseteq', \bot', B', \beta')$ be a further effective cpo and let $\eta' \in Ind(D'_c)$. Then $\Phi : D_c \to D'_c$ is (η,η')-*effective*, if there is some $g \in R^{(1)}$ with $\Phi\eta(i) = \eta' g(i)$. Moreover, Φ is *effectively continuous*, if there is some $s \in R^{(1)}$ such that $\Phi^{-1}(O_{\beta'(j)} \cap D'_c) = L_{s(j)}$.

3. R.E. Inseparability

In this section let $\mathcal{D} = (D,\sqsubseteq,\bot,B,\beta)$ be an effective cpo. Since the index set of a Lacombe set under an indexing $\eta \in \mathrm{Ind}(D_c)$ that satisfies (A1) is r.e., it follows for two sets $S,T \subseteq D_c$ that the r.e. inseparability of $I_\eta(S)$ from $I_\eta(T)$ implies the Lacombe inseparability of S from T. As has been pointed out by Dekker and Myhill (cf.[6,p.45]) the inverse implication is false. For Lacombe inseparability a sufficient condition can be given which is pure topological. We start with the following simple observation.

THEOREM 3.1. Let $S,T \subset D_c$ with $S \cap T = \emptyset$. Then S is inseparable from T by an open set iff $S \cap \overline{T} \neq \emptyset$.

Since every Lacombe set is open in the relative topology on D_c, it follows.

THEOREM 3.2. Let $S,T \subset D_c$ with $S \cap T = \emptyset$ but $S \cap \overline{T} \neq \emptyset$. Then S is Lacombe inseparable from T.

For finite S the condition that $S \cap \overline{T} \neq \emptyset$ is also necessary.

THEOREM 3.3. Let $S,T \subset D_c$ such that S is finite and $S \cap T = \emptyset$. Moreover, let S be Lacombe inseparable from T. Then $S \cap \overline{T} \neq \emptyset$.

PROOF. Let $S = \{u_0,\ldots,u_n\}$ and assume that $S \cap \overline{T} = \emptyset$. Since $D \setminus \overline{T}$ is open, there is some open set O_{b_i} with $u_i \in O_{b_i}$ and $O_{b_i} \cap \overline{T} = \emptyset$, for each $i \leq n$. Set $G = \bigcup \{O_{b_i} \cap D_c \mid i \leq n\}$. Then G is a Lacombe set and $S \subseteq G \subseteq D_c \setminus T$. Thus, S is Lacombe separable from T.

COROLLARY 3.4. Let $\eta \in \mathrm{Ind}(D_c)$ such that (A1) holds. Moreover, let $S,T \subset D_c$ such that S is finite and $S \cap T = \emptyset$. If $I_\eta(S)$ is r.e. inseparable from $I_\eta(T)$, then $S \cap \overline{T} \neq \emptyset$.

In order to obtain a sufficient condition for r.e. inseparability which moreover holds for arbitrary $S \subset D_c$, our assumptions have to be more constructive. As we will see, the modified condition, which seems to be very natural and not too restrictive, already guarantees effective r.e. inseparability. We first present a generalization of a theorem by J. Case [1].

THEOREM 3.5. Let $\eta \in \mathrm{Ind}(D_c)$ such that (A2) holds and the recursion theorem is satisfied. Moreover, let $S,T \subset D_c$ with $S \cap T = \emptyset$. Let $x \in S$ and $p \in R^{(1)}$ such that $\beta p(n) \sqsubseteq \beta p(n+1)$ $(n \in \omega)$ and $x = \bigsqcup_n \beta p(n)$. Finally, let $q \in R^{(2)}$ such that $\beta p(n) \sqsubseteq \beta q(n,0)$, $\beta q(n,m) \sqsubseteq \beta q(n,m+1)$ and $\bigsqcup_m \beta q(n,m) \in T$, for all $n,m \in \omega$. Then $I_\eta(S)$ is effectively r.e. inseparable from $I_\eta(T)$.

PROOF. Let $\sigma,\tau \in R^{(1)}$ such that $\varphi_{\sigma(i)} \in R^{(1)}$ enumerates W_i and $W_{\tau(i)} = \mathrm{range}(\varphi_i)$. Furthermore, set $g(i,j) = \mu m : \varphi_{\sigma(i)}(m) = j$ and let $k \in R^{(2)}$ such that

$$\varphi_{k(i,j)}(n) = \begin{cases} q(g(i,j), n \dot{-} g(i,j)), & \text{if } \exists m \leqslant n \ \ \varphi_{\sigma(i)}(m) = j \\ p(n), & \text{otherwise.} \end{cases}$$

Then $\bigsqcup_n \beta\varphi_{k(i,j)}(n) \in T$, if $j \in W_i$, and $\bigsqcup_n \beta\varphi_{k(i,j)}(n) \in S$, otherwise. Let $d \in R^{(1)}$ be as in condition (A2), then we obtain that

$$\bigsqcup_n \beta\varphi_{k(i,j)}(n) = \bigsqcup \beta(W_{\tau k(i,j)}) = \eta\, d\, \tau\, k(i,j).$$

Since η satisfies the recursion theorem, there is a function $h \in R^{(1)}$ with $\eta d \tau k(i,h(i)) = \eta h(i)$. Thus $I_\eta(S)$ is effectively r.e. inseparable from $I_\eta(T)$.

By Lemma 2.2(iii) the above requirements on η are satisfied by admissible indexings. But as we show in the appendix, there are also non-admissible indexings with these properties. For admissible indexings the above theorem can be formulated in a more topological fashion so that its connection with the results in the beginning of this section becomes more visible.

THEOREM 3.6. Let $\eta \in \mathrm{Ind}(D_c)$ be admissible. Moreover, let $S,T \subset D_c$ such that $S \cap T = \emptyset$ and such that there is an η-r.e. subset T' of T with $S \cap \overline{T'} \neq \emptyset$. Then $I_\eta(S)$ is effectively r.e. inseparable from $I_\eta(T)$.

PROOF. Let $x \in S \cap \overline{T'}$. Then, by Corollary 2.4, there is a $p \in R^{(1)}$ such that $\beta p(n) \ll \beta p(n+1)$ $(n \in \omega)$ and $x = \bigsqcup_n \beta p(n)$. Since $x \in \overline{T'}$, it follows that $O_{\beta p(n)} \cap T' \neq \emptyset$, for all $n \in \omega$. Now, observe that η satisfies (A1) and that T' is η-r.e. Thus there is a function $s \in R^{(1)}$ with $\beta p(n) \ll \eta s(n)$ and $\eta s(n) \in T'$, for all $n \in \omega$. Finally, let $r \in R^{(2)}$ be as in Lemma 2.3 and set $q(m,n) = \varphi_{r(p(n),s(n))}(m)$. Then q has the properties as in Theorem 3.5. Thus this theorem is applicable.

In order to see that for admissible η Theorem 3.5 is implied by this theorem, let x and q be as in Theorem 3.5. Moreover, let $d \in R^{(1)}$ be as in (A2) and let $\tau, s \in R^{(1)}$ such that $W_{\tau(i)} = \text{range}(\varphi_i)$ and $\varphi_{s(n)}(m) = q(n,m)$. Then

$$T' = \{\bigsqcup_m \beta q(n,m) \mid n \in \omega\} = \eta d \tau s(\omega)$$

is an η-r.e. subset of T and by Lemma 2.5 if follows that $x \in S \cap \overline{T'}$, i.e., $S \cap \overline{T'} \neq \emptyset$.

With Lemma 2.6 we now obtain the following consequence which strengthens Theorem 3.2.

COROLLARY 3.7. Let $\eta \in \text{Ind}(D_c)$ be admissible. Moreover, let $S, T \subset D_c$ such that $S \cap T = \emptyset$ and such that there is an η-r.e. subset T' of T with $S \cap \overline{T'} \neq \emptyset$. Then S is effectively Lacombe inseparable from T with respect to η.

In the case that S is a singleton the conditions for T in Theorems 3.5/3.6 are also necessary.

THEOREM 3.8. Let $\eta \in \text{Ind}(D_c)$ be admissible. Moreover, let $T \subset D_c$ and $u \in D_c \setminus T$. Finally, let $p \in R^{(1)}$ such that $\beta p(n) \ll \beta p(n+1)$ $(n \in \omega)$ and $u = \bigsqcup_n \beta p(n)$. Then the following statements are equivalent.

(1) $\exists q \in R^{(2)}\ \forall n \in \omega\ [\beta p(n) \ll \beta q(n,0)\ \wedge$
$\forall m \in \omega\ \beta q(n,m) \ll \beta q(n,m+1)\ \wedge\ \bigsqcup_m \beta q(n,m) \in T]$

(2) $\exists T' \subseteq T\ [u \in \overline{T'}\ \wedge\ T'$ is η-r.e.$]$

(3) $I_\eta(\{u\})$ is effectively r.e. inseparable from $I_\eta(T)$.

PROOF. As we have seen above, (1) implies (2). Moreover, by Theorem 3.6 we obtain that (2) implies (3). Assume now that (3) holds. By (A1) there is a $g \in R^{(1)}$ with $W_{g(n)} = \{j \mid \beta p(n) \ll \eta(j)\}$, and by our assumption there is an $h \in R^{(1)}$ such that either $hg(n) \in I_\eta(\{u\}) \setminus W_{g(n)}$ or $hg(n) \in I_\eta(T) \setminus W_{g(n)}$. In the case that $hg(n) \in I_\eta(\{u\}) \setminus W_{g(n)}$ it follows that $\eta hg(n) = u$ but $\beta p(n) \not\ll \eta hg(n)$, i.e. a contradiction. Thus $hg(n) \in I_\eta(T) \cap W_{g(n)}$. Now, let $r \in R^{(2)}$ be as in Lemma 2.3 and define $q(n,m) = \varphi_{r(p(n),hg(n))}(m)$. Then q satisfies the requirements in (1).

For the cpo of all partial functions from ω into ω this theorem is in essence a result by Ebcioglu [1]. As a further consequence of Theorem 3.6 we obtain a generalized version of a theorem by McNaughton [6, Theorem 4.1].

COROLLARY 3.9. Let $\eta \in \mathrm{Ind}(D_c)$ be admissible. Moreover, let $S,T \subset D_c$ such that $S \cap T = \emptyset$, but such that there is some $x \in S$ and some $y \in T$ with $x \sqsubseteq y$. Then $I_\eta(S)$ is effectively r.e. inseparable from $I_\eta(T)$.

PROOF. Set $T' = \{y\}$. Then T' is η-r.e. and by Lemma 2.5(ii) we have that $x \in \overline{T'}$.

McNaugthon proved this result in the case of the cpo of all subsets of ω. If one is only interested in r.e. inseparability instead of effective r.e. inseparability, then in the above results it suffices to require that η satisfies (A2). In order to see this let $\gamma \in \mathrm{Ind}(D_c)$ be admissible. By Lemma 2.2(ii) there is then a function $f \in R^{(1)}$ with $\gamma = \eta \circ f$. Hence, $I_\gamma(S) = f^{-1}(I_\eta(S))$. From this it follows that if $I_\gamma(S)$ is r.e. inseparable from $I_\gamma(T)$, then $I_\eta(S)$ is also r.e. inseparable from $I_\eta(T)$.

We now use Theorem 3.6 to classify the index sets of certain subsets of D_c. While Lemma 2.6 classifies the index sets of Lacombe sets, the next theorem classifies the index sets of those subsets of D_c that are not open in the relative topology on D_c.

THEOREM 3.10. Let $\eta \in \mathrm{Ind}(D_c)$ be admissible. Moreover, let $S \subset D_c$ such that it is not open in the relative topology on D_c. Then $I_\eta(S)$ is productive.

PROOF. As is easy to see, S is open in the relative topology on D_c if and only if for all $x \in D_c$, $x \in S$ iff there is some $b \in B \cap S$ with $b \ll x$. Thus, if S is not open in the relative topology on D_c, then there is some $\hat{x} \in D_c$ such that either $\hat{x} \in S$ and $b \not\ll \hat{x}$, for all $b \in B \cap S$, or $\hat{x} \notin S$ and $b \ll \hat{x}$, for some $b \in B \cap S$. Let us start with the first case. Set $T = B_{\hat{x}}$. Then $S \cap T = \emptyset$, since for all $b \in B \cap S$ we have that $b \not\ll \hat{x}$, i.e. $b \notin T$. Moreover, $S \cap \overline{T} \neq \emptyset$, because $\hat{x} = \sqcup B_{\hat{x}} \in \overline{T}$, by Lemma 2.5(i). In order to see that T is η-r.e., note that $I_\beta(T)$ is r.e. and that $\beta = \eta \circ f$, for some $f \in R^{(1)}$, by Lemma 2.2(i). With Theorem 3.6 we now obtain that $I_\eta(S)$ is effectively r.e. inseparable from $I_\eta(T)$. Observe that the recursive function which witnesses this r.e.

inseparability is also a productive function for $I_\eta(S)$.
Let us now consider the second case. Define $T = \{\hat{x}\}$ in this case. Then $S \cap T = \emptyset$. Moreover there is some $b \in S$ with $b \sqsubseteq \hat{x}$. By Corollary 3.9 it follows that $I_\eta(S)$ is effectively r.e. inseparable from $I_\eta(T)$. Thus, in the second case $I_\eta(S)$ also possesses a productive function.

For the cpo of all subsets of ω this result is due to Dekker and Myhill [6, Theorem 3.1]. The following consequence of this theorem also generalizes results by Dekker and Myhill [6, pp. 35-36].

COROLLARY 3.11. Let $\eta \in Ind(D_c)$ be admissible. Then the following statements hold.

(i) Let $\emptyset \neq S \subset D_c$. Then at least one of the two sets $I_\eta(S)$ and $I_\eta(D_c \setminus S)$ is productive.

(ii) Let $x \in D_c$ such that $x \notin B$, if x is maximal in D_c with respect to $\sqsubseteq$. Then $I_\eta(\{x\})$ is productive and in case that $x \neq \bot$ it is also coproductive.

(iii) If S is a Lacombe set, then $I_\eta(S)$ is creative.

PROOF. (i) follows since D_c with the Scott-topology is a connected space. For the proof of (ii) suppose that $\{x\}$ is open in the relative topology on D_c. Then there is some $b \in B$ such that $\{x\} = O_b \cap D_c$. If x is not maximal in D_c with respect to $\sqsubseteq$, this is impossible. Assume therefore that x is maximal in D_c with respect to $\sqsubseteq$. Since $x \notin B$, there is then no $b' \in B \setminus \{b\}$ with $b \ll b' \ll x$. Because B_x is directed with respect to $\ll$, this implies for $b' \in B \setminus \{b\}$ with $b' \ll x$ that $b' \ll b$. Thus $x = \sqcup B_x = b$, which contradicts our assumption. Hence, $\{x\}$ is not open in the relative topology on D_c. By the above theorem it follows that $I_\eta(\{x\})$ is productive. If $x \neq \bot$, then $D_c \setminus \{x\}$ is not upwards closed under the restriction of $\sqsubseteq$ to D_c. Thus $D_c \setminus \{x\}$ is not open in the relative topology on D_c, from which it follows that $I_\eta(\{x\})$ is also coproductive. (iii) follows from (i) by Lemma 2.6.

As a consequence of this result, finally, we obtain Rice's theorem for effective cpo-s.

COROLLARY 3.12. Let $\eta \in Ind(D_c)$ satisfy (A2) and let $S \subseteq D_c$. Then $I_\eta(S)$ is recursive iff $S = \emptyset$ or $S = D_c$.

PROOF. Let $\gamma \in Ind(D_c)$ be admissible. Then $I_\gamma(S)$ is recursive iff $S = \emptyset$ or $S = D_c$, by part (i) of the above corollary. By Lemma 2.2(ii) there is some $f \in R^{(1)}$ such that $I_\gamma(S) = f^{-1}(I_\eta(S))$. With this the asserted equivalence follows.

As a further consequence of Theorem 3.10 and hence of Theorem 3.6 we obtain a generalization of two well-known theorems which are moreover shown to be equivalent, namely the non trivial part of the characterization theorem for r.e. index sets of classes of r.e. sets by Rice/Shapiro/McNaughton/Myhill and the theorem by Myhill/Shepherdson on effective operators (cf.[8]). This demonstrates the strength of the above inseparability result. Let us start with the following

COROLLARY 3.13. Let $\eta \in \mathrm{Ind}(D_c)$ satisfy (A2). Moreover, let $V \subseteq D_c$ such that $I_\eta(V)$ is r.e. Then V is open in the relative topology on D_c.

PROOF. Let $\gamma \in \mathrm{Ind}(D_c)$ be admissible. Then $I_\gamma(V)$ is also r.e. Now, assume that V is not open in the relative topology on D_c. Then $I_\gamma(V)$ is productive, by Theorem 3.10. This contradicts the fact that $I_\gamma(V)$ is r.e.

Let now $\mathcal{D}' = (D', \sqsubseteq', \bot', B', \beta')$ be a further effective cpo.

THEOREM 3.14. Let $\eta \in \mathrm{Ind}(D_c)$ and $\eta' \in \mathrm{Ind}(D_c')$ such that η satisfies (A2) and η' satisfies (A1). Then the following statements are equivalent.

(1) Let $V \subseteq D_c$ such that $I_\eta(V)$ is r.e. Then V is open in the relative topology on D_c.

(2) There is a function $k \in R^{(1)}$ with $L_{k(j)} = \eta(W_j)$, for all $j \in \omega$ such that W_j is an index set under η.

(3) Let $\Phi : D_c \to D_c'$ be (η, η')-effective. Then Φ is effectively continuous.

PROOF. Assume that (1) holds and set $A = \beta^{-1}(\eta(W_j))$. Furthermore, let $j \in \omega$ be such that W_j is an index set under η. By Lemma 2.2(i) there is a function $f \in R^{(1)}$ with $\beta = \eta \circ f$. Then $A = f^{-1}(W_j)$. Thus A is r.e. and there is a function $k \in R^{(1)}$ with $A = W_{k(j)}$. By (1) and the definition of A it follows that $\eta(W_j) = \cup\{O_{\beta(i)} \cap D_c | i \in A\}$. Hence $\eta(W_j)$ is a Lacombe set and $\eta(W_j) = L_{k(j)}$.

Next assume that (2) holds. Since Φ is (η, η')-effective, there is a

function $g \in R^{(1)}$ with $\Phi\eta(i) = \eta'g(i)$. Moreover, since η' satisfies (A1), there is a function $q \in R^{(1)}$ with $W_{q(j)} = \{i \mid \beta'(j) \ll \eta'g(i)\}$. Then $\eta(W_{q(j)}) = \Phi^{-1}(O_{\beta'(j)} \cap D_c)$. Since $W_{q(j)}$ is an index set under η, it follows by (3) that $\eta(W_{q(j)}) = L_{kq(j)}$. Thus Φ is effectively continuous.
In [12, Satz 9], finally, it is shown that (3) implies (1).

If D as well as D' is the cpo of all subsets of ω, then statement (3) is a strengthened version of a theorem by Dekker and Myhill [6, Theorem 5.1]. At the end of this section we present a generalization of a sufficient condition for r.e. inseparability of index sets which is due to J. Case [1].

THEOREM 3.15. Let $\eta \in \mathrm{Ind}(D_c)$ satisfy (A2) and the recursion theorem. Moreover, let $S, T \subset D_c$ with $S \cap T = \emptyset$. Let $x_0, \ldots, x_m \in S$ and $p_0, \ldots, p_m \in R^{(1)}$ such that $\beta p_\nu(i) \sqsubseteq \beta p_\nu(i+1)$ $(i \in \omega)$ and $x_\nu = \bigsqcup_i \beta p_\nu(i)$, for $\nu = 0, \ldots, m$. Finally, let $q_0, \ldots, q_n \in R^{(2)}$ such that $\beta q_\kappa(i,j) \sqsubseteq \beta q_\kappa(i,j+1)$, for all $i,j \in \omega$ and $\kappa = 0, \ldots, n$, and such that for each $i \in \omega$ there are $\nu \leq m$ and $\kappa \leq n$ with $\beta p_\nu(i) \sqsubseteq \beta q_\kappa(i,0)$ and $\bigsqcup_j \beta q_\kappa(i,j) \in T$. Then $I_\eta(S)$ is r.e. inseparable from $I_\eta(T)$.

PROOF. Let the pairs $(0,0), \ldots, (m,n)$ be ordered in some fixed way. Moreover, fix some recursive coding function for $(m+1)\cdot(n+1)$-tuples, and for $(\nu,\kappa) = (0,0), \ldots, (m,n)$ let $(\ )_{(\nu,\kappa)} \in R^{(1)}$ such that $(e)_{(\nu,\kappa)}$ is the (ν,κ)-th component of the $(m+1)\cdot(n+1)$-tuple with code number e. Finally, let $\tau \in R^{(1)}$ such that $W_{\tau(e)} = \mathrm{range}(\varphi_e)$. Now, suppose that there is some r.e. set A with $I_\eta(S) \subseteq A \subseteq \omega \setminus I_\eta(T)$. Let $f \in R^{(1)}$ enumerate A and set

$$g(e) = \mu l\colon \forall(\delta,\rho) \leq (m,n)\ \exists j \leq l\ \ f(j) = (e)_{(\delta,\rho)}.$$

Furthermore, let $k_{(\nu,\kappa)} \in R^{(1)}$ such that

$$\varphi_{k_{(\nu,\kappa)}(e)}(i) = \begin{cases} q_\kappa(g(e), i \dot{-} g(e)), & \text{if } \forall(\delta,\rho) \leq (m,n)\ \exists j \leq i\ \ f(j) = (e)_{(\delta,\rho)} \\ p_\nu(i), & \text{otherwise.} \end{cases}$$

If $(e)_{(\delta,\rho)} \notin A$, for some $(\delta,\rho) \leq (m,n)$, then for all $(\nu,\kappa) \leq (m,n)$ we have that $\varphi_{k_{(\nu,\kappa)}(e)} = p_\nu$, and hence that $\bigsqcup_i \beta\varphi_{k_{(\nu,\kappa)}(e)}(i) = x_\nu \in S$. On the other hand, if $(e)_{(\delta,\rho)} \in A$, for all $(\delta,\rho) \leq (m,n)$, then for $\hat{\imath} = g(e)$ there exist $(\bar{\nu},\bar{\kappa}) \leq (m,n)$ with $\beta p_{\bar{\nu}}(\hat{\imath}) \sqsubseteq \beta q_{\bar{\kappa}}(\hat{\imath},0)$. Thus

$\underset{i}{\sqcup}\beta\varphi_{k_{(\bar\nu,\bar\kappa)}(e)}(i) = \underset{i}{\sqcup}\beta q_{\bar\kappa}(\hat{i},i) \in T$. Let $d \in R^{(1)}$ be as in condition (A2), then we obtain that

$$\underset{i}{\sqcup}\beta\varphi_{k_{(\nu,\kappa)}(e)}(i) = \sqcup\beta(W_{\tau k_{(\nu,\kappa)}(e)}) = \eta d\tau k_{(\nu,\kappa)}(e),$$

for all $(\nu,\kappa) \leq (m,n)$. Since η satisfies the recursion theorem, it also satisfies the $(m+1)\cdot(n+1)$-ary recursion theorem (cf. the proof of [8, §11, Theorem X(a)]). Hence there is some $\hat{e} \in \omega$ with $\eta d\tau k_{(\nu,\kappa)}(\hat{e}) = \eta((\hat{e})_{(\nu,\kappa)})$. As we have just seen, we then have that either $\eta((\hat{e})_{(\nu,\kappa)}) \in S$, for all $(\nu,\kappa) \leq (m,n)$, or $\eta((\hat{e})_{(\bar\nu,\bar\kappa)}) \in T$, for some $(\bar\nu,\bar\kappa) \leq (m,n)$.
Let us first consider the case that $\eta((\hat{e})_{(\nu,\kappa)}) \in S$, for all $(\nu,\kappa) \leq (m,n)$. Then $(\hat{e})_{(\nu,\kappa)} \in I_\eta(S)$ and hence $(\hat{e})_{(\nu,\kappa)} \in A$, for all $(\nu,\kappa) \leq (m,n)$. In this case it follows from above that there exist $(\bar\nu,\bar\kappa) \leq (m,n)$ with $\eta((\hat{e})_{(\bar\nu,\bar\kappa)}) \in T$. This is impossible since $S \cap T = \emptyset$.
In the second case there exist $(\bar\nu,\bar\kappa) \leq (m,n)$ with $\eta((\hat{e})_{(\bar\nu,\bar\kappa)}) \in T$. Hence $(\hat{e})_{(\bar\nu,\bar\kappa)} \in I_\eta(T)$, which implies that $(\hat{e})_{(\bar\nu,\bar\kappa)} \notin A$. In this case it follows from above that $\eta((\hat{e})_{(\nu,\kappa)}) \in S$, for all $(\nu,\kappa) \leq (m,n)$. This again contradicts our assumption that $S \cap T = \emptyset$.

In the case that η is admissible the above theorem follows from Theorem 3.6. In order to see this set

$$T' = \{\underset{j}{\sqcup}\beta q_\kappa(i,j) \mid i \in \omega,\ \kappa \leq n\}.$$

Because of (A2) we then have that T' is η-r.e. By Lemma 2.5(ii) and the above assumption it moreover follows that for each $i \in \omega$ there is some $\nu \leq m$ such that $\beta p_\nu(i) \in \overline{T'}$. Thus there is some $\bar\nu \leq m$ with $\beta p_{\bar\nu}(i) \in \overline{T'}$, for infinitely many $i \in \omega$. Since the sequence $\{\beta p_{\bar\nu}(i)\}_{i\in\omega}$ is non-decreasing with respect to $\sqsubseteq$, it follows that $\beta p_{\bar\nu}(i) \in \overline{T'}$, for all $i \in \omega$. Hence $x_{\bar\nu} \in \overline{T'}$, i.e., $S \cap \overline{T'} \neq \emptyset$.

Appendix

THEOREM. Let D be an effective cpo with at least two elements. Then there exists an indexing of D_c that satisfies (A2) and the recursion theorem but is not admissible.

PROOF. Let $K = \{i \mid \varphi_i(i) \text{ converges}\}$. Moreover, let $x \in D_c$ such that there is some $b \in B \setminus \{\bot\}$ with $b \ll x$, and let γ be an admissible indexing of D_c. Finally, let $\nu : \omega \to D_c$ be defined by $\nu(i) = \bot$, if $i \in K$, and $\nu(i) = x$, otherwise. Now define $\Psi \in \mathrm{Ind}(D_c)$ as follows: Set

$$\Psi(3i) = \gamma(i) \quad \text{and} \quad \Psi(3i+1) = \nu(i).$$

The definition of $\Psi(3i+2)$ is carried out in stages $n \geq 0$.

<u>Stage</u> n: Le i be the smallest uncancelled number. Then use the halting problem of φ as an oracle to see whether $\varphi_i(3i+2)$ converges. If $\varphi_i(3i+2)$ diverges, set $\Psi(3i+2) = \bot$, cancel i and go to stage n+1. Otherwise let $m = \varphi_i(3_i+2)$.

(1) If $m = 3j$ or $m = 3j+1$, or if $m = 3j+2$ and j has already been cancelled, then set $\Psi(3i+2) = \Psi(m)$ and cancel i.

(2) If $m = 3j+2$ and j has not yet been cancelled, then set $\Psi(3i+2) = \Psi(m) = x$ and cancel i and j.

Now, let $r \in R^{(1)}$ with $r(i) = 3i$, then $\gamma = \Psi \circ r$, i.e. $\gamma \leq_m \Psi$. By Lemma 2.2(ii) it follows that Ψ satisfies (A2). In order to see that Ψ also satisfies the recursion theorem, let k be a two-place partial recursive function. Then there is some $s \in R^{(1)}$ with $k(i,j) = \varphi_{s(i)}(j)$. Define $g \in R^{(1)}$ by $g(i) = 3s(i)+2$. Then it follows for i with $(i,g(i)) \in \mathrm{dom}(k)$ that $\Psi k(i,g(i)) = \Psi g(i)$. (A1) is not satisfied by Ψ. Assume to the contrary that Ψ satisfies (A1). Then it follows by the choice of ν and x that

$$i \in \omega \setminus K \Leftrightarrow \nu(i) = x \Leftrightarrow \Psi(3i+1) = x \Leftrightarrow 3i+1 \in \{j \mid b \ll \Psi\ (j)\}.$$

Since Ψ satisfies (A1), we obtain that $\omega \setminus K$ is r.e. But as is well-known, $\omega \setminus K$ is not r.e.

References

[1] Case, J.: private communication (1982).

[2] Egli, H.; Constable, R.L.: Computability concepts for programming language semantics. *Theoret. Comp. Sci.* 2, 133-145 (1976).

[3] Eršov, Ju.L.: Theorie der Numerierungen I. *Zeitschr. f. Math. Logik u. Grundl. d. Math.* 19, 289-388 (1973).

[4] Gierz, G. et. al.: *A Compendium of Continuous Lattices.* Berlin: Springer (1980).

[5] Kanda, A.; Park, D.: When are two effectively given domains identical? 4th GI Conference Theoretical Comput. Sci., Aachen 1979, Lec. Notes Comput.Sci. 67, 170-181 (1979).

[6] Myhill, J.; Dekker, J.C.E.: Some theorems on classes of recursively enumerable sets. *Transact. AMS* 89, 29-59 (1958).

[7] Plotkin, G.: Π^{ω} as a universal domain. *J. Computer Syst. Sci.* 17, 209-236 (1978).

[8] Rogers, H., Jr.: *Theory of Recursive Functions and Effective Computability*. New York: McGraw-Hill (1967).

[9] Scott, D.: Outline of a mathematical theory of computation. Techn. Monograph PRG-2, Oxford Uni. Comp. Lab. (1970).

[10] Sciore, E.; Tang, A.: Admissible coherent cpo-s. Automata, Languages and Programming, 5th Colloq., Udine 1978, Lec. Notes Comput. Sci. 62, 440-456 (1978).

[11] Smyth, M.B.: Effectively given domains. *Theoret. Comp. Sci.* 257-274 (1973).

[12] Weihrauch, K.; Deil, Th.: Berechenbarkeit auf cpo-s. Schriften zur Angew. Math. u. Informatik Nr. 63, RWTH Aachen (1980).

ARITHMETICAL DEGREES OF INDEX SETS FOR COMPLEXITY CLASSES

Kenneth W. Regan
Merton College
Oxford, OX1 4JD U.K.

1. Introduction

In what respects are classes such as P, NP, and PSPACE alike? The techniques used today seem unable to distinguish them. We expound upon their similarities instead, hoping to silhouette the techniques that may be needed by blacking in the obstacles to be overcome.

Taking language classes themselves as the basic objects of study, we classify them according to position of associated index sets in the arithmetical hierarchy, and to provability of certain assertions about these classes in first-order logic. We exhibit genera of complexity classes which are structurally more intricate than P, NP, and kin.

Our work extends that of P. Hajek in [Haj79], while our philosophical approach mirrors that of D. Joseph and P. Young in [JY81]. They ask, "How strong must logical theories be to be relevant to Computer Science?" If one regards language classes as standing for concepts, and takes level in the arithmetical hierarchy (of associated index sets) as a measure of difficulty, our question becomes, "How hard must the concepts we work with be for us to settle the important quandaries of Computer Science?"

2. Preliminaries

We associate to each string $x \in \Sigma^*$ (where $\Sigma = \{0,1\}$) the natural number having binary representation 1x, and thence carry over arithmetical operations and the order relation from numbers to strings. The <u>empty string</u> λ corresponds to 1. We let $\langle\ ,\ \rangle$ stand for a (polynomial-time computable) bijection from $\Sigma^* \times \Sigma^*$ to Σ^* s.t. $\forall x,y: |\langle x,y\rangle| \leq 2|x|+2|y|$.

We choose an "acceptable" (in the sense of [Rog67]) recursive enumeration $[M_i]_{i=1}^{\infty}$ of deterministic off-line language-acceptors. We make the indexing uniform for relativized computation by requiring that <u>each</u> M_i have a special <u>query state</u> q? which allows a two-way branch depending on whether the word currently to the right of the worktape head is contained in whatever oracle set is attached. "Real" machines, to be sure, have $\emptyset$ as oracle, and then q? functions in much the same way as any other state. We also assume each machine is M_i for infinitely many i. $\underline{L(M_i)}$ stands for the language accepted by M_i. We limit attention to acceptors in this paper, only touching on the development (in future reports) of analogous results with reference to the more standardly-used list $[T_j]_{j=1}^{\infty}$ of transducers, of which $[M_i]_{i=1}^{\infty}$ may be considered a recursive sublist.

Throughout we let $\underline{F}$ stand for a sound recursively axiomatized first-order theory with equality strong enough to represent all recursive predicates, and which possesses a recursive proof predicate $\underline{FProof}(\psi,d)$ defined to hold exactly when d is a valid derivation of theorem ψ of F. For short we say F is sound etc. Peano Arithmetic and ZF set theory are examples, granting the latter's consistency. We obtain recursive predicates $\underline{Halt}(x,i,n)$ and $\underline{Accept}(x,i,n)$ over the language of F defined to hold when M_i on input x respectively halts or accepts within n steps.

Every formula ψ with $m \geq 1$ unquantified variables defines a language $\underline{L(\psi)}$ of strings $\langle x_1,\langle\ldots,x_m\rangle\ldots\rangle$ which satisfy it. Customarily we write formulas in prenex form, i.e. with all quantifiers preceding a predicate that is quantifier-free. For convenience we always coalesce adjacent like quantifiers into one block, writing e.g. $(\exists x,y)$ for $(\exists x)(\exists y)$. A useful transformation to prenex form is that given by

Lemma 1: For any predicates R,S the formula $\forall x: (\exists y)R(x,y) \Leftrightarrow S(x)$ can be put into the equivalent prenex form

$$(\forall x,y)(\exists y'): [\neg R(x,y) \;\&\; \neg S(x)] \;\vee\; [R(x,y') \;\&\; S(x)],$$

and if T is such that $S(x) \Leftrightarrow (\exists z)T(x,z)$ for all x, further reduced to

$$(\forall x,y,z)(\exists y',z'): [\neg R(x,y) \;\&\; \neg T(x,z)] \;\vee\; [R(x,y') \;\&\; T(x,z')].$$

For $k \geq 0$ a prenex formula is a Σ_k-formula (Π_k-formula) if it has $\leq k$ quantifier blocks on a recursive predicate and the leading one is $\exists$ ($\forall$). Every Σ_k-formula is the negation of a Π_k-formula. We let Σ_k denote the class of languages defined by Σ_k-formulas, and Π_k, their complements. Evidently $\Sigma_0 = \Pi_0 = \underline{REC}$, the class of recursive languages, and for all $k \geq 0$, $(\Sigma_k \cup \Pi_k) \subseteq (\Sigma_{k+1} \cap \Pi_{k+1})$. Taken together, these classes constitute the arithmetical hierarchy, and we set $\underline{AH} := \bigcup_{k=1}^{\infty} \Sigma_k$.

The "weak hierarchy theorem" states that for each k, Σ_k is precisely the class of languages accepted by TM's using oracle sets from Σ_{k-1}. For $k = 1$ this means $\Sigma_1 = \underline{RE}$, the class of recursively enumerable sets. The "strong hierarchy theorem" of Kleene states that the arithmetical hierarchy is infinite, i.e. $\Sigma_k \subsetneq \Sigma_{k+1} \cap \Pi_{k+1} \subsetneq \Sigma_{k+1}$ for all $k \geq 1$. For the polynomial hierarchy (PH) the corresponding weak theorem holds, but whether the strong one holds subsumes $P \neq NP$ and other hard problems.

3. Index Sets And Complexity Classes

The advantage of using index sets is that they allow us to pass easily from classes to sets, and hence describe complexity classes in first-order logic. The recursively invariant definition is given by

Definition 1: A set I of natural numbers is an index set if for some class $\mathcal{C}$ of r.e. languages and "acceptable" recursive enumeration $[M_i]_{i=1}^{\infty}$ of Turing Machine acceptors, $I = \{i \mid L(M_i) \in \mathcal{C}\}$.

We lose no generality by fixing one such enumeration as done above, and denoting by $I_{\mathcal{C}}$ the unique index set it assigns to a class $\mathcal{C}$.

Rice's Theorem states that the only recursive index sets are $I_{\emptyset} = \emptyset$ and $I_{RE} = \Sigma^*$, where $\emptyset$ denotes the empty class. Typical r.e. and Σ_2 index sets, both complete for their respective levels of the arithmetical hierarchy, are $I_{NON-\emptyset} = \{i \mid L(M_i) \neq \emptyset\}$ and $I_{FIN} = \{i \mid L(M_i) \text{ is finite}\}$. We are interested in Σ_3-complete index sets and those placed even higher in the hierarchy. A basic result of H. Rogers (cf. also [Re82]) is

Theorem 2 ([Rog67]): Suppose $J \in \Sigma_3$. Then given a definition of J we can construct a recursive function $f: N \to N$ such that for all i,

(a) $i \in J \Rightarrow L(M_{f(i)})$ is co-finite,
(b) $i \notin J \Rightarrow L(M_{f(i)})$ is not recursive.

[Haj79] notes the corollary for $\mathcal{C}$ = P, DTIME[t(n)], ($t > 1$, recursive).

Corollary 3: Let $\mathcal{C}$ be a class of recursive languages containing some infinite set A together with all sets B such that $A - B$ is finite. Then $I_{\mathcal{C}}$ is Σ_3-hard.

Proof: Let $J \in \Sigma_3$. Since A is recursive and infinite we can find a recursive bijection $\theta: \Sigma^* \to A$, and a recursive function $g: N \to N$ s.t. for all i, $L(M_{g(i)}) = \theta[L(M_i)]$. Composing g with f above reduces J to $I_{\mathcal{C}}$. ∎

The hypothesis that A be infinite is necessary because $I_{FIN} \in \Sigma_2$. From now on we use the slightly stronger condition that $\mathcal{C}$ contain some A together with all finite variations of A, i.e. all languages B such that the symmetric difference $B \Delta A$ is finite. Such a class we call somewhere dense (s.d.). An even stronger condition is that $\mathcal{C}$ be closed under finite variations, i.e. $\forall B \in \mathcal{C}: B^* \subseteq \mathcal{C}$, where $B^* := \{\text{f.v.'s of } B\}$.

We associate another set to complexity classes $\mathcal{C}$ by defining $J_{\mathcal{C}} := \overline{I_{\mathcal{C}}} \cap \underline{TOT} = \{i \mid L(M_i) \notin \mathcal{C}\} \cap \{i \mid M_i \text{ halts for all inputs}\}$. Strictly is not r.p. have been constructed in [LR72]. Now we can characterize recursively presentable classes $\mathcal{C}$ which are s.d. first in terms of $J_{\mathcal{C}}$.

Theorem 7: Let $\mathcal{C}$ be a somewhere-dense class of languages. Then $\mathcal{C}$ is recursively presentable if and only if $\mathcal{C}$ is bounded and $J_{\mathcal{C}} \in \Pi_2$.

Proof: ($\Rightarrow$) Let u be the index of a total machine accepting a recursive universal language U for $\mathcal{C}$, and let s: N $\to$ N be the running-time function of M_u. Then there is a recursive function t: N $\to$ N that majorizes all of $\{\lambda n.s(2n+k) \mid k \in N\}$ such that every language in $\mathcal{C}$ can be accepted by a TM running in time t(n) that first maps inputs x to $\langle x,k\rangle$ for the right fixed value of k and then simulates M_u. Second, for all i: $i \in J_{\mathcal{C}} \Leftrightarrow$

$$i \in TOT \;\&\; (\forall k)(\exists x,n)[Halt(x,i,n) \;\&\; [Accept(x,i,n) \Leftrightarrow \neg U(x,k)]],$$

where U(x,k) stands for '$\langle x,k\rangle \in U$'. As TOT $\in \Pi_2$ this has Π_2 prenex form; and as the last part states $x \in L(M_i) \Delta U_k$, it correctly defines $J_{\mathcal{C}}$.

($\Leftarrow$) Let $J_{\mathcal{C}}$ be defined by $(\forall r)(\exists s)R(i,r,s)$, with R recursive, and let $[M_{i_k}]_{k=1}^{\infty}$ be a recursive presentation of DTIME[t(n)]-containing-$\mathcal{C}$ such that each M_{i_k} is total. Then $L(M_{i_k}) \in \mathcal{C} \Leftrightarrow (\exists r)(\forall s)\neg R(i_k,r,s)$ for all k. Since $\mathcal{C}$ is s.d. there is a recursive A such that $A^* \subseteq \mathcal{C}$. For all k,r let Q_{kr} be a machine that on input $x \in \Sigma^*$ checks whether for all $s \le x$ $\neg R(i_k,r,s)$, and if so accepts x iff M_{i_k} does, otherwise accepting iff $x \in A$. Then $\{\langle x,\langle k,r\rangle\rangle \mid Q_{kr}$ accepts $x\}$ is a recursive universal language for $\mathcal{C}$. ∎

The following criterion often provides a useful way of demonstrating recursive presentability of classes straight from the definitions of their index sets in arithmetic. F has the properties given in Section 2.

Definition 4: A class $\mathcal{C}$ has the recursive refutability property (with respect to F) if for some predicate S over the language of F,

(a) for all i: $L(M_i) \in \mathcal{C} \Leftrightarrow (\exists j)\,S(i,j)$,
(b) for all i,j: M_i total & $\neg S(i,j) \Rightarrow$ '$\neg S(i,j)$' is provable in F.

Theorem 8: For all $\mathcal{C} \subseteq$ REC, $\mathcal{C}$ has the r.r. property $\Leftrightarrow J_{\mathcal{C}} \in \Pi_2$.

Proof: ($\Rightarrow$) Letting FProof(ψ,d) hold when d is a derivation of ψ from axioms of F, and taking S from above, we define $J_{\mathcal{C}}$ for all i by:

$$i \in J_{\mathcal{C}} \Leftrightarrow (\forall x,j)(\exists n,d)\,[Halt(x,i,n) \;\&\; FProof('\neg S(i,j)',d)].$$

($\Leftarrow$) This seems to require showing as a lemma that REC itself has the r.r. property. This is stronger than the well-known statement that REC speaking this is not an index set, but it is useful in giving a finer classification of subrecursive language classes. For now we note,

Proposition 4: Let $\mathcal{C} \subsetneq$ REC. Then $J_{\mathcal{C}}$ is Π_2-hard.

Proof: TOT is itself Π_2-complete. Let A be a recursive language not in $\mathcal{C}$. Then we can construct a recursive function g such that for all i, M_i total $\Leftrightarrow$ $M_{g(i)}$ is total and accepts A. This g reduces TOT to $J_{\mathcal{C}}$. ∎

In the next section we consider a genus of subrecursive classes for which both sets take their lowest possible positions in the hierarchy, and hence which represents the "easiest" notion of complexity class we have.

4. Recursively Presentable Classes

A universal language for a class $\mathcal{C}$ is a set U s.t. $\mathcal{C} = \{U_k \mid k \in N\}$, where $U_k := \{x \mid \langle x,k\rangle \in U\}$ for each k. The following definitions are respectively equivalent to the availability of a recursive enumeration $[Q_k]_{k=1}^{\infty}$ of [total; any] machines accepting precisely the languages in $\mathcal{C}$.

Definition 2: A class $\mathcal{C}$ is recursively presentable (r.p.) if $\mathcal{C}$ has a recursive universal language; $\mathcal{C}$ is r.e.-presentable (i.e. "presentable by r.e. indices") if $\mathcal{C}$ has an r.e. universal language. (And so on.)

Proposition 5: If $\mathcal{C}$ is r.p. or r.e.-presentable, then $I_{\mathcal{C}} \in \Sigma_3$.

Proof: Let u be the index in $[M_i]_{i=1}^{\infty}$ of a machine accepting a universal language for $\mathcal{C}$. Lemma 1 then puts the following in Σ_3 prenex form:

$$\forall i\colon\ i \in I_{\mathcal{C}} \Leftrightarrow (\exists k)(\forall x)\,[(\exists y)\mathrm{Accept}(x,i,y) \Leftrightarrow (\exists z)\mathrm{Accept}(\langle x,k\rangle,u,z)]\,.$$ ∎

The following well-known lemma gives us a basic stock of r.p. classes.

Lemma 6: DTIME[t(n)] is recursively presentable for any recursive, nondecreasing (not necessarily time-constructable) function $t > 1$ a.e.

Definition 3: $\mathcal{C}$ is bounded if $\mathcal{C} \subseteq$ DTIME[t(n)] for some such t.

If any other Blum complexity measure Φ is substituted for TIME above we still obtain the same family of classes, and Lemma 6 holds for Φ so long as $D\Phi[t(n)]$ is s.d. Measures Φ and functions $t > 1$ for which $D\Phi[t(n)]$

has an r.e. "basis" (that is, presentation). We let #Steps(x,i) stand for the (possibly infinite) number of steps M_i has on input x. Then we can informally state a "recursively refutable" predicate defining I_{REC}:

$$S_{REC}(i,\langle k,c\rangle) \ := \ \text{"}L(M_k) = L(M_i) \ \& \ M_k \text{ halts for all inputs } \&$$
$$\forall x,m,n: \ [\#Steps(x,k) = m \ \& \ \sum_{y \leq x} \#Steps(y,i) = n] \Rightarrow m \leq c \cdot n\text{"}.$$

The last clause is superfluous for defining I_{REC} but is needed for the r.r. property. For $L(M_i) \in REC$ one can satisfy it by choosing M_k total and c large enough if M_i is not total, and $k = i$, $c = 1$ if it is. When M_i is total and $S_{REC}(i,\langle k,c\rangle)$ fails then either the last clause fails, as may be checked in F by "running" M_k on a particular x for more than $c \cdot n$ steps, or M_k's running time is bounded by $c \cdot (M_i$'s running time) and it is one of the first two clauses that fails—verifiably so because M_k is then total. Then taking a Π_2-formula $(\forall r)(\exists s)R(i,r,s)$ to define $J_{\mathcal{C}}$ yields

$$i \in I_{\mathcal{C}} \Leftrightarrow (\exists k,c,r)[S_{REC}(i,\langle k,c\rangle) \ \& \ (\forall s) \neg R(k,r,s)]$$

for all i, using the fact that if $S_{REC}(i,\langle k,c\rangle)$ holds then M_k is total. Since R is recursive this definition gives $\mathcal{C}$ the r.r. property. ∎

Corollary 9: Let $\mathcal{C}$ be s.d. Then $\mathcal{C}$ is recursively presentable if and only if $\mathcal{C}$ is bounded and has the r.r. property.

Take away boundedness, and the r.r. property still implies that $\mathcal{C}$ is r.e.-presentable. Theorem 8 shows that it doesn't matter much which F we choose so long as F passes the threshold of strength outlined in Section 2, which is about the same as that for F to be essentially incomplete.

Part (b) of Definition 4 formalizes in one sense the idea of having a "witness" to the failure of a condition S(i,j) defining $\mathcal{C}$. The familiar notion of a witness function for a language L with respect to a presentation $[Q_k]_{k=1}^{\infty}$ such that $L \neq L(Q_k)$ for all k is a map $f: N \to \Sigma^*$ giving $f(k) \in L \,\Delta\, L(Q_k)$ for all k. For r.p. classes $\mathcal{C}$ we can bring the idea of witnesses as proofs in number theory into line with the more common notion of witnesses as individual strings by using the original r.r. predicate for $\mathcal{C}$ to obtain a recursive presentation $[Q_k]_{k=1}^{\infty}$ via Corollary 9, and then redefining $S_{\mathcal{C}}(i,k)$ to be the new r.r. predicate stated informally as "$L(M_i) = L(Q_k)$". Proofs of $\neg S_{\mathcal{C}}(i,k)$ then revolve around strings. Any recursive presentation of $\mathcal{C}$ gives at least one recursive witness function to every language in REC $- \mathcal{C}$.

We invite the reader to verify that the same results can be given for classes $\mathcal{F}$ of total functions with respect to an acceptable enumeration

$[T_j]_{j=1}^{\infty}$ of transducers, under these two modifications: $\mathcal{F}$ is <u>s.d.</u> if it contains some total function f together with all total functions g s.t. g(x) = f(x) for all but finitely many x; $\mathcal{F}$ is <u>recursively presentable</u> if there exists a total recursive function u: $\Sigma^* \times \Sigma^* \to \Sigma^*$ such that $\mathcal{F} = \{\lambda x.u(x,k) \mid k \in N\}$. Although the above proofs still work, there is elsewhere a major difference: the notion of r.e.-presentability does not apply to classes of <u>total</u> functions. (See comment at end of Section C.)

We use these results to describe four types of r.p. and r.r. classes.

<u>(A) Common Complexity Classes</u>. For all i, $L(M_i) \in P$ if and only if $(\exists k,m)[M_k$ runs in time n^m+m and accepts $L(M_i)]$. The predicate in [ ]-brackets satisfies Definition 4, so P is recursively presentable. Moreover, it defines $P^{L(Q)}$ whenever Q is total and L(Q) is the oracle set for <u>both</u> M_i and M_k, and so we can obtain a recursive sublist $[M_{i_k}]_{k=1}^{\infty}$ from our oracle-uniform list of TM's s.t. $[M_{i_k}^{L(Q)}]_{k=1}^{\infty}$ is $P^{L(Q)}$ for any total Q. (To be sure, this can be done more simply by adding to each M_i's program a "clock" that shuts off all computations after n^i+i steps) By the same token, the class <u>FP</u> of polynomial-time computable functions is also r.p.

For any r.p. function class $\mathcal{F}$ and recursive language A, we define $L(M_i) \leq_{\mathcal{F}} A$ informally for all i by $(\exists f \in \mathcal{F})[\forall x: x \in L(M_i) \Leftrightarrow f(x) \in A]$. With $\mathcal{F}$ = FP (so that $\leq_{\mathcal{F}}$ is <u>p</u>olynomial-time <u>m</u>any-one reducibility) and $A = L_{SAT}$ we can conclude that NP is r.p. The roles of A and $L(M_i)$ can be interchanged without affecting recursive refutability of the definition, so that the class of $\leq_m^p$-hard sets for NP may be seen to have the r.r. property (it is unbounded, so not r.p.). Using the fact that the family of classes having the r.r. property is closed under finite intersections (and unions) then tells us that the class of NP-complete languages is recursively presentable. Continuing the process of "chunking" together known r.r. predicates to make new ones, such as was done with "$(\exists f \in \mathcal{F})$" above, eventually lets us conclude that PSPACE, the individual Σ_k^p, Π_k^p, and Δ_k^p classes of the polynomial hierarchy, the whole PH itself, the sets complete for each level, the classes of elementary and primitive recursive functions, and indeed almost all classes considered attractive objects of study, are recursively presentable. (Cf. [Sö82], [Re82].)

<u>(B) Classes of Oracle Sets Having Certain Properties</u>. As an example we take <u>EQ</u> := $\{L \mid NP^L = P^L$; L is recursive$\}$. Using the K-operator of [Sö81], where K(A) is a uniformly-obtainable NP^A-complete set for all A, we may define for all i, $L(M_i) \in EQ \Leftrightarrow (\exists c,k,m,r)[S_{REC}(i,\langle c,k\rangle)$ & M_r with oracle $L(M_k)$ accepts $K[L(M_k)]$ in time $n^m+m]$. We use S_{REC} to state $L(M_i) = L(M_k)$ because we need M_k to be total for subsequent expressions involving $L(M_k)$ to be recursively refutable. ([Haj79] shows $J_{EQ} \in \Pi_2$.)

(C) Classes Corresponding to Provable Properties of Languages. We say a property Π of r.e. languages is provable (in F) if there is some formula ψ over F such that, taking $\mathcal{C} = \{L \mid L \text{ has } \Pi\}$, we have both (1) $\mathcal{C} = \{L(M_i) \mid \psi(i) \text{ holds}\}$ and (2) $\forall L$: $L \in \mathcal{C} \Rightarrow$ for some M_k accepting L, $\psi(k)$ is provable in F. Under this very weak criterion every $L \in P$ is provably in P, for we need only represent L by an n^i+i-clocked machine as in (A) above. This says nothing about the problem of proving $L \in P$ when L is represented "naturally", e.g. as the set of primes or satisfiable Boolean formulae. We ask: For which classes can we prove membership, in this weak sense at least? The answer also has a strong negative side, telling us which $\mathcal{C}$ lack uniform membership tests in any sense, for any F.

Theorem 10: Let F be sound etc., and $\mathcal{C}$ a class of r.e. languages.

(a) $\mathcal{C}$ is r.e.-presentable $\Leftrightarrow$ $\mathcal{C} \neq \emptyset$ and every $L \in \mathcal{C}$ is provably in $\mathcal{C}$.

(b) If $FIN \subseteq \mathcal{C} \subseteq REC$, then $\mathcal{C}$ is r.r. $\Leftrightarrow$ for some ψ as above and all $L \in \mathcal{C}$, $\psi(k)$ is provable for some total machine M_k accepting L.

(c) $\mathcal{C}$ is recursively presentable $\Leftrightarrow$ $\mathcal{C} \neq \emptyset$ and, for some ψ as above and all i, $\psi(i) \Rightarrow M_i$ is total.

Proof: (a) If ψ is some formula such that $L \in \mathcal{C} \Leftrightarrow \psi(k)$ is provable in F for some M_k accepting L, then for each k,d let $Q_{\langle k,d\rangle}$ mimic M_k if FProof('$\psi(k)$',d), and otherwise accept some fixed $A \in \mathcal{C}$. Conversely, suppose $[Q_k]_{k=1}^{\infty}$ is an r.e. presentation of $\mathcal{C}$. Then the sentence $\psi(i)$ can be taken to be "$L(M_i) = L(Q_k)$ for some k". If we represent $L \in \mathcal{C}$ by Q_j for the right j we can prove ψ("j") by going down the list recursively.

(b) With $S_{\mathcal{C}}$ as r.r. predicate for $\mathcal{C}$ we make a special r.e. presentation of $\mathcal{C}$. For all i,j, Q_{ij} mimics M_i on input x if M_i halts on all $y \leq x$ and $(\forall d \leq x) \neg$FProof('$\neg S_{\mathcal{C}}(i,j)$',d). Else Q_{ij} rejects x, possibly by not halting, and so accepts a finite language. Then $[Q_{ij}]_{i,j=1}^{\infty}$ has a total machine accepting each language in $\mathcal{C}$, and so using it as in (a) yields the right ψ. Conversely, an r.r. definition of $\mathcal{C}$ is given by: For all i, $i \in I_{\mathcal{C}} \Leftrightarrow (\exists c,k,d)[S_{REC}(i,\langle k,c\rangle)$ & FProof('$\psi(k)$',d)].

(c) The $\Rightarrow$ part is similar to (a), and assuming F is sound we can do the $\Leftarrow$ part by presenting $\mathcal{C}$ directly. Let $A \in \mathcal{C}$, and for all i,m let Q_{im} be a machine that on all inputs checks FProof('$\psi(i)$',d) for all $d \leq m$. If a satisfying d is found then Q_{im} mimics M_i; else Q_{im} accepts A. ■.

For classes of total functions only (c) works, as per comments above. The standard definition of PROVRECf(F) is $\{f \mid f$ is computed by some T_j whose totality is provable in F$\}$. Applying (c) with $\psi(j)$ = "T_j is total" shows that PROVRECf(F) is r.p. Hence "f is recursive" is undecidable in F for all but a bounded class of recursive functions. (Compare [Haj79].)

(D) Witness Function Classes. Let U be a recursive universal language for some r.p. class $\mathcal{C}$. Then the class $\mathcal{D}$ of recursive languages having recursive witness functions with respect to U equals REC - $\mathcal{C}$, and as we shall see is not even r.e.-presentable (if $\mathcal{C} \supseteq$ FIN). However, if we restrict the witness functions to some r.p. class $\mathcal{F}$, then for all i,

$$i \in I_{\mathcal{D}} \Leftrightarrow (\exists f \in \mathcal{F}; c,k)[S_{REC}(i,\langle k,c\rangle) \;\&\; \forall m: f(m) \in L(M_k) \,\Delta\, U_m],$$

and so $\mathcal{D}$ has the r.r. property. Taking $\mathcal{F}$ = PROVRECf(F), U to present P, and intersecting the class $\mathcal{D}$ above with NP enables us to conclude that if NP $\neq$ P the class of NP languages possessing provably recursive (in F) witness functions with respect to U is recursively presentable.

5. Classes Having Σ_3 Index Sets Which Are Not Recursively Presentable

Faced with such a formidable array of r.p. classes, we might query whether all subrecursive classes having Σ_3 index sets have the r.r. property. This is resolved in the negative, not by a recursion-theoretic result, but by a "structure theorem" in complexity theory independently brought out in slightly different versions by U. Schöning in [Sö82] and by P. Chew and M. Machtey in [CM81]. We combine both forms into one while relaxing the original requirement that $\mathcal{C}_1$ and $\mathcal{C}_2$ be r.p.

Theorem 11: Let $\mathcal{C}_1$, $\mathcal{C}_2$ be classes having the r.r. property and A_1, A_2 recursive languages such that $\mathcal{C}_1$ contains no finite variations of A_1, $\mathcal{C}_2$ no f.v.'s of A_2, and furthermore $A_1 \in$ P, $A_2 \notin \{\emptyset, \Sigma^*\}$. Then we can construct E such that $E \notin \mathcal{C}_1$, $E \notin \mathcal{C}_2$, and yet $E \leq^P_m A_2$—[Sö82]. If $A_1 = \emptyset$ then $\mathcal{C}_1$ need only be r.e.-presentable—[CM81].

The basic example takes $\mathcal{C}_1$ = {NP-complete sets}, $\mathcal{C}_2$ = P, $A_1 = \emptyset$, and $A_2 = L_{SAT}$; if NP $\neq$ P the hypotheses are satisfied, yielding R. Ladner's result that there exist languages in NP - P which are not NP-complete. The following original example strengthens Corollary 4 of [Sö82].

Theorem 12: If NP $\neq$ P, then for any recursive universal language U for P and any sound etc. formal system F, we can construct a language E in NP having no provably (in F) recursive witness function w.r. to U.

Proof: Take $\mathcal{C}_1 = \{L \in NP \mid (\exists f \in PROVRECf(F))(\forall k)[f(k) \in L \,\Delta\, U_k]\}$, which we proved r.p. above, $\mathcal{C}_2$ = P, $A_1 = \emptyset$, and $A_2 = L_{SAT}$. If NP $\neq$ P then applying Theorem 11 successfully constructs the desired E. ∎.

We need not work with P and NP here; Theorem 12 works just as well with EXPTIME (known to contain P properly) in place of NP. If we desire we can throw the class EQ proved r.r. above into the picture and obtain that $NP^E \neq P^E$ as well. We cannot, however, use this method to produce a language E in EXPTIME - P (having no provably recursive witness function w.r. to U presenting P) such that NP^E does equal P^E, owing to a palpable asymmetry between r.r. classes and their complements, between S(i,j) and $\neg$S(i,j) regarding provability in F. The result easily extends to construct $E \in NP$ having no provably recursive witness function with respect to any language which is provably a recursive presentation of P. Oddly enough, if the underlined words above are switched the result seems not to follow! Consider this in tandem with D. Kozen's observation in [Kz80] that a proof of $NP \neq P$ amounts to a proof that every language in NP - P has a recursive witness function w.r. to any recursive presentation of P.

We state the corollary to Theorem 11 drawn by both papers in a more general form, omitting the unnecessary condition that $\mathcal{D}$ itself be r.p.

Corollary 13: Let $\mathcal{C}$ and $\mathcal{D}$ be classes such that $\mathcal{C}$ and $\mathcal{D} - \mathcal{C}$ are both somewhere-dense, $\mathcal{C}$ has the r.r. property, and $\mathcal{D}$ is closed under polynomial-time many-one reducibility. Then $\mathcal{D} - \mathcal{C}$ does not have the r.r. property; and if $FIN \subseteq \mathcal{C}$, $\mathcal{D} - \mathcal{C}$ is not r.e.-presentable either.

Thus EXPTIME - P and NP - P are not r.e.-presentable, and NP - P has the r.r. property iff it is null. The corollary can be read as saying that the topology on REC whose closed sets are generated by the r.r. classes closed under f.v.'s fails to be Hausdorff, in the strong sense that the intersection of any two nonempty open sets is nonempty. Returning at last to what this means in terms of index sets, we can state:

Proposition 14: Let $\mathcal{C}, \mathcal{D} \subseteq REC$ have the r.r. property and be closed under f.v.'s, with $\mathcal{D}$ also closed under $\leq^p_m$, and put $\mathcal{E} = \mathcal{D} - \mathcal{C}$. If $\mathcal{E} \neq \mathcal{D}$ and $\mathcal{E} \neq \emptyset$ then $I_{\mathcal{E}} \in \Sigma_3$ and $J_{\mathcal{E}} \in (\Sigma_3 \cap \Pi_3) - (\Sigma_2 \cup \Pi_2)$.

Proof: Let $S_{\mathcal{C}}$ be an r.r. predicate for $\mathcal{C}$. Then for all i, $i \in I_{\mathcal{D}-\mathcal{C}}$ $\Leftrightarrow i \in I_{\mathcal{D}}$ & $(\exists c,k)[S_{REC}(i,\langle k,c\rangle)$ & $(\forall j)(\exists d)FProof('\neg S_{\mathcal{C}}(k,j)',d)]$. Since $I_{\mathcal{D}} \in \Sigma_3$ this has $\exists\forall\exists$ prenex form, and so $I_{\mathcal{E}}$ is (complete) in Σ_3.

For all i, $i \in J_{\mathcal{E}}$ iff M_i is total and $i \in I_{\mathcal{C}} \vee i \notin I_{\mathcal{D}}$. Formally, $i \in J_{\mathcal{E}} \Leftrightarrow (\forall x)(\exists n)Halt(x,i,n)$ & $[(\exists j)(\forall d)\neg FProof('\neg S_{\mathcal{C}}(i,j)',d) \vee (\forall j)(\exists d)FProof('\neg S_{\mathcal{D}}(i,j)',d)]$. The quantifiers inside the square brackets can be brought out in either order, yielding a Σ_3 and a Π_3 definition for $J_{\mathcal{E}}$. $J_{\mathcal{E}}$ is not in Π_2 because $\mathcal{E}$ does not have the r.r. property, and $J_{\mathcal{E}}$ is not in Σ_2 because it is Π_2-hard. ■.

6. Classes Whose Index Sets Are Not in Σ_3.

The leitmotiv of unprovability noted in the previous three sections now moves from the counterpoint to become the main theme. If a class $\mathcal{C}$ does not have $I_{\mathcal{C}} \in \Sigma_3$, then we cannot generally prove anything either way about membership of given languages in $\mathcal{C}$, neither in the weakish sense of Section 4(C) nor in terms of predicates $S_{\mathcal{C}}(i,j)$ or $\neg S_{\mathcal{C}}(i,j)$ as before. Nor can we represent $\mathcal{C}$ as a finite union or intersection (or difference?) of classes having Σ_3 index sets. We sharpen the demarcation between Σ_3 and non-Σ_3 classes by considering the theory whose axioms are all <u>true</u> closed formulas in the language of arithmetic having Σ_3 prenex form. Since $(\exists n)\psi(n)$ when true logically follows from $\psi(\bar{n})$ for some fixed value $\bar{n}$ making $\psi(\bar{n})$ true, this is the same as the theory obtained from all true Π_2 statements, and so we call it <u>Π_2 Number Theory</u>.

Π_2NT is both weaker than full Peano Arithmetic, because PA's axiom schema $[\psi(\mathbf{0})\ \&\ \forall n{:}\ \psi(n) \Rightarrow \psi(n\mathbf{+1})] \Rightarrow (\forall n)\psi(n)$ for induction is in Σ_3 generally only for closed Σ_1 or Π_1 formulas ψ, and stronger, because its axiom set is not r.e. and includes many powerful unproved statements such as the Riemann Hypothesis (if true). Some authors have opined that Π_2NT includes all constructive methods relevant to Computer Science; for discussion and rebuttal see [JY81]. $P^A \overset{?}{=} NP^A$ is decided in Π_2NT for <u>all</u> recursive A. Most to our point, for all classes $\mathcal{C}$ having Σ_3 index sets and $i \in I_{\mathcal{C}}$, "$L(M_i) \in \mathcal{C}$" is trivially a theorem of Π_2NT. The next result shows that where Π_2NT cannot define a class it cannot for the most part prove membership of the languages contained in it.

<u>Theorem</u> 15: Suppose $I_{\mathcal{C}} \notin \Sigma_3$, with $I_{\mathcal{C}}$ definable by a well-formed formula $\psi_{\mathcal{C}}$ of arithmetic. Then $\mathcal{C}$ can be written as $\mathcal{D} \cup \mathcal{E}$ such that

(a) $\mathcal{D} \cap \mathcal{E} = \emptyset$,

(b) $I_{\mathcal{D}} \in \Sigma_3$ (therefore $I_{\mathcal{E}} \notin \Sigma_3$);

(c) for all $i \in I_{\mathcal{E}}$, $\psi_{\mathcal{C}}(i)$ is not a theorem of Π_2NT.

<u>Proof</u>: Given $\psi_{\mathcal{C}}$ (if none exists the result still holds with $\mathcal{D} = \emptyset$) define

$$D \ := \ \{ i \mid \psi_{\mathcal{C}}(i) \text{ is a theorem of } \Pi_2\text{NT} \}.$$

First we note that D actually is an index set. If M_i and M_k accept the same language then "$L(M_i) = L(M_k)$" is expressed by the true statement $(\forall x)[(\exists n)\,\mathrm{Accept}(x,i,n) \Leftrightarrow (\exists m)\,\mathrm{Accept}(x,k,m)]$. Since this is a Π_2-formula by Lemma 1, $\psi_{\mathcal{C}}(i)$ is provable in Π_2NT if and only if $\psi_{\mathcal{C}}(k)$ is. Thus if we put $\mathcal{D} := \{L(M_i) \mid i \in D\}$, then $D = I_{\mathcal{D}}$. We show that $D \in \Sigma_3$; then setting $\mathcal{E} := \mathcal{D} - \mathcal{C}$ completes the proof of the theorem.

If a formula ψ is derivable in Π_2NT, then it follows from a finite number of Π_2 sentences $\theta_1,\ldots,\theta_k$ (the axioms used in the derivation, including <u>all</u> instances of induction) by elementary rules of inference and properties of arithmetic. We let F_0 be a recursively axiomatized sound etc. formal system encoding these rules and properties, e.g. PA minus induction. F_0 can prove that $\theta_1,\ldots,\theta_k$ are all implied by a single true closed Π_2 formula θ (basically $\theta_1 \& \ldots \& \theta_k$ in prenex form). Let $\underline{W}_2$ denote the set of true well-formed Π_2 formulas. Then for all i,

$$i \in I_{\mathcal{D}} \iff (\exists \theta,d)[\theta \in W_2 \;\&\; F_0\text{Proof}('\theta \Rightarrow \psi_{\mathcal{C}}(i)',d)].$$

Matijasevic-et-al.'s Theorem (see [BM77]) has a corollary stating that for each k the set $\underline{V}_k$ of true Σ_k formulas (encoded over Σ^*) is in Σ_k. Since $\theta \in W_2 \iff '\neg\theta' \notin V_2$ for all θ, the set W_2 is itself definable by a Π_2 formula. So we can put the definition of $I_{\mathcal{D}}$ into Σ_3 prenex form. ∎

The last part shows that the theorem generalizes to non-Σ_k classes and "Π_{k-1}NT" for all $k \geq 3$. We note that $\mathcal{D}$ is maximum among Σ_3 classes $\mathcal{A}$ contained in $\mathcal{C}$ for which '$\forall i: L(M_i) \in \mathcal{A} \Rightarrow \psi_{\mathcal{C}}(i)$' is a theorem of Π_2NT. Unlike classes having intuitively "smooth" contours such as P, NP, and even EXPTIME - P, non-Σ_3 classes have "jagged edges" represented by $\mathcal{E}$ above. Counting arguments show that such classes exist. A model-theoretic result of D. Leivant gives what seems to be strong evidence that certain classes corresponding to immunity properties are not Σ_3.

<u>Definition</u> 5: <u>BM</u> (for <u>bi-immune</u> w.r. to P) is the class of recursive languages B such that no infinite subset of either B or $\overline{B}$ lies in P, i.e. that defined by $Bm(i) \;\&\; i \in I_{REC}$. With U for P, <u>Bm</u>(i) is the Π_3 formula

$(\forall k)[(\exists w)(\forall x \geq w)\neg x \in U_k \;\vee\; (\exists y,z)[y \in U_k \cap L(M_i) \;\&\; z \in U_k \cap \overline{L(M_i)}]]$.

<u>Theorem</u> 16 ([Lei82]): '$\exists i: Bm(i) \;\&\; i \in I_{REC}$' is not a theorem of Π_2NT.

This means that Π_2NT cannot prove that the index set of BM is in Σ_3. However, there could still happen to be a Σ_3 formula whose equivalence to $i \in I_{REC} \;\&\; Bm(i)$ for all i is not provable in Π_2NT, PA, ZF, or whatever. We conjecture that recursion-theoretic methods can show $I_{BM} \notin \Sigma_3$ (ditto for $I_{P\text{-IMMUNE}}$ and $I_{P\text{-COIMMUNE}}$ separately), e.g. by showing I_{BM} Π_3-hard.

We end by returning to the note struck in Sections 1 & 2 on whether methods of logic can separate classes such as P, NP, PH, and PSPACE. The idea is that if $\mathcal{C} \subseteq \mathcal{E} \subseteq \mathcal{D}$, $I_{\mathcal{C}}, I_{\mathcal{D}} \in \Sigma_3$, but $I_{\mathcal{E}} \notin \Sigma_3$, then $\mathcal{C} \neq \mathcal{D}$. By the above discussions, showing $I_{\mathcal{E}} \notin \Sigma_3$ requires ("nonconstructive"?) techniques on a higher level of complexity than those commonly used in current work, but as familiar methods seem impotent here this may well be a plus. To illustrate we give a class possibly separating P from PSPACE.

<u>Definition</u> 6: $\underline{H} := \bigcap_{A \in EQ} P^A = \{L \mid \text{for all } A,\ NP^A = P^A \Rightarrow L \in P^A\}$.

<u>Proposition</u> 17: (a) $PH \subseteq H \subseteq PSPACE$. (b) If $PH = \Sigma_k^p$ for some k, then H = PH. (c) $P^H = NP^H$ (i.e. $\bigcup_{A \in H} P^A = \bigcup_{A \in H} NP^A$). (d) $I_H \in \Pi_4 - \Pi_3$.

<u>Proof</u>: (a) A PSPACE-complete B gives $P^B = NP^B = PSPACE$, while $\forall B \in EQ$, $P^B = PH^B \supseteq PH$. (b) If $PH = \Sigma_k^p$, then for Σ_k^p-complete B, $NP^B = P^B = \Sigma_k^p$. (c) and (d) are clear from the definition of H and Σ_3-hardness of I_H. ∎

If the leading "for all" cannot be removed from the definition of H, then $H \neq PH$, and so by (b), $P \neq NP \neq co\text{-}NP \neq \Sigma_2^p \neq \ldots \neq PH \neq PSPACE$. As this seems too much to ask for (and for other reasons) we conjecture H = PH instead. However, even a proof of H = PH would on condition $PH \neq PSPACE$ resolve an open oracle question by showing that $(\exists A)\ P^A = NP^A \subsetneq PSPACE$.

In sum, we have exhibited different types of language classes whose I and J index sets show their structures to be increasingly more complex, and described what this promises and forbids to future investigations.

References

[BH77] L. Berman and J. Hartmanis. On isomorphisms and density of NP and other complete sets. SIAM J. Comp. 6, 1977, pp 305-321.

[BM77] J. Bell and M. Machover. A First Course in Mathematical Logic. (Amsterdam: North-Holland Pub. Co., 1977).

[CM81] P. Chew and M. Machtey. A note on structure and looking-back (...). J. Comp. Sys. Sci. 22, 1981, pp 53-59.

[Haj79] P. Hajek. Arithmetical hierarchy and complexity of computation. Theor. Comp. Sci. 8, 1979, pp 227-237.

[JY81] D. Joseph and P. Young. Independence results in computer science? J. Comp. Sys. Sci. 23, 1981, pp 205-222.

[Kz80] D. Kozen. Indexings of subrecursive classes. Theoretical Computer Science 11, 1980, pp 277-301.

[Lei82] D. Leivant. Unprovability of theorems of complexity theory in weak number theories. Theor. Comp. Sci. 18, 1982, pp 259-268.

[LR72] L. Landweber and E. Robertson. Recursive properties of abstract complexity classes. J. ACM 19, 1972, pp 296-308.

[Re82] K. Regan. Computability, enumerability, and the polynomial hierarchy. M.Sc. qualif. dissertation, Oxford, 1982.

[Rog67] H. Rogers. Theory of Recursive Functions and Effective Computability. (New York: McGraw-Hill, 1967).

[Sö81] U. Schöning. Untersuchungen zur Struktur von NP... Ph.D dissertation, Stuttgart, 1981.

[Sö82] U. Schöning. A uniform approach to obtaining diagonal sets in complexity classes. Theor. Comp. Sci. 18, 1982, pp 95-103.

RUDIMENTARY RELATIONS AND TURING MACHINES WITH LINEAR ALTERNATION

Hugo Volger
Math.Institut,Universität Tübingen
D7400 Tübingen , Fed.Rep.of Germany

Abstract: Smullyan [61] introduced the class R of rudimentary relations as the smallest class which contains the concatenation relation and which is closed under the boolean operations , explicit transformations and linearly bounded quantification. RUD ,the class of rudimentary languages, consists of the sequential encodings of rudimentary relations. Wrathall [75] has shown that RUD can be described as the union LH of a linear time analogue of the polynomial time hierarchy of PH of Meyer,Stockmeyer [72].

Chandra,Kozen and Stockmeyer [76] introduced the concept of alternating Turing machines (=ATM) . An ATM is a nondeterministic Turing machine with two disjoint sets of states,existential and universal states,which play dual roles in the definition of acceptance . A language L belongs to the alternation class STA(s,t,a) if there exists an ATM M such that for each word w in L there exists a finite accepting computation tree of M for w of depth $\leq t(n)$, alternation depth $\leq a(n)$ and space $\leq s(n)$, where $n = |w|$.

There is a close connection between quantification and linear alternation. Chandra,Kozen and Stockmeyer noted that PH may be described as the union of a hierarchy of bounded alternation . An analogous result will be shown for RUD = LH :

(1) $RUD = \cup\langle STA(-,O(n),k):k \in N\rangle \subseteq ATIME(O(n)) = STA(-,O(n),O(n))$

Extending a result of Nepomnjascii [70] for NLOGSPACE we are able to prove that the alternating LOGSPACE hierarchy of Chandra , Kozen and Stockmeyer is contained in RUD :

(2) $\cup\langle STA(O(n^\alpha),O(n^\beta),k):k \in N\rangle \subseteq RUD$ for $\alpha < 1 \leq \beta$, and hence

(3) $\cup\langle STA(\log(n),-,k):k \in N\rangle \subseteq RUD$

However , the question whether the inclusion $RUD \subseteq ATIME(O(n))$ is proper remains open. A negative answer would solve important open problems in complexity theory.

1. Rudimentary relations

The class R of rudimentary relations on $\{1,2\}^*$ which is based on the concatenation relation was introduced by Smullyan [61] as a string analogue of the class CA of constructive arithmetic relations on N which is based on + and × . Smullyan showed that R forms a base for the recursively enumerable relations i.e. a relation $Q(\overline{x})$ is r.e. iff there exists a (strictly) rudimentary relation $P(\overline{x},y)$ such that $Q(x) \Leftrightarrow \exists y P(\overline{x},y)$. Moreover, Bennett [62] has shown in his thesis that R coincides with CA via the dyadic coding of natural numbers. In the following we shall identify a word in $\{1,\ldots,k\}^*$ with the corresponding natural number under the k-adic coding.

By means of the sequential encoding of relations one obtains the class RUD of rudimentary languages ($\sigma(Q) = \{w_1 c w_2 c \ldots c w_k : (w_1,\ldots,w_k) \in Q\}$, where c is new). It has been studied by several authors (Yu [70,77], Jones [69,75], Nepomnjascii [70,72,78], Wrathall [75,78], Meloul [79]).

The class R (resp. XR) of rudimentary (resp. extended rudimentary) relations is the least class which contains the concatenation relation $x \cdot y = z$ and which is closed under the boolean operations, explicit transformations and linearly (resp. polynomially) bounded quantification. The class R^+ (resp. XR^+) of positive (resp. extended positive) rudimentary relations is the least class which contains the concatenation relation and which is closed under union and intersection, explicit transformations, universal subword quantification and linearly (resp. polynomially) bounded existential quantification. The class R_s of strictly rudimentary relations is the least class which contains the concatenation relation and which is closed under the boolean operations, explicit transformations and subword quantification.

$\exists y : y \subseteq x \wedge \ldots$	subword quantification
$\exists y : \lvert y\rvert \leq k\lvert x\rvert \wedge \ldots$	linearly bounded quantification
$\exists y : \lvert y\rvert \leq \lvert x\rvert^k \wedge \ldots$	polynomially bounded quantification

and analogously for universal quantification. It should be noted that $|y| \leq k|x|$ resp. $|y| \leq |x|^k$ may be replaced by $y \leq x^k$ resp. $y \leq x^{(\log x)^{k+1}}$.

Using the sequential encoding of relations we obtain the corresponding classes of languages: RUD, XRUD, RUD^+, $XRUD^+$, RUD_s. Clearly, these classes are related as follows:

$$\begin{array}{ccccc} & & XRUD^+ & \subseteq & XRUD \\ & & \cup\!\mid & & \cup\!\mid \\ RUD_s & \subseteq & RUD^+ & \subseteq & RUD \end{array}$$

The class RUD_s is quite small, since $RUD_s \subseteq$ LOGSPACE and $\{1^n 2^n : n \in N\} \notin RUD_s$ (cf. Nepomnjascii [78]). However, the NP-complete problem SAT is of the form $\exists y : |y| < |x| \wedge Q(y,x)$ with $Q(y,x)$ strictly rudimentary, as Meloul [79] has shown.

2. Turing machines with linear alternation

Chandra and Stockmeyer [76] and Kozen [76] have extended the concept of nondeterministic Turing machines (NTM's) to alternating Turing machines (ATM's). An ATM M is a NTM which has two disjoint sets of states, the existential and universal states, and a distinguished accepting resp. rejecting state. Configurations and successor configurations are defined as for NTM's. An input w is accepted by M (i.e. $w \in L(M)$), if there exists a finite accepting subtree B of the computation tree of M for w. B is accepting, if (1) the root of B is labeled with the input configuration for w, (2) all leaves of B are labeled with accepting configurations, (3) if C is a node of B labeled with an existential (resp. universal) configuration then at least one (resp. all) successor configuration(s) must appear as labels of the successors of C in B (cf. Chandra, Kozen and Stockmeyer [81]). Note that a NTM is just an ATM with no universal states.

A language L belongs to the alternation class STA(s,t,a), if L is accepted by an ATM M such that each w in L posesses an accepting subtree B of depth $\leq t(n)$ and alternation depth $\leq a(n)$ and each configuration in B uses at most s(n) cells, where $n = |w|$. In particular, one defines ATIME(t) = STA(-,t,-) and ASPACE(s) = STA(s,-,-).

Alternating time bridges the gap between nondeterministic time and deterministic space (cf. Chandra, Kozen and Stockmeyer [81]):

$$\text{NTIME}(t) \subseteq \text{ATIME}(t) \subseteq \text{DSPACE}(t) \quad \text{and} \quad \text{APTIME} = \text{PSPACE} .$$

Moreover, there is a close connection between alternation and quantification. In particular, hierarchies defined by ressource bounded quantification are closely related to hierarchies defined by bounded alternation classes using the same ressources.

3. The linear - and the polynomial time hierarchy

In her thesis Wrathall [75] has shown that the class XRUD is the union of the polynomial time hierarchy of Meyer, Stockmeyer [72] and that the class RUD is the union of a linear time analogue of this hierarchy. There are three different ways of describing the hierarchy for PH resp. LH:

Nondeterministic Oracles:

$$NP_* = \cup\langle NP_k : k \in N\rangle,\ NP_0 = \text{PTIME},\ NP_{k+1} = \underline{NP}(NP_k)$$

$$NL_* = \cup\langle NL_k : k \in N\rangle,\ NL_0 = \text{LTIME},\ NL_{k+1} = \underline{NL}(NL_k),$$

where $\underline{NP}(\underline{A})$ resp. $\underline{NL}(\underline{A})$ is the class of languages accepted by a nondeterministic oracle TM (=NOTM) with a polynomial resp. linear time bound and an oracle for a member of the class $\underline{A}$.

Bounded Quantifiers:

$$PH = \cup\langle P\Sigma_k : k \in N\rangle,\ P\Sigma_0 = \text{PTIME},\ P\Sigma_{k+1} = \exists_p(P\Sigma_k \vee coP\Sigma_k)$$

$$LH = \cup\langle L\Sigma_k : k \in N\rangle,\ L\Sigma_0 = \text{LTIME},\ L\Sigma_{k+1} = \exists_1(L\Sigma_k \vee coL\Sigma_k),$$

where L belongs to $\exists_p(\underline{A})$ resp. $\exists_1(\underline{A})$ if there exists L' in $\underline{A}$ such that x is in L iff $\exists y{:}\,|y| \leq |x|^k \wedge L'(x,y)$ resp. $\exists y{:}\,|y| \leq k|x| \wedge L'(x,y)$.

Bounded Alternation:

$$APH = \bigcup\langle AP_k : k \in N\rangle \ , \ AP_k = \bigcup\langle STA(-,O(n^i),k) : i \in N\rangle$$

$$ALH = \bigcup\langle AL_k : k \in N\rangle \ , \ AL_k = STA(-,O(n),k) \ .$$

The following proposition shows that the three hierarchies coincide in both cases :

Prop.1: (a_1) $NP_k = P\Sigma_k$, $NP_* = PH$, (a_2) $NL_k = L\Sigma_k$, $NL_* = LH$, (b_1) $PH = XRUD$, (b_2) $NP_1 = XRUD^+$, (c_1) $LH = RUD$, (c_2) $NL_1 \subseteq RUD^+$, (d_1) $P\Sigma_k = AP_k$, $PH = APH$, (d_2) $L\Sigma_k = AL_k$, $LH = ALH$.

(a_1) - (c_1) can be found in Wrathall [77,78]. (c_2) follows from the inclusion $CFL \subseteq RUD$ in Yu [70] by means of a result of Book,Greibach [73]. (d_1) was mentioned in Chandra,Kozen and Stockmeyer [81] , but (d_2) seems to be new .

(d_1) and (d_2) can be proved by the same method . Given the syntactical description of L with at most k alternations of bounded quantifiers it is easy to construct an ATM accepting L with the corresponding time bound and at most k alternations . Conversely , given an ATM accepting a language L with at most k alternations , one constructs a DTM accepting a language L' and having k additional tapes , one for each alternation phase . Being in the i-th alternation phase the machine makes use of the content of the i-th tape going from left to right to control the choice of moves as long as necessary and then changes to the (i+1)-th tape if an alternation occurs. Hence L can be obtained from L' by appropriate bounded quantification as desired . This should be compared with the incremental stack automata in Yu [70] .

Meloul [79] presented two different quantifier hierarchies for PH and LH which are based on LOGSPACE in both cases rather than PTIME and LTIME . In addition , he exhibited corresponding oracle hierarchies . He worked with LOGSPACE oracle machines which have a choice tape as source of nondeterminism and required that the length of the choice tape and the oracle tape is polynomially resp. linearly bounded .

The classes XRUD and RUD are related to alternation classes with a linear amount of alternation as follows :

$$APH \subseteq LAPTIME = \bigcup\langle STA(-,O(n^i),O(n)) : i \in N\rangle \subseteq PSPACE$$

$$ALH \subseteq ALTIME = STA(-,O(n),O(n)) \subseteq LSPACE$$

It should be noted that the PSPACE-complete problem QBF is in ALTIME . The question whether these inclusions are proper remains open. A new quantifier seems to be involved . A positive answer would resolve important open problems in complexity theory .

It is known that questions concerning the polynomial time hierarchy may be reduced to questions concerning the linear time hierarchy with the help of polynomial padding. A language L is replaced by the padded language $L_k = \{wc^m : m = |w|^k, w \in L\}$ or $L^k = \{(x,y) : |y| = |x|^k, x \in L\}$. Making use of the fact that SAT and the relation $|y| = |x|^k$ are positive rudimentary (cf. Meloul [79]) one arrives at the following situation:

Prop.2: (i) $\mathrm{LTIME} = \mathrm{RUD}^+$ implies $\mathrm{RUD}^+ \subseteq \mathrm{PTIME}$ (iff $\mathrm{PTIME} = \mathrm{XRUD}^+$)
(ii) $\mathrm{RUD}^+ = \mathrm{RUD}$ implies $\mathrm{RUD} \subseteq \mathrm{XRUD}^+$ (iff $\mathrm{XRUD}^+ = \mathrm{XRUD}$)
(iii) $\mathrm{RUD} = \mathrm{ALTIME}$ implies $\mathrm{ALTIME} \subseteq \mathrm{XRUD}$ (iff $\mathrm{XRUD} = \mathrm{LAPTIME}$)
(iv) $\mathrm{ALTIME} = \mathrm{LSPACE}$ implies $\mathrm{LSPACE} \subseteq \mathrm{LAPTIME}$ (iff $\mathrm{LAPTIME} = \mathrm{PSPACE}$)

4. Alternating LOGSPACE hierarchy

The alternating LOGSPACE hierarchy was defined by Chandra, Kozen and Stockmeyer [81] as follows: $\mathrm{ALOGH} = \cup\langle \mathrm{ALOG}_k : k \in \mathbb{N}\rangle$ with $\mathrm{STA}(\log n, -, k) = \mathrm{ALOG}_k$. In addition, they have shown: $\mathrm{NLOGSPACE} \subseteq \mathrm{ALOGH} \subseteq \mathrm{DSPACE}(\log^2 n)$, $\mathrm{ALOGSPACE} = \mathrm{PTIME}$. Extending the result of Nepomnjascii [70] we are able to show the following:

Prop.3: (*) $\mathrm{NLOGSPACE} \subseteq \mathrm{NST}(O(n^\alpha), O(n^\beta)) \subseteq \mathrm{RUD}^+$ for $\alpha < 1 \leq \beta$
(**) $\mathrm{STA}(\log n, -, k) \subseteq \mathrm{STA}(O(n^\alpha), O(n^\beta), k) \subseteq \mathrm{RUD}$ for all k and $\alpha < 1 \leq \beta$
($\overset{*}{_*}$*) $\mathrm{ALOGH} \subseteq \mathrm{RUD}$.

Nepomnjascii exhibited a binary relation in R^+ which expresses the existence of a sequence of $n^{1-\alpha}$ successive configurations of length n^α. Having iterated this process $t = \lfloor \beta/1-\alpha \rfloor$ times, one is able to describe n^β successive configurations of length n^α. A final existential quantifier with a linear bound yields (*). Now we adapt the method in Berman [80] of expressing acceptance in ATM's with at most k alternations. This introduces k additional quantifiers with linear bounds and yields (**). ($\overset{*}{_*}$*) is an immediate consequence of (**).

Yu [77] has shown that the class 1NSA of languages accepted by 1-way nondeterministic stack automata is contained in RUD. Using the same method as above this result can be extended as follows. The class $\cup\langle 1\mathrm{ASA}_k : k \in \mathbb{N}\rangle$ is contained in RUD, where $1\mathrm{ASA}_k$ is the class of languages accepted by 1-way alternating stack automata with at most k alternations.

References:

Bennett, J.H. [62]: On Spectra, Ph.D. Thesis, Princeton Univ., Princeton N.J., 1962, 135pp.
Berman, L. [80]: The Complexity of Logical Theories, Theoret.Comp.Sci. 11 (1980), 71-77
Book, R., Greibach, S. [70]: Quasirealtime Languages, Math.Systems Theory 4 (1970), 97-111
Chandra, A.K., Stockmeyer, L.J. [76]: Alternation, in: Proc. 17th IEEE Symp. on Found. of Comp.Sci. (1976), 98-108
Chandra, A.K., Kozen, D.C., Stockmeyer, L.J. [81]: Alternation, J.ACM 28 (1981), 114-133

Harrow,K.[78]: The Bounded Arithmetic Hierarchy, Information and Control 36(1978),102-117
Jones,N.D.[69]: Context-free Languages and Rudimentary Attributes, Math. Systems Theory 3(1969)102-109, 11(1977/8),379-380
Jones,N.D.[75]: Space-Bounded Reducibility among Combinatorial Problems, J.Comp.System Sci.11(1975),68-85, 15(1977),241
King,K.N.,Wrathall,C.[78]: Stack Languages and Log n Space, J.Comp.System Sci.17(1978),281-299
Kozen,D.C.[76]: On Parallelism in Turing Machines, in: Proc.17th IEEE Symp. on Found.of Comp.Sci.(1976),89-97
Meloul,J.[79]: Rudimentary Predicates , Low Level Complexity Classes and Related Automata, Ph.D.Thesis, Oxford Univ., Oxford 1979, 210pp.
Meyer,A.R.,Stockmeyer,L.J.[72]: The Equivalence Problem for Regular Expressions with Squaring requires Exponential Space, in: Proc.13th IEEE on Switching and Automata Theory(1972),125-129
Nepomnjascii,V.A.[70a]: Rudimentary Predicates and Turing Computations, Soviet Math.Dokl.11(1970),1462-1465
Nepomnjascii,V.A.[70b]: Rudimentary Interpretation of Two-Tape Turing Computation, Kibernetika 6(1970),29-35
Nepomnjascii,V.A.[78]: Examples of Predicates not expressible by S-RUD Formulae, Kibernetika 12(1978)
Ruzzo,W.L.[80]: Tree-Size Bounded Alternation, J.Comp.System Sci.21 (1980),218-235
Ruzzo,W.L.,Simon,J.,Tompa,M.[82]: Space-Bounded Hierarchies and Probabilistic Computations, in: Proc.14th ACM Symp. on Theory of Computing, San Francisco 1982,215-223
Smullyan,R.[61]: Theory of Formal Systems, Annals of Math.Studies 47 , Princeton Univ.Press 1961, 147pp.
Stockmeyer,L.J.[75]: The Polynomial-Time Hierarchy, IBM Research Report RC 5379(1975)
Stockmeyer,L.J.[77]: The Polynomial-Time Hierarchy, Theoret.Comp.Sci.3 (1977),1-22
Volger,H.[83]: Turing Machines with Linear Alternation,Theories of Bounded Concatenation and the Decision Problem of First Order Theories, Theoret. Comp.Sci.23(1983),333-338
Wrathall,C.[75]: Subrecursive Predicates and Automata, Ph.D.Thesis , Harvard Univ.,Cambridge MA,1975, 156pp.
Wrathall,C.[77]: Complete Sets and the Polynomial-Time Hierarchy, Theoret. Comp.Sci.3(1977),23-33
Wrathall,C.[78]:Rudimentary Predicates and Relative Computation, SIAM J. Computing 7(1978),194-209
Yu,Y.Y.[70]: Rudimentary Relations and Formal Languages, Ph.D.Thesis , Univ.of California,Berkeley 1970, 47pp.
Yu,Y.Y.[77]: Rudimentary Relations and Stack Languages, Math.Systems Theory 10(1977),337-343

A CRITICAL-PAIR/COMPLETION ALGORITHM FOR FINITELY GENERATED IDEALS IN RINGS

B.Buchberger

Mathematisches Institut
Johannes Kepler Universität
A4040 LINZ, Austria

ABSTRACT

In 1965, the author introduced a "critical-pair/completion" algorithm that starts from a finite set F of polynomials in $K[x_1,\ldots,x_n]$ (K a field) and produces a set G of polynomials such that the ideals generated by F and G are identical, but G is in a certain standard form (G is a "Gröbner-basis"), for which a number of important decision and computability problems in polynomial ideal theory can be solved elegantly. In this paper, it is shown how the critical-pair/completion approach can be extended to general rings. One of the difficulties lies in the fact that, in general, the generators of an ideal in a ring do not naturally decompose into a "head" and a "rest" (left-hand side and right-hand side). Thus, the crucial notions of "reduction" and "critical pair" must be formulated in a new way that does not depend on any "rewrite" nature of the generators. The solution of this problem is the starting point of the paper. Furthermore, a set of reduction axioms is given, under which the correctness of the algorithm can be proven and which are preserved when passing from a ring R to the polynomial ring $R[x_1,\ldots,,x_n]$. $Z[x_1,\ldots,x_n]$ is an important example of a ring in which the critical-pair/completion approach is possible.

INTRODUCTION

In /Buchberger 65, 70/ the "critical-pair/completion" approach for solving algorithmical problems in polynomial ideal theory was introduced: Let F be a (finite) set of polynomials in $K[x_1,\ldots,x_n]$ (K a field). In order to solve an algorithmic problem for the ideal Ideal(F) generated by F, first transform F into a certain canonical form G (which was called "Gröbner-basis" in /Buchberger 76/) such that Ideal(F) = Ideal(G) and then solve the problem for G.

Essentially, G is a "Gröbner-basis" iff a certain reduction relation $\rightarrow_G$ induced by G on $K[x_1,...,x_n]$ has the Church-Rosser property (see definition below). It turns out that, in fact, a number of important algorithmic problems for polynomials ideals Ideal(G) can be easily solved as soon as G is a Gröbner-basis, whereas they are extremely complex in general. (Examples: deciding congruence; canonical simplification modulo polynomial "side relations"; computing the multiplication table of the residue class ring modulo polynomial ideals; computation of the elimination ideals of a polynomial ideal; solution of linear diophantine equations in the polynomial ring; computation of the Hilbert function; deciding the solvability of systems of algebraic equations; etc.). Of course, the intrinsic computational complexity of these problems can not be annihilated by this approach either and, in fact, it reappears in the construction of the Gröbner-bases. The advantage of the method, however, is that, as soon as a Gröbner-basis G for F has been constructed once and for all, quite diverse algorithmic problems for Ideal(F) (=Ideal(G)) can be solved easily. Also, for a specific input F, the algorithm for constructing a Gröbner-basis for F may stop quickly (for example, in the extreme case, when F is already a Gröbner-basis) and, then, gives us the legitimation to use the simple algorithmic methods available for Gröbner-bases.

The two basic ideas in our 1965 algorithm for the construction of Gröbner-bases are the consideration of <u>"critical pairs"</u> of polynomials in the basis F and the successive <u>"completion"</u> of the basis by the differences of the reduced forms of critical pairs. More concretely, the algorithm has the following overall structure

```
G:= F
while not all "critical pairs" of G are considered do
     (b1,b2) := a "critical pair" of G which has not yet been considered
     (b1,b2) := the reduced forms of b1,b2 w.r.t. G;
          if b1 ≠ b2 then  "complete" G by b1-b2.
```

In the context of general first order terms instead of polynomials, the same two ideas appeared later (1967) in the well known algorithm of /Knuth-Bendix 67/, which now is widely used in computer algebra and software technology, in particular in the manipulation of abstract data type specifications. In fact, the critical-pair/completion algorithm shown above can as well be read as the Knuth-Bendix algorithm if the appropriate notion of "critical pair" is used and "complete G by b_1-b_2" is replaced by "complete G by the equation b_1=b_2".

In subsequent papers (1976, 1979, 1981, 1982, 1983) the present author has been working on the improvement and the complexity analysis of the 1965 algorithm.

Starting from 1976, also quite a few other authors (R. Schrader, M. Lauer, W. Trinks, D. Spear, M. Bergman, G. Zacharias, F. Winkler, S. Schaller, M. Pohst, D. Yun, F. Mora, J. Guiver, D. Bayer, H. Möller, D. Lazard) worked on the algorithm and gave various applications and generalizations. The special case of the algorithm where all polynomials in F have the form t_1-t_2 (t_1 and t_2 power products) was reinvented two times (in /Ballantyne, Lankford 81/ and in /Bauer 81/). In this case, the algorithm yields a decision procedure for finitely generated commutative semigroups. A fairly complete bibliography on the algorithm with further motivation and hints to applications may be found in /Buchberger, Loos 82/. Some very recent papers on the algorithm are /Buchberger 83/, /Winkler, Buchberger 83/ and /LLopis 83/.

Various authors generalized the algorithm to $Z[x_1,\ldots,x_n]$ (/Lauer 76/) and to $R[x_1,\ldots,x_n]$, where R is a ring that satisfies certain axioms (/Trinks 77/, /Spear 77/, /Zacharias 79/, /Schaller 79/). Roughly, these axioms are:

(R1) the decision problem "$f \in \text{Ideal}(F)$?" ($f \in R$, $F \subseteq R$) must have an algorithmical solution and

(R2) a finite set of generators for the solutions of linear equations in R can be found algorithmically.

In these papers it is also shown that, if R satisfies these axioms, then also $R[x_1,\ldots,x_n]$ satisfies the axioms: if R satisfies the axioms then (a variant of) the above algorithm may be applied to sets $F \subseteq R[x_1,\ldots,x_n]$ yielding Gröbner bases G for which the problems mentioned in (R1) and (R2) are seen to be easily solvable.

/Bergman 78/ is a generalization of the algorithm to the case of associative R-algebras. However, the "left-hand sides" of the generators are supposed to be pure power products. /Bauer 81/ shows also how the commutative semigroup case can be viewed in a much broader perspective.

In the present paper, we present a different generalization of the algorithm, which in various respects is more satisfactory than the generalizations given so far:

(1) The generalization works in general rings (satisfying certain axioms (A)), not only in polynomial rings.

(2) The algorithm preserves its extremely simple structure. A general concept of "critical pair" specializes to a concrete computational step in the various

example domains. In particular, it specializes to the well known procedure in $K[x_1,...,x_n]$. (In the generalizations reported in the literature at the place where "critical pairs" have to be computed in the original 1965 algorithm, linaer equations have to be solved, essentially).

(3) The formulation of the axioms (A) involves only the basic ring operations, a noetherian order relation and variables over ring elements.

(4) The notion of "reduction" is basic (and not the notion of "ideal" or "equation").

(5) The construction of Gröbner-bases in $R[x_1,...,x_n]$ does not presuppose the solution of equally hard algorithmical problems in R (as it is the case in the generalizations cited above, see (R1) and (R2)).

(6) The properties (A), again, carry over from R to $R[x_1,...,x_n]$. However, the proof of this fact does not involve the solvability of algorithmic problems of the type (R1), (R2) either.

(7) The construction has various degrees of freedom: Given R, a wide variety of different sets M of "multipliers", of different noetherian orderings $\succ$ on R and of different notions of reduction "steps" can be chosen. Every choice leads to a correct algorithm. The particular choice can be made in dependence on complexity considerations and other criteria.

(1)-(7) may be conceived as a first attempt to achieve a "<u>constructive ring theory</u>" based on the notion of "<u>reduction</u>" with a "<u>critical-pair/completion algorithm</u>" for enforcing the Church-Rosser property for reduction relations as the basic algorithmic tool.

The specific difficulty of this objective is twofold:

(1) The notion of "reduction" and of "critical pair" in polynomial rings (and also in the general framework of rewrite rules) presupposes the distinction of a "head" and a "rest" ("left-hand side" and a "right-hand side") of the generators (generating equations). In general rings, however, one has to find a way to formulate the concepts of "reduction" and "critical pair" without any appeal to an underlying "left-right" structure of elements.

(2) The axioms (A), at the same time, should be strong and weak. Strong axioms (A) make the correctness proof for the algorithm easier. Weak axioms extend the class of rings, to which the approach is applicable. In the proof of "R satisfies (A) ==> $R[x_1,...,x_n]$ satisfies (A)" strong axioms (A) would be nice in the premise, whereas weak axioms (A) would be nice in the conclusion!

Quite a few attempts were necessary in order to arrive at appropriate concepts of "reducibility" and "critical pair" and to get a feasible balance for the axioms (A). Finally, our notions of "reducibility" and "critical pair", in addition to being appropriate, seem to be fairly natural. Still, most of our axioms (A) seem to be natural. Only (A5) (see below) can not be motivated easily. Because of the limited space available, in this paper, the details of motivation must be omitted. Also, only the proofs for the central lemmas and theorems can be given. For all other details see the technical report /Buchberger 83a/.

From now on, the presentation will be "bottom-up": Section 1. presents the general definition of "reduction" and "critical pairs", which is applicable to arbitrary rings. Section 2. introduces axioms for rings, in which the critical-pair/completion algorithm can be correctly executed. For the moment, such rings will be called "reduction rings". For the sake of notational simplicity, the consideration is restricted to rings without zero divisors. In /Buchberger 83a/ it is shown how the whole program can be carried out in rings with zero divisors. In Section 3., the central theorem and its proof is presented: for checking the Church-Rosser property of the reduction relations in reduction rings, the consideration of the critical pairs is sufficient. Based on this theorem, the general critical-pair/completion algorithm for reduction rings is formulated. In Section 4. it is shown that, if R is a reduction ring, then also $R[x_1,...,x_n]$ is a reduction ring. Finally, in Section 5., the particular example of $Z[x_1,...,x_n]$ is considered in more detail. The reader who is interested in this example only, after having read the definition of the reduction relation $\rightarrow_C$ in Section 1., can immediately pass to the description of the algorithm in Section 3. and the explanation of the subalgorithms in Section 5. He, then, should be able to compute Gröbner-bases for ideals in $Z[x_1,...,x_n]$ and to program the algorithm for this domain.

1. REDUCTION AND CRITICICAL PAIRS

Let R be a commutative ring with 1 (without zero divisors), $<$ a noetherian (partial) order relation on R and M $\subseteq$ R (set of "multipliers"). (Typed variables: a,b,c,d,e for elements in R; C,D for subsets of R; m for elements in M; i,j,k,n for natural numbers. $<$ is noetherian means that there are no infinite sequences $a_1 > a_2 > a_3 > \ldots$)

Definition: $a \rightarrow_c b$ iff $a - m.c = b$ for some m and $a > b$

(read: "a reduces to b modulo c").

Definition: $a \rightarrow_C b$ iff $a \rightarrow_c b$ for some $c \in C$

(read: "a reduces to b modulo C").

(For reduction relations $\rightarrow$, the following additional notation is used: $\leftarrow$ is the inverse relation to $\rightarrow$. $\leftrightarrow$, $\overset{+}{\rightarrow}$, $\overset{*}{\rightarrow}$, $\overset{*}{\leftrightarrow}$ are the symmetric, transitive, reflexive-transitive, and reflexive-symmetric-transitive closure of $\rightarrow$, respectively. Furthermore:

$a \rightarrow$ iff $a \rightarrow b$ for some b

(read: "a is reducible (w.r.t. $\rightarrow$)").

$\underline{a}$ iff $a \rightarrow b$ for no b (i.e. not $a \rightarrow$)

(read: "a is irreducible" or "a is in normal form").

$a \downarrow^* b$ iff $a \overset{*}{\rightarrow} d \overset{*}{\leftarrow} b$ for some d

(read: "a and b have a common successor").

$a \overset{*}{\leftrightarrow}(<d)\ b$ iff for some $e_0,\ldots,e_n$: $a=e_0 \leftrightarrow e_1 \leftrightarrow e_2 \leftrightarrow \ldots \leftrightarrow e_{n-1} \leftrightarrow e_n=b$ and $e_0,e_1,\ldots,e_n < d$

(read: "a and b can be connected below d").

The notation is used quite flexibly: for example, "$a \overset{*}{\rightarrow} d \overset{*}{\leftarrow} b$" is an abbreviation for "$a \overset{*}{\rightarrow} d$ and $d \overset{*}{\leftarrow} b$"; "$\leftarrow_{c_1} a \rightarrow_{c_2}$" is an abbreviation for "$a \rightarrow_{c_1}$ and $a \rightarrow_{c_2}$"; etc.).

Definition: a is a common reducible for c_1 and c_2 iff $\leftarrow_{c_1} a \rightarrow_{c_2}$.

Definition: $c_1 \overset{a}{\Delta} c_2$ iff a is a common reducible for c_1 and c_2 and for no m_1, m_2:

$$a-m_1c_1 \leftarrow_{c_1} a \rightarrow_{c_2} a-m_2c_2 \quad \text{and}$$
$$(a-m_1c_1 \rightarrow_{c_2} a-m_1c_1-m_2c_2 \quad \text{or}$$
$$a-m_2c_2 \rightarrow_{c_1} a-m_2c_2-m_1c_1\ )$$

(read: "a is a non-trivial common reducible for c_1 and c_2").

(It will become clear in Section 3. why, in the above situation, the reducibility of a is called "non-trivial". Roughly, "trivial" common reducibilities present no problem in establishing the Church-Rosser property for the reducibility relations considered.)

Definition: $c_1 \overset{a}{\underline{\Delta}} c_2$ iff $c_1 \overset{a}{\Delta} c_2$ and for no $a' < a$: $c_1 \overset{a'}{\Delta} c_2$
(read: "a is a minimal non-trivial common reducible for c_1 and c_2").

(It will turn out that the minimal non-trivial common reducibles play the crucial role in establishing the Church-Rosser property: the "critical pairs", to which the minimal non-trivial common reducibles can be reduced, are the only pairs of elements where the Church-Rosser property might be injured).

Definition: b_1, b_2 form a critical pair for c_1, c_2 iff
for some a: $c_1 \overset{a}{\underline{\Delta}} c_2$ and $b_1 \leftarrow_{c_1} a \rightarrow_{c_2} b_2$.

Although the reduction $\rightarrow_C$ looks very much as "one directed step" in the ideal theoretical congruence $\equiv_C$, in general, it is not true that $\leftrightarrow_C^* = \equiv_C$. (We use the notation:

$$a \equiv_C b \quad \text{iff} \quad a = b + \sum_{1 \le i \le n} d_i . c_i \quad \text{for some } n,\ d_i \varepsilon R,\ c_i \varepsilon C$$

(read: "a is congruent b modulo (the ideal generated by) C").)

However, under very weak additional assumptions (see (A2) below), $\leftrightarrow_C^* = \equiv_C$ can be derived and, of course, this is what is needed in order that knowledge about the reduction relations (in particular, the knowledge about the Church-Rosser property) can be used for an algorithmic solution to ideal theoretical problems.

2. REDUCTION RINGS

Definition: A commutative ring R with 1 (without zero divisors) together with
a noetherian (partial) order relation on R and
a set $M \subseteq R$ (set of "multipliers")
constitute a reduction ring iff the following axioms are satisfied:

Axioms for multipliers:

(M0) 0 not in M

(M1) $1 \varepsilon M$

(M2) if $m \varepsilon M$ then $-m \varepsilon M$

(M3) if $m_1, m_2 \varepsilon M$ then $m_1.m_2 \varepsilon M$

(M5) if $a \equiv_C b$ then there exist n, $m_i \varepsilon M$, $c_i \varepsilon C$ such that

$$a = b + \sum_{1 \le i \le n} m_i.c_i$$

((M4) requires that no multiplier is a zero divisor. This axiom is not needed in the restricted context of this paper. In some cases, M may be taken equal to R-{0}. Of course, in this case, the axioms (M) are trivially satisfied. However, the consideration of $M \neq R$ provides an additional degree of freedom. For example, in the case of polynomial rings over fields, by taking M:= set of monomials, we obtain the original reduction relation in /Buchberger 65, 70, 76/. In the presence of zero divisors, the availability of sets $M \neq R$ is crucial.)

Axioms for the reduction relation:

(A1) if $a \neq 0$ then $a > 0$

(A2) if $a \rightarrow_C b$ then $a + d \downarrow_C^* b + d$

(A3) if $a \rightarrow_C b$ then $m.a \leftrightarrow_C m.b$

(A4) if $b_1 \leftarrow_C a \rightarrow_C b_2$ then $b_1 \leftrightarrow_C^{*}(<a)\ b_2$

(A5) if $c_1 \overset{a}{\Delta} c_2$ then there exist $a' \leq a$ and m such that
$c_1 \underset{-}{\overset{a'}{\Delta}} c_2$ and
for all c: (if $a'+c < a'$ then $a+m.c < a$)

(These axioms connect the ring operations with reduction. Note that no appeal is made to the solution of "higher" algorithmic problems in R, as was the case with

(R1), (R2). Stronger forms of the axioms, which naturally might come to ones mind, do not even hold in the original example of polynomial rings. For example, (A2) can not be replaced by: if $a \rightarrow_C b$ then $a+d \rightarrow_C b+d$.)

Axioms of effectiveness:

Addition, multiplication, reduction and the formation of critical pairs must be effectively possible in R (as will be explained in more detail in Section 3. after the presentation of the critical-pair/completion algorithm).

Axioms of termination:

(T1) There exists no infinite sequence of subsets $D_1, D_2, \ldots$ of R such that
$Red(D_1) \subset Red(D_2) \subset \ldots$
(where "$\subset$" is strict set inclusion, and $Red(D) := \{ a \mid a \rightarrow_D \}$).

(T2) For all c_1, c_2: $\{ a \mid c_1 \overset{a}{\underline{\Delta}} c_2 \}$ is finite.

(These axioms will guarantee the termination of the general critical-pair/completion algorithm. Note, again, that these axioms are expressed in terms of reduction and not in terms of ideal theoretical notions).

3. CHURCH-ROSSER PROPERTY AND GRÖBNER BASES

The following definition is standard:
$\rightarrow$ has the Church-Rosser property iff for all a,b: (if $a \leftrightarrow^* b$ then $a \downarrow^* b$).

The importance of the Church-Rosser property for reduction relations stems from the following well known fact:
If $\rightarrow$ has the Church-Rosser property then

$a \leftrightarrow^* b$ iff $S(a) = S(b)$

(where S is an arbitrary function (a "simplifier") satisfying:

$$S(a) = \begin{cases} a, & \text{if } \underline{a} \\ S(b), & \text{if } a\rightarrow \text{ and } b \text{ is some element such that } a \rightarrow b). \end{cases}$$

This means that, for Church-Rosser relations $\rightarrow$, "$a \leftrightarrow^* b$" can be decided by simply comparing S(a) and S(b). (A review of known results on Church-Rosser relations may be found, for example, in /Buchberger, Loos 82/). In the case of those rings, in which $\leftrightarrow_C^* = \equiv_C$ holds, one hence has an easy decision procedure for the ideal theoretic congruence $\equiv_C$ as soon as it is known that the reduction relation $\rightarrow_C$ has the Church-Rosser property. (In fact, as was pointed out in the introduction, also a number of other algorithmic ideal theoretical problems can then be solved easily).

Unfortunately, for an arbitrary (finite) C, in general $\rightarrow_C$ is not Church-Rosser. The critical-pair/completion approach sketched in the introduction now suggests to "complete" C by suitable elements (derived from critical pairs) without changing the ideal until a set D is reached, whose coressponding reduction relation $\rightarrow_D$ is Church-Rosser. The feasibility of this approach in the general context of this paper is established by the following theorem.

Presupposition: From now on presuppose that R, $\prec$, M constitute a reduction ring.

Theorem: $\rightarrow_C$ has the Church-Rosser property iff

for all c_1, $c_2 \in C$ and a such that $c_1 \overset{a}{\underline{\Delta}} c_2$
there exists a critical pair b_1, b_2 for c_1, c_2 such that $b_1 \leftrightarrow_C^*(\prec a)\ b_2$.

(Note that the condition formulated in this theorem is even weaker than announced in the rough sketch of the method: we do not require all critical pairs to have a common successor and, also, we do not insist on the existence of common successors but only require that b_1, b_2 may be connected "below" a).

Using this theorem, the following algorithm can be shown to correctly solve the following problem.

Problem: Given: A finite set $C \subseteq R$.
Find: A finite set $D \subseteq R$ such that
$\leftrightarrow_C^* = \leftrightarrow_D^*$ and
$\rightarrow_D$ has the Church-Rosser property.

Algorithm:

D:= C
B:= $\{ (\{c_1,c_2\},a) \mid c_1,c_2 \in C,\ c_1 \overset{a}{\underline{\Delta}} c_2 \}$

<u>while</u> $B \neq \emptyset$ <u>do</u>

$(\{c_1,c_2\},a)$:= one triple out of B
B := $B - \{ (\{c_1,c_2\},a) \}$
(b_1,b_2) := two elements such that $b_1 \leftarrow_{c_1} a \rightarrow_{c_2} b_2$
(b_1,b_2) := $(S_D(b_1),S_D(b_2))$

<u>if</u> $b_1 \neq b_2$ <u>then</u>

$c := b_1 - b_2$
$B := B \cup \{ (\{c,c'\},a) \mid c' \in D, c \overset{a}{\underline{\Delta}} c' \}$
$D := D \cup \{c\}$.

(Here, S_D is a "simplifier" for $\rightarrow_D$ of the kind desribed above. The "algorithm" is effective in the precise sense of algorithm theory provided that the basic operations appearing in the instructions of the algorithm can be effectively executed in the ring considered. This means that, in addition to the axioms (M), (A), and (T) one must stipulate for reduction rings that the ring operations, the determination of a "b such that $a \rightarrow_D b$, in case $a \rightarrow_D$", and the determination of all "a such that $c_1 \overset{a}{\underline{\Delta}} c_2$" are effective operations. Stated differently, the above algorithm is an algorithm "relative" to the effectiveness of the operations mentioned above.)

Due to the following lemma the above problem can also be stated in the following form.

<u>Lemma</u>: $\overset{*}{\leftrightarrow}_C = \equiv_C$.

<u>Problem</u>: Given: A finite set $C \subseteq R$.
Find: A finite set $D \subseteq R$ such that
$\equiv_C = \equiv_D$ and
$\rightarrow_D$ has the Church-Rosser property.

Summarizing, this means that, for an ideal generated by an arbitrary set C, the above algorithm constructs a new basis D for the same ideal such that $\rightarrow_D$ has the Church-Rosser property and, hence, an easy decision procedure for the congruence $\equiv_C = \equiv_D$ is available (namely, $a \equiv_C b$ iff $S_D(a) = S_D(b)$). Extending the terminology introduced in /Buchberger 76/, sets D, whose corresponding reduction relation $\rightarrow_D$ is Church-Rosser, will be called Gröbner bases.

Proof of the lemma:

If $a \leftrightarrow_C^* b$ then $a \equiv_C b$: easy! (Iterate: if $a \rightarrow_C b$ then $a \equiv_C b$).

If $a \equiv_C b$ then $a \leftrightarrow_C^* b$:

Proceed by induction on n in $a = b + \sum_{1 \le i \le n} m_i c_i$ (where the m_i, by (M5), can be chosen in M and the c_i are in C). The case n=0 is clear. Case n≠0:

$$a = b + \sum_{1 \le i \le n} m_i c_i = b + \sum_{1 \le i \le n-1} m_i c_i + m_n c_n \overset{(1)}{\leftarrow_{c_n}^*} b + \sum_{1 \le i \le n-1} m_i c_i \overset{(\text{ind. hyp.})}{\leftrightarrow_C^*} b.$$

(1) is a special instance of the following general law in reduction rings: $d \leftarrow_c^* d+m.c$. (Proof: Case m.c=0 trivial. In case m.c≠0, $m.c \rightarrow_c 0$ and, hence, by (A2) $d+m.c \downarrow_c^* d+0 = d$.)

Proof of the theorem:

The proof can be structured by the following "generalized Newman-Lemma", which seems to be the strongest form of a Newman-type lemma known so far, i.e. it singles out a very weak condition under which the Church-Rosser property of reduction relations can be guaranteed. The author developed this lemma and its proof in /Winkler, Buchberger 83/, where it is used for a different purpose.

Generalized Newman-Lemma: Let $\rightarrow$ and $\succ$ be binary relations on a set T, $\succ$ noetherian and $\rightarrow \subseteq \succ$. Then $\rightarrow$ has the Church-Rosser property iff for all $a,b_1,b_2 \in T$:

if $b_1 \leftarrow a \rightarrow b_2$ then $b_1 \leftrightarrow^*(\prec a)\ b_2$.

For proving the theorem, we take arbitrary $a,b_1,b_2 \in R$, $c_1,c_2 \in C$ and assume

(1) $b_1 \leftarrow_{c_1} a \rightarrow_{c_2} b_2$.

According to the generalized Newman-Lemma it suffices to show

(2) $b_1 \leftrightarrow_C^*(\prec a)\ b_2$.

For notational convenience, $\rightarrow$ etc. will be used instead of $\rightarrow_C$ etc. in this proof. Also, the axioms (M) about the multipliers will be used tacitly.

Case: for some m_1,m_2: $a-m_1c_1-m_2c_2 \leftarrow_{c_2} a-m_1c_1 \leftarrow_{c_1} a \rightarrow_{c_2} a-m_2c_2$:

In this case (where a is a "trivial common reducible" for c_1 and c_2)

$$b_1 \overset{(A4)}{\leftrightarrow_{c_1}^{*}(\prec a)} a-m_1c_1 \overset{(3)}{\leftrightarrow^{*}(\prec a)} a-m_2c_2 \overset{(A4)}{\leftrightarrow_{c_2}^{*}(\prec a)} b_2.$$

(3) needs some more details: $a \rightarrow_{c_1} a-m_1c_1$ implies $a-m_2c_2 \downarrow_{c_1}^{*} a-m_1c_1-m_2c_2$, by (A2). Now, $a-m_1c_1 \rightarrow_{c_2} a-m_1c_1-m_2c_2$ by case assumption. Hence, (3) holds.

<u>Case:</u> for some m_1,m_2: $a-m_1c_1 \leftarrow_{c_1} a \rightarrow_{c_2} a-m_2c_2 \rightarrow_{c_1} a-m_2c_2-m_1c_1$: analogous.

<u>Remaining Case:</u> In this case $c_1 \overset{a}{\Delta} c_2$ holds (this is the "non-trivial" case: a is a "non-trivial common reducible" for c_1 and c_2). Using (A5), choose an $a' \preceq a$ and an m such that

(4) $c_1 \overset{a'}{\underline{\Delta}} c_2$ and

(5) for all c: if $a'+c \prec a'$ then $a+m.c \prec a$.

By the assumption of the theorem, one can choose b_1', b_2' such that

(6) $b_1' \leftarrow_{c_1} a' \rightarrow_{c_2} b_2'$ and

(7) $b_1' \leftrightarrow^{*}(\prec a') \; b_2'$.

Let m_1,m_2 be such that

(8) $b_1' = a'-m_1c_1$, $b_2' = a'-m_2c_2$.

Because of (7) one can choose $e_0,\ldots,e_k$ such that

(9) $b_1'=e_0 \leftrightarrow e_1 \leftrightarrow e_2 \leftrightarrow \ldots \leftrightarrow e_{k-1} \leftrightarrow e_k=b_2'$ and

(10) $e_0,\ldots,e_k \prec a'$.

Now, let $f_0,\ldots,f_k$ be such that

(11) $e_i = a'+f_i$ (for $i=0,\ldots,k$).

(10),(11), and (5), then, imply

(12) $a+m.f_i \prec a$ (for $i=0,\ldots,k$).

Furthermore, using (A3), from (9) one gets

(13) $m.e_0 \leftrightarrow m.e_1 \leftrightarrow \ldots \leftrightarrow m.e_k$

and, using (A2),

(14) $(a-m.a')+m.e_0 \downarrow^{*} (a-m.a')+m.e_1 \downarrow^{*} \ldots \downarrow^{*} (a-m.a')+m.e_k$.

Now, by (11),

(15) $(a-m.a')+m.e_i = a+m.f_i$ (for $i=0,\ldots,k$).

By (8), (9), and (11),

(16) $a+m.f_0 = a-m.m_1.c_1$, $a+m.f_k = a-m.m_2.c_2$.

Hence, by (12),(14),(15), and (16), one finally obtains

(17) $a-m.m_1.c_1 \leftrightarrow^{*}(\prec a) \; a-m.m_2.c_2$.

Because of (12) and (16), one also has

(18) $a \rightarrow_{c_1} a-m.m_1.c_1$, $a \rightarrow_{c_2} a-m.m_2.c_2$.

Hence, one is allowed to use (A4) obtaining

(19) $b_1 \leftrightarrow_{c_1}^{*}(<a)\ a\text{-}m.m_1.c_1, \quad a\text{-}m.m_2.c_2 \leftrightarrow_{c_2}^{*}(<a)\ b_2$.
(17) together with (19) yields (2).

Summary of proof:

One has to "close"

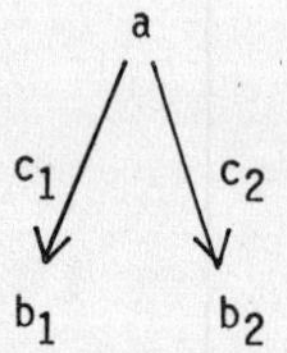

in order to obtain

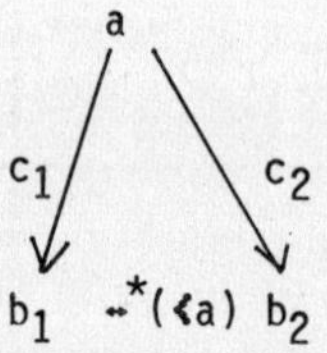

By (A5) and the assumption of the theorem one can find $a' \leq a$, m, m_1, m_2 such that

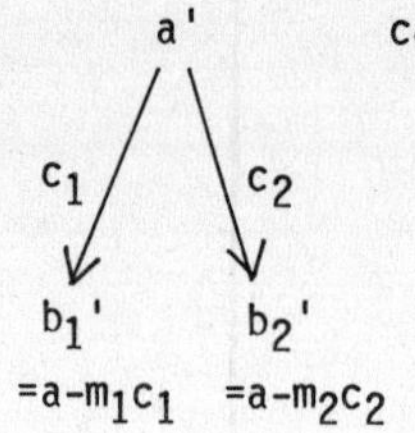

can be closed:

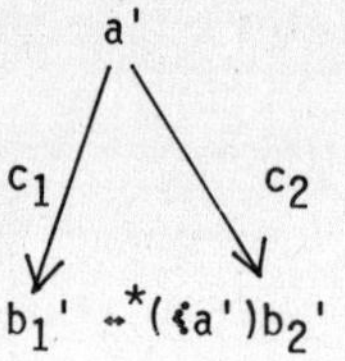

and the closed triangle can be "lifted" to the level of a by an appropriate multiplication with m and an addition of $a\text{-}m.a'$ yielding

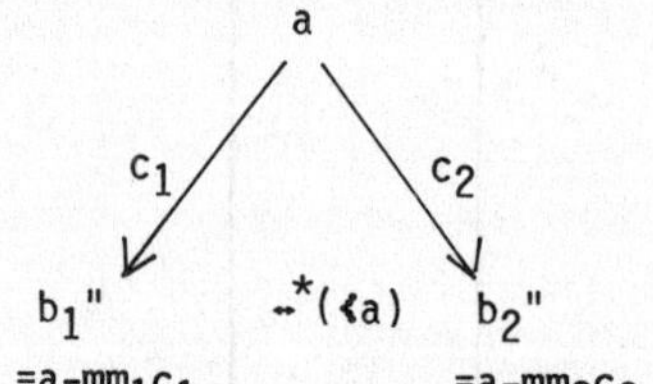

This is not yet what was really wanted. However, using (A4), the interconnection can be established eventually:

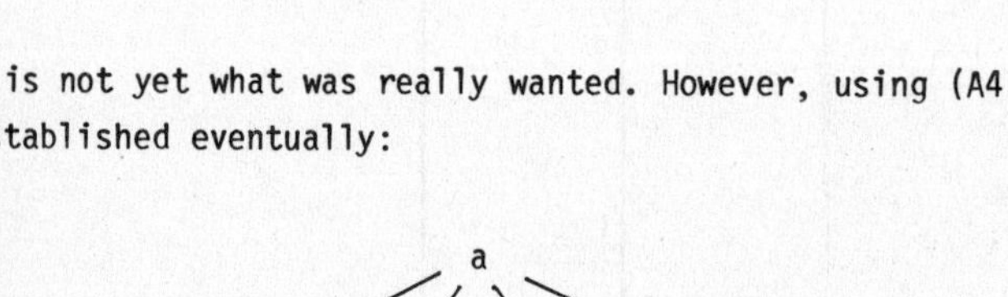

Partial Correctness of the Algorithm:

The partial correctness of the algorithm can be proven by the method of inductive assertions. The inductive assertion for the while-loop is:

$\equiv_C = \equiv_D$,

if $(\{c_1,c_2\},a) \in B$ then $c_1,c_2 \in D$, $c_1 \overset{a}{\Delta} c_2$,

if $c_1,c_2 \in D$, $c_1 \overset{a}{\Delta} c_2$, and $(\{c_1,c_2\},a)$ not in B
then for some b_1, b_2: $b_1 \leftarrow_{c_1} a \rightarrow_{c_2} b_2$ and $b_1 \leftrightarrow_D^*(\lessdot a)\ b_2$.

Knowing the theorem, the other details of the correctness proof are easy, see /Buchberger 83a/.

Termination of the Algorithm:

The termination proof for the algorithm uses the axioms (T1) and (T2). (T2) guarantees that the block within the while-loop is always left after finitely many steps. Furthermore, one can prove:

(1) Let D_i be the value of the variable D after the if-statement in the while-loop has been executed for the i-th time (D_0 = C). Then
$\text{Red}(D_0) \subset \text{Red}(D_1) \subset \text{Red}(D_2) \subset \ldots$

Knowing (1), it is clear that the while-loop can be executed only a finite number of times: Either the if-statement is executed only finitely many times, hence D stabilizes and B will get exhausted, or the if-statement is executed infinitely many times and hence, by (1), a sequence $\text{Red}(D_0) \subset \text{Red}(D_1) \subset \text{Red}(D_2) \subset \ldots$ would exist in contradiction to (T2).

Proof of (1): Surely, $D_i \subseteq D_{i+1}$ and therefore $\text{Red}(D_i) \subseteq \text{Red}(D_{i+1})$ for i=0,1,... One has to show that $\text{Red}(D_i) \neq \text{Red}(D_{i+1})$. Let b_1, b_2 be the elements constructed in the algorithm before the if-statement is entered for the (i+1)-th time. One knows
(2) b_1 and b_2 are irreducible with respect to D_i and
(3) $b_1 \neq b_2$.
One can show:

(4) at least one of b_1, b_2 is reducible with respect to D_{i+1}
and, hence, $Red(D_i)$ and $Red(D_{i+1})$ can not be equal. For proving (4), the following law holding in arbitrary reduction rings may be applied:
(5) if $a \neq b$ then $a \downarrow^*_{a-b} b$.
(Proof: $a \neq b$ implies $a-b \rightarrow_{a-b} 0$. From this, by (A2), one gets $a \downarrow^*_{a-b} b$.)
Applying (5) to (3) yields
(6) $b_1 \rightarrow^*_{b_1-b_2} d \leftarrow^*_{b_1-b_2} b_2$ for some d.
Case $b_1=d=b_2$: not possible because of (3).
Case $b_1 \neq d$: In this case there is a d' such that $b_1 \rightarrow_{b_1-b_2} d' \rightarrow^*_{b_1-b_2} d$. Hence, $b_1 \rightarrow_{D_{i+1}}$ because $b_1-b_2 \in D_{i+1}$.
Case $b_2 \neq d$: Analogous: $b_2 \rightarrow_{D_{i+1}}$.

4. POLYNOMIAL REDUCTION RINGS

(In the following, the typed variables f,p,q,r are used for polynomials and the typed variables s,t,u for n-variate power products (special polynomials of type $x_1^{i_1} \ldots x_n^{i_n}$). $C(p,t)$ will denote the coefficient of the polynomial p at the power product t.)

Theorem: If R, M, $<$ constitute a reduction ring
then $R[x_1,\ldots,x_n]$, $M\{x_1,\ldots,x_n\}$, $\ll$ constitute a reduction ring, where

$R[x_1,\ldots,x_n]$:= set of polynomials with coefficients in R,
$M\{x_1,\ldots,x_n\}$:= $\{ m.s \mid m \in M, s \text{ is a power product} \}$,
$p \ll q$ iff there is a t such that
$C(p,t) < C(q,t)$ and
for $s >_T t$: $C(p,s)=C(q,s)$

(Here, $>_T$ is an arbitrary total ordering on the power products satisfying
(PP1) for all $s \neq 1$: $1 <_T s$
(PP2) for all s,t,u: if $s <_T t$ then $s.u <_T t.u$).

Remark: Of course, one also could formulate the theorem, first, for the transition from R to $R[x]$ and then obtain the above theorem by iteration. However, the direct transition from R to $R[x_1,\ldots,x_n]$ opens the possibility to use a big variety of different orderings $<_T$ and, hence, orderings $\ll$ as an additional degree of freedom, which may prove useful for the algorithmic aspects pursued in this paper. The iteration of the transition from R to R[x] implies that, finally, the power products would be ordered according to the "purely lexicographical" ordering.

Proof: The proof of this theorem is quite tedious. We only can give some hints at the crucial points. The details are contained in /Buchberger 83a/.

Proof that $\ll$ is a noetherian partial ordering: It is easy to show that $\ll$ is a partial ordering. It is well known that (PP1) and (PP2) imply that $<_T$ is noetherian. From this and the assumption that $<$ is noetherian it can be proven that $\ll$ is noetherian. Similar proofs are contained in /Lauer 76/, /Trinks 78/, /Zacharias 78/, /Schaller 79/.

Proof of (M1),...,(M5): Easy. For (M5) note that, by (M5) in R, all elements $a \in R$ can be represented in the form $a = \sum m_i$ (Take C:= {1}, b:= 0. Then $a \equiv_C b$ and, hence, by (M5), $a = \sum m_i c_i$ with $c_i \in C$, i.e. $a = \sum m_i$.)

Proof of (A1): Clear.

Proof of (A2), (A3), (A4): All these proof are based on the fact that

(RED) $p \to_f q$ iff there exist $a \neq 0$, t, m, u such that

$$t = u.LP(f),$$
$$p = H(p,t) + a.t + L(p,t),$$
$$q = H(p,t) + (a-m.LC(f)).t + L(p,t) - m.u.R(f)$$
$$\text{and } \quad a-m.LC(f) < a.$$

(Here, H(p,t) and L(p,t) are the "higher" and "lower" part of p w.r.t. t, LC(f), LP(f) and R(f) are the "leading coefficient", the "leading power product" and the "residuum of f".

More formally, H(p,t) and L(p,t) are two polynomials such that

$$p = H(p,t) + C(p,t).t + L(p,t),$$
$$\text{for all } s: \text{ if } C(H(p,t),s) \neq 0 \text{ then } s >_T t \text{ and}$$
$$\text{for all } s: \text{ if } C(L(p,t),s) \neq 0 \text{ then } s <_T t,$$

and the element LC(f) in R, the power product LP(f) and the polynomial R(f) such that

$$f = LC(f).LP(f) + R(f) \quad \text{and}$$
$$\text{for all } s: \text{ if } C(R(f),s) \neq 0 \quad \text{then } s <_T LP(f).)$$

Having established property (RED), the properties (A2), (A3), (A4) can be "lifted" from R to $R[x_1,\ldots,x_n]$ in a manner which is more or less straightforward. One should note at this point that this is in sharp contrast to the procedures reported in the literature so far, which need to apply the critical-pair/completion algorithm itself (in $R[x_1,\ldots,x_n]$) in order to "lift" the axioms of the type (R1), (R2) from R to $R[x_1,\ldots,x_n]$.

Proof of (A5): This proof is based on the following two properties (whose proof is very tedious):

(CR1) $f_1 \Delta^p f_2$ iff there do not exist t_1, t_2 such that $t_1 \neq t_2$ and
$LP(f_1)$ divides t_1, $C(p,t_1) \to_{LC(f_1)}$,
$LP(f_2)$ divides t_2, $C(p,t_2) \to_{LC(f_2)}$,
but there exists a t such that
$LP(f_1)$ divides t,
$LP(f_2)$ divides t and
$LC(f_1) \Delta^{C(p,t)} LC(f_2)$

(CR2) $f_1 \underline{\Delta}^p f_2$ iff $p = LC(p).LCM(LP(f_1),LP(f_2))$ and
$LC(f_1) \underline{\Delta}^{LC(p)} LC(f_2)$.

(Here, $LCM(t_1,t_2)$ denotes the least common multiple of t_1, t_2).

Again, having (CR1), (CR2), the property (A5) can be "lifted".

Proof of the axioms of effectiveness: The effectiveness of addition and multiplication, of course, easily carries over from R to $R[x_1,\ldots,x_n]$. (RED) and (CR2) show that reduction and formation of critical pairs are effective in $R[x_1,\ldots,x_n]$ if they are effective in R.

Proof of (T1): This proof, again, is tedious. The techniques involved, however, are similar to those used in showing that $\gg$ is noetherian.

Proof of (T2): (T2) in $R[x_1,\ldots,x_n]$ follows immediately from (CR2) and (T2) in R.

5. EXAMPLE: $Z[x_1,\dots x_n]$

It is easy to see that all fields K are reduction rings. In fact, every set C in a field K is a Gröbner basis. Hence, also all polynomial rings $K[x_1,\dots,x_n]$ (K a field) are reduction rings and, in fact, the general critical-pair/completion algorithm presented in Section 3., in this case, specializes to our original 1965 algorithm. To see this, one has to specialize the characterizations of reduction and of critical pairs given in (RED) and (CR2), respectively. This is left to the reader (see /Buchberger 83a/).

Rather, the example of $Z[x_1,\dots,x_n]$ will be considered in detail. As in the general case, also in this example the algorithm of Section 3. yields ideal bases for which ideal congruence can be decided by reduction and, furthermore, the Gröbner bases constructed are "invariants" in the sense that the ideals generated by C_1, C_2 are equal iff the Gröbner bases constructed for C_1, C_2 are equal. (The latter statement is true if, after the construction of the Gröbner bases according to the algorithm in Section 3., every polynomial in the Gröbner basis is reduced to normal form with respect to the other polynomials in the basis. This is totally analogous to the situation in $K[x_1,\dots,x_n]$, see /Buchberger 76a/.) The problem of deciding congruence and of constructing invariants for ideals in $Z[x_1,\dots,x_n]$ has a long and interesting history involving work of /Kronecker, Hensel 01/, /Szekeres 52, 65, 75/, /Redei 56/, /Trotter 69, 78/, /Simmons 70/, /Hurd 70/, /Richman 74/, /Lauer 76/, /Trinks 77/, /Zacharias 78/, /Sims 78/, /Schaller 79/, /Ayoub 81/. For some of the details of the history, see /Ayoub 81/. The first general solution of both problems, based on the critical-pair completion approach, but needing two different versions of "critical pairs" at the same time, was given in /Lauer 76/. Other solutions based on the critical-pair/completion approach using (R1), (R2) were given in /Trinks 77/, /Zacharias 78/, /Schaller 79/. The first general solution based on a different approach was given in /Ayoub 81/. Our solution seems to be much more concise than the solutions given so far.

In order to get the specialization of the general critical-pair completion algorithm of Section 3. for $Z[x_1,\dots,x_n]$, according to (RED), (CR1) and (CR2), one has to fix a suitable M, $<$ and $<_T$, and to give effective procedures for finding b if $a \to_C$ ($a \in Z$, $C \subseteq Z$) and for finding (all) a such that $c_1 \overset{a}{\Delta} c_2$ (c_1, $c_2 \in Z$). One possible choice is (proofs see /Buchberger 83a/):

$M := Z-\{0\}$; $0 < -1 < 1 < -2 < 2 < -3 < 3 < \ldots$. If $a \to_c$ then $b := a - MP(a,c).c$ is such that $a \to_c b$, where $MP(a,c)$, for example, may be determined by the following algorithm (which, in practice, may be replaced by a modified integer division algorithm):

(1) M(a,c):= <u>if</u> a and c have the same sign
<u>then</u> <u>if</u> a-c < a <u>then</u> M(a-c,c)+1
<u>else</u> 0
<u>else</u> <u>if</u> a+c < a <u>then</u> M(a+c,c)-1
<u>else</u> 0.

Furthermore, $c_1 \overset{a}{\Delta} c_2$ iff $a = LCR(c_1,c_2)$, where $LCR(c_1,c_2)$, the "least common reducible" of the integers c_1, c_2, can be computed as follows:

(2) $LCR(c_1,c_2) = \max_{<}(L(c_1),L(c_2))$,
$L(c) := abs(c)/2$, if c is even
$- (abs(c)+1)/2$, if c is odd.

Hence, for example, 0, -1, 1, -2, 2, -3, 3, -4, 4 are the values of L for the arguments 0, 1, 2, 3, 4, 5, 6, 7, 8, respectively, and LCR(3,1)=-2, LCR(7,8)=4. Note that L(c)=L(-c).

Consider now the following set $F := \{f_1,f_2,f_3\}$ of polynomials (the same example is considered in /Ayoub 81/):

$f_1 := 3x^2y+2xy+y+9x^2+5x-3$
$f_2 := 2x^3y-xy-y+6x^3-2x^2-3x+3$
$f_3 := x^3y+x^2y+3x^3+2x^2$.

We fix the "purely lexicographical" ordering for the bivariate power products: $1 <_T$ $<_T x <_T x^2 <_T x^3 <_T \ldots <_T y <_T xy <_T x^2y <_T x^3y <_T \ldots <_T y^2 <_T xy^2 <_T < x^2y^2 <_T \ldots$ Then, $LP(f_1)=x^2y$, $LP(f_2)=LP(f_3)=x^3y$. The polynomial x^3y may be reduced by f_2 in the following way (compare (RED)):

Take $a:=1$, $t:=x^3y$, $m:=1$, $u:=1$.
Hence, $x^3y \to_{f_2} -x^3y+xy+y-6x^3+2x^2+3x-3 =: g$.

g may be further reduced modulo f_3:

Take $a:=1$, $t:=x^3y$, $m:=-1$, $u:=1$.
Hence, $g \to_{f_3} x^2y+xy+y-3x^3+4x^2+3x-3 =: g'$.

g' is irreducible with respect to F. The polynomial x^3y may also be reduced by f_3:

Take $a:=1$, $t:=x^3y$, $m:=1$, $u:=1$.
Hence, $x^3y \to_{f_3} -x^2y-3x^3-2x^2 =: h$.

Also h is irreducible with respect to F, i.e. we have the following situation

$$\underline{g'}_F \leftarrow_F^* x^3y \to_F^* \underline{h}_F,\quad g' \neq h,$$

which shows that $\to_F$ does not possess the Church-Rosser property.

In order to "complete" F by the critical-pair/completion algorithm one has to consider "critical pairs" of polynomials in F (in any order). For example, one can start with f_2, f_3, and compute (all) p such that $f_2 \;\underline{\Delta}^p\; f_3$. This can be done by (CR2): $LC(f_2)=2$, $LC(f_3)=1$, hence by (2) above $LC(p)=LCR(2,1) = 1$. Furthermore, $LCM(LP(f_2),LP(f_3)) = x^3y$. Thus, $p=x^3y$. One reduction step of x^3y modulo f_2 and f_3 yields (the "critical pair") g and h, respectively. (Actually, in $Z[x_1,\ldots,x_n]$, there exists only one critical pair for a given pair of polynomials). Further reduction of g and h to normal forms modulo F yields g' and h, respectively. $g' \neq h$ and, hence,

$$f_4 := g'-h = 2x^2y+xy+y+6x^2+3x-3$$

must be adjoined to the basis. Similarly, one now has to consider the next critical pair, for example, the one corresponding to f_1, f_4: $p=-2x^2y$. Reduction of p modulo f_1 and f_4 yields $x^2y+2xy+y+9x^2+5x-3$ and $xy+y+6x^2+3x-3$, respectively. Reduction to normal forms yields $-x^2y+xy+3x^2+2x$ (using f_4) and $xy+y+6x^2+3x-3$. Thus, the difference of these two polynomials must be adjoined to the basis:

$$f_5 := -x^2y-y-3x^2-x+3.$$

Similarly, the consideration of the critical pair of f_4 and f_5 leads to

$$f_6 := -xy+y-x-3.$$

The consideration of the critical pair of f_5 and f_6 leads to

$$f_7 := 2y+2x^2-3x-6.$$

Finally, the consideration of the critical pair of f_6 and f_7 leads to

$$f_8 := 2x^3-5x^2-5x.$$

The consideration of all the other critical pairs leads to identical normal forms. (Actually, the consideration of most of these critical pairs can be avoided a priori by applying "criteria" of the type described in /Buchberger 79/ and /Winkler, Buchberger 83/. Furthermore, some of the polynomials in the basis can also be canceled in the course of the algorithm, see /Buchberger 70/). Hence, $G := \{f_1,\ldots,f_8\}$ is a Gröbner basis corresponding to F, i.e. F and G generate the same ideal in $Z[x_1,\ldots,x_n]$ and $\rightarrow_G$ has the Church-Rosser property. Reduction of all the f_i modulo $G-\{f_i\}$ leaves us with $G' := \{f_6', f_7, f_8\}$, where

$$f_6' := -xy-y-2x^2+2x+3.$$

Summarizing, the algorithm produced G' such that F and G' generate the same ideal, $\rightarrow_{G'}$ has the Church-Rosser property (hence, $S_{G'}$ decides the congruence $\equiv_F$) and, given the ideal generated by F, G' is uniquely determined.

CONCLUSIONS

We have shown that an axiomatic approach to a constructive ring theory based on the notion of reduction and on a critical-pair/completion is possible. The general critical-pair completion algorithm presented in this paper specializes to our 1965 algorithm in $K[x_1,\ldots,x_n]$ (K a field) (which, again, specializes to Euclid's algorithm in $K[x]$ and to Gauß' algorithm in the case of linear polynomials in $K[x_1,\ldots,x_n]$), it specializes to (a whole spectrum) of Euclidean algorithms in Z and it yields an elegant algorithm for deciding congruence and related problems in $Z[x_1,\ldots,x_n]$.

The major weakness of this paper is that, at the present stage of the investigations, only a few other examples of rings satisfying the axioms are known. Further research is necessary in order to clarify whether a sufficiently wide class of rings satisfies the axioms or whether a further attempt at weakening the axioms must be made. Both tasks are challenging: The first task leads to the investigation of new properties of rings that so far have not received any attention. The second task requires to develop new ideas for the proof of the main theorem in Section 3. and will clarify the insight, which properties of rings lie "really" at the basis of the critical-pair/completion approach.

REFERENCES

Ayoub, C. W., 81:
On Constructing Bases for Ideals in Polynomial Rings over the Integers.
The Pennsylvania State University, Dept. of Mathematics, Report Nr. 8184, 1981, submitted to publication.

Ballantyne, A. M., Lankford, D. S., 81:
New Decision Algorithms for Finitely Presented Commutative Semigroups.
Comp. and Maths. with Appls. 7 (1981), 159-165.

Bauer, G., 81:
The Representation of Monoids by Confluent Rule Systems.
University of Kaiserlautern, FRG, Fachbereich Informatik, Ph.D. Thesis, 1981.

Bergman, G. M., 78:
The Diamond Lemma for Ring Theory.
Adv. Math. 29, (1978), 178-218.

Buchberger, B., 65:
An Algorithm for Finding a Basis for the Residue Class Ring of a Zero-Dimensional Polynomial Ideal (German).
Univ. of Innsbruck, Austria, Math. Inst., Ph.D. Thesis, 1965.

Buchberger, B., 70:
An Algorithmical Criterion for the Sovability of Algebraic Systems of Equations (German).
Aequationes mathematicae 4/3 (1970), 374-383.

Buchberger, B., 76:
A Theoretical Basis for the Reduction of Polynomials to Canonical Form.
ACM SIGSAM Bull. 10/3 (1976), 19-29.

Buchberger, B., 76a:
Some Properties of Gröbner Bases for Polynomial Ideals.
ACM SIGSAM Bull. 10/4 (1976), 19-24.

Buchberger, B., 79:
A Criterion for Detecting Unnecessary Reductions in the Construction of Gröbner-Bases.
Proc. EUROSAM 79, Marseille (Ng, W., ed.), Lecture Notes in Computer Science 72 (1979), 3-21.

Buchberger, B., 83:
A Note on the Complexity of Constructing Gröbner-Bases.
Proc. of the EUROCAL 83 Conf., London, Lecture Notes in Computer Science, Sringer, 1983, to appear.

Buchberger, B., 83a:
A Critical-Pair/Completion Algorithm in Reduction Rings.
University of Linz, Austria, Math. Institute, Technical Report Nr. CAMP 83-21.0, 1983.

Buchberger, B., Loos, R., 82:
Algebraic Simplification.
In: Computer Algebra (B. Buchberger, G. Collins, R. Loos eds.), Springer, Wien - New York, 1982, 11-43.

Guiver, J. P., 82:
Contributions to Two-Dimensional System Theory.
Univ. of Pittsburgh, Math. Dept., Ph.D. Thesis, 1982.

Hurd, C. B., 70:
Concerning Ideals in $Z[x]$ and $Z_{p^n}[x]$.
The Pennsylvania State University, Department of Mathematics, Ph.D. Thesis, 1970.

Knuth, D. E., Bendix, P. B., 67:
Simple Word Problems in Universal Algebras.
Proc. of the Conf. on Computational Problems in Abstract Algebra, Oxford, 1967, (Leech, J., ed.), Oxford: Pergamon Press, 1970.

Kronecker, L., Hensel, K., 01:
Lectures on Number Theory (German).
Leipzig, 1901.

Lauer, M., 76:
Canonical Representatives for the Residue Classes of a Polynomial Ideal.
University of Kaiserslautern, FRG, Dept. of Mathematics, Diploma Thesis, 1976.

Llopis de Trias, R., 83:
Canonical Forms for Residue Classes of Polynomial Ideals and Term Rewriting Systems.
Univ. Aut. de Madrid, Division de Matematicas, submitted to publication, 1983.

Redei, L., 56:
Equivalence of the Theorems of Kronecker-Hensel and Szekeres (German).
Acta Sci. Math. Szeged 17 (1956), 198-202.

Richman, F., 74:
Constructive Aspects of Noetherian Rings.
Proc. AMS 44/2 (1974), 436-441.

Schaller, S., 79:
Algorithmic Aspects of Polynomial Residue Class Rings.
University of Wisconsin, Madison, Ph.D. Thesis, Comput. Sci. Tech. Rep. 370, 1979.

Simmons, H., 70:
The Solution of a Decision Problem for Several Classes of Rings.
Pac. J. Math. 34 (1970), 547-557.

Sims, C., 78:
The Role of Algorithms in the Teaching of Algebra.
In: Topics in Algebra (Newman, M. F. ed.), Lecture Notes in Mathematics 697 (1978), Springer, 95-107.

Spear, D., 77:
A Constructive Approach to Commutative Ring Theory.
Proc. of the MACSYMA Users' Conference, Berkeley, 1977, (Fateman, R.J., ed.), published by MIT, 369-376.

Szekeres, G., 52:
A Canonical Basis for the Ideals of a Polynomial Domain.
Am. Math. Monthly 59 (1952), 379-386.

Szekeres, G., 65:
Metabelian Groups with Two Generators.
Proc. Internat. Conf. Theory of Groups (Canberra 1965), Gordon & Breach, 1967, 323-346.

Szekeres, G., 75:
Homogeneous Ideals in K[x,y,z].
Acta Math. Acad. Sci. Hungar. 26 (1975), 355-367.

Trinks, W., 78:
On B. Buchberger's method for Solving Algebraic Equations (German).
J. Number Theory 10/4 (1978), 475-488 (preprint 1977).

Trotter, P. G., 69:
A Canonical Basis for Ideals of Polynomials in Several Variables and With Integer Coefficients.
Univ. of New South Wales, Ph.D. Thesis, 1969.

Trotter, P. G., 78:
Ideals in Z[x,y].
Acta Math. Acad. Sci. Hungar. 32 (1-2) (1978), 63-73.

Winkler, F., Buchberger, B., 83:
A Criterion for Eliminating Unnecessary Reductions in the Knuth-Bendix Algorithm.
Coll. on Algebra, Combinatorics and Logic in Computer Science, Györ, Sept. 12-16, 1983, 21 p.

Zacharias, G., 78:
Generalized Gröbner Bases in Commutative Polynomial Rings.
MIT, Dept. Comput. Sci., Bachelor Thesis, 1978.

ACKNOWLEDGEMENT: The research described in this paper is supported by the Austrian Research Fund, Project Nr. 4567. Part of the work was carried out while the author was visiting the University of Delaware (Dept. of Computer and Information Sciences), the IBM Thomas Watson Research Center (Dept. of Mathematics), Yorktown Heights, and the University of Wisconsin-Madison (Dept. of Computer Science), January-September 1982.

EXTENSIBLE ALGORITHMS

Hans-Georg CARSTENS and Peter PÄPPINGHAUS

Department of Mathematics, University of Bielefeld, Universitätsstr. 1, D-4800 Bielefeld, Fed. Rep. of Germany, and Institute of Mathematics, University of Hannover, Welfengarten 1, D-3000 Hannover, Fed. Rep. of Germany

§1 INTRODUCTION

There are various measures for the performance of an algorithm, e.g. the time needed in the worst case or in the average, the space needed, numerical accuracy, on line performance etc. We add to these another one: extensibility.

Before going into details, let us look at a practical problem: time tables. There are computer programs computing time tables, but generally they meet strong opposition by the people involved. Why? First of all: the computer does not argue with people. Secondly and more seriously: the new time table in most cases has almost no similarities with the old one used last period. So people have to change their habits. Things would be easier with a computer program making as few changes in the old time table as possible. Hence one likes to have an algorithm which extends the solution of the old time table problem (i.e. the old time table) to the new situation.

Such a property of an algorithm is implicitly employed in many classical algorithms. Let us demonstrate this by a rather old one: The Newton interpolation schema. Within this algorithm you have to compute the divided differences as follows.

x	k=o	k=1	k=2	k=3
x_o	$f[x_o]$			
x_1	$f[x_1]$	$f[x_o,x_1]$		
x_2	$f[x_2]$	$f[x_1,x_2]$	$f[x_o,x_1,x_2]$	
x_3	$f[x_3]$	$f[x_2,x_3]$	$f[x_1,x_2,x_3]$	$f[x_o,x_1,x_2,x_3]$

The underlined terms are the coefficients in the interpolating polynomial. If you add more values at $x_4, x_5, \ldots$, you are allowed to use the old results and continue the computation of the new interpolating polynomial in the next line of the schema.

As data for our extensible algorithms we admit arbitrary finite relational structures, but for simplicity let us concentrate for a moment on algorithms for graphs. To be specific consider e.g. a 5-coloring algorithm for planar graphs as mentioned below. Our idea to make precise the concept of extensibility in this setting is as follows.

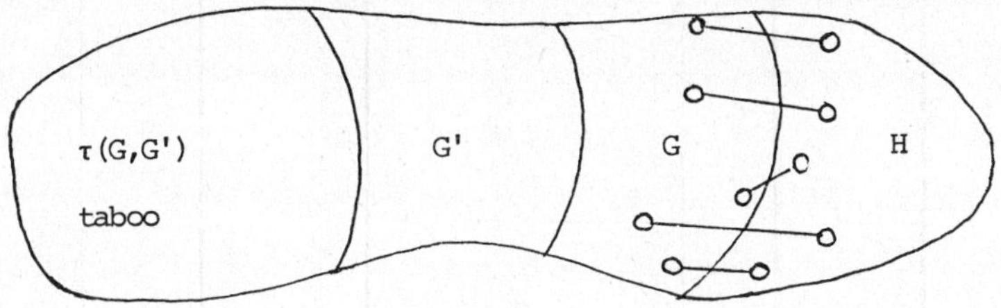

If G is a finite planar graph and G' a subgraph of G, then there should be an induced "taboo" subgraph $\tau(G,G')$ of G' s.t. the extensible coloring algorithm "extends" the coloring of G to any finite planar supergraph H of G, which joins no new lines to G', making color changes if necessary in $G \setminus \tau(G,G')$, but leaving the coloration of $\tau(G,G')$ completely unchanged. Furthermore $\tau(G,G')$ should be big if G' and G are very big.

Let us mention three examples of extensible coloring algorithms:

- An extensible algorithm 5-coloring planar graphs [C].
- An extensible algorithm coloring graphs of maximal degree Δ with Δ many colors, provided the graph contains no exceptional subgraph mentioned in Brooks' theorem [CP1], and [Sch 2] .
- An extensible algorithm for finding a one-way/endless Euler path in graphs having such a path [B].

Further extensible coloring algorithms can also be found in [K] and [Sch 1].

One can easily imagine variants of this concept. Our choice is motivated by the following pleasing fact. There is a connection between extensible algorithms for a class of finite objects and uniform algorithms for the class of canonically formed limits of chains of objects in the class of finite things i.e. infinite objects. (For the class of finite planar graphs the limits are the locally finite countable planar graphs.) So extensibility fits to recursion theory. This is implied by the following theorem. There is an extensible algorithm solving a certain problem for a class of finite objects iff there is a corresponding uniform algorithm for the limits. E.g. 5-coloring finite planar graphs is extensibly solvable iff 5-coloring countable locally finite planar graphs is uniformly solvable. In regard to the continuity of uniform algorithms the theorem comes not as a surprise. It simply states that things are as nice as they should be. What we gain from this theorem is an interplay of themes in computer science and logic. We illustrate this by a discussion of problems mainly from the area of combinatorial algorithms, e.g. paths in a tree, coloring,, matching, etc. In one direction we employ recursive unsolvability results to show extensible unsolvability of certain problems. In the other direction we conclude that a problem for infinite data has a uniform recursive solution, for there is an extensible algorithm for the corresponding finite data. For this latter purpose Kierstead introduced a notion, which can be viewed as a special case of extensibility, see [K, Lemma 1.1, p. 1281]. We give an example for extensible unsolvability. A theorem of Manaster and Rosenstein [MR 1] says that there is a highly recursive bipartite graph having a matching, but no recursive one. From this we conclude that matching is not extensibly solvable. This will be seen to imply that the maximal flow problem for networks and linear integer programming are not extensibly solvable.

The facts from mathematical logic used in this paper can be found in Shoenfields book [Sh],the notions from graph theory are as in Hararys book [H].

§2 EXTENSIBILITY

In this section we give a mathematical formulation of extensibility in terms of recursive partial functions.

We think of inputs as finite structures and of outputs as finite sequences, e.g. an input is a graph and an output is the sequence of points of a path in the graph. For simplicity we require that the domain D(S) of a structure $S=\langle D(S);R_1,\ldots,R_n;c_1,\ldots,c_m\rangle$ is a set of natural numbers, and we consider relational structures only. In the following we let always the set of inputs I be a class of finite structures of a fixed signature, and the set of outputs be the set $\mathbb{N}^{<\omega}$ of all finite sequences of natural numbers.

Now we consider a relation $Q \subseteq I \times \mathbb{N}^{<\omega}$ thinking of $Q(S,r)$ as expressing "the finite sequence r answers question Q for the I-structure S".
We intend that $r_i = 0$ codes the "null information" corresponding to undefined values of partial functions. Therefore we use the relation
$r \prec r' :\Leftrightarrow \forall i < \ell h(r): (r_i \neq 0 \Rightarrow r'_i = r_i)$, which means that r' extends the information contained in r. $\ell h(r)$ denotes the length of the sequence r.

DEFINITION 1. $P = (I,Q)$ is called a *problem* iff the following is satisfied:

(i) I is a class of finite relational structures of a fixed signature, with domain $\subseteq \mathbb{N}$.

(ii) $Q \subseteq I \times \mathbb{N}^{<\omega}$, and

(iii) $Q(S,r) \wedge r \prec r' \Rightarrow Q(S,r')$.

The last condition reflects the fact, that we want to allow answers to contain superfluous information.

Let us first define the type of algorithm we need.

DEFINITION 2. α is called a *P-solving algorithm* iff the following hold:
$\alpha : I \to \mathbb{N}^{<\omega}$ is a recursive partial function s.t.

$$S \in I \Rightarrow \alpha(S)\downarrow \wedge Q(S,\alpha(S)).$$

$\downarrow$ means "is defined", and we assume that finite objects in I, $\mathbb{N}^{<\omega}$, etc. are coded as natural numbers in some fixed way.

The idea of a P-solving extentible algorithm is captured in the following. There is a function ι selecting a part of the solution $\alpha(S)$ of problem P for S, s.t. an invariance condition holds for this part, and s.t. the invariant part $\iota(S,S')$ grows arbitrarily big, if the input S,S' is growing.

In order to express this invariance condition, we use the following relation $S \underset{M}{=} T$ saying "for points of M, structure S looks like T".

DEFINITION 3. For I-structures S,T and $M \subseteq \mathbb{N}$:
$S \underset{M}{=} T :\Leftrightarrow \forall a \in M: (a \in D(S) \Leftrightarrow a \in D(T))$ and for all corresponding relations R_S in S and R_T in T the following holds:
$\forall a_1,\dots,a_n \in \mathbb{N}: (\exists i: a_i \in M \Rightarrow (R_S(a_1,\dots,a_n) \Leftrightarrow R_T(a_1,\dots,a_n)))$, and the corresponding constants of S and T are equal.

To express the growth condition mentioned in the above description of an invariance selector, we use ascending sequences of I-structures. In the following section such sequences will also serve the purpose of constructing infinite structures as unions of ascending chains. We want to allow also finite structures to occur as such unions of chains. Therefore "ascending" should not necessarily mean "strictly ascending". On the other hand to obtain some growth condition, we need proper growth of the ascending sequences in some sense. This problem is solved by allowing different codes for the same structure, and so we can require the *length* of the coding sequence to grow properly.

More specifically, we code a finite structure S by sequences so that the i-th member s_i of the sequence s contains the information, whether i is in the domain of S, plus the information, in which relations of S i occurs (the "relational neighbourhood" of i). The different codes of S are thus distinguished by the amount of information on non-members of $D(S)$ they contain.

Precisely speaking, I has to be taken to be the set of these codes, and $S,T,\dots$ have to denote such a code. But we will continue to use these letters somewhat ambiguously for codes of structures and also for the structures themselves. As a rule, when we speak of $S \subseteq T$, $S = T$ and $Q(S,r)$, we mean the structures themselves, but when writing $\ell h(S)$, we refer to the length of a sequence, which codes the structure.

DEFINITION 4. $(S_n)_{n \geq 0}$ is called a *canonically ascending* I*-sequence* iff the following is satisfied:

$\forall n$: $S_n \in I \wedge S_n \subseteq S_{n+1} \wedge \ell h(S_n) < \ell h(S_{n+1}) \wedge S_{n+1} \underset{[0,\ell h(S_n)[}{=} S_{n+2}$, where $[0,k[$ denotes the interval $\{i \in \mathbb{N} \mid i < k\}$.

We are ready to make precise the notion of an extensible algorithm.

DEFINITION 5. α is called a P-*solving extensible algorithm* if the following hold: α is a P-solving algorithm for which there is an *invariance selector* ι satisfying:

(ι) $\iota: I \times I \to \mathbb{N}^{<\omega}$ is a recursive partial function s.t.

(ι1) $S,S' \in I \wedge S' \subseteq S \wedge \ell h(S') \leq \ell h(S) \Rightarrow \iota(S,S')\downarrow \wedge \iota(S,S') \prec \alpha(S)$

(ι2) $S,S',T,T' \in I \wedge S' \subseteq S \wedge \ell h(S') \leq \ell h(S) \wedge T' \subseteq T \wedge \ell h(T') \leq \ell h(T)$

$\wedge\, S \underset{[0,\ell h(S')[}{=} T \wedge \ell h(S') \leq \ell h(T') \wedge S' \subseteq T' \Rightarrow \iota(S,S') \prec \iota(T,T')$

(ι3) for every canonically ascending I-sequence $(S_n)_{n\geq 0}$:

$\forall n\, \exists k\, \exists T \in I$: $S_n \subseteq T \subseteq S_k \wedge Q(T, \iota(S_{k+1},S_k))$.

Now let us come back to our intuitive description of an extensible algorithm given in the introduction. There we considered a function τ exhibiting a taboo substructure of $S' \subseteq S$, where an extensible algorithm never changes the solution. Furthermore $\tau(S,S')$ has to become arbitrarily big, if the input S,S' is growing. This concept corresponds to a slight strengthening of our requirements for a well shaped problem, which can easily be fulfilled in many cases as e.g. 5-coloring planar graphs. We make this precise in the following. Λ denotes the empty structure.

DEFINITION 6. (i) ρ is called a *restrictor* for P iff the following is satisfied:

(ρ) $\rho: I \times \mathbb{N}^{<\omega} \to I \cup \{\Lambda\}$ is a recursive partial function s.t.

$Q(S,r) \wedge I \ni S' \subseteq S \wedge r' \prec r \Rightarrow$

(ρ1) $\rho(S',r')\downarrow \wedge \rho(S',r') \subseteq S' \wedge (\rho(S',r') \neq \Lambda \Rightarrow Q(\rho(S',r'),r'))$,

(ρ2) $Q(S',r') \Rightarrow \rho(S',r') = S'$,

(ρ3) $I \ni S'' \subseteq S' \wedge r'' \prec r' \Rightarrow \rho(S'',r'') \subseteq \rho(S',r')$.

(ii) Let α be an extensible P-solving algorithm with an invariance selector ι.

τ is called a *taboo selector* for α iff the following is satisfied:

(τ) $\tau: I \times I \to I \cup \{\Lambda\}$ is a recursive partial function s.t.

(τ 1) $S,S' \in I \wedge S' \subseteq S \wedge \ell h(S') \leq \ell h(S) \Rightarrow \tau(S,S')\downarrow \wedge \tau(S,S') \subseteq S' \wedge$

$(\tau(S,S') \neq \Lambda \Rightarrow Q(\tau(S,S'), \iota(S,S')))$

(τ 2) for every canonically ascending sequence $(S_n)_{n \geq 0}$:

$\forall n \exists k \geq n: S_n \subseteq \tau(S_{k+1},S_k)$ and $\forall n: \tau(S_{n+1},S_n) \subseteq \tau(S_{n+2},S_{n+1})$.

THEOREM. *A P-solving extensible algorithm has a taboo selector if* P *has a restrictor.*

PROOF. Let α be a P-solving extensible algorithm with invariance selector ι, and let ρ be a restrictor for P. We define $\tau(S,S') :\simeq \rho(S', \iota(S,S'))$. Obviously τ is a recursive partial function and $\tau: I \times I \to I \cup \{\Lambda\}$. For $S,S' \in I$ s.t. $S' \subseteq S \wedge \ell h(S') \leq \ell h(S)$ we have $Q(S,\alpha(S))$ and $\iota(S,S') \prec \alpha(S)$. So ($\rho$1) implies ($\tau$1). As to ($\tau$2) let $(S_n)_{n \geq 0}$ be a canonically ascending sequence, and let n be a natural number. By (ι3) there is a k and a $T \in I$ s.t. $S_n \subseteq T \subseteq S_k$ and $Q(T, \iota(S_{k+1},S_k))$, and by (ρ3) and (ρ2) we see that the following is true.
$\tau(S_{k+1},S_k) = \rho(S_k, \iota(S_{k+1},S_k)) \supseteq \rho(T, \iota(S_{k+1},S_k)) = T \supseteq S_n$.

We remark that for a P-solving extensible algorithm with invariance selector ι having a taboo selector τ, the property (ι3) follows from (τ).

DEFINITION 7. P is called *extensibly solvable* iff there exists a P-solving extensible algorithm.

To illustrate these notions we devote the remainder of this section to the description of a simple example. For this example fix $k \geq 2$, and let I be the class of all finite connected graphs of chromatic number k having 0 as a point. We consider the question Q asking for a 2k-1 coloring for graphs in I. More precisely a sequence r answers question Q for a graph G in I, if r_i is a color for point $i \in D(G)$ in a 2k-1 coloring of G with colors from $\{1,\ldots,2k-1\}$. We show that P=(I,Q) is extensibly solvable and describe a P-solving extensible algorithm as follows.

STEP 1. α colors point o and all points of distance ≤ 2 to point o with k colors by trial and error.

STEP s+1. We assume that all points of distance $\leq 2s$ to point o are already colored in such a way, that the points of distance 2s or 2s-1 are colored by k many colors. We remove one color, say c_1, from the points having distance 2s to point o and color the now uncolored points of distance 2s to point o and the points having distance 2s+1 or 2s+2 with c_1 and the remaining k-1 colors not used for points of distance 2s. Obviously the assumption needed in step s+2 holds.

α assigns to finite connected graphs G with chromatic number k a sequence of colors of a 2k-1 coloring of G in an effective way s.t. the sequence is an answer to question Q. So α is a P-solving algorithm. Let d be maximal s.t. all points of G having distance $\leq$ d to o are contained in G'. We define $\iota(G,G') :\simeq \alpha\restriction$the subgraph of G' spanned by the points having distance $\leq$d-3 to o. Obviously ι is a recursive partial function and $\iota: I\times I \to \mathbb{N}^{<\omega}$. Let $G' \subseteq G$, then $\iota(G,G')\downarrow$ and $\iota(G,G') \prec \alpha(G)$. So ($\iota$1) is satisfied. Now ($\iota$2) follows from the fact that for even d the colors assigned to the points of distance $\leq$d-1 to point o are not changed any more when all points of distance $\leq$d to point o have received a color. Let $(S_n)_{n\geq o}$ be a canonically ascending sequence and n a natural number. We define k as minimal s.t. there is a d s.t. S_n is contained in the subgraph of $\bigcup_n S_n$ spanned by the points having distance $\leq$ d-3 to point o, and S_k contains the subgraph of $\bigcup_n S_n$ spanned by the points having distance $\leq$ d to point o. As the structure T we are allowed to choose S_n. Now the following holds: $S_n \subseteq T \subseteq S_k$ and $Q(T, \iota(S_{k+1}, S_k))$. This implies ($\iota$3). Hence ι is an invariance selector for α and α is a P-solving extensible algorithm. We define $\rho(G,r) :\simeq$ the maximal connected subgraph $H \subseteq G$ containing o s.t. r is an answer to Q for H. Obviously ρ is a restrictor for P. Hence by the theorem α has a taboo selector τ, which can be defined as follows: $\tau(G,G') \simeq \rho(G', \iota(G,G'))$. If $G' \subsetneq G$, then we see that $\tau(G,G')$ is the subgraph of G' spanned by the points having distance $\leq$ d-3 to o, where d is from the definition of ι.

§3 Approximation of infinite problems by finite problems

In this section we prove a theorem connecting extensible algorithms on finite structures with uniform algorithms on canonically formed limits of finite structures.

We think of I as approximating the class of infinite structures consisting of the unions of canonically ascending I-sequences. Conversely, by an approximation of an infinite structure S we mean a canonically ascending sequence $(S_n)_{n\geq 0}$ of substructures s.t. $S = \bigcup_n S_n$.

Definition 1. $\lim I := \{ S \mid \exists (S_n)_{n\geq 0} : (S_n)_{n\geq 0}$ is a canonically ascending I-sequence and $S = \bigcup_n S_n \}$

Let us look at the nature of the structures in $\lim I$. As mentioned in the previous section, $\lim I$ contains besides countably infinite structures also the finite structures of I. (Simply take a sequence of codes of increasing length of one and the same structure $S \in I$). For the infinite structures in $\lim I$ we make the following observations. Consider a canonically ascending I-sequence $(S_n)_{n\geq 0}$. The requirement $S_{n+1} \underset{[0,\ell h(S_n)[}{=} S_{n+2}$ amounts to demanding that the new data of S_{n+2} do not interfere the data of S_n as it is embedded in S_{n+1}. This implies the following for $S := \bigcup_n S_n$. If $a \in S_n$ and b is in the relational neighbourhood of a w.r.t. S, then b is an element of S_{n+1}, and b is in the relational neighbourhood of a w.r.t. S_{n+1}. It follows that S is *locally finite*. Moreover for $b \in \mathbb{N}$ s.t. $b < \ell h(S_n)$, we have that b is in S iff b is in S_{n+1}. This implies that S is highly recursive, when the sequence $(S_n)_{n\geq 0}$ is recursive.

A locally finite, countable structure S is called *highly recursive* iff its representing function is recursive. We represent S by a total function f_S as follows. If $S = \langle D(S); R_1,\ldots,R_m; c_1,\ldots,c_k\rangle$ and $n \in D(S)$, then $f_S(n)$ is a code $\neq 0$ for the finite sequence $\langle R_1[n],\ldots,R_m[n],c_1,\ldots,c_k\rangle$. And $R_i[n]$ is the finite set of those tuples of elements of S, which are related to n by R_i. If $n \notin D(S)$, then

we let $f_S(n) := 0$. The definitions are set up so that we obviously have
$S \underset{M}{=} T \Leftrightarrow f_S \upharpoonright M = f_T \upharpoonright M$.

Apart from generating only locally finite countable structures in $\lim I$, the canonical requirements lead to no further restrictions for $\lim I$. This follows from the fact that from every sequence $S_0 \subseteq \ldots \subseteq S_n \subseteq \ldots$ in I one can easily extract a subsequence $S_{k_0} \subseteq \ldots \subseteq S_{k_n} \subseteq \ldots$ which - when adequately coded - is canonically as-ascending. So we obtain the following lemma.

LEMMA. $\lim I = \{S \mid S$ locally finite and there is a sequence $S_0 \subseteq \ldots \subseteq S_n \subseteq \ldots$ of I-structures s.t. $S = \bigcup_n S_n\}$.

PROOF. Let $S_0 \subseteq \ldots \subseteq S_n \subseteq \ldots$ be a sequence in I, and w.l.o.g. let the coding be such that $\ell h(S_n) < \ell h(S_{n+1})$. We define k_n inductively. Let $k_0 := 0$, and assume $k_0, \ldots, k_n$ to be defined already. Then we define k_{n+1} to be the least $k > k_n$ s.t. the following finite set is contained in $D(S_k)$: $\{i \mid i < \ell h(S_{k_n}) \wedge i \in D(S)\} \cup$
$\cup \{i \mid i$ is in the relational neighbourhood of some $j \in S_{k_n}\}$. It is obvious that $(S_{k_n})_{n \geq 0}$ is a canonically ascending I-sequence, and so
$S := \bigcup_n S_n = \bigcup_n S_{k_n} \in \lim I$.

Like I approximates $\lim I$, so a problem P determines a problem $\lim P$. Before defining $\lim P$, we first look at infinite problems in general. Here we let the set of inputs I^* be a class of countable and locally finite structures of a fixed signature, and the set of outputs be the set of all partial functions from $\mathbb{N}$ to $\mathbb{N}$. Again we want answers to possibly contain superfluous information, which can be expressed using the following relation: $r \prec^* r' :\Leftrightarrow \forall i: (r(i)\downarrow \Rightarrow r(i) \simeq s(i))$.

DEFINITION 2. $P^* = (I^*, Q^*)$ is called a *problem for countable structures* iff the following is satisfied:

(i) I^* is a class of countable and locally finite relational structures of a fixed signature with domain $\subseteq \mathbb{N}$,

(ii) $Q^* \subseteq I^* \times \{r \mid r: \mathbb{N} \rightarrow \mathbb{N}$ partial function$\}$, and

(iii) $Q^*(S,r) \wedge r \prec^* r' \Rightarrow Q^*(S,r')$.

The idea to define the problem $\lim P = (\lim I, \lim Q)$ from $P = (I,Q)$ is as follows. We think of a partial function r as answering question $\lim Q$ for a $\lim I$-structure S, if answers to Q for the finite substructures of S are contained in suitable finite initial segments of r. To make this precise, we define for a partial function $r:\mathbb{N} \to \mathbb{N}$ and $k \in \mathbb{N}$ the finite sequence $r \restriction k$ of length k as follows: for $i < k$: $(r \restriction k)_i := r(i) + 1$, if $r(i)\downarrow$, and $(r \restriction k)_i := 0$ otherwise.

DEFINITION 3. Let $P = (I,Q)$ be a problem.

$$\lim P := (\lim I, \lim Q) ,$$

where $\lim Q \subseteq \lim I \times \{r \mid r:\mathbb{N} \to \mathbb{N} \text{ partial function}\}$ is defined by

$$\lim Q(S,r) : \Leftrightarrow \forall T' \in I: (T' \subseteq S \Rightarrow \exists k\, \exists T \in I: T' \subseteq T \subseteq S \wedge Q(T, r \restriction k)) .$$

Since we are going to define an algorithm for structures in $\lim I$, we will need an effective operation λ, called an approximator, assigning to every $\lim I$-structure S an approximating sequence of I-structures. By an effective operation we mean a recursive partial type-2-functional.

DEFINITION 4. $\lim P$ is *effectively based* on P via an *approximator* λ, iff the following is satisfied:

(i) Q is decidable, and

(ii) $\lambda:\lim I \times \mathbb{N} \to I$ is a recursive partial functional s.t. $S \in \lim I \Rightarrow \lambda(S,\cdot)$ is total $\wedge\ S = \bigcup_n \lambda(S,n) \wedge (\lambda(S,n))_{n \geq 0}$ is a canonically ascending I-sequence.

To illustrate these notions, we give an example. For this example let I be the class of finite rooted trees. We consider the question to find a path of maximal length in such a tree starting with the root. More precisely, the finite sequence r

is an answer to question Q for T if r has an initial segment r' s.t. for all $i < \ell h(r')$: $r'_i \neq 0$, and the numbers $r'_i - 1$, $i < \ell h(r')$ are the points of a rooted path of maximal length in T in the order given by the path. Obviously $P = (I,Q)$ is a problem, and $\lim I = \{T \mid T$ is countable, locally finite rooted tree s.t. $D(T) \subseteq \mathbb{N}\}$. $\lim Q(T,r)$ iff for every finite rooted tree $S' \subseteq T$ there is a k and a finite rooted tree S s.t. $S' \subseteq S \subseteq T$ and $r \restriction k$ is a rooted path of maximal length in S, but this is equivalent to the fact that r is a rooted infinite path in T. Hence lim P is the natural infinite analogon of P.

$\lambda(T,n)$ is recursively defined as follows. $\lambda(T,0)$ is a code of length 1 of the tree consisting only of the root. Assume $\lambda(T,n)$ to be defined. Let T_{n+1} be the subtree of T generated by $D(\lambda(T,n)) \cup \{b \in T \mid b$ adjacent to a point of $\lambda(T,n)\} \cup$ $\cup \{b \in T \mid b \leq \ell h(\lambda(T,n))\}$, and let $\lambda(T,n+1)$ be that code of T_{n+1}, which is of length $\max(D(T_{n+1})) + 1$.

Obviously $\lambda: \lim I \times \mathbb{N} \to I$ is a partial functional, computable using T as an oracle. For $T \in \lim I$, $\lambda(T,\cdot)$ is total and $T = \bigcup_n \lambda(T,n)$. Clearly $(\lambda(T,n))_{n \geq 0}$ is a canonically ascending I-sequence. Moreover Q is decidable. Hence lim P is effectively based on P via λ.

DEFINITION 5. (i) α is called a lim P-*solving uniform algorithm* iff the following is satisfied: $\alpha: \lim I \times \mathbb{N} \to \mathbb{N}$ is a recursive partial functional s.t.

$$S \in \lim I \Rightarrow \lim Q(S, \alpha(S,\cdot)) .$$

(ii) lim P is called *uniformly solvable* iff there exists a lim P-solving uniform algorithm.

The desired theorem connecting extensible algorithms on finite structures in I with uniform algorithms on countable structures in lim I reads as follows.

THEOREM. *For a problem* P *effectively based on its limit* lim P *the following holds. There is a* P-*solving extensible algorithm iff there is a* lim P-*solving uniform algorithm.*

PROOF. Let P be effectively based on its limit $\lim P$ via an approximator λ, and let α be a P-solving extensible algorithm with an invariance selector ι. We define a $\lim P$-solving uniform algorithm as follows.

$$\lim \alpha(S,\cdot) :\simeq \bigcup_k (\iota(S_{k+1},S_k))^*$$

where for $S \in \lim I$, $S_k := \lambda(S,k)$, and for a finite sequence r, r^* is defined as follows: $r^*(i) := r_i - 1$, if $i < \ell h(r) \wedge r_i \neq 0$, and $r^*(i)\uparrow$ otherwise. $(S_k)_{k \geq 0}$ is a canonically ascending sequence, hence for all k: $\iota(S_{k+1},S_k)\downarrow$, and for $k \geq m$, $(\iota(S_{m+1},S_m))^* \prec^* (\iota(S_{k+1},S_k))^*$. So $\lim \alpha(S,\cdot)$ is a partial function. The value of $\lim \alpha(S,n)$ can be computed as follows. First compute the least k s.t. $\ell h(\iota(S_{k+1},S_k)) > n$. Then compute the n-th member of $\iota(S_{k+1},S_k)$, say j. If $j \neq 0$, then output $j-1$. Thus $\lim \alpha$ is a recursive partial functional. Now we use the definition of $\lim Q$ to verify that $\lim Q(S, \lim \alpha(S,\cdot))$. For $T' \in I$ s.t. $T' \subseteq S$ we find an m s.t. $T' \subseteq S_m$, and by $(\iota 3)$ there is a k and a $T \in I$ s.t. $S_m \subseteq T \subseteq S_k \wedge Q(T, \iota(S_{k+1},S_k))$. Obviously for some n, $\iota(S_{k+1},S_k) \prec \lim \alpha(S,\cdot)\restriction n$ holds, and so $Q(T, \lim \alpha(S,\cdot)\restriction n)$. Hence $\lim \alpha$ is a $\lim P$-solving uniform algorithm. For the converse let α be a $\lim P$-solving uniform algorithm. To define a P-solving extensible algorithm $\hat{\alpha}$, we use Kleene's normal form theorem for functionals. Thereby α has an index $\ulcorner \alpha \urcorner \in \mathbb{N}$ so that it can be written as follows:

$$\alpha(f,i) \simeq U(\min k.\ T(\ulcorner \alpha \urcorner, \langle f(0),\ldots,f(k-1)\rangle, i, k)).$$

W.l.o.g. U,T can be taken to satisfy:

$$k_0 < k_1 \wedge T(\ulcorner \alpha \urcorner, \langle f(0),\ldots,f(k_0-1)\rangle, i, k_0)$$
$$\Rightarrow T(\ulcorner \alpha \urcorner, \langle f(0),\ldots,f(k_1-1)\rangle, i, k_1) \wedge U(k_0) = U(k_1).$$

We define $\beta(f,n) := \langle r_0,\ldots,r_{n-1}\rangle$, where $r_i := U(n) + 1$, if $T(\ulcorner \alpha \urcorner, \langle f(0),\ldots,f(n-1)\rangle, i, n)$, and $r_i := 0$ else. Note that β is recursive and total and satisfies the following properties:

(i) $n < k \Rightarrow \beta(f,n) \prec \beta(f,k) \prec \alpha(f,\cdot)\restriction k$,

(ii) $\forall k\ \exists n$: $\alpha(f,\cdot)\restriction k \prec \beta(f,n)$, and

(iii) $f \upharpoonright n = g \upharpoonright n \Rightarrow \beta(f,n) = \beta(g,n)$.

We observed that I is contained in lim I. Literally speaking this is, however, not true, since I consists of natural numbers representing finite structures, whereas lim I consists of total functions from $\mathbb{N}$ to $\mathbb{N}$ representing countable structures. At this point we want to be precise about this, so we assume to be given a recursive partial functional $\eta: I \to \lim I$, which is defined for every $S \in I$ and produces the infinite representation of $S: \eta(S) \in \lim I$.

Now we are ready to define a P-solving algorithm as follows.

$\hat{\alpha}(S) :\simeq \beta(\eta(S), n_0(S))$, where $n_0(S) :\simeq \min n.\ Q(S, \beta(\eta(S), n)) \wedge n \geq \ell h(S)$.

(Remember that S is the code of a sequence, so this makes sense). We show that $\hat{\alpha}$ is a P-solving algorithm. Let $S \in I$ be given. Then $\eta(S) \in \lim I$, and so $\lim Q(\eta(S), \alpha(\eta(S), \cdot))$. By definition of lim Q, there are $k \in \mathbb{N}$ and $T \in I$ s.t. $S \subseteq T \subseteq \eta(S)$ and $Q(T, \alpha(\eta(S), \cdot) \upharpoonright k)$. Clearly $T = S$, and so by (ii) above $\exists n: Q(S, \beta(\eta(S), n))$. We see that $n_0(S)$ and $\hat{\alpha}(S)$ are defined, and $Q(S, \hat{\alpha}(S))$ holds. So $\hat{\alpha}$ is a P-solving algorithm.

Now we define an invariance selector as follows.

$\iota(S,S') :\simeq \beta(\eta(S), \ell h(S'))$.

To verify the axioms for an invariance selector, let $S, S' \in I$ be given s.t. $S' \subseteq S$ and $\ell h(S') \leq \ell h(S)$. Clearly $\iota(S,S')$ is defined. From $\ell h(S') \leq \ell h(S) \leq n_0(S)$ and (i) above it follows that $\iota(S,S') \prec \hat{\alpha}(S)$, and so ($\iota$1) holds.

We turn to the invariance condition (ι2). Let S,S',T, and T' satisfy the assumptions of (ι2). Then $\eta(S) \underset{[0,\ell h(S')[}{=} \eta(T)$, and so by (iii) and (i) above:

$\iota(S,S') = \beta(\eta(S), \ell h(S')) = \beta(\eta(T), \ell h(S')) \prec \beta(\eta(T), \ell h(T')) = \iota(T,T')$, and so ($\iota$2) holds.

As to the growth condition (ι3), let $(S_n)_{n \geq 0}$ be a canonically ascending I-sequence and $S := \bigcup_n S_n$. Then $S \in \lim I$, and $\lim Q(S, \alpha(S, \cdot))$. Now we fix an n. By definition of lim Q there are m_0 and $T \in I$ s.t.

$S_n \subseteq T \subseteq S \wedge Q(T,\alpha(S,\cdot)\restriction m_0)$. By (ii) above we find an m_1 s.t. $Q(T,\beta(S,m_1))$. Further we find - using the definition of a canonically ascending I-sequence - a number k s.t. $T \subseteq S_k$ and $\ell h(S_k) \geq m_1$. Clearly $S_{[0,\ell h(S_k)[} = \eta(S_{k+1})$, and so by (iii) and (i) above we see that

$\beta(S,m_1) = \beta(\eta(S_{k+1}),m_1) \prec \beta(\eta(S_{k+1}), \ell h(S_k)) = \iota(S_{k+1},S_k)$, and hence $Q(T,\iota(S_{k+1},S_k))$ holds. This proves $(\iota 3)$ and concludes the proof of the theorem.

§4 APPLICATIONS

In this section we show that matching, max-flow, and linear integer programming are not extensibly solvable.

We consider the problem to find a matching for a bipartite graph. For simplicity we represent bipartite graphs as structures $G = \langle D(G);A,B,L\rangle$ where A and B are disjoint sets s.t. $A \cup B = D(G)$, and there are no lines joining points in A or joining points in B . A matching for such a graph G is a set of mutually disjoint lines of G s.t. every point of A is incident with some line in this set.

Now fix a $k \geq 2$ and let I be the class of finite bipartite "almost k-regular" graphs in the following sense. A bipartite graph $G = \langle D(G);A,B,L\rangle$ is in I iff every $a \in A$ has degree k in G, and every $b \in B$ has degree $\leq k$ in G . It is easy to see that every graph $G \in I$ satisfies a well-known criterion of Hall, whereby every $G \in I$ has a matching.

The finite sequence r is an answer to question Q for $G \in I$ iff for all $a \in A$: $a < \ell h(r) \wedge r_a \neq 0$, and $\{\{a,r_a-1\} \mid a \in A\}$ is a matching for G .

Let Matching(k) be the problem (I,Q) .

LEMMA 1. *(i)* $\lim I$ *is the set of locally finite bipartite graphs with domain* $\subseteq \mathbb{N}$, *which are "almost* k*-regular".*

(ii) The partial function r *is an answer to* lim Q *for a graph* $G \in \lim I$ *iff* $\forall a \in A$: $r(a)\downarrow$ *, and* $\{\{a,r(a)\} \mid a \in A\}$ *is a matching for* G.

LEMMA 2. lim *Matching(k) is effectively based on Matching(k).*

PROOF. Let $G = \langle D(G);A,B,L\rangle$ be an element of lim I. $\lambda(G,n)$ is defined recursively as follows. G_0 is the subgraph of G generated by the smallest $a \in A$ and all its G-neighbours. $\lambda(G,0)$ is that code of G_0 having length $\max(D(G_0)) + 1$.
Assume $\lambda(G,n)$ to be defined. First we form the set
$C_{n+1} := D(\lambda(G,n)) \cup \{c \in G \mid c$ is adjacent in G to some $d \in \lambda(G,n)\} \cup$
$\cup \{c \in G \mid c \leq \ell h(\lambda(G,n))\}$, then we let $D_{n+1} := C_{n+1} \cup \{b \in B \mid b$ is adjacent in G to some $a \in C_{n+1}\}$. Now G_{n+1} is the subgraph of G generated by D_{n+1}, and $\lambda(G,n)$ is that code of G_{n+1} having length $\max(D_{n+1}) + 1$. It is easy to verify that λ is an approximator.

To show that Matching(k) is not extensibly solvable, we use the following theorem of Manaster and Rosenstein [MR 2]:

- For every $k \geq 2$ there is a highly recursive bipartite and k-regular graph having a matching, but no recursive one.

Hihgly recursive graphs are locally finite. So in our terminology a graph G is highly recursive iff f_G is recursive.

THEOREM 1. *For every* $k \geq 2$*: Matching(k) is not extensibly solvable.*

PROOF. Let $k \geq 2$, and suppose Matching(k) is extensibly solvable. By lemma 2 and the theorem of §3 lim Matching(k) is uniformly solvable. Hence there is an algorithm α computing a matching for every G in lim I, using G as an oracle. Hence for recursive f_G, $\alpha(f_G,\cdot)$ has to be recursive. By the theorem of Manaster and Rosenstein and lemma 1 there is a G in lim I s.t. f_G is recursive but has no recursive matching. We have met a contradiction.

To show that max-flow and linear integer programming are not extensibly solvable we use well-known techniques for simulation. What we have to show is that these simulations respect extensibility. For the following let $P_1=(I_1,Q_1)$ and $P_2 = (I_2,Q_2)$ be problems.

DEFINITION. P_2 *simulates extensibly* P_1 iff there are recursive partial functions $\sigma: I_1 \to I_2$ and $o: \mathbb{N}^{<\omega} \to \mathbb{N}^{<\omega}$ s.t. σ is defined on I_1 and o is total and s.t. for every P_2-solving extensible algorithm α with invariance selector $\iota, o(\alpha(\sigma(\cdot)))$ is a P_1-solving extensible algorithm with invariance selector $o(\iota(\sigma(\cdot), \sigma(\cdot\cdot)))$.

LEMMA 3. *If the following conditions are satisfied, then* P_2 *simulates extensibly* P_1.

(i) $S,T \in I_1 \wedge S \subseteq T \Rightarrow \sigma(S) \subseteq \sigma(T)$.

(ii) *There is a recursive, total, and strictly increasing function* ε *s.t.* $S \in I_1 \Rightarrow \ell h(\sigma(S)) = \varepsilon(\ell h(S))$.

(iii) $S,T \in I_1 \wedge k \leq \ell h(S) \wedge k \leq \ell h(T) \wedge S \underset{[0,k[}{=} T \Rightarrow \sigma(S) \underset{[0,\varepsilon(k)[}{=} \sigma(T)$.

(iv) $S \in I_1 \wedge Q_2(\sigma(S),r) \Rightarrow Q_1(S,o(r)) \wedge (r_1 \prec r_2 \prec r \Rightarrow o(r_1) \prec o(r_2))$.

(v) *For every canonically ascending* I_1 *-sequence* $(S_n)_{n \geq 0}$*:* $\forall n\, \exists k\, \exists T \in I_1: S_n \subseteq T \subseteq S_k \wedge Q_1(T, o(\iota(\sigma(S_{k+1}), \sigma(S_k))))$.

PROOF. We check the definitions of an extensible algorithm and its invariance selector. Observe that a canonically ascending I_1-sequence $(S_n)_{n\geq 0}$ is by virtue of (i)-(iii) mapped to a canonically ascending I_2-sequence $(\sigma(S_n))_{n\geq 0}$.

Now we consider the problem to find a flow of maximal value in a network. By a network we mean a directed graph with capacities at its lines (i.e. an assignment of natural numbers to the lines), a source and a sink. A flow in a network is an assignment of currents to the lines s.t. the current of a line is less or equal to its capacity, and for every point which is not the source or the sink, the sum of currents going in equals the sum of currents going out, i.e. Kirchhoffs rule is satisfied. The value of a flow is the sum of currents coming out of the source (or equivalently going into the sink). That we are allowed to consider integer flows only is due to a theorem of Ford and Fulkerson [FF] saying that in a network with integer capacities there is an integer flow having maximal value.

To meet our requirements for a problem we have to change this setting a little bit. We give every line coming out of the source its own source and every line going into the sink its individual sink. We represent a network N having capacities $\leq c$ as follows: $N = \langle D(N); So, Si, L, C_1, \ldots, C_c \rangle$ where D(N) is the set of points in N, So is the set of sources, Si the set of sinks, L the set of directed lines, and C_i the set of lines having capacity i.

Now let I(c) be the class of finite networks having a flow, capacities $\leq$ c, and every source is connected with a sink. The sequence r is an answer to question Q for a network N in I(c) iff for all $i \in D(N)$: $i < lh(r) \wedge r_i \neq 0$, and $r_i - 1$ is a code of the set of points j s.t. there is a line directed from i to j together with a current of (i,j), and these currents yield a flow of maximal value in N. Let *Max-flow*(c) be the problem (I(c),Q).

THEOREM 2. *Max-flow(1) is not extensibly solvable.*

PROOF. Suppose *Max-flow*(1) is extensibly solvable. To get a contradiction it suffices to show that *Max-flow*(1) extensibly simulates *Matching* (2). First we define a function $\sigma: I_1 \to I_2$ where I_1 is the input set of *Matching* (2) and I_2 the input set of *Maxflow* (1). Let $G = \langle D(G); A, B, L(G) \rangle$ be an element of I_1. We obtain $\sigma(G) = \langle D; So, Si, L, C_1 \rangle$ as follows. We adjoin to every $a \in A$ a source so(a) and a directed line from so(a) to a, having capacity 1, and to every $b \in B$ a sink si(b) and a directed line from b to si(b) having capacity 1. Finally we direct the lines of G from A to B. More precisely we put: $D := 2(A \cup B) \cup (2(A \cup B)+1)$, $So := 2A+1$, $Si := 2B+1$, $(a,b) \in L$ iff $(a, b \in 2(A \cup B) \wedge \{a/2, b/2\} \in L(G)) \vee (b \in 2A \wedge a = b+1) \vee (a \in 2B \wedge b = a+1)$, $C_1 := L$.

Further, if G is coded with length n, we let $\sigma(G)$ be coded with length 2n.

Next we define $o: \mathbb{N}^{<\omega} \to \mathbb{N}^{<\omega}$ as follows: $\ell h(o(r)) := \lceil \ell h(r)/2 \rceil$ and for

$i < \ell h(o(r))$, $o(r)_i$ is 1+ the minimal j s.t. $r_{2i} \neq 0 \wedge (2j,1) \in r_{2i}-1$ if it exists, and o otherwise. Now we use lemma 3. Obviously (i)-(iii) of lemma 3 are satisfied. (iv) follows from the definition of o and the fact that the lines in $\sigma(G)$ corresponding to lines in G and having current 1 in a flow of maximal value yield a matching of G. For a canonically ascending I_1-sequence $(G_n)_{n\geq 0}$ we fix an n. By $(\iota 3)$ there is a k and a network N s.t. $\sigma(G_n) \subseteq N \subseteq \sigma(G_k)$ and $Q_2(N, \iota(\sigma(G_{k+1}), \sigma(G_k)))$. So we have for $G_n = \langle D(G_n); A_n, B_n, L_n \rangle$ the following. As we have already shown via (iv) of lemma 3, $o(\alpha(\sigma(G_{k+1})))$ yields a matching for G_{k+1}. By the nature of the input class of Matching (2), this induces a matching for $G_n \subseteq G_{k+1}$. Also by (iv) of lemma 3, we see that $o(\iota(\sigma(G_{k+1}), \sigma(G_k))) \prec o(\alpha(\sigma(G_{k+1})))$. So it remains only to see that enough information is preserved in $o(\iota(\sigma(G_{k+1}), \sigma(G_k)))$ to describe this matching for G_n. For this we observe that for $a \in A_n$ obviously $2a < \ell h(\iota(\sigma(G_{k+1}), \sigma(G_k))) \wedge (\iota(\sigma(G_{k+1}), \sigma(G_k)))_{2a} \neq 0$.

Finally we consider linear integer programming. A linear program is a schema of the following type: $Ax \lesssim b$, $x \geq 0$, $c^t x = \text{Max!}$, where A is a matrix, b and c are vectors, and $\lesssim$ means that $\leq$ and $=$ are allowed for the rows of the schema. For an integer program all entries in the matrix or vectors are integers, and one has to find a vector x of integers s.t. the equalities and inequalities of the program are satisfied and $c^t x$ is maximal. As is easily verified, we are allowed to restrict our attention to programs with equalities and inequalities having at most three variables, and where a single variable x_m has to be maximized. So we represent a finite linear integer program PR having coefficients of absolute value $\leq c$ as follows: $PR = \langle D(PR); ,\ldots, C_m^{\leq}, C_w^{=}, \ldots \rangle$, where $D(PR)$ is the set of indices of variables in the program (we may assume that the variables are indexed by natural numbers), w is a sequence of length ≤ 4 of natural numbers a s.t. $-c \leq a \leq c$, and $C_w^{\lesssim}$ codes restrictions of the program, e.g. $(i,j,k) \in C^{\lesssim}_{a_0 a_1 a_2 a_3}$ iff $a_0 x_i + a_1 x_j + a_2 x_k \lesssim a_3$ is a restriction of the program. The variable with the greatest index is to be maximized.

Now let $I(c)$ be the class of finite linear integer programs having coefficients of absolute value $\leq c$. The sequence r is an answer to question Q for a program PR in $I(c)$ iff for all $i \in D(PR)$: $i < \ell h(r) \wedge r_i \neq 0$, and $r_i - 1$ is the value of x_i in a fixed solution of the program PR. Let *Linear integer programming*(c) be the problem $(I(c),Q)$.

To show that *Linear integer programming*(c) is not extensibly solvable we again extensibly simulate Matching (2).

THEOREM 3. *Linear integer programming*(1) *is not extensibly solvable.*

PROOF. Suppose *Linear integer programming*(1) is extensibly solvable. To get a contradiction it suffices to show that *Linear integer programming*(1) extensibly simulates *Matching*(2). First we define a function $\sigma : I_1 \to I_2$, where I_1 is the input set of *Matching*(2), and I_2 is the input set of *Linear integer programming*(1). Let $G = \langle D(G);A,B,L\rangle$ be an element of I_1. We obtain $\sigma(G)$ as follows.

The variables of the program correspond to the lines in G. To meet the formal requirements, we use an elementary recursive coding of pairs satisfying $\langle k,k\rangle = k^2$, and $\forall i,j$: $(\langle i,j\rangle < k^2 \Leftrightarrow i < k \wedge j < k)$. Now we number each line $\{a,b\}$ of G with $a \in A$ and $b \in B$ by $\langle a,b\rangle$, and let $D(\sigma(G)) := \{\langle a,b\rangle \mid a \in A \wedge b \in B \wedge \{a,b\}$ is a line of $G\}$. If G is coded with length k, we let $\sigma(G)$ be coded with length k^2.

The relations of $\sigma(G)$ are given by the following program:

$x_{\langle a,b_1\rangle} + x_{\langle a,b_2\rangle} = 1$ for $a \in A$ and $b_1 < b_2$ adjacent to a in G,

$x_{\langle a_1,b\rangle} + x_{\langle a_2,b\rangle} \leq 1$ for $b \in B$ of degree 2 and $a_1 < a_2$ adjacent to b in G.

Since the variables $x_{\langle a,b\rangle}$ correspond to the lines $\{a,b\}$ of G, it is obvious that a solution of the program $\sigma(G)$ yields a matching for G. Conditions (i)-(iii) of lemma 3 follow easily.

Now we define $o(r)$ for a finite sequence r as follows. $\ell h(o(r)) := \lceil\sqrt{\ell h(r)}\rceil$, and for $a < \ell h(o(r))$: $o(r)_a := 1+$ the least b s.t. $\langle a,b\rangle < \ell h(r) \wedge r_{\langle a,b\rangle} = 2$, if such a b exists, and $o(r)_a := 0$ otherwise. We leave it to the reader to verify condition (iv) of lemma 3, and also condition (v) following the lines of the proof of theorem 2.

REFERENCES

[B] Bean, D.R., *Recursive Euler and Hamilton paths*, Proceedings of the American Mathematical Society, 55 (1976), 385-394.

[C] Carstens, H.-G., *Recursive colorings on surfaces*, to appear.

[CP 1] Carstens, H.-G., Päppinghaus, P., *Recursive coloration of countable graphs*, Annals of Pure and Applied Logic 25 (1983), 19-45.

[CP 2] Carstens, H.-G., Päppinghaus, P., *Abstract construction of counterexamples in recursive graph theory*, to appear.

[FF] Ford, L.R., Fulkerson, D.R., *Flows in networks*, Princeton University Press, N.J., 1962.

[H] Harary, F., *Graph Theory*, Reading (Mass.), 1969.

[K] Kierstead, H.A., *Recursive colorings of highly recursive graphs*, Canadian Journal of Mathematics, 33 (1981), 1279-1290.

[MR 1] Manaster, A.B., Rosenstein, J.G., *Effective match making (Recursion theoretic aspects of a theorem of Phillip Hall)*, Proceedings of the London Mathematical Society, 25 (1972), 615-645.

[MR 2] Manaster, A.B., Rosenstein, J.G., *Effective match making and k-chromatic graphs*, Proceedings of the American Mathematical Society, 39 (1973), 371-379.

[Sch 1] Schmerl, J.H., *Recursive colorings of graphs*, Canadian Journal of Mathematics, 32 (1980), 821-830.

[Sch 2] Schmerl, J.H., *The effective version of Brook's theorem, Canadian Journal of Mathematics, 34 (1982), 1036-1046.*

[Sh] Shoenfield, J.R., *Mathematical Logic*, Reading (Mass.), 1967.

SOME REORDERING PROPERTIES

FOR INEQUALITY PROOF TREES

Michael M. Richter
Technische Hochschule Aachen

1. Preliminaries

In automatic theorem proving the restriction to a certain class of structures has the advantage that additional knowledge is available. Such knowledge can be added in the form of new axioms or in the form of new rules; it seems that the latter is more promising. An example has been provided by W. W. Bledsoe for the theory of total dense linear order without end points and arbitrary functions (cf. [Bl - Hi]).
In order to describe it shortly we recall some definitions and notations.
Variables are denoted by x,y,z,..; a,b,c,... stand for terms and C,D,... for clauses. Literals are of the form $a \leq b$ or $b < a$ where $b < a$ is $\neg(a \leq b)$; a and b are also called half-literals. Clauses will alternatively be regarded as disjunctions or sets. Substitutions are regarded as mappings and $\sigma(a)$ is the result of applying σ to a. Throughout the paper substitutions which sepatate variables are neglected; so we assume all clauses have disjoint sets of variables. A variable x which occurs in a term of the form $f(t_1,\ldots,t_n)$ is called shielded in that term. A variable which occurs in a clause C but does not occur shielded in any term in C is called eligible.
The following rules of derivation are used by W. W. Bledsoe:

(I) Variable elimination VE:

This is a one premis rule. Assume $\leq_i$ is either $\leq$ or $<$ and x is eligible in C but does not occur in Q.

a) From $C = \bigvee_{i=1}^{n} a_i \leq_i x \vee x \leq \bigvee_{j=1}^{m} x \leq_i b_j \vee Q$

one can deduce

$$D = \bigvee_{i,j} a_i \leq_{ij} b_j \vee Q$$

where $\leq_{ij}$ is $\leq$ iff both $\leq_i$ and $\leq_j$ are $\leq$.

b) From $C = \bigvee_{i=1}^{n} a_i \leq_i x \vee Q$

and from $C = \bigvee_{j=1}^{m} x \leq_j b_j \vee Q$

one can deduce Q.

(II) Chaining CH:

We have a two premis and a one premis rule.

a) From $C = a \leq_1 b \vee P$, $D = b' \leq_2 c \vee Q$, σ a most general unifier (m. g. u.) of b and b' one can infer $R = \sigma(a \leq_3 c \vee P \vee Q)$ where $\leq_3$ is $\leq$ iff both $\leq_1$ and $\leq_2$ are $\leq$. This form of CH is called chaining.

b) From $C = a < a' \vee P$, σ a m. g. u. of a and a' one can infer $\sigma(P)$. This form of the rule is called self-chaining.

(III) The usual rule of factoring Fac.

Using VE, CH and Fac various calculi have been defined, cf. [Bl - Ne - Sh] or [Ku]. To indicate derivations in such a calculus we use the symbol "$\vdash$" with or without subscripts; proofs are thought to be written in tree form with clauses attached to the nodes. The completeness of a calculus means that $S \vdash \square$ iff the set S of clauses is unsatisfiable in a totally dense ordered structure without end points. For completeness proofs cf. [Bl - Ne - Sh], [Bl - Ku - Sh].

The difference between these calculi consists in the conditions for the application of the rules. Such conditions are of two different types:

1) Conditions which restrict the application of a rule in certain situations;

2) Conditions which enforce the application of a rule in certain situations.

Both types of conditions are intended to improve the efficiency of the theorem proving system. The intention is not to watch out for a short proof but to keep the search space as small as possible. One such calculus has been implemented quite successfully by W. W. Bledsoe and L. Hines in Austin (Texas).

We consider here the following calculus $\mathscr{K}$ defined by the restrictions:

1) In (IIa) the rule CH is restricted to the case where neither b nor b' is a variable.
2) VE is mandatory whenever possible.

This is not the most restrictive calculus but it contains all the essentials. The usual techniques do not lead to a completeness proof; such a proof is the main purpose of this paper. As a by-product we get in our opinion a better understanding of the different effects when two non-commuting rules are applied in a different order. The technique we apply is to compare the calculus in question (which is supposed to be efficient) with another calculus where to completeness is known. The comparison is given by a syntactic transformation using a rule which is not sound. This means intuitively that the first calculus is so much better that one has to add an inconsistent rule to the second calculus in order to get the same intermediate results.

This research was done on the suggestion of Woody Bledsoe. The author wants to thank him for many fruitful discussion; the author is also indebted to Woody Bledsoe and K. Kunen for pointing out several errors in earlier versions.

2. Some commutative and non-commutative situations

In most of the rest of the discussion we will no longer distinguish between $\leq$ and $<$ and will carry out the formulations only for $\leq$. Because we often need to chain several times on "the same literal" we introduce a new rule.

<u>1. Def.:</u> Suppose τ = m. g. u. (a,d). The rule MCH of multiple chaining is defined by

(i) From $\bigvee_i b_i \leq a \vee P$ and $d \leq c \vee Q$

one can infer

$$\tau(\bigvee_i b_i \leq c \vee P \vee Q).$$

(ii) From $\bigvee_i a \leq b_i \vee P$ and $c \leq d \vee Q$

one can infer

$$\tau(\bigvee c \leq b_i \vee P \vee Q).$$

It is clear that MCH can up to a renaming of variables be simulated by CH and Fac.

A first observation is a commutativity property.

2. Proposition: Suppose

(1) x is eligible in C;

(2) R is obtained from C and D by CH;

(3) C' is obtained from C by VE of x;

(4) R' is obtained from R by VE of x.

Then one can infer up to a renaming of variables R' by MCH from C' and D.

Proof: We first observe that (4) is meaningful because an eligible variable remains eligible after an application of CH; chaining is not allowed on the variable itself and because it does not occur shielded the substitution in the chaining has no effect on it. For the rest we consider only one (typical) case:

$$C = \bigvee_i a_i \leq x \vee x \leq b_j \vee x \leq b \vee P$$

$$D = b' \leq c \vee Q$$

$$R = \sigma(\bigvee_i a_i \leq x \vee \bigvee_j x \leq b_j \vee x \leq c \vee P \vee Q)$$

$$R' = \sigma(\bigvee_{i,j} a_i \leq b_j \vee \bigvee_i a_i \leq c \vee P \vee Q)$$

$$C' = \bigvee_{i,j} a_i \leq b_j \vee \bigvee_i a_i \leq b \vee P)$$

The result follows immediately.

Next we consider two consecutive steps of VE.

3. Proposition: Suppose x and y are both eligible in C. Then VE on x commuts with VE on y.

The proof is done simply by inspection.

These are, however, not the only cases of two consecutive VE-steps.

Example 1: $C = x \leq f(y) \vee a \leq y$

Here y is shielded but becomes eligible after the elimination of x. This leads to the next definition.

4. Def.: A variable y in C is conditionally eligilbe with respect to x iff y is shielded in C and x is eligible in C and VE of x makes y eligible.

The next example is the first of a series of non-commuting cases; some of these examples are due to K. Kunen.

Example 2: $C = x \leq f(y) \vee a \leq y \vee y \leq b$

$D = f(g(z)) \leq d.$

By VE we get from C

$\hat{C} = a \leq b;$

CH on C and D yields

$$R = x \leq d \vee a \leq g(z) \vee g(z) \leq b,$$

which gives

$$\hat{R} = a \leq g(z) \vee g(z) \leq b \qquad \text{by VE.}$$

There is, however, no way to obtain from $\hat{R}$ a clause which is subsumed by $\hat{C}$; in fact, the term g(z) cannot be removed. In order to get control over such a non-commutative situation we introduce a very general rule.

<u>5. Def.:</u> (i) The rule TE ("term elimination") allows to deduce from $\bigvee_i a_i \leq b_i \vee \bigvee_j c_j \leq d_j \vee Q$ the clause $\bigvee_{i,j} a_i \leq d_j \vee Q$,

i. e. arbitrary terms can be eliminated without any restriction (e. g. they can occur in Q); the clause of all term eliminants of C is denoted by TE(C).

(ii) If σ is a substitution then $C \leq_{q(\sigma)} D$ means $\sigma(C) \subseteq D \cup TE(D)$.

(iii) $C \subseteq_q D$ ("C quasi subsumes D") iff $C \subseteq_{q(\sigma)} D$ for some σ.

<u>Remark:</u> The rule TE is not sound; its purpose is not to be used in proofs but to compare proofs. In one important exception it is sound, however: If $C \vdash_{TE} \square$ then $C = \square$. Also, TE is strictly a two premis rule, the one premis analogue of VE is not defined (see also the remark at the end of the paper). First we derive some useful closure properties of TE. C, D and E are arbitrary clauses and σ, τ, λ arbitrary substitutions.

<u>6. Proposition:</u>

(i) $C \cup TE(C)$ is closed unter TE.

(ii) If $C \subseteq_{q(\sigma)} D$ and $D \subseteq_{q(\tau)} E$ then $C \subseteq_{q(\tau \bullet \sigma)} E$.

(iii) Quasisubsumption is transitive.

(iv) $\lambda(C \cup TE(\sigma)) = \lambda(C) \cup TE(\lambda(C))$.

<u>Proof:</u>

(i) Assume $C = \bigvee_i a_i \leq b_i \vee \bigvee_j c_j \leq d_j \vee \bigvee_k a'_k \leq b'_k \vee \bigvee_l c'_l \leq d'_l \vee Q$;

the most general case is to consider the term eliminants

$\bigvee_{i,j} a_i \leq d_j$ and $\bigvee_{k,l} a'_k \leq d'_l$

and from this the term eliminant

$\bigvee_{i,l} a_i \leq d'_l$;

this can be obtained directly from C.

(ii) If $\sigma(C) \subseteq D \cup TE(D)$

and

$\tau(D) \subseteq E \cup TE(D)$,

then

$$\begin{aligned}\tau\circ\sigma(C) &\subseteq \tau(D) \cup \tau(TE(D))\\ &\subseteq E \cup TE(E) \cup TE(\tau(D))\\ &\subseteq E \cup TE(E) \cup TE(TE(E)) = E \cup TE(E).\end{aligned}$$

(iii) is immediate from (ii).

(iv) The easy inclusion is $TE(\lambda(C)) \subseteq \lambda(TE(C))$.

For the other direction we assume

$$C = \bigvee_i a_i \leq b_i \vee \bigvee_j c_j \leq d_j \vee Q$$

and have to show that $\lambda(\bigvee_{i,j} a_i \leq d_j)$ is in $\lambda(C) \cup TE(\lambda(C))$.

Omitting the trivial case, λ does not unify any literals from C, we proceed inductively and can assume that λ unifies exactly two literals $a \leq b$ and $a' \leq b'$.

The only case which provides some difficulties is that neither $a \leq b$ nor $a' \leq b'$ is in Q and

$a \leq b$ is some $a_{i_o} \leq b_{i_o}$

and $a' \leq b'$ is some $c_{j_1} \leq d_{j_1}$.

We split up:

$$\bigvee_{i,j} a_i \leq d_j = \bigvee_i \bigvee_{j \neq j_1} a_i \leq d_j \vee \bigvee_{i \neq i_o} a_i \leq d_{j_1} \vee a_{i_o} \leq d_{j_1}$$

$$= C_1 \vee C_2 \vee C_3.$$

From

$$\begin{aligned}\lambda(C) &= \lambda(\bigvee_i a_i \leq b_i \vee \bigvee_{j \neq j_1} \vee c_j \leq d_j \vee Q)\\ &= \lambda(\bigvee_{i \neq i_o} a_i \leq b_i \vee \bigvee_j c_j \leq d_j \vee Q)\end{aligned}$$

we get

$\lambda(C_1) \subseteq TE(\lambda(C))$, $\lambda(C_2) \subseteq TE(\lambda(C))$

and $\lambda(a_{i_o} \subseteq d_{j_1}) = \lambda(a_{i_o} \subseteq b_{i_o}) \in \lambda(C)$,

as required.

Next we come to a first use of quasisubsumption.

7. Proposition: Assume:

(1) x is eligible in C and y is conditionally eligible with respect to x;

(2) VE of x and y in C gives $\hat{C}$;

(3) CH on C and some clause D gives R;

(4) VE of x in R gives $\hat{R}$.

Then we have $\hat{C} \subseteq_q \hat{R}$.

<u>Proof:</u> Recall that we neglect separation of variables. Using the same substitution for $\subseteq_q$ as in the previous application of CH the result follows by direct comparison of $\hat{C}$ and $\hat{R}$.

In the light of example 2 the definiton of TE and $\subseteq_q$ seems to be unnecessarily liberal. Additional examples will show that this is not the case; in fact we will have to refine our definition.

<u>Example 3:</u> We continue example 2 and apply CH to $\hat{R}$ and $E = g(z) \leq c$ to obtain

$$\tilde{R} = a \leq c \vee g(z) \leq b.$$

This shows that elimination of just one term is insufficient.

<u>Example 4:</u> Put $C = a_1 \leq x \vee a_2 \leq x \vee x \leq b_1 \vee x \leq b_2 \vee t \leq b_2$.

VE gives

$$C_1 = \bigvee_{i=1,2} \bigvee_{j=1,2} a_i \leq b_j \vee t \leq b_2$$

while factoring with the unification of x and t provides us

$$C_2 = \bigvee_{i=1,2} a_i \leq t \vee \bigvee_{j=1,2} t \leq b_j.$$

This does not only show that VE and Fac in general do not commute. It shows also that we need besides TE on C_2 the clause C_2 itself in order to get a clause which is subsumed by C_1.

<u>Example 5:</u> Continuing example 3 we consider the following derivation:

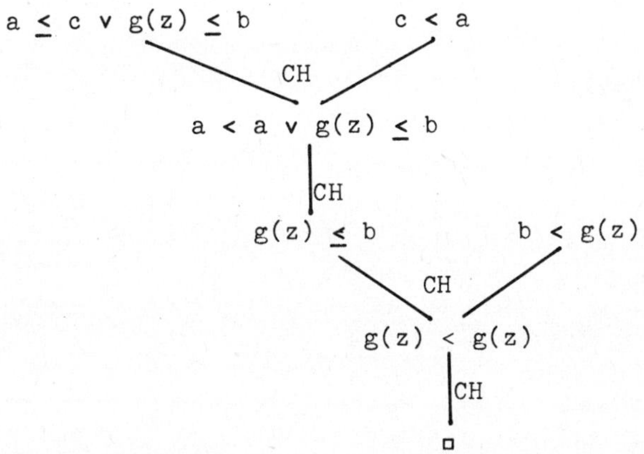

In order to obtain □ from $a \leq b$ we need the clauses $g(z) \leq c$, $c < a$, $b < g(z)$ which were used in the derivation of □. Therefore we have to introduce a bookkeeping procedure in order to have access to previously used clauses.

Example 6: Postponing VE may have the effect of applying a substitution. Take $C = x \le f(y) \vee u \le g(y) \vee a \le y \vee y \le b$.
Chaining with $f(c) \le d$ yields

$$D = x \le d \vee u \le g(c) \vee a \le c \vee c \le b$$

and one obtains by VE on x

$$E = u \le g(c) \vee a \le c \vee c \le b$$

while immediate VE in C gives

$$\hat{C} = u \le g(y) \vee a \le y \vee y \le b;$$

E is a substitution instance of $\hat{C}$. Further chaining on $a \le c$ in E with $c \le d$ could of course be imitated in $\hat{C}$ but only on the cost of chaining on variables.

3. Non-commutative situations

This leads to our final subsumption property. Let $C, C_1, \ldots, C_n, D$ be clauses.

8. Def.: $\neg D$ is the set of unit clauses containing the negations of literals in D.

9. Def.: $(C \mid C_1, \ldots, C_n) \subseteq_q D$
iff
there is some substitution σ s. t.
(i) $C \subseteq_{q(\sigma)} D$;
(ii) for each $D' \subseteq D$ there is some C' s. t.
α) $C' \subseteq_q D \setminus D'$;
β) there is some -derivation
$C, C_1, \ldots, C_n, \neg D' \vdash C'$ such that in this derivation:
1) For all occurring substitutions σ' there is some μ with $\sigma = \mu \circ \sigma'$
2) self-chaining is restricted to the descendants of C and the unit clauses in $\neg D'$, it occurs at most $n = |D'|$ times.

The next task is to establish that this general quasi-subsumption is in some sense compatible with the rules of derivation.

10. Proposition: If $(C \mid C_1, \ldots, C_n) \subseteq_q D$ and $C \vdash_{VE} E$
then $(E \mid C_1, \ldots, C_n) \subseteq_q D$ is true.

Proof: We restrict ourselves here and in the following propositions to the case that E is obtained from C in a one-step derivation.
Suppose $\sigma(C) \subseteq D \cup TE(D)$ and assume furthermore

$$C = \bigvee_i a_i \le x \vee \bigvee_j x \le b_j \vee Q,$$

$$E = \bigvee_{i,j} a_i \le b_j \vee Q.$$

(i) Because D ∪ TE(D) is closed under TE we have

$$E \subseteq_{q(\sigma)} D.$$

(ii) Assume $D' \subseteq D$; we have for some $C' \subseteq_q D \setminus D'$ a derivation

$$C, C_1, \ldots, C_n, \neg D' \vdash C'$$

which satisfies the conditions in (ii)β) of definition 9.

This derivation will be modified in the following way:

a) C is replaced by E.

b) CH-steps with $a_i \leq x$ (res. $x \leq b_j$) are replaced by multiple chainings with the $a_i \leq b_j$.

If E' is the result of this derivation then E' differs from C' only in so far as literals of the form $\bar{a}_i \leq x$ and $x \leq \bar{b}_j$ in C' are replaced by literals of the form $\bar{a}_i \leq \bar{b}_j$. Then we have

$$C' \vdash_{VE} E'$$

in a derivation of at most one step. By (i) we get $E' \subseteq_q D \setminus D'$. Furthermore all the required properties in the derivation of E' remain true.

<u>11. Proposition:</u> If $(C \mid C_1, \ldots, C_n) \subseteq_q D$ and $D \vdash_{VE} E$ then $(\hat{C} \mid C_1, \ldots, C_n) \subseteq_q E$ for some $\hat{C}$ s. t. $C \vdash_{VE} \hat{C}$.

<u>Proof:</u> Suppose $\sigma(C) \subseteq D \cup TE(D)$

and assume $D = \bigvee_i a_i \leq x \vee \bigvee_j x \subseteq b_j \vee Q$,

$$E = \bigvee_{i,j} a_i \leq b_j \vee Q.$$

(i) The only members of TE(D) which are not automatically in TE(E) are of the form

$u \leq x$ (from $a_i \leq x$ and some $u \leq v$ in Q)

and $x \leq u$ (from $x \leq b_j$ and some $u \leq v$ in Q).

If x does not occur in the image under σ we put $\hat{C} = C$; otherwise we take $\hat{C}$ as the result of applying VE to $y := \sigma^{-1}(x)$.

(ii) We take some $E' \subseteq E$ and define $D' \subseteq D$ by replacing each $a_i \leq b_j$ by $a_i \leq x \vee x \leq b_j$. By assumption there is some $C' \subseteq_q D \setminus D'$ s. t.

$$C, C_1, \ldots, C_n, \neg D' \vdash C'.$$

This derivation will be modified in the following way:

a) C is replaced by $\hat{C}$.

b) Chaining on a unit clause $(x < a_i)$ for $a_i \leq x$ in D_1 (yielding some $x < v$) is replaced by multiple chaining on all $b_j < a_i$ (yielding $b_j < v$); similar for the $b_j < x$.

c) Chaining on $\sigma^{-1}(a_i \leq x)$ in C is replaced with the corresponding multiple chaining on $\sigma^{-1}(a_i \leq b_j)$ for all j; similar for $\sigma^{-1}(x \leq b_j)$.

As in the previous proposition we see that for the result C'' in this derivation $C'' \subseteq_q E\ E'$ holds.

All required properties of the new derivation remain true.

12. Proposition: If $(C \mid C_1,\dots,C_n) \subseteq_q D$ and $D \vdash_{Fac} E$ then $(\hat{C} \mid C_1,\dots,C_n) \subseteq_q E$ for some $\hat{C}$ s. t. $C \vdash_{Fac} \hat{C}$.

Proof: Suppose $\sigma(C) \subseteq D \cup TE(D)$ and assume

$$E = \lambda(D).$$

(i) Observing that $\lambda\circ\sigma$ is a unifier we find a m. g. u. ξ and some substitution τ such that the following diagram commutes:

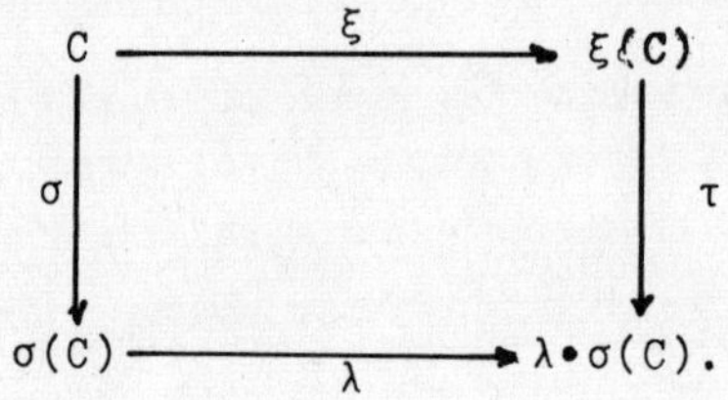

We have $\lambda\circ\sigma(C) \subseteq \lambda(D \cup TE(D))$

$$= TE(\lambda(D)) \cup \lambda(D) = TE(E) \cup E$$

by proposition 6.

Hence we can take $\hat{C} = \xi(C)$ for which

$$\hat{C} \subseteq_{q(\tau)} E \text{ holds.}$$

(ii) We take some $E' \subseteq E$ and define $D' = \lambda^{-1}(E')$; there is $C' \subseteq_q D \setminus D'$ for which we have a derivation

$$C, C_1,\dots,C_n, \neg D' \vdash C'.$$

If in this derivation a substitution σ' is applied, then by assumption $\sigma = \sigma''\circ\sigma'$ for some σ''. Therefore we can project the given derivation to a new one using $\hat{C} = \xi(C)$ instead of C which again fulfills the required conditions. The result is a clause C'' s. t. for some τ' we have

$$\tau'(C'') \subseteq \lambda((D \setminus D') \cup TE(D \setminus D'))$$
$$\subseteq (E \setminus E') \cup TE(E \setminus E')$$

which shows $C'' \subseteq_q E \setminus E'$.

13. Proposition: Suppose

(1) $(C^1 \mid C^1_1,\dots,C^1_{n_1}) \subseteq_q D^1$,

(2) $(C^2 \mid C^2_1,\dots,C^2_{n_2}) \subseteq_q D^2$

(3) R is a chain resolvent of D^1 and D^2.

Then at least one of the following is true:

(a) There is a multiple chain resolvent C of C^1 and C^2 s. t.

$$(C \mid C_1^1,\dots,C_{n_1}^1,C_1^2,\dots,C_{n_2}^2) \subseteq_q R$$

or (b) $(C^1 \mid C_1^1,\dots,C_{n_1}^1,C_1^2,\dots,C_{n_2}^2,C^2) \subseteq_q R$

or (c) $(C^2 \mid C_1^1,\dots,C_{n_1}^1,C_1^2,\dots,C_{n_2}^2,C^1) \subseteq_q R$.

Proof: We assume $\sigma(C^1) \subseteq D^1 \cup TE(D^1)$,

$$\tau(C^2) \subseteq D^2 \cup TE(D^2),$$

$$D^1 = \bigvee_i a_i \le b_i \vee \bigvee_j c_j \le d_j \vee P$$

$$D^2 = \bigvee_k e_k \le f_k \vee \bigvee_l g_l \le h_l \vee Q.$$

Let furthermore λ be the unifier used in the application of CH. We recall again that D_1 and D_2 have no variables in common.

(i) We have to consider the different possibilities for the chaining and the term elimination. Assume

$$R = \mu(a_{i_o} \le h_{l_o} \vee \bigvee_{i \neq i_o} a_i \le b_i \vee \bigvee_j c_j \le d_j \vee \bigvee_k e_k \le f_k$$

$$\vee \bigvee_{l \neq l_o} g_l \le h_l \vee P \vee Q)$$

and $\mu(b_{i_o}) = \mu(g_{l_o})$.

Typical literals from $\mu(D^1)$ and $\mu(D^2)$ which are not automatically in TE(R) are of the form

$$c_j \le b_{i_o} \quad \text{resp.} \quad g_{l_o} \le f_k.$$

If $c_j \le b_{i_o}$ is not in $\sigma(C^1)$, then $C_1 \subseteq_{q(\mu\circ\sigma)} R$;

if $g_{l_o} \le f_k$ is not in $\tau(C^2)$, then $C_2 \subseteq_{q(\mu\circ\tau)} R$.

In the remaining cases the pre-images b_{i_o} and g_{l_o} under σ resp. τ are unifiable, hence we get by multiple chaining literals which are up to substitution of the form $c_j \le f_k$, i. e. term eliminants of R. We observe that we might be forced here to chain on variables (cf. example 6).

(ii) Let $R' \subseteq R$; we split $R' = R_1' \cup R_2' \cup S$ where the R_i' are inherited from the D^i resp. and S is the unit clause containing the chain resolvent. For the R_i' we can use the hypothesis of the proposition; the hypothesis also to the removal of S if one replaces the original literals by the chain resolvents.

<u>14. Proposition:</u> Suppose

1) $(C \mid C_1,\ldots,C_n) \subseteq_q D$

2) $\hat{D}$ is obtained from D by self-chaining.

Then there is some $\hat{C}$ s. t.

a) $(\hat{C} \mid C_1,\ldots,C_n) \subseteq_q \hat{D}$

b) $\hat{C}$ is obtained from $C, C_1,\ldots,C_n$ in a derivation in which no shielded variable becomes eligible except possibly after the last step.

<u>Proof:</u> Assume $\sigma(a) = \sigma(b)$ and $\hat{D} = \sigma(D) \setminus (\sigma(a) < \sigma(b))$.
By assumption we can derive some $\hat{C} \subseteq_q D \setminus D'$, where $D' = \sigma^{-1}(\sigma(a) < \sigma(b))$, from $C, C_1,\ldots,C_n$; here no VE took place and w. l. o. g. at most one step of self-chaining. This step can be done at the end, there no shielded variable becomes eligible at an earlierstep. It is immediate that $(\hat{C} \mid C_1,\ldots,C_n) \subseteq_q \hat{D}$ holds.

<u>15. Proposition:</u> Suppose S is a set of clauses. The for each proof $\mathcal{D}$ of $\square$ from S we can construct a proof $\mathcal{B}$ (not necessarily in $\mathcal{K}$) of $\square$ from S in which VE is mandatory.

<u>Proof:</u> We will first construct by top-down induction on the nodes of $\mathcal{D}$ a certain tree $\hat{\mathcal{D}}$. At each node of $\hat{\mathcal{D}}$ an expression of the form $(C_o \mid C_1,\ldots,C_n)$ will be attached such that the following conditions are satisfied:

a) At top nodes we have expressions $(C \mid \emptyset)$ with $C \in S$,

b) In the one parent case

$$\begin{array}{c} (C_o \mid C_1,\ldots,C_n) \\ | \\ (C'_o \mid C'_1,\ldots,C'_n) \end{array}$$

we have either

(1) $C'_i = C_i$, $0 \le i \le n$;

or

(2) $C_i = C'_i$, $1 \le i \le n$, and C'_o is obtained in a derivation from $C_o,\ldots,C_n$ in which VE is not possible;

or

(3) $C_i = C'_i$, $1 \le i \le n$, and $C_o \vdash_{VE} C'_o$.

c) In the two parent case

$$\begin{array}{ccc} (C_o \mid C_1,\ldots,C_n) & & (D_o \mid D_1,\ldots,D_m) \\ & \searrow \quad \swarrow & \\ & (E_o \mid E_1,\ldots,E_k) & \end{array}$$

we have either

(1) $(E_o \mid E_1,\ldots,E_k) = (C_o \mid C_1,\ldots,C_n,D_1,\ldots,D_m,D_o)$

or

(2) $(E_o \mid E_1,\ldots,E_k) = (D_o \mid D_1,\ldots,D_m,C_1,\ldots,C_n,C_o)$

or

(3) $(E_1,\dots,E_k) = (C_1,\dots,C_n,D_1,\dots,D_m)$
and E_o is obtained from C_o and D_o by MCH.

d) For each clause D occurring in $\mathcal{D}$ there is an expression
$$(C \mid C_1,\dots,C_n) \subseteq_q D$$
occurring in $\hat{\mathcal{D}}$.

The inductive definiton of $\hat{\mathcal{D}}$ is done according to the derivation steps in $\mathcal{D}$ using the propositions 11-13.
There is one exception: If a VE step on C in $(C \mid C_1,\dots,C_n)$ is possible then it is inserted; here we apply proposition 10.
We consider the bottom clause $\square$ in $\mathcal{D}$:
There is some $(C \mid C_1,\dots,C_n) \subseteq_q \square$ in $\hat{\mathcal{D}}$, in particular $C \subseteq_q \square$ which implies $C = \square$. The first components of the expressions in $\hat{\mathcal{D}}$ then define the proof $\mathcal{B}$ of $\square$ from S in which VE is mandatory.

The proof obtained by the last proposition may contain applications of CH to variables which we have to remove.

<u>16. Proposition:</u> Suppose S is a set of clauses and $\mathcal{D}$ a proof of $\square$ from S in which VE is mandatory. Then we can construct a proof $\mathcal{B}$ of $\square$ from S in the calculus $\mathcal{K}$.

<u>Proof:</u> Suppose we have applied CH to a variable at a certain node in the proof $\mathcal{D}$. W. l. o. g. we have the following situation:

$C = a \leq x \vee Q \qquad D = b \leq c \vee P$

$\sigma_o = \text{m. g. u. } \sigma(x,b),\ \sigma_o(x) = b,$

$R = \sigma(a \leq b \vee Q \vee P).$

Because VE was mandatory Q must contain half-literals $t_1,\dots,t_n$ in which x occurs; if b = y was a variable too then $t_{n+1},\dots,t_m$ are the half-literals in which y occurs. We call $t_1,\dots,t_m$ the shielding list for x (resp. x and y).
We define the shielding list now for all nodes below the given one by top-down induction. It shall contain the descendants of the half-literals on the previous shielding list; if chaining takes place on a half-literal of the shielding renaming the variable x (resp. its descendant) as z then all half-literals in which z occurs are added.
We also have the sequence $\sigma_o,\sigma_1,\dots$ of the applied substitutions on the path down. Because we are in a proof of the empty clause $\square$ one of the following cases happens:

(1) There is a first node where $\sigma_n(\sigma_{n-1}(\dots(\sigma_o(x))\dots)$ is not a variable and the applied rule is either factoring or chaining on a half--literal of the shielding list.

(2) Such a node does not exist but there ist a first node where no half--literal from the shielding list is anymore in the clause (they can disappear by VE, self-chaining or chaining).

The given proof is now changed as follows. Suppose at the node in case 1) resp. 2) we have derived a clause E.

In case 1): We omit all applications of CH on the half-literal x and all steps depending on that step. After carrying out the remaining steps the chaining on the half-literal from the shielding list (resp. the factoring) is done (using the substitution $\sigma_n \circ \sigma_{n-1} \ldots \circ \sigma_o$). Then the omitted steps are inserted; this is possible because the involved substitutions are compatible and it is allowed because we no longer chain on a variable. The clause we obtain is again E, up to renaming of variables.

In case 2): We omit the same steps as in (1). The variable x is renamed but it becomes eligible; therefore we can remove it by VE. The result is a clause $\hat{E} \subseteq_q E$.

We have described how to avoid chaining on a specific variable x. Proceeding inductively on the number of variables which take place in chainings and using transitivity of quasi-subsumption we get a deduction $\mathscr{B}$ of a clause $F \subseteq_q \square$, i. e. we get $\square$.

Using that we have completeness when VE is optional (cf. [Bl - Ne - Sh] or [Bl - Ku - Sh]) and applying propositions 15 and 16 we get finally:

17. Theorem: If a set S of clauses is unsatisfiable in any totally ordered dense structure without endpoints then there is a proof of $\square$ from S in the calculus $\mathscr{K}$.

Finally we want to remark that the rule of term elimination has also a practical aspect. Suppose a set S of clauses is not inconsistent; then the derivation procedure might not terminate but it might come to an end using TE. If it stops that way we have two possibilities:

a) We have not derived $\square$; then we definitely cannot derive $\square$ without TE and have shown that S is satisfiable.

b) We have derived $\square$; in this case we are left with a subgoal, namely to show that the applications of TE in this case were correct (which may or may not be the case).

This means, TE could also be used as a heuristic.

References

[Bl - Hi]: W. W. Bledsoe - L. M. Hines: Variable elimination and chaining in a resolution - based prover for inequalities. Proc. 5th Conference on Automated Deduction. Springer LNCS 87, ed. W. Bibel - R. Kowalski, p. 70 - 87.

[Bl - Ku - Sh]: W. W. Bledsoe - K. Kunen - R. Shostak: Completeness Results for Inequality Provers. Manuscript 1982.

[Bl - Ne - Sh]: W. W. Bledsoe - R. Neveln - R. Shostak: Completeness Results for a Class of Inequality Provers, Univ. of Texas Math. Dept. Memo ATP 60, March 1981

[Ku]: K. Kunen: A Completeness Theorem for Multiple Chaining, Univ. of Texas Math. Dept. Memo ATP 61, Aug. 1980.

[Lo]: D. W. Loveland: Automated Theorem Proving: A Logical Basis. Amsterdam 1978.

MODULAR DECOMPOSITION OF AUTOMATA

Anne Brüggemann[+], Lutz Priese[*]

Dieter Rödding[+], Rüdiger Schätz[+]

Abstract

This paper deals with some aspects of Normed Networks (NN) and Asynchronous Parallel Automata nets (APA). A NN is a canonical composition of sequential machines and NN theory deals thus with the classical theory of sequential computation. An APA net is also a straightforward composition of concurrent automata and APA net theory thus handles asynchronous parallel computations. APA nets are very closely related to Vector Addition Systems but allow only local conflicts within the component machines while the network structure is a modular one and thus conflict-free. As a main aspect the paper asks for optimal realisations of sequential automata by NN and APA nets, respectively.

Fachbereich Mathematik-Informatik[*]
Warburger Straße 1oo

D-479o Paderborn

Institut für Mathematische Logik[+]
und Grundlagenforschung
Einsteinstraße 62

D-44oo Münster

0. Introduction

The intention of this paper is twofold: it firstly gives an overview on some research about a) sequential and b) concurrent automata and automata nets and secondly gives some estimations about 'size' and 'time' of smooth network constructions.

A sequential automaton (s.a.) is defined as usual in automata theory as a (possibly non-completely defined and deterministic) Mealy-automaton. Also we use the classical definitions of simulation (or realization) of such automata with the result that our research is about classical standard automata theory. As a major difference to classical approaches we encode input and output signals by a 1-out-of-n-code, where each input line of an automaton presents one letter of its input alphabet and at most one unit signal may pass an automaton via at most one input line at any time (analogously for outputs). This modification leads in a natural way to the conception of so-called Normed Networks (NN) over s.a., which are used to obtain normal forms for s.a.

This approach has proven quite fruitful and led to various results about decomposition and realization of automata in Normed Networks and to applications in different disciplines. As a s.a. is a quite general model of (sequential) finite computations we have been able to apply our results in logic and computation theory, systems theory with applications in theoretical biology and economics, fault-tolerant switching theory, etc. For an overview and a list of references see [Rö 83].

As s.a. are a smooth model for sequential computations, an attempt is made to introduce analogously 'concurrent automata' (c.a.) as a model for concurrent computations. Realization of c.a. leads to Asynchronous Parallel Automata nets (APA-nets) and Locally Sequential Nets (LSN). In this paper we use APA-nets and LSN to present a best possible decomposition result for s.a. with respect to size and time.

NN have been introduced by D. Rödding in the 60s and led to a research group at the Institut für Mathematische Logik und Grundlagenforschung at the University of Münster. APA-nets and LSN have been introduced by L. Priese in the 70s (or 80s, respectively) and led to a further group at the University of Paderborn. Both groups cooperate quite closely and often it is hard to tell whom to blame for some result.

1. Definitions

1.1. Normed Network Theory

Let's start with the basic definitions of Normed Network Theory.

Definition 1:

1. A sequential automaton (s.a.) is a tuple $A=(I_A, O_A, S_A, \delta_A, \lambda_A)$, consisting of finite, non-empty, mutually disjoint sets I_A, O_A, S_A and two partial functions $\delta_A: I_A \times S_A \to S_A$ and $\lambda_A: I_A \times S_A \to O_A$ with $dom(\delta_A)=dom(\lambda_A)$.
2. A s.a. A is totally defined iff $dom(\delta_A)=dom(\lambda_A)=I_A \times S_A$.
3. $\mathcal{A}$ is the class of all s.a.

Let $A=(I,O,S,\delta,\lambda)$ be a s.a. I is called the set of input lines, O the set of output lines, S the set of internal states, δ the transition function, and λ the output function. The input and output signals of A are encoded by a 1-out-of-n-code: In the internal state $s \in S$, A may receive an (unit) input signal on its input line $i \in I$, change its state to $\delta(i,s)$, and put an (unit) output signal on its output line $\lambda(i,s)$ (provided that $\delta(i,s)$ and $\lambda(i,s)$ are defined). A must be used sequentially only, i.e. the environment of A may give another input signal to A only after having removed the old output signal from the line $\lambda(i,s)$. Nevertheless no clock is involved: We don't know anything about processing time of A or the resting interval between an output signal and a new input signal. Finally s.a. which are isomorphic in the usual sense (i.e. identical besides the names of input lines, output lines, and states) are identified.

Next we give some examples of s.a. Here the transition function and the output function of a s.a. A are given by a list Q_A with the elements $i,s \to \lambda_A(i,s), \delta_A(i,s)$ $(i \in I_A,\ s \in S_A)$.

Definition 2:

$H = (\{t,u,d\},\ \{t^u, t^d, u', d'\},\ \{up, down\},\ Q_H)$,

$E = (\{t,c\},\ \{t^u, t^d, c'\},\ \{up, down\},\ Q_E)$,

$D = (\{t,c\},\ \{t^u, t^d\},\ \{up, down\},\ Q_D)$,

$T = (\{t,c\},\ \{t^u, t^d\},\ \{up, down\},\ Q_T)$,

$F = (\{c\},\ \{c^u, c^d\},\ \{up, down\},\ Q_F),$

$K = (\{i_1, i_2\},\ \{o\},\ \{0\},\ Q_K)$ with

Q_H: $t\ up \rightarrow t^u\ up$
$t\ down \rightarrow t^d\ down$
$u\ up \rightarrow u'\ up$
$u\ down \rightarrow u'\ up$
$d\ up \rightarrow d'\ down$
$d\ down \rightarrow d'\ down$

Q_E: $t\ up \rightarrow t^u\ up$
$t\ down \rightarrow t^d\ down$
$c\ up \rightarrow c'\ down$
$c\ down \rightarrow c'\ up$

Q_D: $t\ up \rightarrow t^u\ up$
$t\ down \rightarrow t^d\ down$
$c\ up \rightarrow t^d\ down$
$c\ down \rightarrow t^d\ up$

Q_T: $t\ up \rightarrow t^u\ down$
$t\ down \rightarrow t^d\ down$
$c\ up \rightarrow t^d\ down$
$c\ down \rightarrow t^d\ up$

Q_F: $c\ up \rightarrow c^u\ down$
$c\ down \rightarrow c^d\ up$

Q_K: $i_1\ 0 \rightarrow o\ 0$
$i_2\ 0 \rightarrow o\ 0$

The automata defined above are graphically represented in Figure 1:

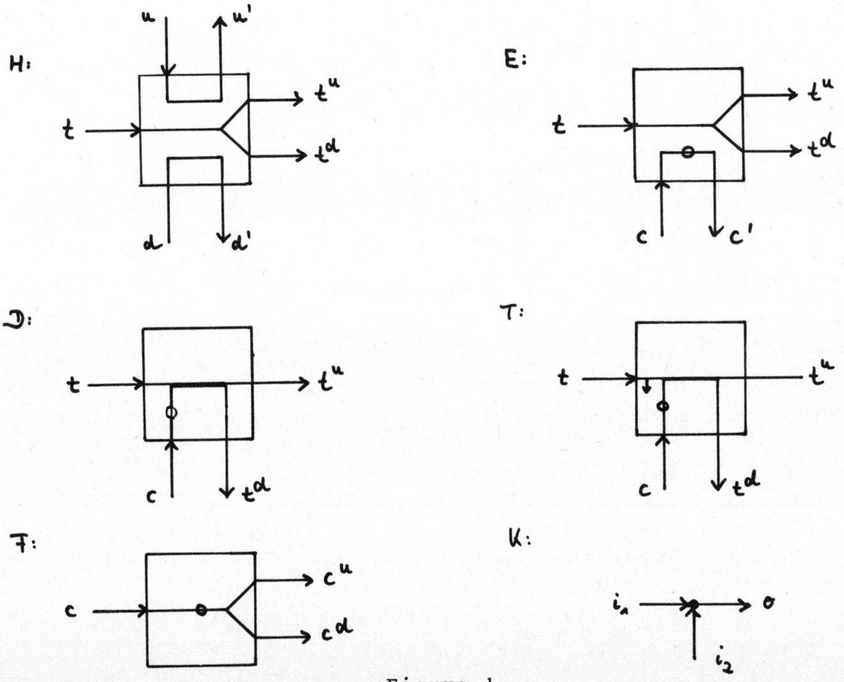

Figure 1

H is a s.a. with two states up and down. An input signal on line t works as a test: The state is not changed and the output signal gives information about the state. An input signal on u(d) results in the state up (down) and an output signal on u' (d').
E has a test device like H and a signal on c changes the state.
D works as E but two output lines are identified.
T works as D but a t-input always resets T to state down.
F is a flip-flop and K a union of wires.

Next we define Normed Networks over a set $\mathcal{B}$ of s.a. A Normed Network over $\mathcal{B}$ consists of some copies of elements of $\mathcal{B}$, the components of the network, and interconnections between them, where each interconnection leads from exactly one output line of one component to exactly one input line of the same or another component. Each input line or output line may be used by at most one interconnection. Such networks can be defined inductively by the two processes "parallel composition" and "feedback". This definition tells us also how to interpret such a network as a s.a. again.

<u>Definition 3:</u>

Let A,B be s.a., $I_A \cap I_B = O_A \cap O_B = \emptyset$ (without loss of generality, see above)
$A \| B := (I_A \cup I_B,\ O_A \cup O_B,\ S_A \times S_B, \delta, \lambda)$ is the s.a. with

$$\delta(i,(s,t)) \cong \begin{cases} (\delta_A(i,s),t) & \text{if } i \in I_A \\ (s,\delta_B(i,t)) & \text{if } i \in I_B, \end{cases}$$

$$\lambda(i,(s,t)) \cong \begin{cases} \lambda_A(i,s) & \text{if } i \in I_A \\ \lambda_B(i,t) & \text{if } i \in I_B. \end{cases}$$

$A \| B$ is called the <u>parallel composition</u> of A and B.

<u>Definition 4:</u>

Let A be a s.a. with $|I_A|, |O_A| > 1$, $i_o \in I_A, o_o \in O_A$.

$A_{i_o}^{o_o} := (I_A \setminus \{i_o\},\ O_A \setminus \{o_o\}, S_A, \delta, \lambda)$ is the (uniquely determined) s.a. with
$\delta(i,s) \tilde{=} s' \wedge \lambda(i,s) \tilde{=} o$

$$\Longleftrightarrow \exists n \in \mathbb{N}_+ : \exists s_o, \ldots, s_n \in S_A :$$
$$[s_o = s \wedge s_n = s'$$
$$\wedge\ \delta_A(i,s_o) \tilde{=} s_1 \wedge \forall 1 \leq \nu \leq n-1 : \delta_A(i_o,s_\nu) \tilde{=} s_{\nu+1} \wedge (n=1 \rightarrow \lambda_A(i,s_o) \tilde{=} o)$$
$$\wedge\ (n>1 \rightarrow (\lambda_A(i,s_o) \tilde{=} o_o \wedge \forall 1 \leq \nu < n-1 : \lambda_A(i_o,s_\nu) \tilde{=} o_o \wedge \lambda_A(i_o,s_{n-1}) \tilde{=} o))]$$

$A_{i_o}^{o_o}$ is called the <u>feedback</u> of A concerning i_o and o_o.

Parallel composition and feedback are graphically presented in figure 2.

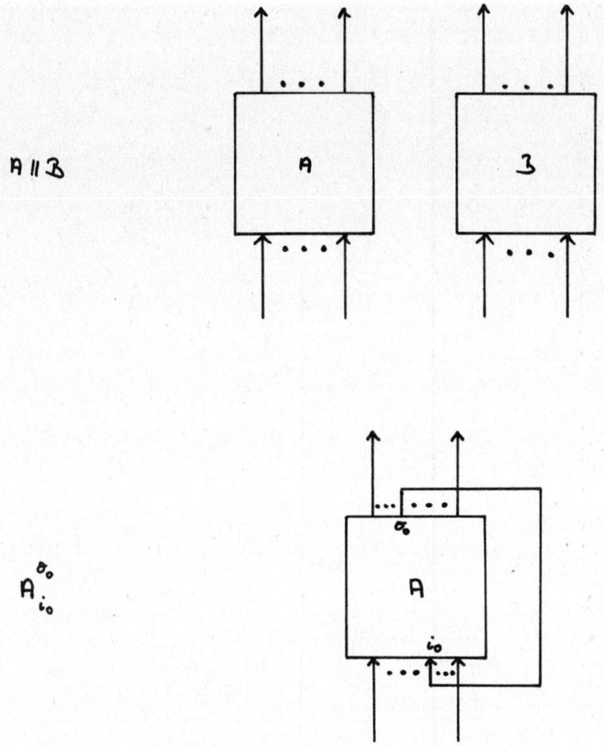

Figure 2

<u>Definition 5:</u>

Let $\mathcal{B}$ be a class of s.a.
The class NN($\mathcal{B}$) of Normed Networks over $\mathcal{B}$ is the smallest class of s.a. which contains $\mathcal{B}$ and is closed under parallel composition and feedback.

The term "normed" is due to the fact, that in a Normed Network only one signal runs during a computation process.

Definition 6:

Let A,B be s.a. with $I_A \subseteq I_B$, $O_A \subseteq O_B$, $\mathcal{B}$ be a class of s.a.

1. A is isomorphically embedded in B (B simulates A isomorphically) iff there is a non-empty set $S \subseteq S_B$ such that A and $B_{|S} := (I_A, O_A, S, \delta_{B|I_A \times S}, \lambda_{B|I_A \times S})$ are isomorphic.

2. B simulates A (behaviouristically) iff there holds:
$\forall s \in S_A \; \exists \bar{s} \in S_B : s \sim \bar{s}$
("$s \sim \bar{s}$" denotes equivalence of states in the usual sense that $\lambda^*_B(w,s) = \lambda^*_A(w,\bar{s})$ for all $w \in I^*_A$).

3. $\mathcal{B}$ is called a NN-basis for s.a. iff for all s.a. C there exists a $N \in NN(\mathcal{B})$ which simulates C.

We illustrate the above definitions by embedding H in a Normed Network N_H over E and K:

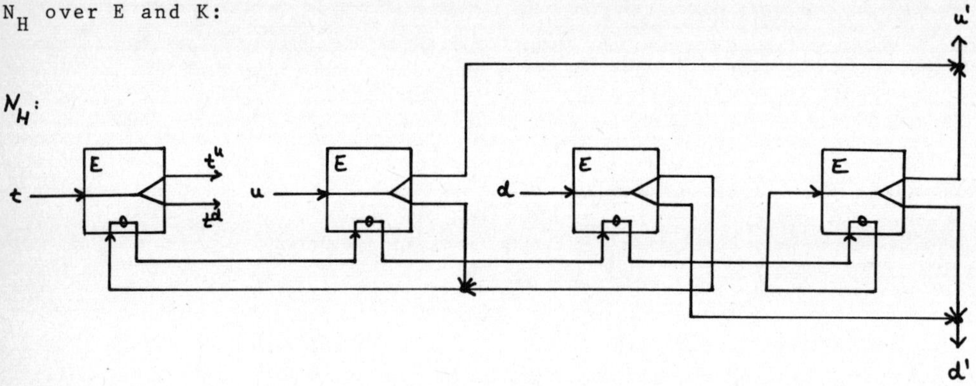

Figure 3

The state up (down) of H corresponds to the state of N_H in which all E-components are in the state up (down).

1.2. Concurrent automata and APA-nets

For the investigation of concurrent automata and parallel computations sometimes it becomes convenient to switch to a slightly more abstract level regarding S-systems.

Definition 7:

An S-system (state-system) A is a tuple $A=(S,\rightarrow)$ of a set S (of states) and a relation $\rightarrow \subseteq S \times S$ (of transitions). For $(s,s') \in \rightarrow$ we also write $s \rightarrow s'$. $\overset{*}{\rightarrow}$ denotes the reflexive and transitive closure of $\rightarrow$.

A <u>labeled S-system</u> (or <u>transition-system</u>, or briefly <u>T-system</u>) A is a tuple $A=(S,\Sigma,\rightarrow)$ of an S-system $(S,\rightarrow)$ and a finite set Σ (of names for transitions), such that for any $e\in\Sigma$ there exists a relation $\rightarrow_e \subseteq S\times S$ with $\rightarrow \; = \bigsqcup_{e\in\Sigma} \rightarrow_e$.

Obviously, all models for computations may be regarded to be S-systems, most to be T-systems.

Example: Any Petri-net N is a T-system in a canonical sense: Let P_N denote the set of places and E_N the set of events of N. Then N may be regarded as the T-system $(\mathbb{N}_o^d, E_N, \rightarrow)$ with $d=|P_N|$, where $s\rightarrow_e s'$ denotes that in state (marking) s event e is fireable and may fire with the new resulting state s'.

Definition 8:

An S-system (or T-system) $A=(S,\rightarrow)$ (or $(S,\Sigma,\rightarrow)$) has a <u>dimension</u> d iff $S \subseteq \mathbb{N}_o^d$.

P denotes the set of coordinates of $\mathbb{N}_o^d$, $P=\{p_1,\ldots,p_d\}$, and we also regard a state $s\in S\subseteq\mathbb{N}_o^d$ as a mapping $s:P\rightarrow\mathbb{N}_o$, i.e. we identify $\mathbb{N}_o^d = \mathbb{N}_o^P$. An element $p\in P$ is also called a <u>place</u> of A.

For any relation $\rho\subseteq\mathbb{N}_o\times\mathbb{N}_o$ and subset P' of P we define

$$s\rho s' \text{ on } P' :\Leftrightarrow \forall p\in P': s(p)\,\rho\, s'(p).$$

If one analyses a model of computation, say an S-System A, A usually is decomposed into simple sub-parts and their connections. Without any restriction we regard in this paper only S-systems with a countable state set, i.e. $S\subseteq N_o$. Thus, a state of a decomposed S-system contains all states of the sub-parts and of the connections which results in a dimensional S-system.

Obviously, such important concepts as liveness, hang-up-freeness, local determination, commutativeness, persistency, confluency, safeness etc. may be defined on the abstract level of (dimensional) T-systems.

Definition 9:

A <u>finite concurrent automaton</u> <u>(c.a.)</u> A is a tuple $A=(I,O,Z,R)$ of mutually disjoint finite sets I,O,Z of "inputs", "outputs", and "states", respectively, and a relation $R\subseteq(2^I\times Z)\times(2^O\times Z)$ of "transitions". Without any restriction we assume $Z\subseteq\mathbb{N}_o$, but are free to use mnemonic names for elements $z\in Z$.

An (<u>infinite</u>) <u>c.a.</u> is defined as above but Z is allowed to be infinite.

We again use the model that any $x \in I \cup O$ refers to one input or output line. A transition $(M,z)R(N,z')$ means that in state z the c.a. A may take off one signal from each input line x for all $x \in M$, send out one signal onto each output of N, and switch to state z'. Thus, in order that a transition $(M,z)R(N,z')$ is 'activated' there must be at least one signal on each input $x \in M$ and A has to be in state z. In such a case we also call the transition $(M,z)R(N,z')$ 'fireable'. We thus have dropped the '1-out-of-n-code' for c.a., but still we introduce no synchronizing clock. That we shall dispense with any clock is expressed by the fact that a fireable transition is never forced to fire. For example, a c.a. A may possess the transitions $(\{x_1,x_2\},s_1)R(\{y_1\},s_2)$ and $(\{x_1\},s_1)R(\{y_1,y_2\},s_3)$, be in state s_1 and one signal may be on line x_1 and x_2. Nevertheless A may choose the second transition $(\{x_1\},s_1)R(\{y_1,y_2\},s_3)$, ignoring x_2. Thus, instead of talking about the "arrival-time" of a signal, we imagine that a signal will pass a line in zero time, <u>but</u> the automaton may ignore its presence. By this simple trick we model the absence of any clock. Nothing is known about delays, durations of signals on wires, accumulation time within gates, etc. This heuristics is wellknown in the conception of Petri-nets. It will be made precise in the definition of APA-nets. This is done on the level of S-systems. Therefore we identify c.a. with S-systems according to Definition 10:

<u>Definition 10:</u>

A c.a. $A=(I,O,Z,R)$ is identified with the following S-system, also named A: $A=(\mathbb{N}_o^n \times \mathbb{N}_o^m \times Z, \rightarrow)$ with

1. $n=|I|$, $m=|O|$
2. $\forall s,s' \in \mathbb{N}_o^n \times \mathbb{N}_o^m \times Z$:

$$(s \rightarrow s' \Leftrightarrow \exists M \subseteq I : \exists N \subseteq O : \exists z,z' \in Z : \exists u,v \in \mathbb{N}_o^n \times \mathbb{N}_o^m :$$
$$(s=(u,z) \wedge s'=(v,z') \wedge (M,z)R(N,z')$$
$$\wedge\ u \geq 1 \text{ on } M$$
$$\wedge\ v=u-1 \text{ on } M$$
$$\wedge\ v=u+1 \text{ on } N$$
$$\wedge\ v=u \quad \text{ on } (I \cup O) \setminus (M \cup N))).$$

<u>Examples 11:</u>

$V=(\{1\},\{2,3\},\{0\},\{((\{1\},0),\ (\{2,3\},0))\})$ is a c.a.
V acts as a FORK-element (or parbegin), as V takes off one signal via

input 1 and sends out two signals, one onto each output 2 and 3. The state 0 is only formally needed.

Its counterpart, namely JOIN (or parend), is given by W:

$W=(\{1,2\},\{3\},\{0\},\{((\{1,2\},0),(\{3\},0))\})$.

$S=(\{1,2,e\},\{1',2'\},\{on,off\},R)$ with R consisting of the transitions

$\{1\}$, on R$\{1'\}$, off
$\{2\}$, on R$\{2'\}$, off
$\{e\}$, off R $\emptyset$, on

is a c.a. counterpart of a so-called 'simple Petri-net'. In Figure 4 S is graphically presented as a c.a. and as a PN (PN stands for "Petri-net" in the seque l).

$R=(\{1,2,3,4\},\{5,6,7\},\{0\}, \{((\{1,2\},0),(\{5\},0)),$
$((\{2,3\},0), (\{6\},0)), ((\{3,4\},0), (\{7\},0))\})$

is a "universal" PN-element (see Theorem 25).

The automata defined above are graphically represented in Figure 4:

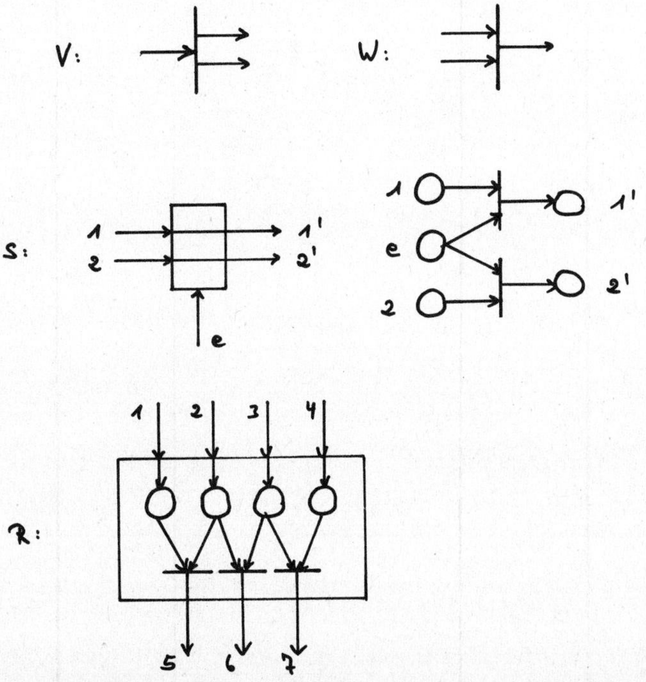

Figure 4

Obviously any s.a. can be interpreted as a c.a. E, e.g., may be regarded as the c.a. $E=(\{t,c\},\{t^u,t^d,c'\}, \{up,down\},R)$ with R consisting of the transitions

$(\{t\},up)R(\{t^u\},up)$
$(\{t\},down)R(\{t^d\},down)$
$(\{c\},up)R(\{c'\},down)$
$(\{c\},down)R(\{c'\},up).$

R may be increased by the transitions

$(\{t,c\},up)R(\{t^u,c'\},down)$
(*) $(\{t,c\},up)R(\{t^d,c'\},down)$
$(\{t,c\},down)R(\{t^u,c'\},up)$
$(\{t,c\},down)R(\{t^d,c'\},up).$

The transitions (*) may be regarded as results of a (nondeterministic) concurrent application of a t- and a c-signal. If we regard E as an S-system according to Definition 10, it is of no consequence for the $\rightarrow^*$-relation of E whether we add or drop the (*)-transitions.

Definition 12:

An APA-net-graph (asynchronous, parallel automata-net-graph) N over some given c.a. $A_1,\ldots,A_n$ is a graph with copies of $A_1,\ldots,A_n$ as nodes and connections of inputs and outputs as directed edges with the restriction that any edge connects exactly one output of some automaton copy with one input of the same or another copy. Let $B_1,\ldots,B_m$ be the set of all copies of $A_1,\ldots,A_n$ in N. Any B_i is called a component of N. $I_N(O_N)$ is the set of all inputs (outputs) of components that are not connected via edges and is regarded as the set of all inputs (outputs) of N.

$S_C := Z_{B_1}\times\ldots\times Z_{B_m}$ is the set of all component-states of N.

$S_W := \mathbb{N}_o^l$ is the set of all wire states, where l is the number of edges inside N.

An APA-net (also named 'N') is an APA-net-graph N together with a set $S\subseteq \mathbb{N}_o^{|I_N|}\times \mathbb{N}_o^{|O_N|}\times S_W\times S_C$ of admissible states. We regard N to be the following T-system of dimension $|I_N|+|O_N|+ l+m$:

$N=(S,\{B_1,\ldots,B_m\},\rightarrow)$ where for all $s,s'\in S$ there holds:

$s \underset{B_i}{\rightarrow} s' \;:\Leftrightarrow\; \exists M\subseteq I_{B_i} : \exists M'\subseteq O_{B_i} : \exists z,z'\in Z_{B_i} :$

$(M,z)R_{B_i}(M',z')$

$\wedge\ s=z$ on $\{Z_{B_i}\}$

$\wedge\ s'=z'$ on $\{Z_{B_i}\}$

$\wedge\ s\geq 1$ on M

$\wedge\ s'=s-1$ on $M\setminus M'$

$\wedge\ s'=s+1$ on $M'\setminus M$

$\wedge\ s'=s$ on $M\cap M'$

$\wedge\ s'=s$ on $P_N\setminus(\{Z_{B_i}\}\cup M\cup M')$,

where M and M' are canonical coordinates of I_N, O_N and of S_W depending on the interconnections of B_i in N.
For any class $\mathcal{B}$ of c.a. APA($\mathcal{B}$) denotes the class of all APA-nets over c.a. of $\mathcal{B}$. APA_f denotes the class of all APA-nets over finite c.a.

Next we have to generalize the conception of "simulation" that was defined only for s.a. As we want to compare various different models of computing systems via simulation we meet the problem to find one general conception. Fortunately such a general definition of simulating can be defined on the level of S-systems. Briefly, an S-system B simulates an S-system A iff for any state $s\in S_A$ there exist equivalent states $\bar{s}\in S_B$ such that any computation of A ($s\overset{*}{\underset{A}{\rightarrow}}s'$) can be refound in some computation $\bar{s}\overset{*}{\underset{B}{\rightarrow}}\bar{s}'$ of B for any $\bar{s}$ equivalent to s and some $\overline{s'}$ equivalent to s'. Further, any computation $\bar{s}\overset{*}{\underset{B}{\rightarrow}}\tilde{s}$ in B starting with any state $\bar{s}$ equivalent to some state s in A can be continued to reach a state $\overline{s'}$ equivalent to some state s' of A, i.e. $\tilde{s}\overset{*}{\underset{B}{\rightarrow}}\overline{s'}$ holds, and also $s\overset{*}{\underset{A}{\rightarrow}}s'$ is a correct computation in A. Thus, any computation of A can be refound in B, and B, started with correct states, simulates only computations of A (but possibly with some intermediate states not equivalent to some states of A). Such a simulation of a computation must not introduce dead-locks in B and a (k-) safe computation in A must become a (k'-) safe computation in B (k'>k is sometimes unavoidable). A formal definition of simulation is somewhat involved. Therefore and for a deeper discussion we refer to [Pr 83].

It should be noted that the net N_H of Figure 3, regarded as an APA-net, does not simulate H as a c.a., for if you send in state up one input signal on each line u and d, N_H can send out two output signals on output line u', but H cannot.

Definition 13:

Let B_1, B_2 be two classes of S-systems, A be some S-system.

$B_1 \mathrel{\tilde{\subseteq}} B_2 \;:\Leftrightarrow\; \forall B \in B_1 : \exists C \in B_2 :$ C simulates B.

$B_1 \mathrel{\tilde{=}} B_2 \;:\Leftrightarrow\; B_1 \mathrel{\tilde{\subseteq}} B_2$ and $B_2 \mathrel{\tilde{\subseteq}} B_1$.

$B_1 \mathrel{\tilde{\subset}} B_2 \;:\Leftrightarrow\; B_1 \mathrel{\tilde{\subseteq}} B_2$ and not $B_1 \mathrel{\tilde{=}} B_2$.

$A \mathrel{\tilde{\in}} B_1 \;:\Leftrightarrow\; \exists B \in B_1 :$ B simulates A.

By k-safe B_1 we denote all k-safe S-systems in B_1.
B_1 is called an APA-basis for B_2 iff $B_2 \mathrel{\tilde{\subseteq}} APA(B_1)$.

1.3. Locally sequential nets

As mentioned above, "sequential behaviour" or "sequential usage" of an automaton A=(I,O,S,R) means that the environment of A communicates with A in such a way that a further signal will be sent to some input of A only after an output signal has been received as a reaction of A to the previous input signal. In this context it is presupposed that a "sequentially used" automaton is restricted to such input-state-combinations to which it can react with exactly one output signal only. As a simple consequence, any sequentially used automaton must be 1-safe.

Definition 14:

Let B_1 be a class of s.a., B_2 be a class of c.a.
LSN(B_1;B_2), the class of all locally sequential nets over B_1 and B_2, is the class of all nets N∈APA($B_1 \cup B_2$) with such state sets that ensure the following:

1. Any A∈B_1 may be used sequentially only within N.
2. Elements of B_2 may be used concurrently within N.

Example 15:

Regard the net N over E,K,V and S of Figure 5 with the wires 1 up to 15. A state s=(2,3,0,1,0,0,3,0,2,0,0,1,0,0,0,off,up), e.g., means that there are exactly 2 signals on wire 1, 3 on wire 2, 1 on wire 4, 3 on wire 7, 2 on wire 9, and 1 on wire 12, S is in its weal state off, and E in up. We may drop the trivial states for V- and K-elements.

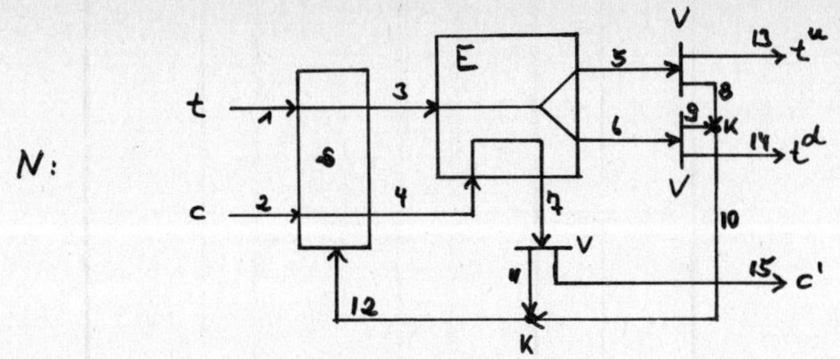

Figure 5

Let us define

$S_N \subseteq \mathbb{N}_o^{15} \times Z_S \times Z_E$ with

$s \in S_N \Leftrightarrow ((\sum_{i=3}^{12} s(i)=1$ and $s(Z_S)=\text{off})$ or $(\sum_{i=3}^{12} s(i)=0$ and $s(Z_S)=\text{on}))$

as the state set of N, i.e. for $s \in S_N$ there is at most one signal on the wires 3 up to 12. Then N becomes a net of LSN({K,E},{V,S}).

For any class $\mathcal{B}_1$ of s.a., LSN($\mathcal{B}_1$;{V,W}) describes computations with parbegin and parend, but also a non structured interplay of these elements is allowed as long as the net ensures that all $\mathcal{B}_1$-elements are used only sequentially.

The environment may use a LSN sequentially or concurrently: Whenever we use LSN in context with s.a. (such as NN, e.g.) the LSN must be used only sequentially by its environment but when we use LSN in context with c.a. or APA-nets it may also be used concurrently by its environment. With this restriction the following holds:

Lemma 16: LSN($\mathcal{B}$;$\emptyset$) $\cong$ NN($\mathcal{B}$)
LSN($\emptyset$;$\mathcal{B}$) = APA($\mathcal{B}$) for any set $\mathcal{B}$ of s.a.

2. Basic results

2.1. Some results on Normed Networks

First we give a simple proof for the wellknown fact that {H,K} is a basis for s.a. ([Ott 73], [PrRö 74]).

Theorem 17:

{H,K} is a NN-basis for s.a.

Proof:

Let $A=(I,O,S,\delta,\lambda)$ be a s.a. with $I=\{x_o,\dots,x_{i_o}\}$, $O=\{y_o,\dots,y_{k_o}\}$, $S=\{z_o,\dots,z_{j_o}\}$, $i_o,j_o,k_o \in \mathbb{N}_o$. Let H^1_{ij}, H^2_{ij}, H^3_{ij} $(0\leq i\leq i_o,\ 0\leq j\leq j_o)$ be s.a. isomorphic to H. The names for the input and output lines of these automata can be taken from the following graphical representation:

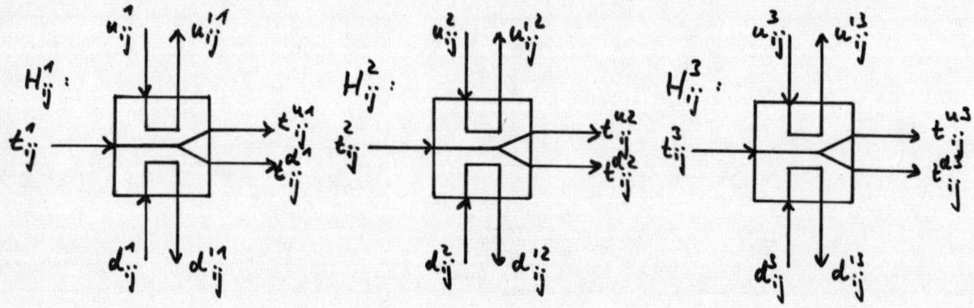

Figure 6

Now we compose the automata H^1_{ij}, H^2_{ij}, H^3_{ij} to a Normed Network N' by parallel composition. We imagine the components to be ordered in three matrices:

$$N':\quad \begin{matrix} H^1_{oo} & H^1_{o1} & \dots H^1_{oj_o} \\ H^1_{1o} & H^1_{11} & \dots H^1_{1j_o} \\ \vdots & \vdots & \vdots \\ H^1_{i_oo} & H^1_{i_o1} & \dots H^1_{i_oj_o} \end{matrix} \qquad \begin{matrix} H^2_{oo} & \dots & H^2_{oj_o} \\ \vdots & & \vdots \\ H^2_{i_oo} & \dots & H^2_{i_oj_o} \end{matrix} \qquad \begin{matrix} H^3_{oo} & \dots & H^3_{oj_o} \\ \vdots & & \vdots \\ H^3_{i_oo} & \dots & H^3_{i_oj_o} \end{matrix}$$

Next we introduce feedback-connections between the components of N'. A connection between an output line u and an input line v of H-components in N' is indicated by writing $u\to v$. We may write $u\to w$ and $v\to w$ for different output lines u and v. This instruction is realized in the network by using a K-component: 

An instruction 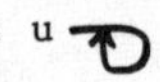 indicates a K-component, too: 

By an instruction $x_i\to u$ $(0\leq i\leq i_o)$ we indicate that the input line u of the net is renamed with x_i, by $u\to y_k$ $(0\leq k\leq k_o)$ we indicate that the output line u of the net is renamed with y_k (possibly after having identified several output lines by K-components).
Naturally, by definition of Normed Networks, instructions like $u\to v$ and $u\to w$ with $v\neq w$ are forbidden.
Now we give the list of connections in N':

1. $x_i \rightarrow t^1_{io}$

2. $t^{d1}_{ij} \rightarrow t^1_{ij+1}$ if $j<j_o$

3. $t^{d1}_{ij_o} \circlearrowleft$

4. $t^{u1}_{ij} \rightarrow u^2_{ij}$

5. $u'^2_{ij} \rightarrow u^3_{ij}$

6. $u'^3_{ij} \rightarrow d^1_{oj}$

7. $d'^1_{ij} \rightarrow d'^1_{i+1j}$ if $i<i_o$

8. $d'^1_{i_oj} \rightarrow t^2_{oo}$

9. $t^{d2}_{ij} \rightarrow t^2_{ij+1}$ if $j<j_o$

10. $t^{d2}_{ij_o} \rightarrow t^2_{i+1o}$ if $i<i_o$

11. $t^{d2}_{i_oj_o} \circlearrowleft$

12. $t^{u2}_{ij} \rightarrow d^2_{ij}$

13. $d'^2_{ij} \rightarrow u^1_{oj'}$ if $\delta(x_i,z_j) \cong z_{j'}$

14. $d'^2_{ij} \circlearrowleft$ if $\delta(x_i,z_j)$ is undefined

15. $u'^1_{ij} \rightarrow u^1_{i+1j}$ if $i<i_o$

16. $u'^1_{i_oj} \rightarrow t^3_{oo}$

17. $t^{d3}_{ij} \rightarrow t^3_{ij+1}$ if $j<j_o$

18. $t^{d3}_{ij_o} \rightarrow t^3_{i+1o}$ if $i<i_o$

19. $t^{d3}_{i_oj_o} \circlearrowleft$

20. $t_{ij}^{u3} \rightarrow d_{ij}^{3}$

21. $d_{ij}^{'3} \rightarrow \lambda(x_i,z_j)$ if $\lambda(x_i,z_j)$ is defined

22. $d_{ij}^{'3}$ ↻ if $\lambda(x_i,z_j)$ is undefined

$(0 \leqq i \leqq i_o,\ 0 \leqq j,\ j' \leqq j_o)$

The resulting Normed Network over H and K is called N.

A state z_j of A corresponds to the state of N where all H-components of N have the state down, except the components H_{ij}^1 $(0 \leqq i \leqq i_o)$, which have the state up.

Now let N be in the state corresponding to z_j $(0 \leqq j \leqq j_o)$ and let an input signal enter the net by line x_i. We'll describe the way the signal takes through the net:
First the signal passes through the test devices in the i-th row of the first matrix and reaches the component H_{ij}^1, which is the first with state up (connections 1.-3.). Now the signal has the information (i,j) and stores it in the 2. and 3. matrices by forcing the state up in H_{ij}^2 and H_{ij}^3 (4.-5.). Next the information about the state is extinguished in the 1. matrix by resetting the j-th column to state down (6.-7.). Then the signal returns to the second matrix and searches row by row for the component H_{ij}^2, which is in the state up (8.-11.). Then the signal resets H_{ij}^2 (12.) and switches the $\delta(x_i,z_j)$ – column of the 1. matrix into the state up (if $\delta(x_i,z_j)$ is defined , otherwise the signal gets into an infinite loop and the transition of the network is undefined, too)(13.-15.). Next the signal returns to the third matrix and searches for the component H_{ij}^3, which is in the state up (16.-19.). Finally the signal resets H_{ij}^3 (20.) and leaves the network by output line $\lambda(x_i,z_j)$ (21.-22.) (if $\lambda(x_i,z_j)$ is defined, see above). When the output is produced, the network N is in the state corresponding to $\delta(x_i,z_j)$.
Therefore A is isomorphically embedded into N.

<u>Remark 19:</u>
Constructions like

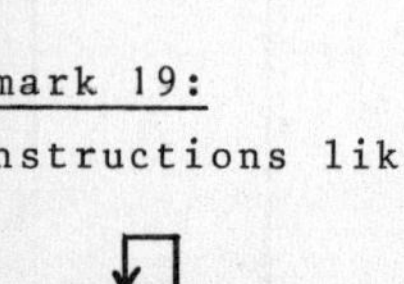

and

Figure 7

show how to construct "free" output and input lines that may be required in the net N.

Corollary 20:

{E,K} is a NN-basis for s.a.

Proof:

Figure 3.

{H,K} and {E,K} are bases even with respect to isomorphic embedding. But this is not the general case, as the following theorem shows ([BrKl 83]):

Theorem 21:

1. {F,D} is a NN-basis for s.a.
2. There doesn't exist any network $N \in NN(F,D)$, such that K is isomorphically embedded into N.

In this context the question arises how complicated a set $\mathcal{B}$ of s.a. has to be to form a basis. The full decidability problem remains open, but there are some approximative results:

Definition 22:

Let $A=(I,O,S,\delta,\lambda)$ be a s.a., $\mathcal{B}$ a finite set of s.a.

1. $\deg(A) := (|I|-1)\cdot(|O|-1)\cdot(|S|-1)$.
2. $\deg(\mathcal{B}) := \max\{\deg(B) \mid B \in \mathcal{B}\}$.
3. A is called monotonous iff there holds:
   $\forall s \in S \quad \forall p,q,r \in I^*:$
   $[(\delta^*(p,s)=s=\delta^*(qr,s) \wedge L(q) \subseteq L(p)$
   $\Rightarrow L(\lambda^*(q,s) \subseteq L(\lambda^*(p,s))]$.
4. A is called antimonotonous iff A is not monotonous.

Here $\delta^*: I^* \times S \to S$ and $\lambda^*: I^* \times S \to O^*$ are the canonical extensions of δ and λ, and L(p) denotes the set of letters which occur in a word p.

[Kö 76] and [Vo 80] have proven the following theorem:

Theorem 23:

1. Monotony of s.a. is decidable.
2. All s.a. of degree 0 are monotonous.

3. Every NN-basis for s.a. contains at least one antimonotonous automaton.
4. For every antimonotous s.a. A there exists a monotonous automaton B, such that {A,B} is a NN-basis for s.a.
5. For all totally defined s.a. A with $|S_A| \leq 2$ there holds: {A,F,K} is a NN-basis for s.a. <u>iff</u> A is antimonotonous.

<u>Application 24:</u>

{D,F,K} and {T,F,K} are NN-bases for s.a. of <u>minimal</u> degree 1.

2.2. Some results on APA-nets

To study the power of APA-nets, a comparison with PN - as the most well known conception of concurrent computations - is necessary. There is one important difference: APA-nets possess a <u>modular</u> structure with components with welldefined interfaces without possible conflicts: a signal on a wire can reach at most <u>one</u> input-line of one component. In contrast conflicts in PN are possible as a place may be input-place of several events. Such events in conflict have to 'agree' via some invisible logical structure which one fires. On the other hand, events in PN are locally deterministic, but c.a. in APA-nets generally are not. Nevertheless both conceptions are equivalent via simulation:

<u>Theorem 25:</u>

For any NN-base $\mathcal{B}$ for s.a. and for any $k \in \mathbb{N}$ there holds:

$$\begin{aligned} PN &\cong APA_f \\ &\cong APA(\{K,V,W,S\} \cup \mathcal{B}) \\ &\cong APA(\{K,V,R\}) \\ &\cong APA(\{K,V,W,E\}) \end{aligned}$$

$$\begin{aligned} k\text{-safe PN} &\cong 1\text{-safe PN} = 1\text{-safe } APA_f \\ &\cong 1\text{-safe } APA(\{K,V,W,S\} \cup \mathcal{B}) = 1\text{-safe } APA(\{K,V,R\}) \\ &\cong 2\text{-safe } APA(\{K,V,W,E\}). \end{aligned}$$

For a proof we refer to [Pr 83].

Up to now there exists no theorem for c.a. analogous to theorem 23. There is some progress to generalize "monotony" for c.a., but the conception of "degree", as defined in definition 22 for s.a., is not reasonable for c.a.:

Theorem 26:

There exists an APA-basis for c.a. of degree 0.
The proof is a trivial observation that the conception of states in c.a. is avoidable. Simply substitute any c.a. $A=(I,O,Z,R)$ by $A' := (I \cup Z, O \cup Z, \{0\}, R')$ and feed back any output $z \in Z$ to the input with the same name z. A transition $(M,z)R(N,z')$ now becomes a transition $(M \cup \{z\}), \{0\}\ R'(N \cup \{z'\}, \{0\})$. Thus, a state z in A is transformed in A' into a signal on a wire named also z. A' has degree 0 as it possesses only one formally needed state, 0.

2.3. Some results on LSN

Theorem 25 for APA-nets can be reformulated for LSN to a slightly deeper result:

Theorem 27:

For any NN-base $\mathcal{B}$ for s.a. and any $k \in N$ there holds:

$$\begin{aligned} PN &\cong APA_f \\ &\cong LSN(\mathcal{B} \cup \{K\}; \{V,W,S\}) \\ &\cong LSN(\{K\}; \{V,W,E\}) \end{aligned}$$

$$\begin{aligned} k\text{-safe } PN &\cong 1\text{-safe } APA_f \\ &\cong 1\text{-safe } LSN(\mathcal{B} \cup \{K\}; \{V,W,S\}) \\ &\cong 2\text{-safe } LSN(\{K\}; \{V,W,E\}). \end{aligned}$$

This is an immediate consequence of the proofs in [Pr 83]. Although E is also a s.a., it must be regarded as a c.a. in the last result.

Theorem 28:

For any set $\mathcal{B}$ of totally defined s.a. there holds:

$$\begin{aligned} &LSN(\emptyset; \mathcal{B} \cup \{K,V,W,S\}) \\ &\cong LSN(\mathcal{B} \cup \{K\}; \{V,W,S\}). \end{aligned}$$

Proof:

If $A=(I,O,Z,R)$ is a totally defined s.a. used in some APA-net $N \in LSN(\emptyset; \mathcal{B} \cup \{K,V,W,S\})$, $A \in \mathcal{B}$, A may be used concurrently. We have to ensure that A is used only sequentially without changing the

computations of N. This is done by help of some c.a.
S^n : S^n := ({1,...,n,e},{1',...,n'},{on, off},R) with the transitions

({i},on)R({i'},off) $(1 \le i \le n)$
({e},off)R(∅,on).

Thus, in state 'on' one signal on the input-lines 1,...,n of S^n may pass S^n and switch S^n to state 'off'. By using the input e, S^n may be switched back to 'on'. Figure 8 shows how to protect A by S^n, $n=|I_A|$, such that A is used only sequentially but the behaviour of A remains unchanged.

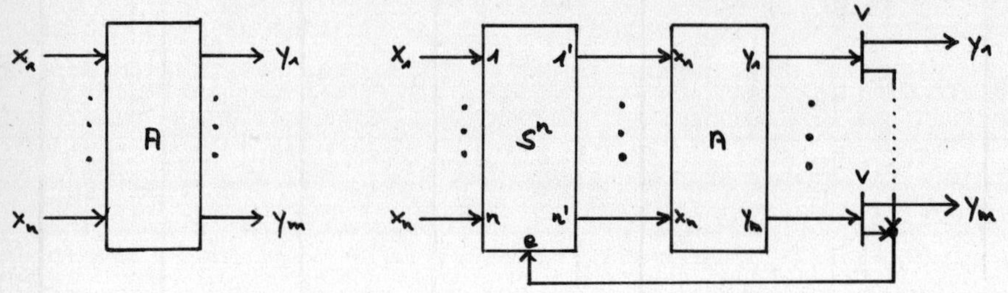

Figure 8

To complete the proof we show

$S^n \tilde{\in} LSN(\{K\};\{V,W,S\})$.

A net $\tilde{S}^n \in LSN(\{K\};\{V,W,S\})$ simulating S^n is presented in figure 9 with n=8. State 'on' of S^n is normally represented by the following state of S^n: All S-components of the inleading "left tree" are in state 'on' and all S-components of the outleading "right tree" are in state 'off'. Any signal entering S^n via some input line i $(1 \le i \le n)$ must pass the center element S_c in order

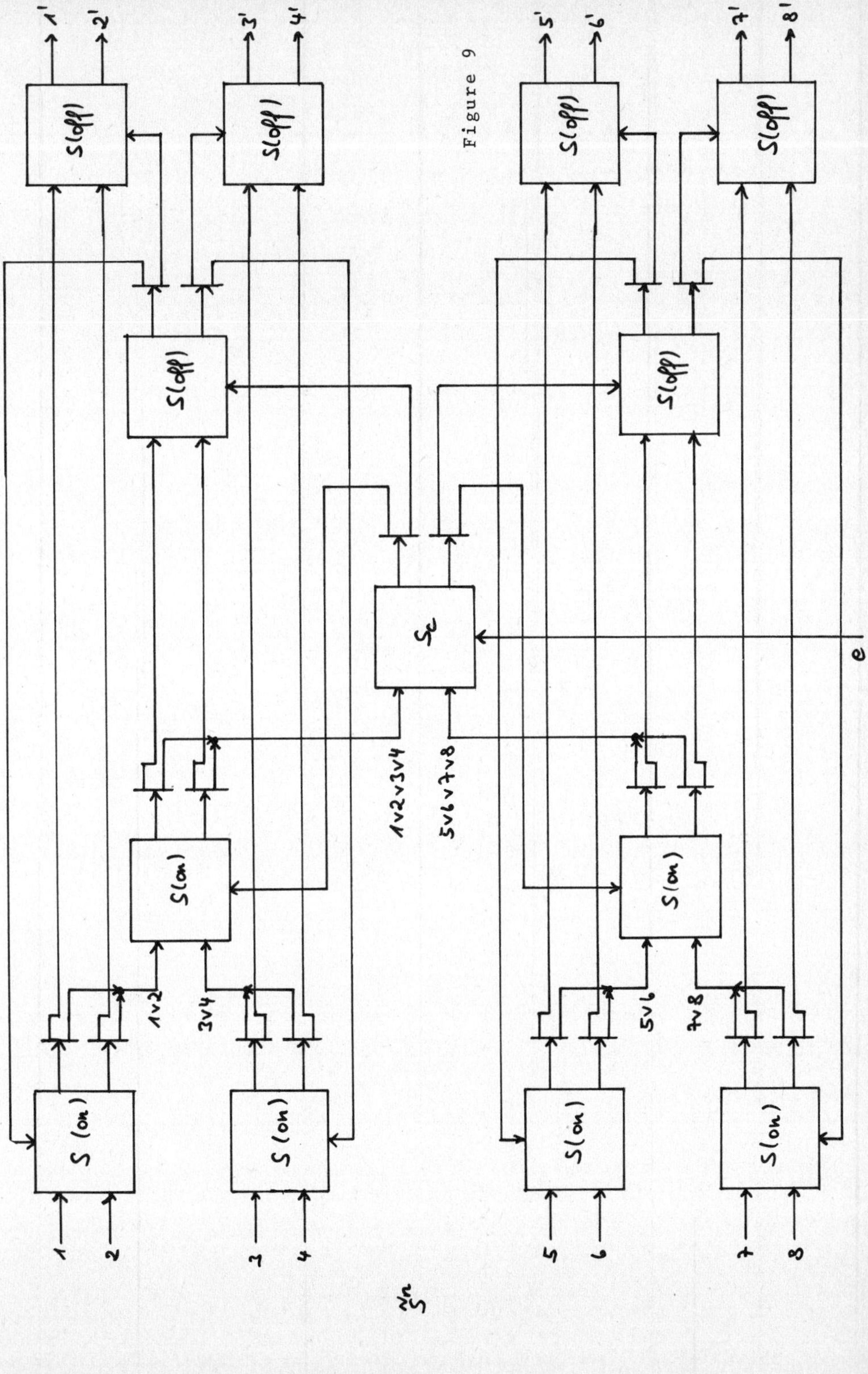
1'
2'
3'
4'
5'
6'
7'
8'
S(off)
S(on)
Sc
e
1v2
3v4
5v6
7v8
1v2v3v4
5v6v7v8
1
2
3
4
5
6
7
8

Figure 9

to pass $\tilde{S}^n$. It should be obvious that $\tilde{S}^n$ simulates the behaviour of S^n in quite a strong sense. For later purpose it should be noted that any signal passing S^n passes $O(\lg n)$ components in $\tilde{S}^n$.

Theorem 26 is incorrect for LSN. There holds a result similar to theorem 23 for s.a.:

<u>Theorem 29:</u>

1. For any NN-basis $\mathcal{B}$ for s.a. there holds:
   $A \tilde{\subset} LSN(\mathcal{B};\{V,W,S\})$
2. For any set $\mathcal{B}$ of s.a. there holds:
   $A \tilde{\subset} LSN(\mathcal{B};\{V,W,S\}) \Rightarrow deg(\mathcal{B})>0.$

<u>Proof:</u>

1. is trivial as $A \tilde{\subset} LSN(\mathcal{B};\emptyset) \tilde{=} NN(\mathcal{B})$ holds for any NN-base $\mathcal{B}$ for s.a.

2. is also quite simple. The idea is to prove that any sequentially used s.a. of degree 0 can be replaced by a <u>simple</u> PN. "Simple" means that no event possesses more than one input place that is also an input place of another event. V, W and S are also simple PN. Further it is known that simple PN don't form a basis for c.a., and by theorem 27 they thus also don't form a basis for s.a.

There arises the question, whether any set $\mathcal{B}$ of s.a. with $A \tilde{\subset} LSN(\mathcal{B};\{V,W\})$ is also a NN-basis for s.a. This ist not true:

<u>Theorem 30:</u>

1. $NN(\{T,K\}) \tilde{\not\subset} A$
2. $A \tilde{\subset} LSN(\{T,K\};\{V,W\})$

<u>Proof:</u>

1. We leave 1. as a (non-trivial) exercise for the reader. Hint: Find a property C of s.a. such that the following conditions hold:

a) T and K fullfil C
b) C is invariant under the operations feedback and parallel composition (i.e. <u>any</u> $A \in NN(\{T,K\})$ fullfils C).
c) C is invariant under simulation.
d) F doesn't fullfil C.

2. As $\{T,K,F\}$ is a NN-Basis for s.a., it suffices to show that $F \in LSN(\{T,K\};\{V,W\})$ holds. F actually is simulated by the following

network N :

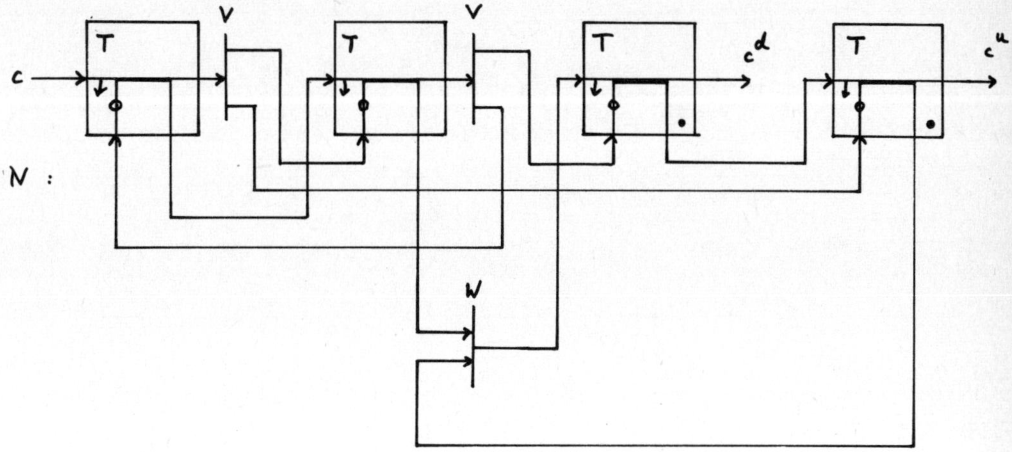

Figure 10

There state up (down) of T corresponds to state (up, down, down, down) ((down, up, down, down)) of N .

3. Computing time

In this section we sharpen our basis-results for s.a. with respect to the computing time, i.e. the number of components which are passed by the signal during a simulation process.
In the standard proof given above the network N over H and K, which simulates the s.a. $A=(I,O,S,\delta,\lambda)$, has $3\cdot|I|\cdot|S|$ H-components and $\leq 4\cdot|I|\cdot|S|$ H-components are passed by a signal during a simulation process. The number of components needed to simulate A, namely $O(|I|\cdot|S|)$, cannot be improved, just because the simulating network must possess sufficiently many output lines and $|O|=|I|\cdot|S|$ may hold in A. But we will see in this section that the computing time can be improved essentially.

Definition 31:

For $n,l\in\mathbb{N}_o$ we define a network Δ_l^n over H and K.
The H-components of Δ_l^n are called H_p^ν $(1\leq\nu\leq n,\ p\in\{u,d\}^{\leq l} :=$ $:= \{w\in\{u,d\}^*\,|\,lh(w)\leq l\})$. The input and output lines of H_p^ν are renamed according to figure 11:

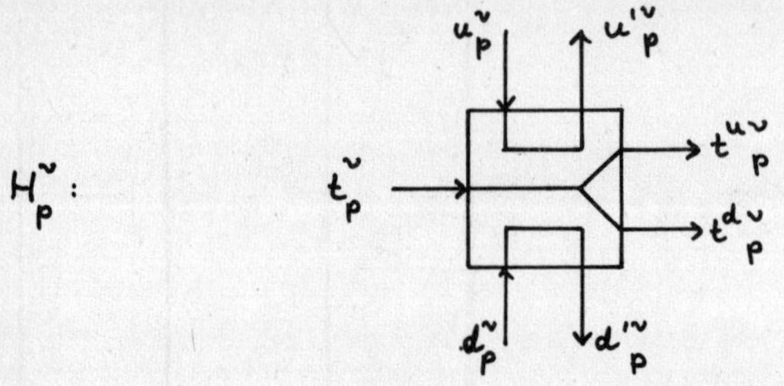

Figure 11

The input lines of Δ^n_l are called $t^\sim (1 \leq \sim \leq n)$ and $s_p (p \in \{u,d\}^{l+1} :=$
$:= \{w \in \{u,d\}^* \mid lh(w)=l+1\})$ and the output lines are $t^\sim_p (0 \leq \sim \leq n, p \in \{u,d\}^{l+1})$ and s'. The interconnections between the H-components of Δ^l_n are listed below:

1. $t^\sim \rightarrow t^\sim_\Lambda$
2. $t^{u\sim}_p \rightarrow t^\sim_{pu}$
3. $t^{d\sim}_p \rightarrow t^\sim_{pd}$
4. $s_{pu} \rightarrow u^1_p$ if $lh(p)=l$
5. $s_{pd} \rightarrow d^1_p$ if $lh(p)=l$
6. $u'^\sim_p \rightarrow u^{\sim+1}_p$ if $1 \leq \sim < n$
7. $d'^\sim_p \rightarrow d^{\sim+1}_p$ if $1 \leq \sim < n$
8. $u'^n_{pu} \rightarrow u^1_p$ if $lh(p)<l$
9. $d'^n_{pu} \rightarrow u^1_p$ if $lh(p)<l$
10. $u'^n_{pd} \rightarrow d^1_p$ if $lh(p)<l$
11. $d'^n_{pd} \rightarrow d^1_p$ if $lh(p)<l$
12. $u'^n_\Lambda \rightarrow s'$
13. $d'^n_\Lambda \rightarrow s'$

$(p \in \{u,d\}^{\leq l},\ 1 \leq \sim \leq n)$.

Example:

Δ_1^2:

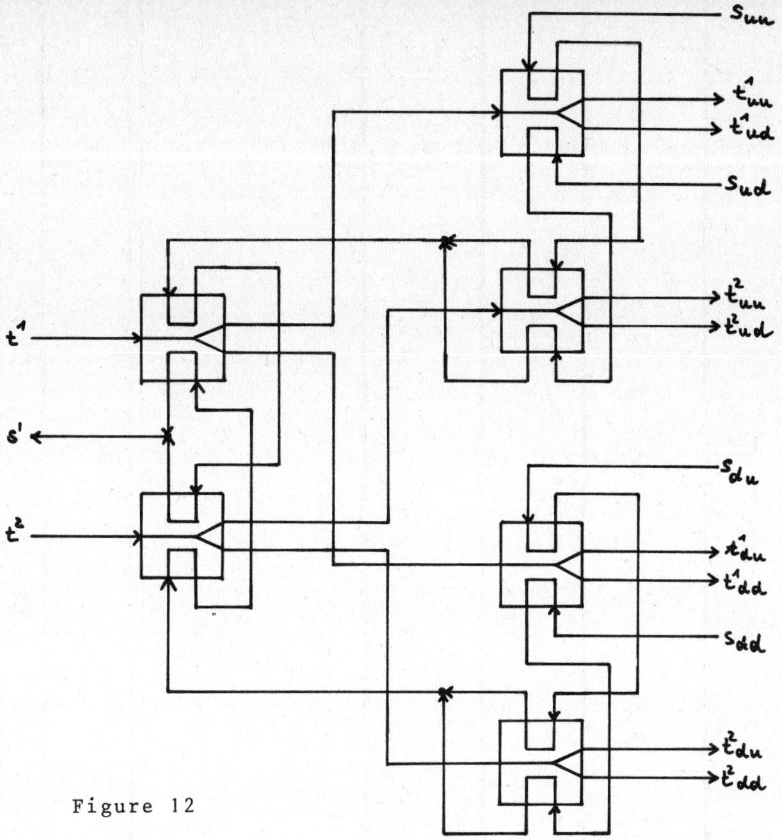

Figure 12

Description 32:

1. Δ_l^n is a binary tree of depth $l+1$, whose nodes are occupied with n H-components.
2. Assume $p \in \{u,d\}^{l+1}$. Every state of Δ_l^n, which fullfils the following condition (*), is associated to p:

   (*) $\forall q \in \{u,d\}^{\leq l}: \forall 1 \leq \nu \leq n:$

   $[qu \subseteq p \Rightarrow H_q^\nu$ has the state up

   $qd \subseteq p \Rightarrow H_q^\nu$ has the state down].

   ("$\subseteq$" means "is initial segment of").
3. If Δ_l^n is in a state associated to $p \in \{u,d\}^{l+1}$, a signal on t^ν will force an output signal on t_p^ν (without changing the state). A signal on s_p switches Δ_l^n into a state associated to p and is followed by an output signal on s'.
4. The number of H-components in Δ_l^n is $n \cdot (2^{l+1}-1)$ and the computing

time of Δ_1^n is $n \cdot (1+1)$.

Therefore we can encode and decode 2^{1+1} states in Δ_1^n in $\leq 1+1$ steps, and Δ_1^n has n independent decoding devices.

The construction of Δ_1^n is near at hand, see for example [Ke 74], but we can iterate it to speed up the basis-construction for s.a.:

Theorem 33:

For all s.a. $A=(I,O,S,\delta,\lambda)$ there exists a Normed Network N over H and and K that simulates A. N has $\leq 3 \cdot |I| \cdot |S|$ many H-components, and a signal passes $\leq 7 \cdot \mathrm{ld}\ |I| \cdot \mathrm{ld}|O|$ many H-components during a simulation process.

Proof:

Without loss of generality assume $S=\{u,d\}^{1+1}$, $I=\{u,d\}^{k+1}$, $1 \neq k$.

The network N consists of Δ_1^2-components $(\Delta_1^2)_r$ $(r \in \{u,d\}^{\leq k-1})$, Δ_1^4-components $(\Delta_1^4)_r$ $(r \in \{u,d\}^k)$ and H-components H_r^1, H_r^2 $(r \in \{u,d\}^{\leq k})$.

An input line or output line v of Δ_1^n is called $(v)_r$ in $(\Delta_1^n)_r$, and the lines of $H_r^{\sim}$ are renamed according to figure 13:

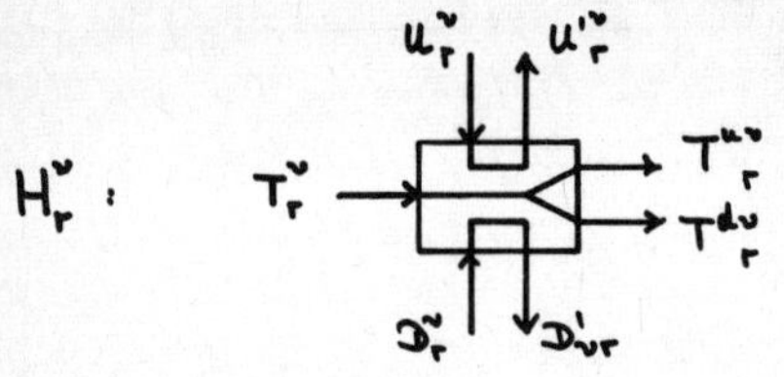

Figure 13

We imagine the components to be ordered in the following way:

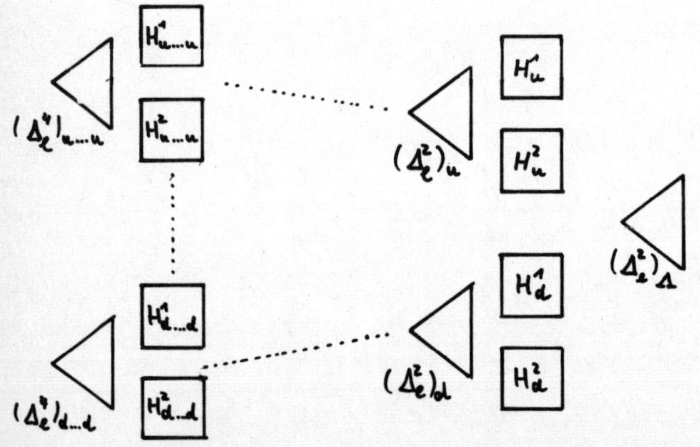

Figure 14

The sets of input lines and output lines of N are the same as of A. To obtain the network N we make the following interconnections between the components:

1.	$ru \to U^1_r$	if $lh(r)=k$
2.	$rd \to D^1_r$	if $lh(r)=k$
3.	$U'^1_r \to U^2_r$	
4.	$D'^1_r \to D^2_r$	
5.	$U'^2_{ru} \to U^1_r$	if $lh(r)<k$
6.	$D'^2_{ru} \to U^1_r$	if $lh(r)<k$
7.	$U'^2_{rd} \to D^1_r$	if $lh(r)<k$
8.	$D'^2_{rd} \to D^1_r$	if $lh(r)<k$
9.	$U'^2_\Lambda \to T^1_\Lambda$	
10.	$D'^2_\Lambda \to T^1_\Lambda$	
11.	$T^{u1}_r \to (t^1)_r$	
12.	$T^{d1}_r \to (t^2)_r$	
13.	$(t^1_p)_r \to (s_p)_{ru}$	if $lh(r)<k$
14.	$(t^2_p)_r \to (s_p)_{rd}$	if $lh(r)<k$
15.	$(s')_r \to T^1_r$	if $lh(r)>0$
16.	$(t^1_p)_r \to (s_{\delta(ru,p)})_\Lambda$	if $lh(r)=k$, $\delta(ru,p)$ def.
	$(t^1_p)_r$ ⊙	if $lh(r)=k$, $\delta(ru,p)$ undef.
17.	$(t^2_p)_r \to (s_{\delta(rd,p)})_\Lambda$	if $lh(r)=k$, $\delta(rd,p)$ def.
	$(t^2_p)_r$ ⊙	if $lh(r)=k$, $\delta(rd,p)$ undef.
18.	$(s')_\Lambda \to T^2_\Lambda$	
19.	$T^{u2}_r \to T^2_{ru}$	if $lh(r)<k$
20.	$T^{d2}_r \to T^2_{rd}$	if $lh(r)<k$
21.	$T^{u2}_r \to (t^3)_r$	if $lh(r)=k$
22.	$T^{d2}_r \to (t^4)_r$	if $lh(r)=k$
23.	$(t^3_p)_r \to \lambda(ru,p)$	if $lh(r)=k$, $\lambda(ru,p)$ def.
	$(t^3_p)_r$ ⊙	if $lh(r)=k$, $\lambda(ru,p)$ undef.

24. $(t_p^4)_r \to \lambda(rd,p)$ if $lh(r)=k$, $\lambda(rd,p)$ def.

$(t_p^4)_r$ ↻ if $lh(r)=k$, $\lambda(rd,p)$ undef.

$(r \in \{u,d\}^{\leq k},\ p \in \{u,d\}^{l+1})$.

Assume $p \in \{u,d\}^{l+1}$. All states of N, for which the state of $(\Delta_1^2)_\Lambda$ is associated to p, are associated to p.
Let now N be in a state associated to $p \in \{u,d\}^{l+1}$ and let an input signal enter N by input line $r \in \{u,d\}^{k+1}$.

First the input line r is encoded in the components $H_{r_1}^1$, $H_{r_1}^2$ $(r_1 \in \{u,d\}^{\leq k})$ like in a Δ_k^2-module (1.-8.). Then for $r_1 \subseteq r$, $lh(r_1) \leq k$ the following process iterates: First the state of $H_{r_1}^1$ is tested. State up (down) gives us the information $r_1 u \subseteq r$ $(r_1 d \subseteq p)$.
Then $(\Delta_1^2)_{r_1}$ is tested via t^1 resp. t^2 (according to the state of $H_{r_1}^1$). In the previous iteration step (resp. in the previous simulation process for $r_1=\Lambda$) $(\Delta_1^2)_{r_1}$ has been switched to a state accociated to p, and so the signal leaves $(\Delta_1^2)_{r_1}$ via $(t_p^1)_{r_1}$ resp. $(t_p^2)r_1$. If $lh(r_1) \leq k-1$ holds, $(\Delta_1^{2(4)})_{r_1 u}$ resp. $(\Delta_1^{2(4)})_{r_1 d}$ is switched to a state associated to p. The internal wires passed during the iteration process are described by lines 9.-15.
Finally the signal enters line $(t_p^1)_{r_1}$ (if $r=r_1u$) or $(t_p^2)_{r_1}$ (if $r=r_1d$). Next $(\Delta_1^2)_\Lambda$ is switched to a state accociated to $\delta(r,p)$ (if $\delta(r,p)$ is defined; 16.-17.). Then the input line r is decoded from the $H_{r_1}^2$-components $(r_1 \in \{u,d\}^{\leq k})$, and the signal reaches line $T_{r_1}^{u2}$ (if $r=r_1u$) or $T_{r_1}^{d2}$ (if $r=r_1d$), (18.-20.). Finally $(\Delta_p^4)r_1$ is tested via t^3 (if (if $r=r_1u$) or t^4 (if $r=r_1d$), the signal reaches line $(t_p^3)_{r_1}$ resp. $(t_p^4)_{r_1}$, and the output signal $\lambda(r,p)$ is given, 21.-24.).

Remark 34:

1. The previous construction can be slightly modified to obtain a network N, that simulates the s.a. A isomorpically.
2. The previous result holds for every NN-basis $\mathcal{B}$ for s.a. Only the constants 3 und 7 have to be modified.

We can speed up this construction again, if we allow parallelism in the simulating network. In this context, s.a. are simulated by LSN with concurrent components V and W only. The simulating networks have the additional property that in the "time interval" between an output signal and the next input signal all internal wires of the network are free. Such networks evidently are 1-safe and determinated and can be regarded as s.a. Furthermore, in order to understand the next construction one only has to know how V and W work, but nothing else about the theory of APA or LSN.

When we calculate computation time of a LSN we take into account all steps which can be done simultaneously, as a single one only. For example, the V-tree of depth 3.

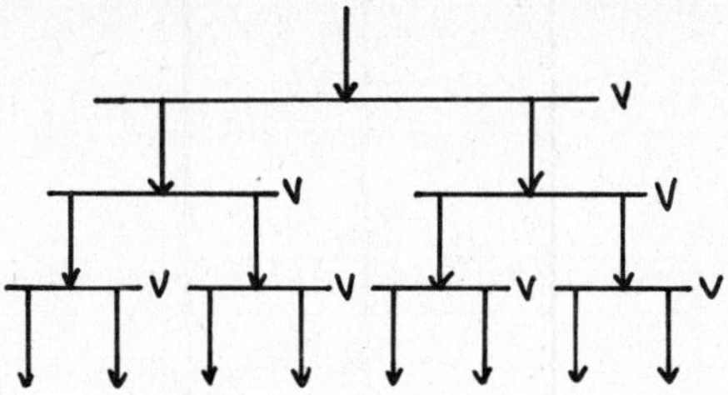

Figure 15

which increases one input signal 8-fold, needs computation time 3.

In the following construction we need generalized V-modules with 1 output lines and W- and K-modules with 1 input lines ($1 \in N_+$). It is clear how to simulate such generalized modules by V- resp. W- resp. K-trees, which require 1-1 basic modules and computation time ld(1).

Theorem 35:

Let $A=(I,O,S,\delta,\lambda)$ be a s.a. . There exists a network $N \in LSN(\{H,K\};\{V,W\})$, which simulates A. N consists of $\leq 3\cdot|I|\cdot|S|$ copies of H, $\leq 6\cdot|I|\cdot|S|$ copies of V, $\leq 3\cdot|I|\cdot|S|$ copies of W, $\leq 5\cdot|I|\cdot|S|$ copies of K, and requires a computation time $\leq 9(\mathrm{ld}|I|+\mathrm{ld}|S|)$.

Proof:

Assume $I=\{x_o,\ldots,x_{i_o}\}$, $S=\{z_o,\ldots,z_{j_o}\}$, $O=\{y_o,\ldots,y_{l_o}\}$. The network N consists of the following components:

1. Networks H_{ij} over H,V and K ($0\leq i\leq i_o$, $0\leq j\leq j_o$):

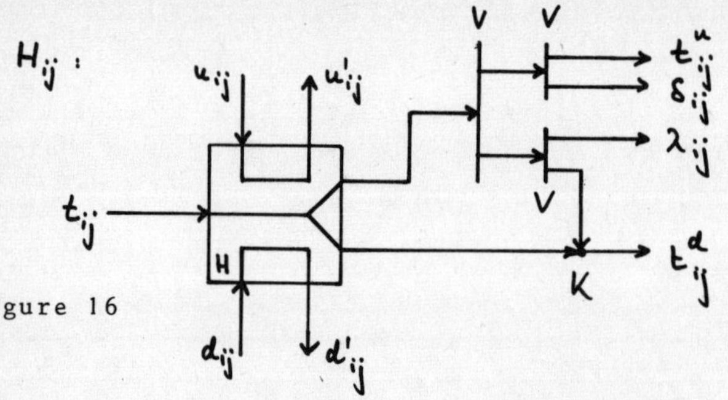

Figure 16

2. Generalized V-components V_i^1, V_j^2, V_j^3 $(0 \leqq i \leqq i_o,\ 0 \leqq j \leqq j_o)$:

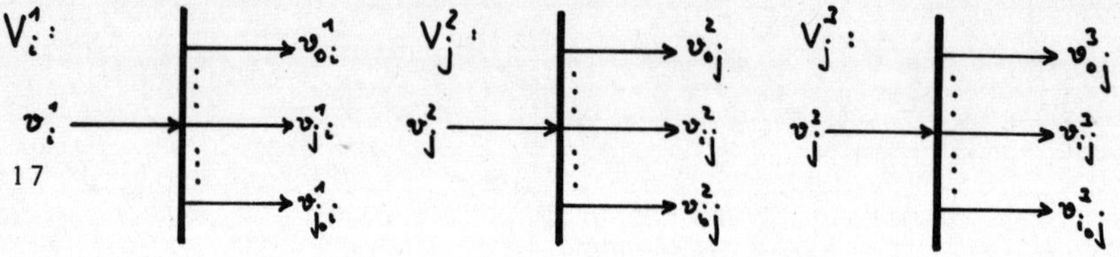

Figure 17

3. Generalized W-components W_i^1, W_j^2, W_j^3, W^4 $(0 \leqq i \leqq i_o,\ 0 \leqq j \leqq j_o)$:

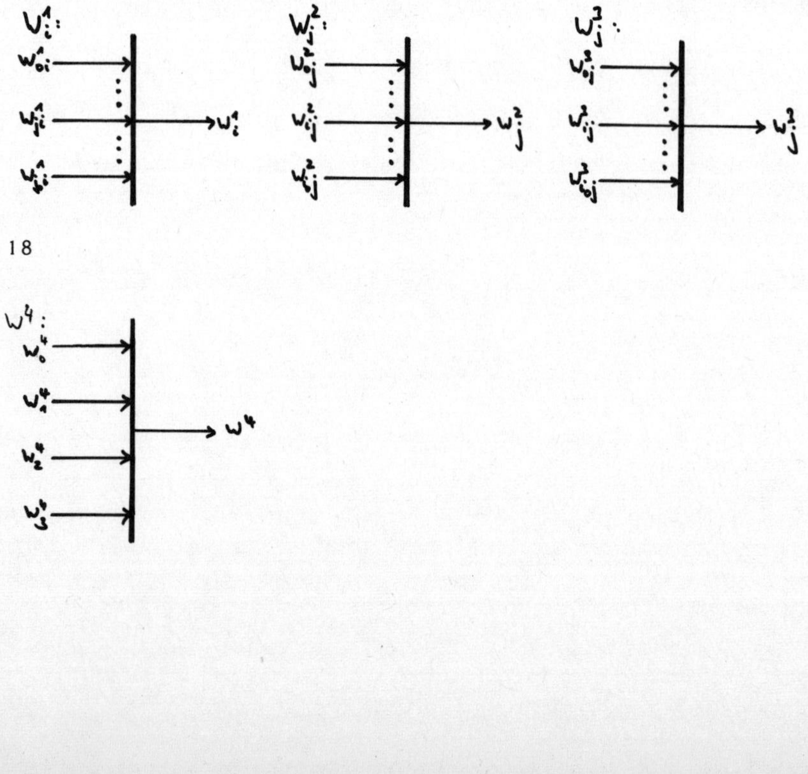

Figure 18

4. Generalized K-components K^1, K^2, $K^3_{j'}$, K^4_l, K^5_j, K^6 $(0 \leq j, j' \leq j_o,\ 0 \leq l \leq l_o)$:

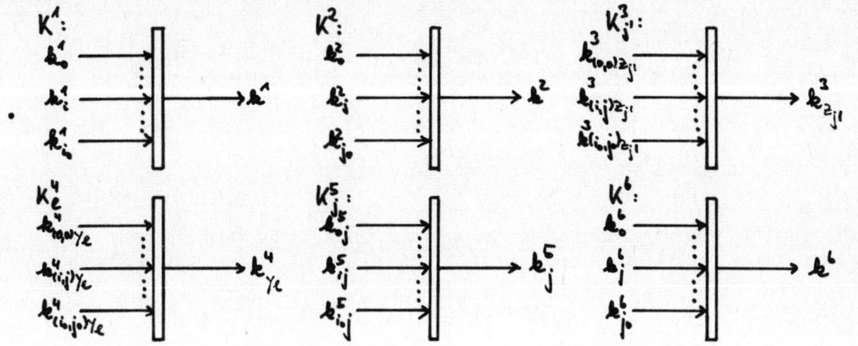

Figure 19

5. Modules $\Delta^1_{ld(|S|)}$, $\Delta^1_{ld(|O|)}$:

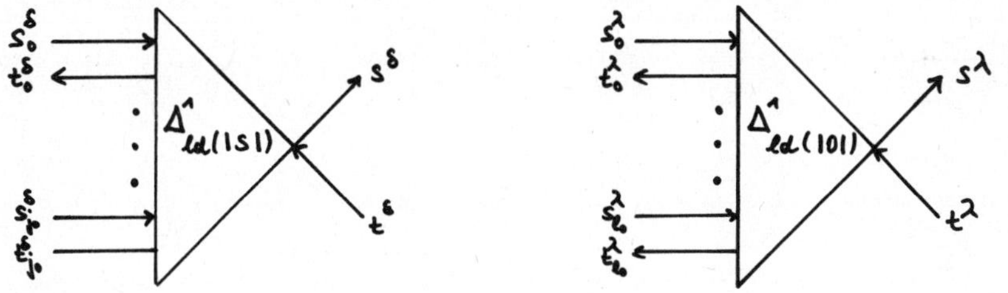

Figure 20

Between these components we make interconnections according to the following rules:

1. $x_i \rightarrow v^1_i$
2. $v^1_{ji} \rightarrow t_{ij}$
3. $t^d_{ij} \rightarrow w^1_{ji}$
4. $w^1_i \rightarrow k^1_i$
5. $k^1 \rightarrow w^4_2$
6. $t^u_{ij} \rightarrow k^5_{ij}$
7. $k^5_j \rightarrow v^2_j$
8. $v^2_{ij} \rightarrow d_{ij}$
9. $d^!_{ij} \rightarrow w^2_{ij}$
10. $w^2_j \rightarrow k^2_j$

11. $k^{2} \to w^{4}_{3}$

12. $\delta_{ij} \to k^{3}_{(i,j)j'}$ if $\delta(x_i,z_j)=z_{j'}$,

δ_{ij} $\circlearrowleft$ if $\delta(x_i,z_j)$ is undefined

13. $k^{3}_{j} \to s^{\delta}_{j}$

14. $s^{\delta} \to w^{4}_{o}$

15. $\lambda_{ij} \to k^{4}_{(i,j)l}$ if $\lambda(x_i,z_j)=y_l$

λ_{ij} $\circlearrowleft$ if $\lambda(x_i,z_j)$ is undefined

16. $k^{4}_{l} \to s^{\lambda}_{l}$

17. $s^{\lambda} \to w^{4}_{l}$

18. $w^{4} \to t^{\delta}$

19. $t^{\delta}_{j} \to v^{3}_{j}$

20. $v^{3}_{ij} \to u_{ij}$

21. $u'_{ij} \to w^{3}_{ij}$

22. $w^{3}_{j} \to k^{6}_{j}$

23. $k^{6} \to t^{\lambda}$

24. $t^{\lambda}_{l} \to y_l$,

$(0 \leq i \leq i_o,\ 0 \leq j,j' \leq j_o,\ 0 \leq l \leq l_o)$

The network N is sketched on the next page.

Assume $0 \leq j \leq j_o$. All states of N, for which the components H_{ij} $(0 \leq i \leq i_o)$ are in the state up, all components $H_{ij'}$ $(0 \leq i \leq i_o,\ 0 \leq j' \leq j_o,\ j' \neq j)$ are in the state down, and all internal wires carry no signal, are associated to z_j.

Let now N be in a state associated to z_j and let a signal enter N by input line x_i $(0 \leq j \leq j_o,\ 0 \leq i \leq i_o)$. The signal first reaches V^{1}_{i} and is increased $|S|$-fold. Then each component $H_{ij'}$ $(0 \leq j' \leq j_o)$ receives a t-signal (1.-2.).

N:

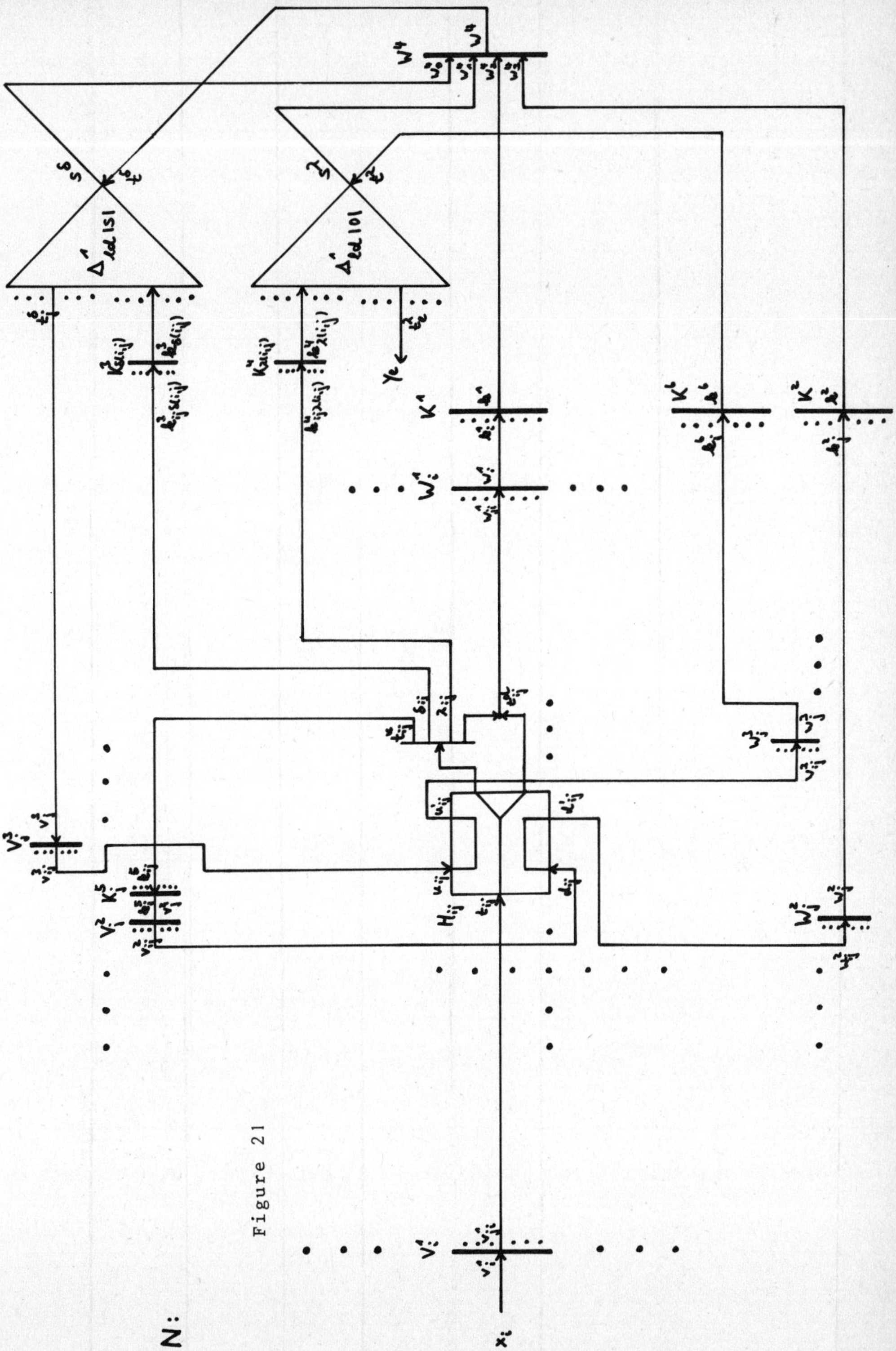

Figure 21

Then the following parallel processes run:

1. Each $H_{ij'}$-component ($0 \leq j' \leq j_o$) puts an output signal on line t^d_{ij}. These signals enable W^1_i. W^1_i enables K^1, and K^1 puts an output signal on line w^4_2 of W^4 (3.-5.).
2. H_{ij} puts an output signal on line t^u_{ij}, too. This signal enters V^2_j via K^5_j, which increases it $|I|$-fold. These signals parallelly reset the j-th column of the H-matrix and are collected again by W^2_j. W^2_j enables K^2, which puts a signal on line w^4_3 of W^4 (6.-11.).
3. H_{ij} puts an output signal on line δ_{ij}, too. This signal reaches line $s^\delta_{j'}$ of $\Delta^1_{1d|S|}$ by passing $K^3_{\delta(x_i,z_j)}$ (if $\delta(x_i,z_j)=j'$ otherwise the signal gets into an infinite K-loop), switches $\Delta^1_{1d(|S|)}$ into a state associated to $\delta(x_i,z_j)$, leaves $\Delta^1_{1d(|S|)}$ by line s^δ, and reaches line w^4_o of W^4 (12.-14.).
4. H_{ij} puts an output signal on line λ_{ij}, too. Like in process 3., this signal switches $\Delta^1_{1d|O|)}$ into a state associated to $\lambda(x_i,z_j)$ if $\lambda(x_i,z_j)$ is defined, and reaches line w^4_1 of W^4 (15.-17.).

Finally W^4 is enabled if and only if $\delta(x_i,z_j)$ and $\lambda(x_i,z_j)$ are defined and stored in the Δ-components and if the whole net is "clean". In this case W^4 sends a signal to t^δ. This signal reaches line $t^\delta_{j'}$, and is increased $|I|$-fold by $V^3_{j'}$. These signals now set the j'-th column of the H-matrix into state up, before they are collected again by $W^3_{j'}$ (18.-21.). $W^3_{j'}$ sends a signal to t^λ via K^6 and the output signal $\lambda(x_i,z_j)=y_1$ is given on line t^λ_1, (22.-24.). By this description it should be clear that N simulates A.

<u>Remark 36:</u>

Again, theorem 35 holds with respect to isomorphic embedding and for every NN-basis $\mathcal{B}$ for s.a. instead of {H,K}.
In the above construction the number of components in the simulating network $O(|I|\cdot|S|)$ <u>and</u> the computing time $O(1d|I|\cdot 1d|S|)$ are optimal.

4. Self-correction

In this section we switch to another part of Normed Network Theory, namely self-correction. Normed Networks with self-correcting properties first have been investigated in [PrRö 74]. In this section we indicate a much more lucid method of fault-tolerant simulation and show how to transfer our results about computing time to this topic.

Definition 37:

1. A fault in a net of s.a. is a spontaneous change (i.e. caused by no signal) of the state of one of its components.
2. For a s.a. A we define "period of using" to be the "time interval" between an input signal entering A and the next input signal entering A, and "resting interval" the "time interval" between an output signal leaving A and the next input signal entering A.
3. A network N simulates a s.a. A k-self-correcting (stable-k-self-correcting) iff N acts like A even if in every resting interval (period of using) at most k faults occur in N.

Definition 38:

For $k \in N_+$ let us define a network $H_k^* \in NN(H,K)$. H_k^* shall consist of $(k+1)^2$ H-components $H_{\nu\mu}$ $(0 \leq \nu, \mu \leq k)$ with input and output lines named according to figure 22:

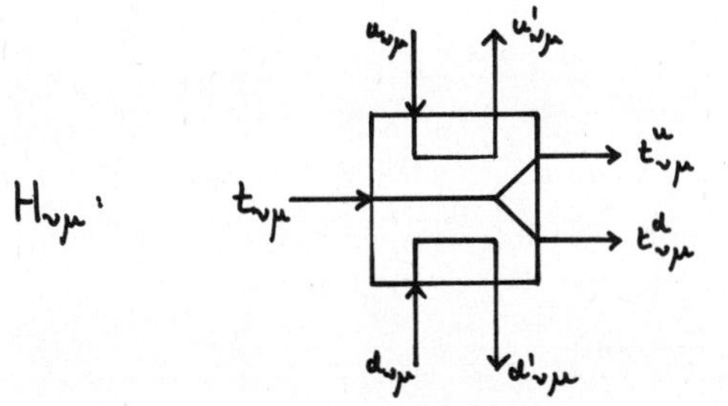

Figure 22

H_k^* has input lines t,u,d and output lines t^u, t^d, u', d'. Between the H-components of H_k^* there shall be drawn the following connection lines:

1. $t \to t_{oo}$
2. $t^u_{\nu\mu} \to t_{\nu+1\mu}$ $\qquad (0 \leq \nu < k)$
3. $t^u_{k\mu} \to t^u$

4. $t^d_{\sim\mu} \rightarrow t_{\sim\mu+1}$ $(0 \leq \mu < k)$

5. $t^d_{\sim k} \rightarrow t^d$

6. $u \rightarrow u_{oo}$

7. $u'_{\sim\mu} \rightarrow u_{\sim\mu+1}$ $(0 \leq \mu < k)$

8. $u'_{\sim k} \rightarrow u_{\sim+1o}$ $(0 \leqq \sim < k)$

9. $u'_{kk} \rightarrow u'$

10. $d \rightarrow d_{oo}$

11. $d'_{\sim\mu} \rightarrow d_{\sim\mu+1}$ $(0 \leq \mu < k)$

12. $d'_{\sim k} \rightarrow d_{\sim+1o}$ $(0 \leqq \sim < k)$

13. $d'_{kk} \rightarrow d'$

Figure 23, for example, represents H^*_2:

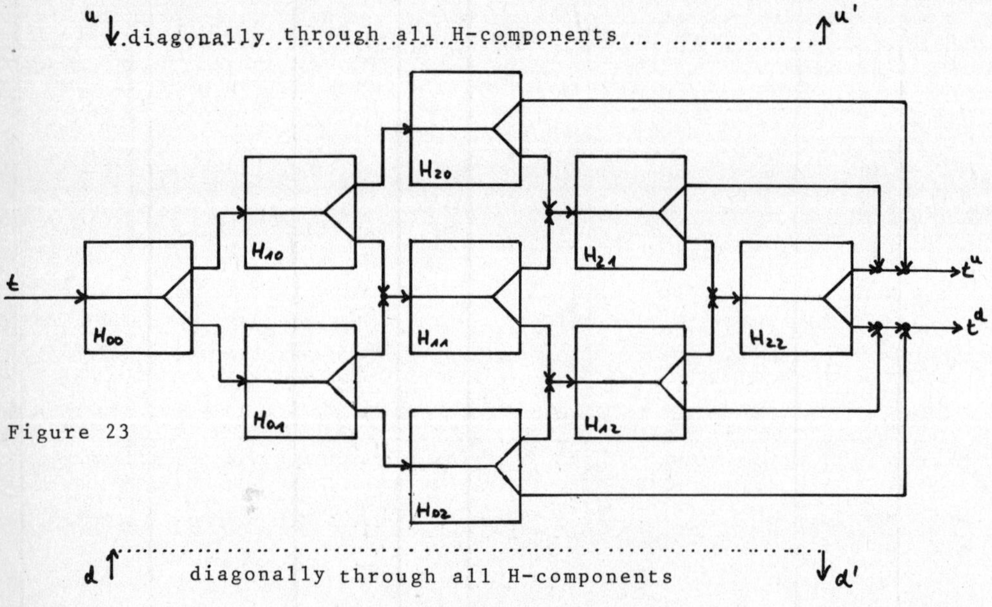

Figure 23

<u>Remark 39:</u>

Let state up (down) of H be associated to the state of H^*_k, where all H-components of H^*_k are in state up (down). Then there holds:

1. If H^*_k is in a state associated to up or down, then H^*_k acts like H during an experiment (i.e. a sequential usage by the environment),

also if k faults occur during the experiment.

2. An input signal on line u or d causes a correction of arbitrary many faults in H_k^*.

Theorem 40:

For every s.a. A there exists a network $N \in NN(H,K)$, which simulates A k-self-correcting.

Proof:

Let us start with a network $N' \in NN(H,K)$ which simulates the s.a. A and whose H-components can be divided into two sets N_1 and N_2, such that there holds for every simulation process:

1. Every H-component in N_1, which is used during the simulation process, is used first by input line u or d.
2. Every H-component in N_2, which is used during the simulation process, has been used in the simulation process before by input line u or d.

To obtain such a network N', one can slightly modify the network of theorem 17 or simply take the network of theorem 32. In the last case N_2 consists of the H-components of $(\Delta_1^2)_\Lambda$ and N_1 of all the others. Then simply substitute in N' each H-component of N_2 by H_k^*, and the resulting network N simulates A k-self-correcting (apply remark 39).

Theorem 41:

For every s.a. A there exists a network $N \in NN(H,K)$, which simulates A stable-k-self-correcting.

Proof:

Let us start again with a network $N' \in NN(H,K)$ as in the previous proof, but now replace the H-components of N_1 by H_k^* and the H-components of N_2 by H_{2k}^*. Again remark 39 ensures that the resulting network N simulates A stable-k-self-correcting.

Remark 42:

From the previous proofs it can be seen, that a network $N \in NN(H,K)$, which simulates a s.a. A (stable-)k-self-correcting, needs at most $O(k^2 \cdot ld(|I_A|) \cdot ld(|S_A|))$ many steps for a simulation process.
If one considers self-correcting networks $N \in LSN(\{H,K\};\{V,W\})$, one can transfer Theorem 35 in the same way to obtain a computation time of

$O(k^2 \cdot (ld|I_A|+ld|S_A|))$.

5. References

[BrKl 83] Brüggemann, A./Klein, R.: Über die Minimalität von {K,F,D}. Arbeitspapier Nr. 8, Inst. f. Math. Logik u. Grundlagenforschung, Münster (1983)

[Ke 74] Keller, R.M.: Towards a Theory of Universal Speed-Independent Modules, IEEE Transactions on Computers, Vol. C-23, No. 1, Jan. 1974

[Kö 76] Körber, P.: Untersuchungen an sequentiellen, durch normierte Konstruktionen gewonnenen Netzwerken endlicher Automaten. Dissertation, Inst. f. Math. Logik u. Grundlagenforschung, Münster (1976)

[Ott 73] Ottmann, Th.: Über Möglichkeiten zur Simulation endlicher Automaten durch eine Art sequentieller Netzwerke aus einfachen Bausteinen. ZmLG 19 (1973)

[PrRö 74] Priese, L./Rödding, D.: A Combinatorial Approach to Self-Correction. J. Cybern. 4 (1974)

[Pr 83] Priese, L.: Automata and Concurrency. TCS, Vol. 25, Number 3, July 1983

[Rö 83] Rödding, D.: Modular Decomposition of Automata (Survey). in: Karpinski, M. (ed.): Proceedings of the 1983 International FCT-Conference, LNCS 158

[Vo 80] Vobl, R.: Komplexitätsuntersuchungen an Basisdarstellungen endlicher Automaten. Diplomarbeit, Inst. f. Math. Logik u. Grundlagenforschung, Münster (1980)

MODULAR MACHINES, UNDECIDABILITY AND INCOMPLETENESS

Daniel E. Cohen
Queen Mary College
Mile End Rd.
London E1

0. Introduction

In [1,2] a new kind of machine, called a modular machine, was introduced. In that paper we gave a very simple proof of the unsolvability of the word problem for finitely presented groups. We commented that these machines fit in with standard constructions and simplify some of the proofs.

In this paper we show how the Undecidability and Incompleteness Theorems for elementary number theory follow easily from the existence of a modular machine with unsolvable halting problem. After modular machines have been introduced the proof follows well-known lines. However, I feel that the use of modular machines considerably simplifies the details, and that they form a significant addition to our factory of machines for obtaining partial recursive functions. The main purpose of this paper is not to prove new results, but to act as an introduction to modular machines and as propaganda for their use elsewhere.

1. Modular Machines.

We look at a Turing machine T defined by quintuples qaa'q'D (where D is one of the direction symbols L and R). We may consider its alphabet to consist of the integers 0(blank),1,...,n and its states to consist of integers less than m, with at most one state among 0,...,n (for this paper we could take the state set and the alphabet to be disjoint; in other contexts it is useful to consider machines with 0 as their unique halting state). Minsky [5] associates to the configuration $C = \ldots b_1 b_0 q a c_0 c_1 \ldots$ the quadruple u,v,a,q where $u = \sum b_i n^i$ and $v = \sum c_i n^i$. We shall represent C by either of the pairs (um+a,vm+q) or (um+q,vm+a), which we refer to as the pairs left- and right-associated to C respectively. Plainly we can tell whether or not an element of $\mathbf{N}^2$ is associated to some configuration, and, if so, we can find the configuration (if two integers could be both letters and states this would not hold). We use whichever associate is most convenient at the time, as indicated below (if each state is always entered from a particular direction we need only consider one associate for each configuration, but it will still be sometimes the left associate and sometimes the right). Suppose T has a quintuple qaa'q'R. Then C yields $C' = \ldots b_1 b_0 a' q' c_0 c_1 \ldots$ whose right associate is $(um^2+a'm+q', v)$. Similarly if T has a quintuple qaa'q'L, then C yields a configuration C' whose left associate is $(u, vm^2+a'm+q')$. It is this procedure which requires the use of both associates and which motivates the definition of modular machines given below.

A modular machine M consists of an integer m>1, an integer n with $0<n<m$, and quadruples (a,b,c,D) where a,b,c are integers with $0\leq a,b<m$, $0\leq c<m^2$, and D is one of the direction symbols L and R; we require that for each pair a,b at most one quadruple begins with a,b (if this condition is omitted we obtain a non-deterministic modular machine). If n is 1 we call the modular machine special.

The modular machine corresponding to a Turing machine T of the form above has as its data the integers m,n as above and two quadruples (q,a,a'm+q',D) and (a,q,a'm+q',D) for each quintuple qaa'q'D of T.

A configuration of M is an element (α,β) of $\mathbf{N}^2$. Write α,β as $\alpha=um+a$, $\beta=vm+b$, where $0\leq a,b<m$. If no quadruple begins with a,b we call (α,β) terminal and define $\mathrm{Next}_M(\alpha,\beta)$ to be (α,β). If there is such a quadruple we say (α,β) yields (α',β'), and define $\mathrm{Next}_M(\alpha,\beta)$ to be (α',β'), where $\alpha'=um^2+c$, $\beta'=v$ if there is a quadruple (a,b,c,R), while $\alpha'=u, \beta=vm^2+c$ if there is a quadruple (a,b,c,L). The name "modular machine" was given because the behaviour of M on (α,β) depends on the

remainders of α,β modulo m.

Let M be the modular machine corresponding to a Turing machine T and let (α,β) be an associate of the configuration C. Plainly if C is terminal so is (α,β), while if C yields C' then (α,β) yields (α',β') where (α',β') is one of the pairs associated with C'.

To say when a function $f:\mathbf{N}\to\mathbf{N}$ (or more generally a function from $\mathbf{N}^r$ to $\mathbf{N}$) is computed by M we need to define the input function $In_M:\mathbf{N}\to\mathbf{N}^2$ and the output function $Out_M:\mathbf{N}^2\to\mathbf{N}$ (we usually omit the subscript M) . We then say f is computed by M provided fx is defined if and only if M started on In x reaches a terminal configuration whose output is fx. The definitions should be made in such a way that, if M is the modular machine corresponding to the Turing machine T, then In_Mx is an associate of the input of T corresponding to x, and so that if (α,β) is an associate of the terminal configuration C, then $Out_M(\alpha.\beta)$ is the output of C. If this is done we will see at once that M simulates T, and, in particular, that M and T compute the same function. The precise definition of the input and output functions in the general case is straightforward (depending only on m and n), but is messy in detail, so we only give the details for special modular machines.

The input function for a special modular machine M is given by $In_Mx=(2,(m^{x+1}-1)/(m-1))$, while the output is $Out_M(\alpha,\beta)=(\mu y\leqslant\beta)(m^{y+1}>\beta)$. If M corresponds to the Turing machine T(with alphabet 0,1), then In_Mx is an associate of $q_2 01^x$ and $Out_M(\alpha,\beta)=y$ if (α,β) is an associate of $q01^y\}$. If we say that f is computed by T provided that fx=y if and only if T started on $q_2 01^x$ halts on $q01^y$ for some q, we see that the function computed by T is also computed by the corresponding modular machine M (the similar result holds for Turing machines on the alphabet 0,...,n for any n).Since every partial recursive function can be computed by such a Turing machine we see also that partial recursive functions are computable by modular machines.

Conversely it is clear from the formulae that, for any (special) modular machine M, the set of terminal configurations is primitive recursive, as are In_M, Out_M, $Next_M$, and the iterate $Comp_M:\mathbf{N}^3\to\mathbf{N}^2$ of $Next_M$. So we see at once that any function computable by a modular machine is partial recursive, and uses only one minimisation which is applied to a primitive recursive function. To obtain Kleene's Normal Form Theorem we need only find a Godel numbering for (special) modular machines and check that the above set and functions remain primitive recursive when the Godel number is part of the initial data. We define the Godel number of a special modular machine with data m and the quadruples (a_i,b_i,c_i,D_i) to be

$$2^m \prod p_{4i+1}{}^{a_i} \, p_{4i+2}{}^{b_i} \, p_{4i+3}{}^{c_i} \, p_{4i+4}{}^{d_i},$$

where d_i is 1 if D_i is L and d_i is 2 if D_i is R, and p_i are the primes. For any integer k we define primitive recursive functions m, a_i, etc. of k and i by the above formula, and let $r = (\mu s \leqslant k)(\exp_{4s+5} k=0)$. For any$(\alpha,\beta)$ let $j=(\mu i \leqslant r)(\text{rem}(m,\alpha)=a_i$ and $\text{rem}(m,\beta)=b_i)$ (where rem denotes the remainder function). Then j is a primitive recursive function of α,β, and k (it is most convenient to make the bounded minimisation have value r+1 if there is no such i). Next, as a function of α,β, and k is defined as follows (this definition, which is clearly primitive recursive, may be used whether or not k is a Godel number, and agrees with Next_M if k is the Godel number of M):

$\text{Next}(\alpha,\beta,k)=(\alpha',\beta',k)$ where

if $j>r$ then $\alpha'=\alpha$ and $\beta'=\beta$,

if $j\leqslant r$ and $d_j=1$, then $\alpha'=u$ and $\beta'=vm^2+c_j$, where u,v are the quotients when α,β are divided by m,

$\alpha'=um^2+c_j$ and $\beta'=v$ otherwise.

Comp is now defined as the iterate of Next, and In, Out are defined as before, as is the terminal set. These allow us to define the function corresponding to k whether or not k is a Godel number. If we wish to obtain an indexing of all partial recursive functions it is probably most convenient to define the function of index k as that function corresponding to the integer 4k+4 (so that m is at least 2).

Although this definition is somewhat complicated, it is still much less so than the definition required for Godel numbering of Turing machines. The advantage seems to be that modular machines are simultaneously Godelisations of Turing machines and machines in their own right, and all their data are numerical. The existence of a Turing machine with unsolvable halting problem (or of a non-recursive r.e. set)leads immediately to a modular machine with unsolvable halting problem. It now takes only two pages (see [1]) to exhibit, using Britton's Lemma, a finitely presented group with unsolvable word problem, and a further one page, using the unsolvability of the confluence problem for modular machines [3] ,to obtain a finitely presented group with solvable word problem but unsolvable conjugacy problem.We now proceed to obtain the Undecidability and Incompleteness Theorems , with less effort than usual, from the existence of a modular machine with unsolvable halting problem.

We now show how the s-m-n Theorem, the Recursion Theorem and Rogers' Isomorphism Theorem can be obtained with the indexing of functions given above. As is well-known (for instance, the treatment in chapter 3 of [4]) to obtain the s-m-n Theorem and the Recursion Theorem it is enough to prove the following simple case of the s-m-n Theorem:

Let $f:\mathbf{N}^3\to\mathbf{N}$ be partial recursive.Then there is a primitve recursive function $g:\mathbf{N}^2\to\mathbf{N}$ such that, for all x,y, the function with index g(x,y) sends z to f(x,y,z) for all z.

Let T be a Turing machine on alphabet 0,1 with states 3(starting state),...,n which computes f when started on the first 0 of $01^x01^y01^z$. Let P(x,y) be a Turing machine with starting state 2, halting state 3, and remaining states integers greater than n which started on the first 0 of 01^z ends on the first 0 of $01^x01^y01^z$. The required function g has 4g(x,y)+4 as the Godel number of the modular machine corresponding to P(x,y)T.

For Rogers' Isomorphism Theorem it is enough to prove the following Padding Lemma:

There is a one-one primitive recursive function $p:\mathbf{N}^2\to\mathbf{N}$ such that, for all k,r, the functions of indices k and p(k,r) are the same.

Let m,n,a be integers with n>m>1 and let $a=\sum\alpha_i m^i$,where $\alpha_i<m$.The integer $\sum\alpha_i n^i$ will be called a transferred from m to n. It is easy to check that if k is a non-zero integer and 4k'+4 is obtained by transferring every exponent of 4k+4 from m (= $\exp_0(4k+4)$) to n then k and k' are indices of the same function.Thus we need only define p(k,r) to be obtained in this way with n=J(m,r),J being the standard pairing function.

2.Elementary Number Theory.

We consider the language with equality L which has one constant symbol O, one unary function symbol s and two binary function symbols + and . and the binary predicate symbol $\leqslant$. To each $n\in\mathbf{N}$ there is the associated numeral n defined inductively. We use informal notation, writing (u+v) or u+v instead of the more formal +uv (where u,v are terms), etc.

Let A be the following set of sentences:

(i) m+n=p for all $m,n,p \epsilon N$ with m+n=p;

(ii) m.n=q for all m,n,q with mn=q;

(iii) $\neg(m=n)$ for all m,n with $m \neq n$;

(iv) $\forall x(x \leqslant n \rightarrow (x=0 \vee x=1 \vee \ldots \vee x=n))$ for all n.

Plainly **N** is a model of A, and A forms a very weak set of axioms for number theory. There are finite sets of which **N** is a model and from which A can be derived, for instance (as can be seen by an easy induction) the following set B:

(1) $\forall x(x+0=x)$;

(2) $\forall x \forall y(x+sy=s(x+y))$;

(3) $\forall x(x.0=0)$;

(4) $\forall x \forall y(x.sy=x.y+x)$;

(5) $\forall x \forall y(\neg(x=x+sy))$;

(6) $\forall x(x \leqslant 0 \rightarrow x=0)$;

(7) $\forall x \forall y(x \leqslant sy \rightarrow x=sy \vee x \leqslant y)$.

The subset S of $\mathbf{N}^k$ will be called <u>weakly expressible</u> iff there is a wff P such that

if $(n_1,\ldots,n_k) \epsilon S$ then $A \vdash P(n_1,\ldots,n_k)$,

while if $(n_1,\ldots,n_k) \epsilon S$ then $P(n_1,\ldots,n_k)$ is false in the model **N**. A (total) function $f:\mathbf{N}^p \rightarrow \mathbf{N}^q$ is <u>weakly representable</u> if its graph in $\mathbf{N}^{p+q}$ is weakly representable. These definitions are similar to, but weaker than, the more usual definitons of expressibility and representability.

Let $J:\mathbf{N}^2 \rightarrow \mathbf{N}$ be the bijection given by $J(a,b)=((a+b)^2+3a+b)/2$, and let $\gamma:\mathbf{N}^3 \rightarrow \mathbf{N}$ be the Godel sequencing function, i.e. $\gamma(k,t,u)$ is the remainder when (k+1)t+1 divides u.

<u>Lemma 1.</u> J, J^{-1} and γ are weakly representable.

<u>Proof</u> J is plainly weakly represented by the following wff φ:

$2.z=x.x+y.y+2.(x.y)+3.x+y$.

Consequently J^{-1} ia weakly represented by the wff

$\exists u \exists v(\varphi(y,u,x) \wedge \varphi(v,z,x))$.

Also γ is weakly represented by

$\exists y(x_3=y.(s(x_2.sx_1)+x_4 \wedge x_4 \leqslant x_2.sx_1)$.

<u>Lemma 2.</u> If $f:\mathbf{N}^p \rightarrow \mathbf{N}^q$ and $g:\mathbf{N}^q \rightarrow \mathbf{N}^r$ are weakly representable, so is gf.If

S is a weakly expressible subset of N^q then $f^{-1}S$ is weakly expressible.

Proof. Let f and g be weakly represented by the wffs φ and ψ and let S be weakly expressed by η. Then it is easy to see that gf is weakly represented by

$$\exists y_1 \ldots \exists y_q(\varphi(x_1,\ldots,x_p,y_1,\ldots,y_q) \wedge \psi(y_1,\ldots,y_q,z_1,\ldots,z_r))$$

while $f^{-1}S$ is weakly expressed by

$$\exists y_1,\ldots \exists y_q(\varphi(x_1,\ldots,x_p,y_1,\ldots,y_q) \wedge \eta(y_1,\ldots,y_q)).$$

Lemma 3. Let $F:N^2 \to N$ be the iterate of $f:N \to N$. If f is weakly representable, so is F.

Proof. By definition $F(r,s)=r'$ iff there is a sequence $a_0,\ldots,a_s,a_{s+1}$ such that $a_0=r, a_s=r'$ and, for all $i \leq s$, $a_{i+1}=fa_i$ (it is convenient, though not necessary, to require a_{s+1} to exist and equal fa_s). In turn, this holds iff there are p,q with $\gamma(k+1,p,q)=f\gamma(k,p,q)$ for all $k \leq s$ and with $\gamma(0,p,q)=r$ and $\gamma(s,p,q)=r'$. If φ and ψ represent f and γ, then F is weakly represented by the following wff:

$\exists u \exists v(\psi(0,u,v,x_1) \wedge \psi(x_2,u,v,x_3) \wedge \forall y(y \leq x_2 \to \theta)$, where θ is $\exists z \exists z'(\psi(y,u,v,z) \wedge \psi(sy,u,v,z') \wedge \varphi(z,z'))$.

Corollary. Let $G:N^3 \to N^2$ be the iterate of $g:N^2 \to N^2$. If g is weakly representable, so is G.

Proof. By Lemmas 1 and 2, JgJ^{-1} is weakly represntable. If H is the iterate of this, then G is $J^{-1}HJ'$, where $J'(p,q,r) = (J(p,q).r)$.

Proposition 4. The halting set of any modular machine is weakly expressible.

Proof. Let I,J,K be index sets such that the modular machine has data (a_i,b_i,c_i,R) for $i \in I$, (a_j,b_j,c_j,L) for $j \in J$ and has no quadruples beginning with a_k,b_k for $k \in K$.

(α,β) is terminal iff there are u,v,and k with $\alpha=um+a_k$ and $\beta=vm+b_k$. So the terminal set is weakly expressed by

$$\exists y \exists z(\bigvee_k (x_1=y.m+a_k \wedge x_2=z.m+b_k).$$

Similarly the function Next is weakly represented by

$$\exists y,z,y',z'(x_1=y.m+y' \wedge x_2=z.m+z' \wedge \xi),$$

where ξ is $\bigvee_i \varphi_i \vee \bigvee_j \psi_j \vee \bigvee_k \theta_k$, and φ_i is $y'=a_i \wedge z'=b_i \wedge x_3=y.m.m+c_i \wedge x_4=z$,

ψ_j is $y'=a_j \wedge z'=b_j \wedge x_3=y \wedge x_4=z.m.m+c_j$, and

θ_k is $y'=a_k \wedge z'=b_k \wedge x_3=x_1 \wedge x_4=x_2$.

By the corollary to Lemma 3, Comp, which is the iterate of Next, is weakly representable, and then, by lemma 2, $\{(\alpha,\beta,n);\mathrm{Comp}(\alpha,\beta,n)$ is terminal$\}$ is weakly expressible, by ω, say. Then the halting set is weakly expressed by $\exists y\omega(x_1,x_2,y)$.

Proposition 5. A subset of **N** is weakly expressible iff it is recursively enumerable.

Proof Let S be expressed by φ. By Church's Thesis, the map sending n to $\varphi\mathbf{n}$ is recursive. Also the set of consequences of A is recursively enumerable. Since $n\in S$ iff $\varphi\mathbf{n}$ is a consequence of A, we see S is r.e.

Let S be r.e. Its partial characteristic function is computed by a special modular machine M, and so $n\in S$ iff $In_M n$ is in the halting set of M. So we need only show In_M is weakly representable, by Proposition 4 and Lemma 2. Now $In_M n = ((m^{n+1}-1)/(m-1),2)$. Multiplication by m is plainly weakly representable, and hence so is its iterate E. Since $m^{x+1}=E(m,x)$, the function sending x to m^{x+1} is weakly representable, and, as is easily seen, so is In_M.

It follows immediately that any partial recursive function is weakly representable. We could use similar techniques to show that any recursive function is representable, but , unlike the weak properties, the proof shows no saving over conventional proofs. To obtain Godel's Second Incompleteness Theorem and the Undefinability of Truth it seems necessary to have the stronger properties.

3. Undecidability and Incompleteness.

Let T be the set of sentences valid in **N**, and F the set of sentences not valid in **N**; then $\sigma\in F$ iff $\neg\sigma\in T$. For any set Σ of sentences, let $Cn\Sigma$ be its set of consequences.

Theorem 1 (General Undecidability Theorem). CnA is recursively inseparable from F.

Proof Let M be a modular machine with unsolvable halting problem, and let φ weakly express its halting set. If C separates CnA and F, then (p,q) is in the halting set of M iff $\varphi(\mathbf{p},\mathbf{q})$ is in C. By Church's Thesis, the map sending (p,q) to $\varphi(\mathbf{p},\mathbf{q})$ is recursive. Since the halting set of M is not recursive, it follows that C cannot be recursive.

<u>Theorem 2(Undecidability of Number Theory).</u> Let Σ be a set of sentences in a countable language L' containing L. Let **N** be made into an L' structure which is a model of Σ and suppose $\Sigma \vdash A$. Then $Cn\Sigma$ is not recursive.

<u>Proof.</u> The set Δ of sentences of L which are in $Cn\Sigma$ plainly separates CnA from F, and Δ would be recursive if $Cn\Sigma$ were recursive.

<u>Theorem 3(Strong Undecidability).</u> Let Σ be a set of sentences in a countable language L' containing L and let **N** be made into an L' structure which is a model of Σ. Then $Cn\Sigma$ is not recursive.

<u>Proof.</u> Take the finite set B of sentences which was defined in Section 2, and let β be the conjunction of the sentences in B. Then A is a consequence of $\Sigma \cup \{\beta\}$. So by Theorem 2 $Cn(\Sigma \cup \{\beta\})$ is not recursive. Now σ is in this set iff $\beta \to \sigma$ is in $Cn\Sigma$. Since the map sending σ to $\beta \to \sigma$ is recursive it follows that $Cn\Sigma$ cannot be recursive.

<u>Theorem 4(Undecidability of Predicate Logic).</u> The theory of L is not recursive.This holds whether L is regarded as a language with equality or a pure language.

<u>Proof.</u> In Theorem 3 take L' to be L and Σ to be empty. This deals with the case where L is regarded as a language with equality. Now the axioms of equality may be taken to be a finite set, and we let their conjunction be the sentence η. Then σ is in the theory of L regarded as a language with equality iff $\eta \to \sigma$ is in the theory of L regarded as a pure language. The result follows as above.

<u>Theorem 5 (Undecidability of Truth).</u> T is not r.e.

<u>Proof.</u> Since σ is in F iff $\neg \sigma$ is in T, we see that F would be r.e. if T were r.e. But F is the complement of T, so that T would then be recursive. This is impossible by Theorem 1.

<u>Theorem 6(Godel's First Incompleteness Theorem; weak form).</u> Let Σ be an r.e. set of sentences in a countable language L' containing L. Let **N** be made into an L' structure and suppose it is a model of Σ. Then there is a sentence in T but not in $Cn\Sigma$.

<u>Proof.</u> By Church's Thesis (or by an explicit proof) $Cn\Sigma$ is r.e., and so the set Δ of those sentences of L which are in $Cn\Sigma$ will also be r.e.. As **N** is a model of Σ , we see that T contains Δ and they cannot

be equal as T is not r.e.

Theorem 7(Godel's First Incompleteness Theorem;strong form). With the above notation, a sentence σ in T but not in $Cn\Sigma$ can be found explicitly.

Remark. Actually finding σ is difficult. Among other results, it requires obtaining $Cn\Sigma$ from Σ, which would require a very detailed analysis.

We will assume that A is contained in $Cn\Sigma$. As above, this can be ensured by replacing Σ by $\Sigma \cup \{\beta\}$.

Proof. For each integer n, obtain a wff δ_n as follows. If n is the Godel number of a wff φ with one free variable, then δ_n is $\varphi(n)$; if not, δ_n is 0=0. By Church's Thesis, the map sending n to δ_n is recursive. As $Cn\Sigma$ is r.e., so is $\{n; \delta_n \text{ is in } Cn\Sigma\}$.Let this set be weakly expressed by a wff ψ and let φ be $\neg\psi$.

Then if $\delta_n \in Cn\Sigma$ we have $\psi(n)$ in CnA, and so it is in T and $\varphi(n)$ is in F.while if $\delta_n \notin Cn\Sigma$ then $\psi(n)$ is in F and $\varphi(n)$ is in T. Now φ is a wff with one free variable; let its Godel number be k. As **N** is a model of Σ, and by definition δ_k is the same as $\varphi(k)$, we see that δ_k would be in T if it were in $Cn\Sigma$, and so find at once that δ_k must be in T but not in $Cn\Sigma$.

References

1. S.Aanderaa and D.E.Cohen, Modular machines, the word problem for finitely presented groups and Collins' theorem, Word Problems II;the Oxford Book (North-Holland,1980),1-16.

2. S.Aanderaa and D.E.Cohen, Modular machines and the Higman-Clapham-Valiev embedding theorem, Word Problems II;the Oxford Book (North-Holland,1980),17-28.

3. D.E.Cohen, Degree problems for modular machines, J.Symbolic Logic 45(1980),510-528.

4.M.Machtey and P.Young, An Introduction to the General Theory of Algorithms, North-Holland,1978.

5. M.L.Minsky, Recursive unsolvability of Post's problem of 'Tag' and other topics in the theory of Turing machines, Annals of Math. 74(1961),437-455.

UNIVERSAL TURING MACHINES (UTM) AND JONES-MATIYASEVICH-MASKING

G. Hasenjaeger
Seminar für Logik und Grundlagenforschung
der Universität Bonn

Ø. Background. When I learned from reports given by BÖRGER (spring 1982) and JONES (fall 1982) about a new combination of coding of sequences with coding of Boolean algebras, as a tool to describe the behavior of register machines, I tried to apply this tool on my earlier variants of small universal Turing machines hoping these application should lead to some sufficiently simple solutions for exponential diophantine predicates universal for r.e. sets.

1. The grid of attachment and local masks. For a suitable $q = 2^{\pi}$ ($\pi \in \mathbb{N}$ the period) let $A = \sum_{i \leq h} q^{i}$ the grid of Attachment, $M = \sum_{i+1<\pi} 2^{i} = q/2 - 1$ a basic Mask used to derive several masks, for instance $A \cdot M$. (If inverse operations are to be avoided, use "$2(M+1) = q$".)

A sequence attached to the "points" of A, can be coded as

$r = \sum_{i \leq h} r_{i} q^{i}$ for $2r_{i} < q$. Thus one digit gaps are to make sure that adjacent r_{i} do not interfer.

MATIYASEVICH's and JONES' observation: that several essential properties and relations can be expressed as elementary boolean conditions for the related finite sets S_r, will be somehow extended, where

$S_r = \{j \mid \underline{dig}(r,2,j) = 1\}$ and

dig is the digit-function known to be exponential diophantine:

$\underline{dig}(a,b,c) = d \leftrightarrow \exists xy(a = x \cdot b^{c+1} + d \cdot b^{c} + y \wedge d < b \wedge y < b^{c})$

2. Boolean algebra I. As a simple consequence of

$S_a \cap S_b = \emptyset \leftrightarrow \binom{a+b}{a} \equiv 1 \ \underline{mod}\ 2$, ("odd")

MATIYASEVICH (1975/80) after KUMMER (1852) and

$S_a \subseteq S_b \leftrightarrow \binom{b}{a} \equiv 1 \ \underline{mod}\ 2$

JONES/MATIYASEVICH (1980) after LUCAS (1877)

I shall also apply

$S_a \cap S_b = S_c \leftrightarrow \binom{a}{b} \cdot \binom{a+b-c}{a}$ odd, $\leftrightarrow$: $a \underset{\sim}{\cap} b = c$

$S_a \cup S_b = S_c \leftrightarrow \binom{c}{a} \cdot \binom{a}{c-b}$ odd, $\leftrightarrow$: $a \underset{\sim}{\cup} b = c$

$S_a \setminus S_b = S_c \leftrightarrow \binom{a}{c} \cdot \binom{b+c}{a}$ odd, $\leftrightarrow$: $a \underset{\sim}{\setminus} b = c$

The first two conditions could be made symmetric by some canceling, but there seems to be no characterisation of the resulting <u>tri</u>nomial coefficient like

$$\binom{n}{k} = \underline{\text{dig}}((b+1)^n,\ b,\ k) \quad \text{for } b > \text{all } \binom{n}{k}.$$

As only the parity of $\binom{n}{k}$ is used $b = 2^{n+1}$ seems a better choice than $b = 2^n+1$, as this choice makes

$$\binom{n}{k} \underline{\text{mod}}\ 2 := \underline{\text{dig}}((2^{n+1}+1)^n,\ 2,\ (n+1)\cdot k)$$

For "subsets" $X \underset{\sim}{\subseteq} A$ the mask $X \cdot M$ can be used to pick some parts $X \cdot M \underset{\sim}{\cap} R$ of a given sequence coded by $R \underset{\sim}{\subseteq} A \cdot M$.

<u>3. Boolean algebra II.</u> As the relational definitions of $\underset{\sim}{\cap}$ and $\underset{\sim}{\cup}$ yield no immediate extension to multiple operations, a variant for "subsets" of A seems useful. As the case of <u>seven</u> (in general 2^k-1) arguments is ample for present applications this case should suffice. The fact that some space can be assumed at the "upper" side of each "point" of A

$$Y = \bigvee_{1 \leq i < 8} X_i \quad \text{for } X_i \underset{\sim}{\subseteq} A,\ Y \underset{\sim}{\subseteq} A$$

can be described by

$$8A - \sum_{1 \leq i < 8} X_i \underset{\sim}{\subseteq} 8A - Y$$

As negation (or complementation) is $A - X_i$ or $A - Y$ respectively, for instance $Y = X_1 \vee X_2 \vee X_3 \vee \neg X_4 \vee \neg X_5$ leeds to

$$6A - X_1 - X_2 - X_3 + X_4 + X_5 \underset{\sim}{\subseteq} 8A - Y$$

or $Y = X_1 \wedge X_2 \wedge X_3 \wedge \neg X_4 \wedge \neg X_5$, hence $\neg Y = \neg X_1 \vee \neg X_2 \vee \neg X_3 \vee X_4 \vee X_5$ leeds to $5A + X_1 + X_2 + X_3 - X_4 - X_5 \underset{\sim}{\subseteq} 7A + Y$.

Some cases which have solutions simpler than the result of expressing conjunctive or disjunctive normal forms are

$$A - Y \underset{\sim}{\subseteq} X_1 + X_2 \underset{\sim}{\subseteq} 3A - Y \quad \text{for } Y = X_1 \leftrightarrow X_2 \text{ , and}$$

$$Y \underset{\sim}{\subseteq} X_1 + X_2 \underset{\sim}{\subseteq} 2A + Y \quad \text{for } Y = X_1 \leftrightarrow X_2$$

but there seems to be no general method without composition described by quantifiers.

<u>4. Universal Turing machines.</u> As decoding a number (introduced as an additional argument indicating a program or a particular machine) certainly needs more states, smaller solutions are obtained by introducing an additional program tape or program loop. Instead of targets for conditional jumps (or "gotos") an additional register to count the "in between" for <u>suspended</u> operations seems adequate. As besides conditional halt and jump three operations on <u>the</u> tape are sufficient (l, r, print, or l, r, change sign) hence are to be transcribed, a similar code should also serve for a multiple counter concept: Just <u>one</u> counter is on duty; operations are: add 1, take 1, change of counters in a given cyclic order.

<u>5. From Turing tapes to counting registers.</u> As 25 years ago finding it harder to materialize a Turing machine with an operation: change symbol (instead of: print 1, no erazing) we introduced a multitape version: one 1 on each tape, and moving for counting. By changing the tape "on duty" in a cyclic order all tapes can be operated; (i.e. I materialized "early bronze age" model and D. RÖDDING had finished a theory when MINSKY[s] became known). Though 2 counters are known to be sufficient a 3 counter version should indicate the general case. Let R_0, R_1, $R_2 \underset{\sim}{\subset} A\cdot M$ indicate the sequences of the register contents, R_0 being the working register (Formally the contents not the register will "circulate").Let $\arg(x) = x$ or $\arg(x) = 2^x$ with $\arg(x) \leq M$ (instead of the more "expensive" $\arg(x) \underset{\sim}{\subset} M$) the initial argument of the procedure; but here there is no need for coding the "moving for counting" as multiplication by 2 or by 1/2 respectively. Now the conditions for R_0, R_1, R_2 can be given as:

$$R_2 = A\cdot M \cap q\cdot(R_2 - R_2 \cap (C\cdot M) + R_0 \cap (C\cdot M))$$
$$R_1 = A\cdot M \cap q\cdot(R_1 - R_1 \cap (C\cdot M) + R_2 \cap (C\cdot M))$$
$$R_0 = A\cdot M \cap q\cdot(R_0 - R_0 \cap (C\cdot M) + R_1 \cap (C\cdot M) + E - D) + \arg(x)$$

where C, D, E, all $\underset{\sim}{\subset} A$ and determined by the coded program, decide wich operation is on:

C for the cyclic changes,

E for enlarging by one,

D for decreasing by one.

Any two-bit-code on the program tape offers an additional code for "if...". Two states $S \underset{\sim}{\subset} A$, and $A-S$ respectively, are sufficient to read such a four letter assembler loop.

<u>Remark on push down stores.</u> This kind of registers would need 3 basic operations <u>besides</u> C (in the above sense) thus no coding of "if..." would be available in 2 bits. Thus their basic operations, 2x, 2x+1, $[x/2]$, might better be programmed on counting registers, with one additional counter (as this is about small solutions, only the 2 symbols case is considered).

<u>Remark on Turing tapes.</u> Handling one Turing tape like two particularly connected pushdown stores brings it into the scope of the previous remark. - The decision to put the tape between the even points of A and to use the odd points of A as the sequence of working fields of the tape seems to offer no simpler solution.

<u>6. Reading the program loop.</u> Zero-value of S and zero-position of an additional counter whose history is coded by J indicates that three two bit words (c, e, d) are read as C, E, D respectively. The remaining word f changes S into 1. If now the halt or jump condition is not satisfied the next sign other than f just changes S into 0. If the

jump condition is satisfied each f of an additional sequence of f's enlarges the counter J. A terminal d returns S to 0, but J still being $\neq 0$, execution of c,e,d is still suspended; but each c, e, d read reduces the temporal value of J. If the jump condition is satisfied but instead of f...fd the next sign is c the evaluation of this situation in the sense of Boolean algebra II is to connect it with the top of A.

As it can he supposed that the counter J is 0 at start and halt, its history is given as

$J = q\cdot(J+J_+-J_-)$ where $J_+, J_- \underset{\sim}{\subseteq} A$ exhibit the times when there are the changes indicated by the respective subscript. A program coded as $p \underset{\sim}{\subseteq} M$ combined with a length λ (to be sure about zeroes at the top) into $m = 4^\lambda+p$ could be read by moving the program (cyclically) or moving a pointer appropriately. As scanning "at A" (for 2 bits at 3A respectively) seems simpler, circular moving of the program is preferred. A seqence of program codings where uniformly 2 bits from the lower end are moved to the upper end is described by:

$$P = A\cdot M \underset{\sim}{\cap} q/4\cdot(P+(4^\lambda-1)\cdot(3A \underset{\sim}{\cap} P)) + p$$

or in order to avoid denominators

$$4P = 4A\cdot M \underset{\sim}{\cap} q\cdot(P+(4^\lambda-1)\cdot(3A \underset{\sim}{\cap} P)) + 4p$$

Now the 2 bits S_1, S_2 of $3A \underset{\sim}{\cap} P$ can be described as

$$S_1 = A \underset{\sim}{\cap} P \quad \text{and} \quad 2S_2 = 2A \underset{\sim}{\cap} P,$$

as the temporal state of J $(=0, \neq 0)$ by

$$S_3 \underset{\sim}{\subseteq} A \quad \text{and} \quad A\cdot 2^\lambda-J \underset{\sim}{\subseteq} A\cdot 2^\lambda-S_3,$$

and the "signum" $S_0 \underset{\sim}{\subseteq} A$ of the register R_0 by

$$(M+1)A-R_0 \underset{\sim}{\subseteq} (M+1)A-S_0.$$

7. Comparisons. Whereas the use of non-uniform registers (here the program loop besides counters) yields an obvious gain of simplicity, compared e.g. with GREGUŠOVÁ/KOREC (1979) applied to register machines as arithmetized in JONES/MATIYASEVICH (1980), it is far from being optimal in terms of prefix length as given in JONES/MATIYASEVICH (1981). It seems to offer a fairly short way to the universal version compared with the earlier results as reported in DAVIS (1973) albeit no gain with respect to the harder step from "exponential diophantine" to "diophantine", see in JONES (1982). It may just be hoped that the applied techniques might prove useful, e.g. handling of multiple scanning by Boolean algebra II. As for this short contribution an extended bibliography seems out of proportion, besides the used sources some instances with sufficient bibliography will be given, - and SINGMASTER (1974), as quoted by MATIYASEVICH (1975), for the early result on binomial coefficients.

8. The machine table. A machine table that shows the result of the previous stipulations is given by:

Registers:	R_0	P		J	state	operations on			
scanned:	S_0	S_1	S_2	S_3	S	R_0	J	S_{next}	
		1	1	0	0	c		0	
	1	1	0	0	0	d		0	
		0	1	0	0	e		0	
		0	0	0	0			1	(f for if)
	0	1	1		1			0	NOOP on R_0, but
	0	1	0		1			0	end of suspension
	0	0	0		1			1	progression on P only
	1	1	1		1			0	fc is read
	1	0	0		1		j_+	1	f...f is being read
	1	1	0		1			0	f...fd is read
		0	0	1	0			0	progression on P only
		1	1	1	0		j_-	0	
		1	0	1	0		j_-	0	count down on J
		0	1	1	0		j_-	0	

Note that the less simple boolean functions are

$$J_- = (S_1 \vee S_2) \wedge (S_3 \wedge \neg S) \quad \text{and} \quad S_{next} = \neg((S_1 \vee S_2) \wedge (S_3 \wedge \neg S))$$

9. An exponential diophantine universal r.e. predicate.

In order to keep somehow recognizable the ideas behind this solution the reductions of $<$ and $\underset{\sim}{\subsetneq}$ as well as of $\wedge$ to one equation is not carried out.

$$x \in W_m \leftrightarrow \exists p\lambda q\pi MAhS_0R_0PS_1S_2S_3JUVSJ_-J_+CDER_0R_1R_2[x,\ m, \ldots],$$

where $[x, m, \ldots]$ is

$$\begin{aligned}
[m = p+4^{\lambda} &\wedge \pi \geq 4 \wedge q = 2^{\pi} \wedge 2(M+1) = q \wedge \arg(x) \leq M \\
&\wedge A\cdot q+1 = q^{h+1}+A \wedge S_0 \underset{\sim}{\subsetneq} A \wedge (M+1)\cdot A-R_0 \underset{\sim}{\subsetneq} (M+1)A-S_0 \\
&\wedge 4P = 4AM \underset{\sim}{\cap} q\cdot(P+(4^{\lambda}-1)\cdot(3A \underset{\sim}{\cap} P)) + p \\
&\wedge S_1 = P\underset{\sim}{\cap}A \wedge 2S_2 = P\underset{\sim}{\cap}2A \wedge S_3 \underset{\sim}{\subsetneq} A \wedge A\cdot 2^{\lambda}-J \underset{\sim}{\subsetneq} A\cdot 2^{\lambda}-S_3 \\
&\wedge U = S_1\underset{\sim}{\cup}S_2 \wedge V = S_3\underset{\sim}{\setminus}S \wedge J_- = U\underset{\sim}{\cap}V \wedge S = q\cdot(A - (U\underset{\sim}{\cup}V)) \\
&\wedge J_+ \underset{\sim}{\subsetneq} A \wedge 6A+S_0+S-S_1-S_2 \underset{\sim}{\subsetneq} 7A+J_+ \wedge C \underset{\sim}{\subsetneq} A \wedge 6A+S_1+S_2-S_3-S \underset{\sim}{\subsetneq} 7A+C \\
&\wedge D \underset{\sim}{\subsetneq} A \wedge 6A+S_0+S_1-S_2-S_3-S \underset{\sim}{\subsetneq} 7A+D \wedge E \underset{\sim}{\subsetneq} A \wedge 7A+S_2-S_1-S_3-S \underset{\sim}{\subsetneq} 7A+E \\
&\wedge R_0 = AM \underset{\sim}{\cap} q\cdot(R_0-(R_0 \underset{\sim}{\cap} CM) + (R_1 \underset{\sim}{\cap} CM) + E - D) + \arg(x) \\
&\wedge R_1 = AM \underset{\sim}{\cap} q\cdot(R_1-(R_1 \underset{\sim}{\cap} CM) + (R_2 \underset{\sim}{\cap} CM)) \\
&\wedge R_2 = AM \underset{\sim}{\cap} q\cdot(R_2-(R_2 \underset{\sim}{\cap} CM) + (R_0 \underset{\sim}{\cap} CM)) \\
&\wedge 4A+S_0+S_1+S_2+S \underset{\sim}{\subsetneq} 7A+q^{h}]
\end{aligned}$$

References

DAVIS (1973) Hilbert's tenth problem is unsolvable. Amer. Math. Monthly 80 (1973) 233-269.

SINGMASTER (1974) Notes on binomial coefficients I-III. J. London Math. Soc. 8 No 3, (1974) 545-560.

MATIYASEVICH (1975) A new proof of the theorem on exponential diophantine representations of enumerable sets. J. Soviet Math. 14 (1980), 1475-1486; Result announced 1975.

GREGUŠOVÁ/KOREC (1979) Small universal Minsky machines, Mathem. Foundations of Computer Science 1979. Lecture Notes in Computer Science 74. Berlin-Heidelberg-New York, 1979.

JONES/MATIYASEVICH (1980) Direct translation of register machines into exponential diophantine equations, Preprint 1980, available in the Report on the 1st GTI-workshop Lutz Priese (Ed.) Paderborn March 1983.

JONES/MATIYASEVICH (1981) Exponential diophantine representation of recursively enumerable sets, Proceedings of the Herbrand colloquium '81. Amsterdam-New York-Oxford 1982.

JONES (1982) Universal diophantine equation, J. Symbolic Logic 47 (1982) 549-571.

COMPLEXITY OF LOOP-PROBLEMS IN NORMED NETWORKS

H. Kleine Büning
Institut für Angewandte Informatik
und Formale Beschreibungsverfahren
Universität Karlsruhe
Postfach 6380
7500 Karlsruhe

Introduction. In this paper we will discuss some loop and related problems for normed networks. The decision problem whether a finite set of automata forms a basis or not is still open. So we try to classify the property to be a basis by the complexity of the mentioned problems assuming the unequality of P, NP and PSPACE. We will give a short description of the theory of normed networks, for details see [5] (in this Lecture Note).

A normed network N of a finite set $A_1,\dots,A_t$ of Mealy automata - $N \in NN(A_1,\dots,A_t)$ - consists of copies of A_i, where each output-line is connected with at most one input-line. An external input x means that one signal on the input-line x enters the net and may change the states of the components, where as a reaction of one input-signal of a component one output-signal appears. The state of the net is given by the states of its components.

In the following $x, s_1 \underset{N}{\to} s_2, y$ means that an signal on the input-line x in state s_1 of the net N appears on the output-line y and the final state of N is s_2.

A basis B is a finite set of Mealy automata, such that for each finite automaton A there is a net $N \in NN(B)$ simulating A.

Reachability Problem.

At first we will consider the reachability problem, which is defined for a fixed set A of automata as follows

Instance: Net $N \in NN(A)$, input $x \in I_N$
states $s_1, s_2 \in S_N$, output $y \in Q_N$.

Question: $x, s_1 \underset{N}{\to} s_2, y$?

<u>Theorem 1.</u> The reachability problem is PSPACE-complete for each basis.

<u>Proof:</u> The reachability problem belongs to PSPACE, because a net of size n consists of at most c^n states (perform at most c^n steps in the net and compare with x, $s_1 \underset{N}{\rightarrow} s_2$, y). Let be given a deterministic linear bounded Turing machine M.Then this DTM M with n+1 tape squares can be simulated by a net of size O(n) with components of B in polynomial time.

Each tape square can be considered as an finite automaton and therefore simulated by a net N $\in$ NN(B). Combining these nets we obtain a tape of the DTM M, where each tape square contains the program M [4,5]. Now we associate to a finite string W of length n over the input alphabet of M in polynomial time an initial state s_1 and an input x of the net (start of the computation of M). Then it holds that M accepts x iff the net reaches an accepting state s_2 and an output follows. Since there is a fixed deterministic linear bounded automata for which the acceptance problem is PSPACE-complete ([2], page 265) we have shown that the reachability problem of each basis is PSPACE-complete.

q.e.d.

For non-basis sets of automata it is still open whether or not there exists an PSPACE-complete reachability problem. Naturally there are non-basis sets with polynomial reachability problem. For two often used automata F(flip-flop) and K(connection) we will show that the problem is decidable in polynomial time.

F = 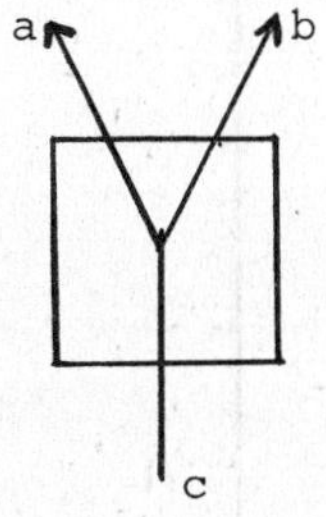

$S_F = \{l,r\}$

c,l $\rightarrow$ r,a

c,r $\rightarrow$ l,b

K = 

$S_K = \{1\}$

a,1 $\rightarrow$ 1,d

b,1 $\rightarrow$ 1,d

<u>Theorem 2.</u> The reachability problem for the non-basis set $\{F,K\}$ belongs to P.

<u>Proof:</u> In the first step we will mark the inputs and outputs of the components of the net with variables and finally solve linear equations in rational numbers. If the marking consists only of natural numbers we have got the number of signals which locally correct run on this lines during the computation of the net.
In a second step we have to decide whether or not the local solution is a global solution too.

<u>Local solution</u>

Let be given the net as follows

$(F_1,\ldots,F_t,K_{t+1},\ldots,K_r)^{b_1\ldots b_m}_{a_1\ldots a_m}$, where the names of inputs and outputs of the components are

F_i = 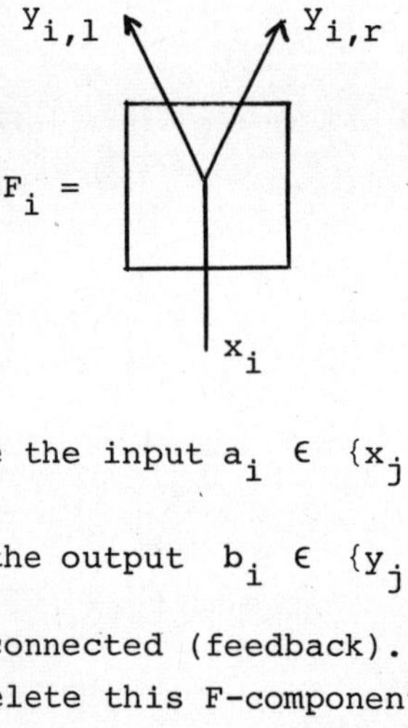

and K = 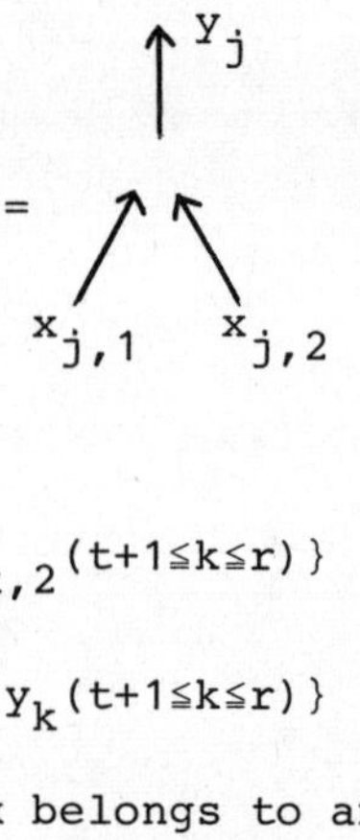

and

where the input $a_i \in \{x_j(1 \leq j \leq t), x_{k,1}, x_{k,2}(t+1 \leq k \leq r)\}$

and the output $b_i \in \{y_{j,l}, y_{j,r}(1 \leq j \leq t), y_k(t+1 \leq k \leq r)\}$

are connected (feedback). If the input x belongs to an F-component, we delete this F-component and associate input x to the component with F depending on the state of F. So we can suppose that the external input x is $x_{t+1,1}$.

The states are given as $(S^1_1,\ldots,S^1_t)$ and $(S^2_1,\ldots,S^2_t)$.

For $i := 1$ to t do {mark F-component}

begin

mark x_i with variable k_i;

if $s_i^1 = s_i^2$ then mark $y_{i,l}$, $y_{i,r}$ with $k_i / 2$;

if s_i^1 = right, s_i^2 = left then begin

mark $y_{i,e}$ with $(k_i-1)/2$;

mark $y_{i,r}$ with $(k_i+1)/2$

end;

if s_i^1 = left, s_i^2 = right then begin

mark $y_{i,l}$ with $(k_i+1)/2$;

mark $y_{i,r}$ with $(k_i-1)/2$

end

end

The interpretation for example of the last case is that for k_i inputs on x_i we obtain $(k_i+1)/2$ outputs on $y_{i,l}$ and $(k_i-1)/2$ on $y_{i,r}$ if both expressions have values in $\mathbb{N}$ for some $k_i \in \mathbb{N}$.

For $i := 1$ to $(r-t)$ do {mark K-components}

begin

mark $y_{t+i,1}$ with variable k_{t+2i-1};

mark $x_{t+i,2}$ with variable k_{t+2i};

mark y_{t+i} with expression $k_{t+2i}+k_{t+2i-1}$

end

For $j := 1$ to m do { feedback }

{output b_j is connected with input a_j}

b_j is marked with $f_{b_j} := \Sigma q_r k_r + q_o$ for some $q_r, q_o \in \mathbb{Q}$;

a_j is marked with k_w for some $w \in \{1,..,2r-t\}$

begin

if $q_w \neq 1$

then for all occurences of k_w substitute the variable k_w by the expression $(\sum_{r \neq w} q_r k_r + q_o)(1-q_w)^{-1}$

{f_{b_j} signals on b_j are now k_w inputs on a_j}

<u>else</u> delete the feedback b_j-a_j and for all occurences of k_w substitute k_w b_y O

{$k_w + \sum_{r \neq w} q_r k_r + q_o$ outputs are now k_w inputs,
these k_w inputs never leave the net (local loop) and therefore we cut the connection}

<u>end.</u>

For all inputs $x' \in I_N (x \neq x')$ and all outputs $y' \in Q_N (y' \neq x)$ we set the corresponding variables to $\mathbb{Q}$ and compute the expressions as long as possible.
After this we have only expressions with one variable corresponding to input x or rational numbers on the outputs and inputs. Because of one input x and the desired result output y we set the variable $k_{t+1} := 1$ and output y to 1.

In the next step we must control whether or not the marking of the net is locally correct, that means for each output and input of the components the numbers are natural numbers and there isn't a local contradiction according to components and connections.
Now set every value $\mathbb{Q}$-$\mathbb{N}$ to 0 and check the values on the components and connections.
In case of a contradiction we can follow $x, s_1 \not\to_N s_2, y$.

But unfortunately a correct solution in $\mathbb{N}$ isn't sufficient for $x, s_1 \to_N s_2, y$ as the following example shows:

Net 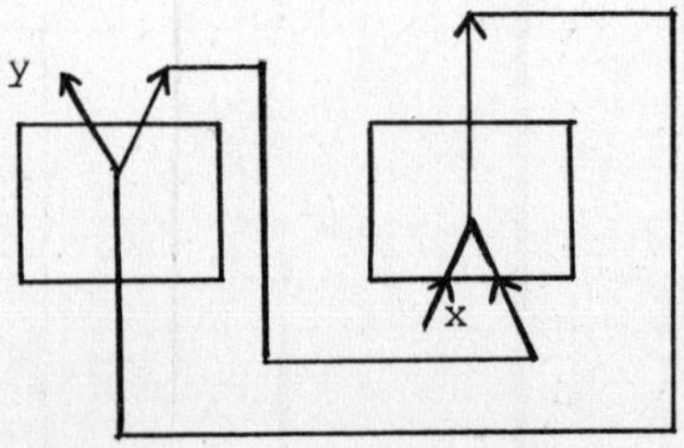

$= (F_1, K)^{y_1, y_2}_{x_3\ x_1}$ $s_1 = s_2 = \text{left}$

Question: $x, s_1 \underset{N}{\rightarrow} s_2, y$?

1)

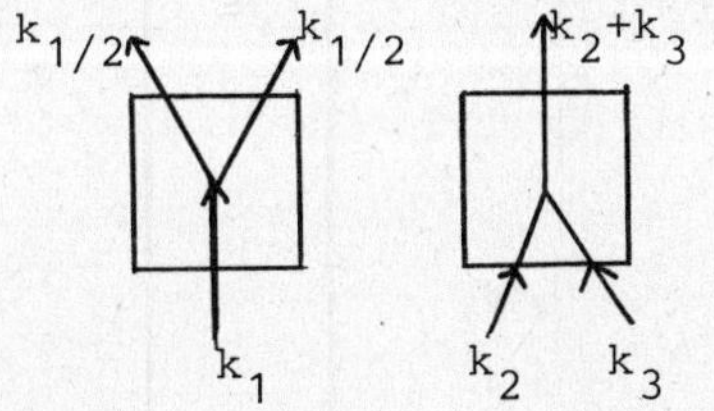

2) {feedback $y_{1,r} - x_3$}

$k_{1/2} = k_3$

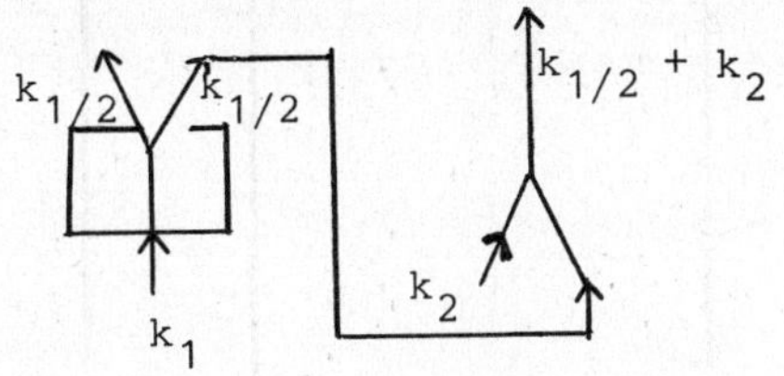

3) {feedback $y_2 - x_1$}

$k_1 = k_1/2 + k_2 \rightarrow k_1/2 = k_2$

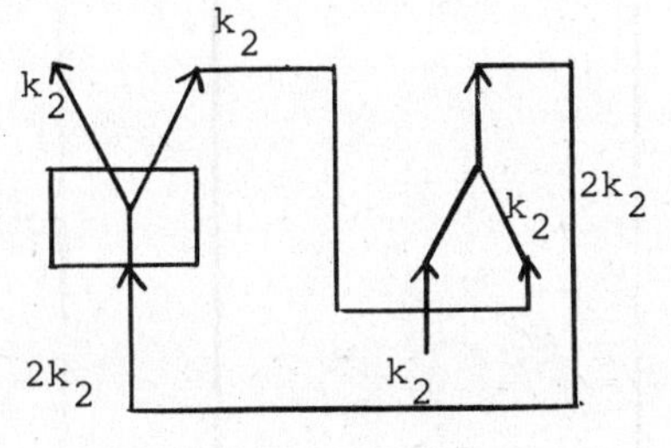

4) $k_2 = 1$ (input x), correct solution in $\mathbb{N}$ (local),

but $x, \text{left} \underset{N}{\rightarrow} \text{right}, y$ in contradiction to our desired result $x, \text{left} \underset{N}{\rightarrow} \text{left}, y$.

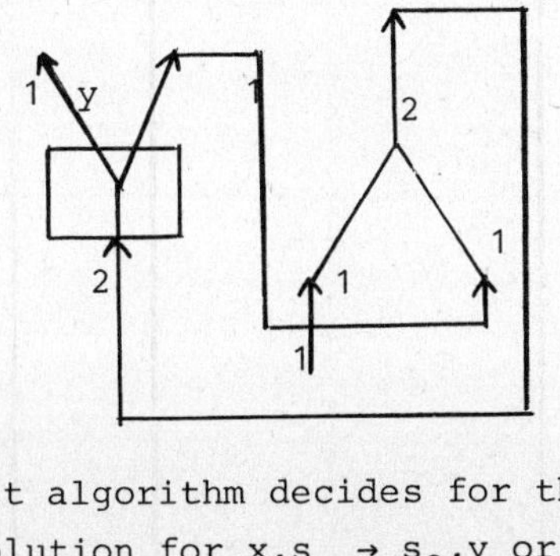

Now the last algorithm decides for the solution in $\mathbb{N}$ whether this is a global solution for $x, s_1 \underset{N}{\rightarrow} s_2, y$ or not.

Now we will mark the last output for each component with *.

<u>For</u> i:=1 <u>to</u> t <u>do</u> {F_i-components}

<u>if</u> s_i^2 = right <u>then</u> mark $y_{i,l}$ with * if the value on $y_{i,l}$ is greater than 0.

<u>else</u> mark $y_{i,r}$ with * if the value on $y_{i,r}$ is greater than 0;

<u>For</u> i:=1 <u>to</u> (r-t) do {K-components}

<u>if</u> one of the inputs of K_{t+i} has a * <u>then</u> mark the output with *.

Finally we check:

(1) Is there a cyclic path marked with * ?

(2) Is there a path marked with * from input x to the output y?

In case of a cyclic path the signal never reaches the output. If there isn't a cyclic path then (1) garanties that an signal on input x lead to an output y, such that $x,s_1 \underset{N}{\rightarrow} s_2,y$.

(1) and (2) can be decided in polynomial time. Altogether we have constructed a polynomial time algorithm.

q.e.d.

<u>Input-Loop Problem.</u>

In this chapter we will discuss the complexity to decide whether for a net an input doesn't produce an output.

<u>Definition.</u> For a fixed set A of automata the input-loop-problem - called IL(A) - is defined as:

Instance: Net $N \in NN(A)$, input x.

Question: $\exists s \in S_N \;\; \exists y \in Q_N : x,s \underset{N}{\rightarrow} y$?

For the input-loop problem the complexity doesn't depend on to be a basis as we will see in the following theorems.

We introduce a further automaton E which forms together with K a basis [5].

$I_E = \{t,s\}$, $Q_E = \{t_u, t_d, s'\}$, $S_E = \{up, down\}$

t, up → up, t_u

t, down → down, t_d

s, up → down, s'

s, down → up, s'

E = 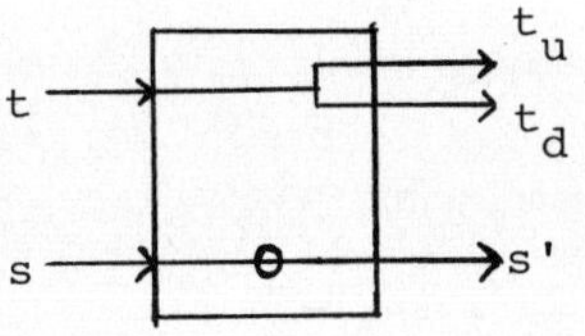

<u>Theorem 3.</u> The input-loop problem for {F,K} and the basis {E,K} belongs to P.

<u>Proof:</u> At first we will show that IL({E,K}) belongs to P. For {F,K} the theorem follows immediately because of the following simulation

F ≅ 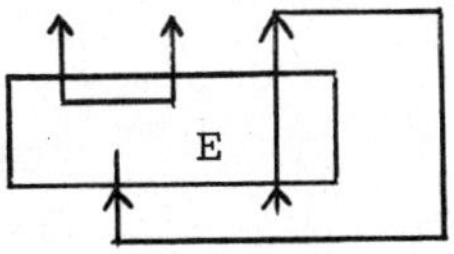

,which shows that the polynomial algorithm for E can be applied to nets N ∈ NN(F,K).

The algorithm for {E,K} reduces the size of the given net as long as the set of external outputs isn't empty or we have found a direct connection (for one component) of the input x with some output of the reduced net.
For technical purpose we need a further automaton T defined as

$I_T = \{t\}$, $Q_T = \{t_u, t_d\}$, $S_t = \{up, down\}$

t, up → up, t_u

t, down → down, t_d

T = 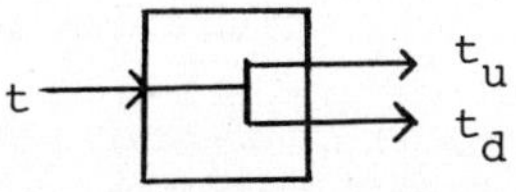

Algorithm IL({E,K}): {net N $\in$ NN({E,K}), input x $\in I_N$}

result:= FALSE; Q := Q_N;

while Q not empty and result = FALSE do

begin

if $\exists$y $\in$ Q :

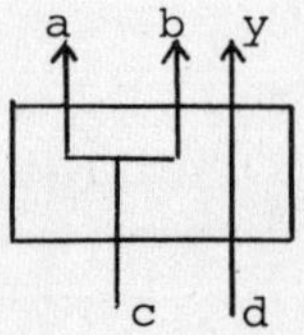

{component E of N}.

then begin

if d = x then result := TRUE;

if d $\in I_N$-{x} then Q := Q-{y} and

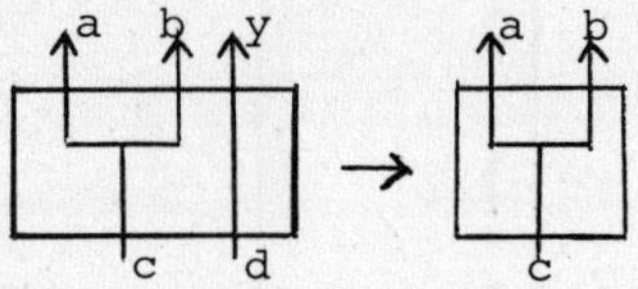

{delete $\uparrow^{y}_{d}$};

if d $\notin I_N$ then indentify d := y and

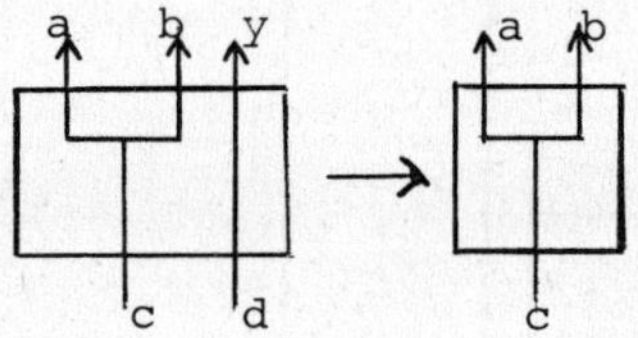

{if output y then d must be reached before, the behavior of E with input c doesn't depend on the state changed by input d }

end

else

if $a_1 \in$ Q and

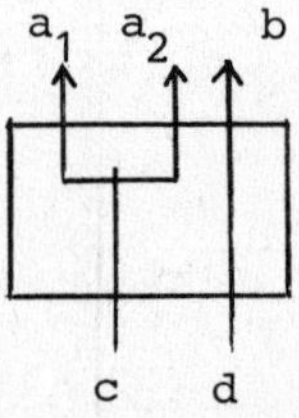

then begin

if c = x then result := TRUE;

{initial state can be choosen, s.th. input c leads to a_1}

if c $\in I_N$-{x} then Q := Q - {a_1,a_2} and

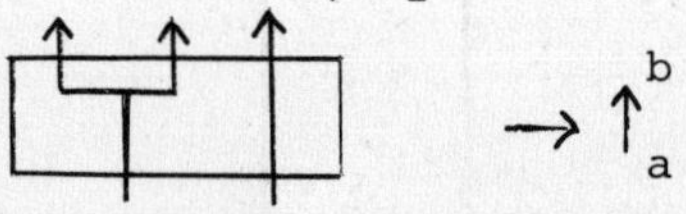

{delete input c};

if $c \notin I_N$ then identify $c := a_1$ and $Q := Q - \{a_2\}$

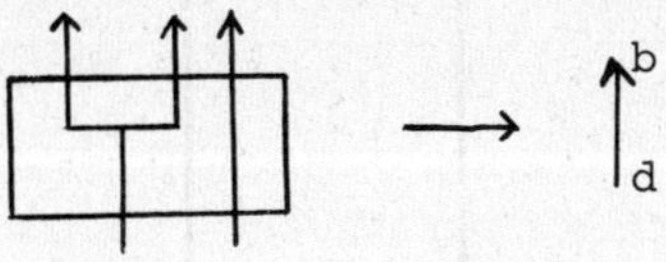

{if output a_1 then c must be reached before, the initial state can be choosen, s.th. input c leads to a_1}

end

else

if 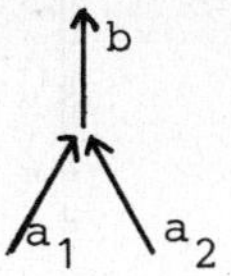

and $b \in Q$

then begin

if $a_1 = x$ or $a_2 = x$ then result := TRUE;

if $a_1, a_2 \in I_N - \{x\}$ then $Q := Q - \{b\}$ and delete  ;

if $a_1, a_2 \notin I_N$ then $Q := Q \cup \{a_1, a_2\} - \{b\}$;

{if output y then a_1 or a_2 must be reached before}

if $a_1 \in I_N$, $a_2 \notin I_N$ then 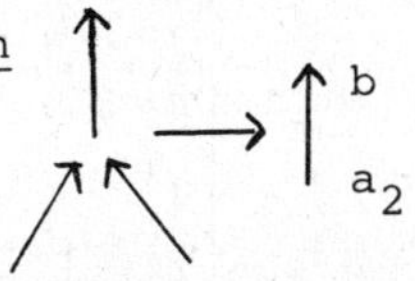

else 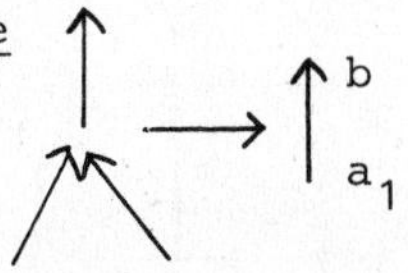

{delete input different from x}

end

else

if 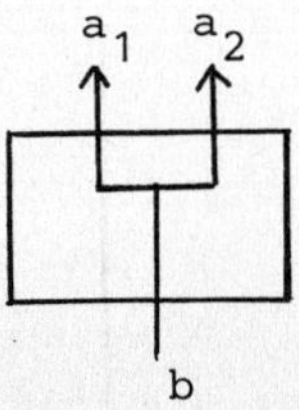

and $a_1 \in Q$

<u>then</u> <u>begin</u>
    <u>if</u> b = x <u>then</u> result := TRUE;
    {initial state can be choosen, s.th. b leads to a_1}
    <u>if</u> $b \in I_N - \{x\}$ <u>then</u> Q := Q-$\{a_1, a_2\}$ and delete T;
    {delete inputs different from x }
    <u>if</u> $b \notin I_N$ <u>then</u> identify b := a_1 and delete T;
    {if output a_1 then b must be readed before}
    <u>end</u>
<u>else</u>
<u>if</u> $\uparrow^{a}_{b}$ and $a \in Q$ <u>then</u>
    <u>begin</u> <u>if</u> b = x <u>then</u> result := TRUE;
        <u>if</u> $b \in I_N - \{x\}$ <u>then</u> delete $\uparrow^{a}_{b}$;
        if $b \notin I_N$ <u>then</u> identify b := a
    <u>end</u>
<u>end</u>; {end of while}
<u>if</u> result = TRUE <u>then</u> $\exists s \in S_N \ \exists y \in Q : x,s \underset{N}{\rightarrow} y$.

The algorithm IL({E,K}) reduces in each step (while-block) the size of the net and therefore gives in polynomial time the answer TRUE or Q is empty and the value of result remains FALSE.

q.e.d.

The next theorem gives us an example of a non-basis set {B,K}, for which the input-loop problem is NP-complete.

The automaton B is defined as
$I_B = \{u,d,t\}$, $Q_B = \{t_u, t_d, u', d', e\}$, $S_B = \{s_u, s_d, s, s*\}$

$t, s_u \rightarrow s*, t_u$
$t, s_d \rightarrow s*, t_d$
$d, s \rightarrow s_d, d'$
$u, s \rightarrow s_u, u'$
rest $\rightarrow s*, e$

B =

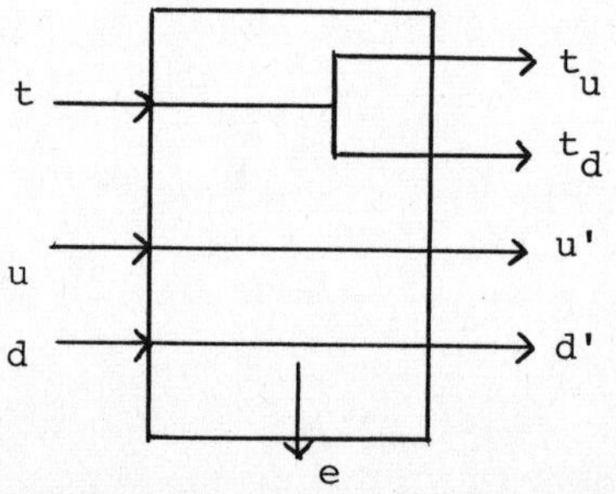

The set {B,K} cannot be a basis, because after at most 3 uses of the component B the state becomes constant (s*) and the output is e for every input.

<u>Theorem 4:</u> The input-loop for the non-basis set {B,K} is NP-complete.

<u>Proof</u>: That IL({B,K}) belongs to NP can be seen by guessing a state s of the net and a polynomial test whether input x and state s produces an output (at most 3 uses per component B). In order to show taht IL ({B,K}) is complete we specify a polynomial transformation from 3-SAT to IL({B,K}).

Let be given a formula $C = \bigwedge_{1 \leq i \leq n} C_i$ in 3-conjunctive normalform

with variables $a_1,...,a_k$ and literals $d_1,...,d_{3n}$. Now we will construct to C a net $N_C \in NN(\{B,K\})$. For each literal d_j we have a component B- called B_j-connected as follows

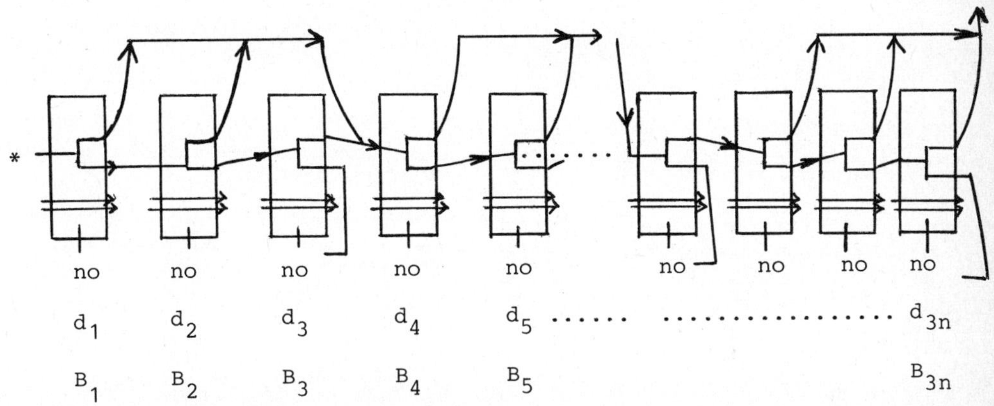

$*$ = test for a truth assignment

Furthermore we associate to each variable $a_i (1 \leq i \leq k)$ a component B, called B^i. All outputs and inputs of $B^i (1 \leq i \leq k)$ different from t, t_u and t_d get the name "no". The net N_C has only input t of B^1 and one output y.

The outputs "no" are collected

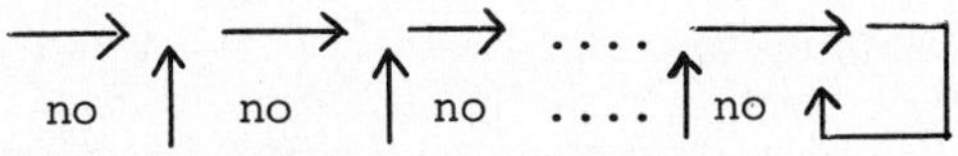

and an signal on one of the "no"-line never leaves the net N_C.

For the literals $d_{i_1},\ldots,d_{i_l}$ with $d_{i_p} = (1 \leq p \leq l)$ t_u resp. t_d of B^i is connected with input u resp. d of B_{i_1} and output u'resp.d' of B_{i_p} $(1 \leq p \leq l)$ with u resp. d of $B_{i_{p+1}}$.

For the literals $d_{i_{l+1}},\ldots,d_{i_q}$ with $d_{i_p} = \bar{a}_i$ $(l+1 \leq p < q)$ u'resp.d' B_{i_l} is connected with input d resp. u of $B_{i_{l+1}}$ and output d'reps.u' of B_{i_p} $(l+1 \leq p < q)$ with input d resp. u of $B_{i_{p+1}}$. The output u' and d' of B_{i_q} is connected with t of B^{i+1} for $i < k$. For $i = k$ the outputs u' and d' of B_{k_q} are connected with $*$-test for the truth assignment. The states s_u resp. s_d of B_i are copied to s_u resp. s_d of $B_{d_{i_j}}$.

If there is a truth assignment satisfying C, then we guess state s for $B_j (1 \leq j \leq 3n)$ and state s_u for $B^i (1 \leq i \leq k)$ if a_i becomes true, otherwise state s_d. For this state of N_C after input x we obtain output y(true). Now we assume starting with input y that we reach output y(true) for some initial state of N.

Then $B^i (1 \leq i \leq k)$ must have initial state s_u or s_d and all components $B_j (1 \leq j \leq 3n)$ have initial state s, otherwise an input x would lead to "no" (loop).
After state s_u resp. s_d of $B^i (1 \leq i \leq k)$ has been copied, such that for all $j (1 \leq j \leq 3n)$ if d_j is a_i then B_j has state s_u resp. s_d and if d_j is $\bar{a}_i$ then B_j has state s_d resp. s_u, we obtain after the test of the truth assignment (a_i=true iff B^i has state s_u) output y(true) iff the truth assignment satisfies C.
Hence we have construct a polynomial transformation from 3SAT to IL{B,K}). Altogether we have proved our desired result.
q.e.d.

Several attempts have failed to find a non-basis set for which the input-loop problem is PSPACE-complete whereas the basis {H,K} has PSPACE-complete IL-problem.

The automation H is defined as

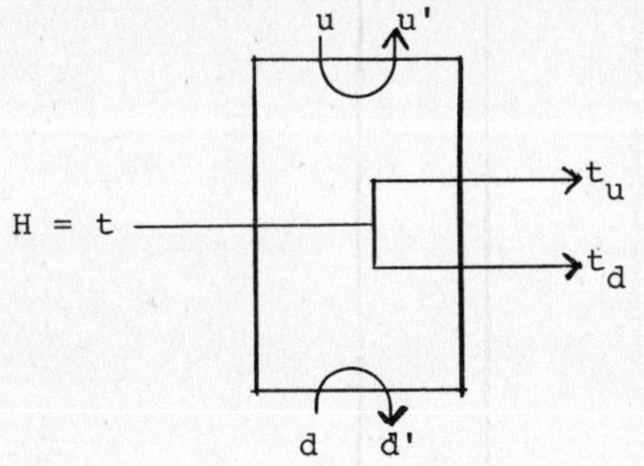

$S_H = \{up, down\}$

$t, up \rightarrow up, t_u$
$t, down \rightarrow down, t_d$
$u, up \rightarrow up, u'$
$u, down \rightarrow up, u'$
$d, up \rightarrow down, d'$
$d, down \rightarrow down, d'$

For the proof is very helpfull to present another loop problem.

Definition. For a fixed set A of automata the input-state-loop problem - called ISL(A) - is defined as:

Instance: Net $N \in NN(A)$, $x \in I_N$, $s \in S_N$.

Question: $\exists y \in Q_N$: $(x,s) \underset{N}{\rightarrow} y$?

Theorem 5. The input-state-loop problem for the basis $\{H,K\}$ is PSPACE-complete.

Proof: Since $\{H,K\}$ is a basis (4), the proof of theorem 4 can be adopted without attention to the final state. The net has in this case only one output which can be reached iff the linear bounded automata accepts the string.

q.e.d.

Theorem 6: The input-loop problem for the basis $\{H,K\}$ is PSPACE-complete.

Proof: We will extend the construction of the net N in proof of theorem 5, such that at the beginning of the computation the net reaches the desired state s and then works like the net N. Suppose that the H-components of N are denoted as $H_1^1, \ldots, H_1^t$. Now we substitute each H_1^i by a simple net consisting of two additional H-components H_2^i and H_3^i as follows.

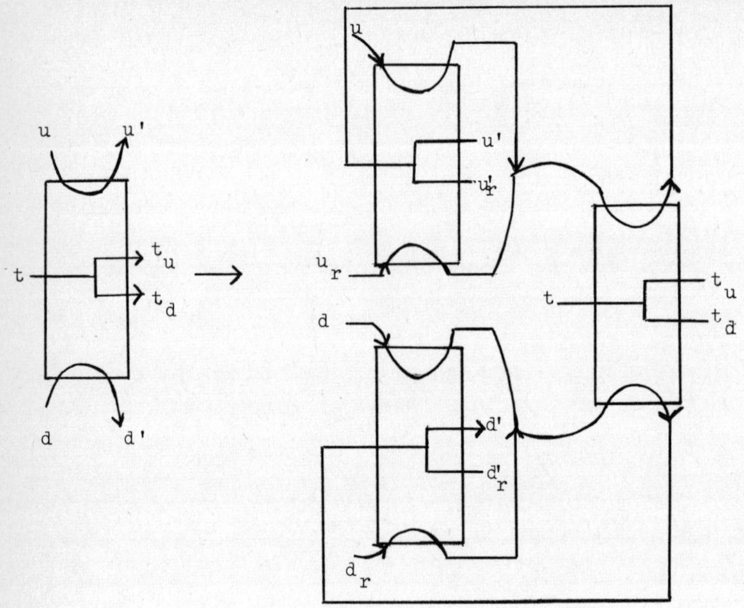

The inputs t,d,u and outputs t_u, t_d, u' and d' are connected as before. In theorem 5 we had to decide whether $\exists\, y: (x,s) \underset{N}{\rightarrow} y$.

Let $s = (s_1,..,s_t)$ be the state of N. Now we connect the inputs d_r, u_r and outputs u'_r, d'_r getting a net N_1, such that an input x_1 leads in a first run to state $(s_1,...,s_t)$ for the components $(H_1^1,...,H_1^t)$ and the signal leaves the net N_1 iff

$\exists\, y \in Q_N : (x,s) \underset{N}{\rightarrow} y.$

{connection of u_r, d_r, u'_r, d'_r}

<u>If</u> $s_1 = u$ <u>then</u> $x_1 := u_r(H_2^1)$
<u>else</u> $x_1 := d_r(H_3^1)$;

<u>for</u> i := 1 <u>to</u> t-1 <u>do</u>

<u>if</u> $s_i = u$ <u>then</u> <u>if</u> $s_{i+1}=u$ <u>then</u> $u_r(i+1) = u'_r(i)$
<u>else</u> $d_r(i+1) = u'_r(i)$

<u>else</u> <u>if</u> $s_{i+1}=u$ <u>then</u> $u_r(i+1) = d'_r(i)$
<u>else</u> $d_r(i+1) = d'_r(i)$;

<u>if</u> $s_{i+1}=u$ <u>then</u> u_r' is connected with $x \in I_N$.

<u>else</u> d_r' is connected with $x \in I_N$.

Obviously for N_1 the state at the beginning doesn't play any role, because state s will occur after some steps in the subnet N. Hence we have a polynomial transformation $x \to x_1$; $s_1, N \to N_1$ for which it holds

$\exists y \in Q_N : (x,s) \underset{N}{\to} y$ iff

$\exists s \in S_{N_1}$ $\exists y \in Q_{N_1}$: $(x_1,s) \underset{N_1}{\to} y$. Since ISL({H,K}) is PSPACE complete we have proved our desired result.

q.e.d.

<u>References:</u>

(1) Karp, R.M. 1972 , "Reducibility among combinatorial problems", in R.E. Miller, J.W. Thatcher (eds.) Complexity of Computer Computations, Plenum Press, New York, 85-103.

(2) Garey, M.R., Johnson, D.S. 1979 "Computer and Intractability", W.H. Freeman and Company, San Francisco

(3) Kleine Büning, H., Ottmann, Th. 1977 , "Kleine universelle mehrdimensionale Turingmaschinen", EIK 13(1977)

(4) Ottmann, Th., 1973 , "Über Möglichkeiten zur Simulation endlicher Automaten durch eine Art sequentieller Netzwerke aus einfachen Bausteinen", ZMLG 19 (1973)

(5) Priese,L., Rödding, D., Schätz, D., Brüggemann, A., "Modular Decomposition of Automata", in Börger, E., Hasenjäger, G., Rödding, D. (eds), Proceedings of the Symposium "Rekursive Kombinatorik", Münster 1983, LNCS

On the solvability of the extended
$\forall\exists \wedge \exists\forall^*$ - Ackermann class with identity.

by

Stål Aanderaa.

Abstract

Dreben and Goldfarb define the class 5.7 to be the class of all schemata of the form

$$\forall y\exists xPy \wedge \forall y_1\forall y_2\ldots\forall y_n G$$

where G is quantifier-free and contains, apart from the dyadic predicate letter P, only monadic predicate letters. Dreben and Goldfarb prove the docility of the class 5.7, but the solvability of this class is left open (see Dreben and Goldfarb(1979) p. 136 and p.264). We shall give an outline of a proof of the solvability of the class 5.7. In fact, we shall give an outline of a proof of the solvability of an extension of this class.

1. Introduction

In 1928 Ackermann solved the decision problem for formulas with prefix of the form $\forall\exists^*$. (Here $\exists^*$ means $\exists\exists\exists\ldots\exists$ and $\forall^*$ means $\forall\forall\ldots\forall$.) This class is usually called the Ackermann class. We shall not in this paper discuss this class. But in a paper of 1936 Ackermann proposed to study another class, and the decision problem for this class is still open, according to Dreben and Goldfarb, p. 135. A class is called docile if there exists a method for deciding, given F in the class, whether F has a finite model. Ackermann showed that his class of 1936 is docile. Dreben and Goldfarb prove in their book of 1979 pp. 136-137 that an extension of this class is docile. This extended result was first shown by Denton (see Dreben and Denton, 1963; Denton 1963) and by Kostyrko (1962). We shall define an extension 5.7b of this class and prove that this class is solvable. Since I would like to publish this result now, and since I do not have the time to work out every detail, the proof will be somewhat sketchy. But I hope that an experienced reader will be able to fill in the gaps if he chooses to spend some time on it. Some of the lemmas are stated without proof, but the basic steps in the construction are stated explicitly. A more complete proof will perhaps be published later. Suppose $\mathfrak{A}$ is a model for a formula ϕ and suppose $\mathfrak{A}$ is in some sense periodic. Then ϕ also has a finite model. We shall define a new kind of model which we will call a skew-periodic model. The main difference between a periodic model and a skew-periodic model is that in the skew-periodic model the period will not always continue on the other side of the diagonal.

We shall prove that all formulas in the class 5.7b have skew-periodic models. Suppose that ϕ has a skew-periodic model, this does not imply that ϕ has a finite model. But skew-periodic models have many properties in common with finite models. A class is called finitely controllable if each formula in it that has a model at all has a finite model. A finitely controllable class is solvable. In nearly the same way we shall prove that a class where each formula in it that has a model at all has a skew-periodic model, is solvable. This solves the class 5.7b. The main steps in the proof are as follows. With regard to the consistent formulas in the class 5.7b, three cases arise.

1. Those formulas that have finite models.
2. Those formulas where P may be interpreted as the successor function i.e. $P\bar{m}\bar{n} \leftrightarrow m+1=n$.
3. The last case is the case where long chains may be shortened, i.e. it is consistent to assume that a long chain $Paa_2, Pa_2a_3, \ldots, Pa_{r-2}a_{r-1}Pa_{r-1}a$ may be shortened to the chain $Pab_2, Pb_2b_3, Pb_3b_4 \ldots Pb_{k-2}b_{k-1}, Pb_{k-1}b$ where k is less than a fixed constant. The prove of the existence of skew-periodic models in case 1 and 2, is a more or less standard procedure. But in the case 3 we shall apply a combinatorial principle called Ramsey's partition theorem. As far as we know this is the first time Ramsey's partition theorem has been used in order to prove that a class is solvable. Ramsey used his partition theorem to prove that the spectrum of a Bernays-Schönfinkel formula with identity is either finite or cofinite.

2. Basic definitions and the main theorem

Dreben and Goldfarb (1979) defines the class 5.7 as follows:

$$\forall y \exists x Pyx \wedge \forall y_1 \ldots \forall y_n G,$$

where G is quantifier-free and contains, apart from the predicate letter P, only monadic predicate letters. We shall define the class 5.7b as follows.

Definition 1.

The class 5.7b consists of all formulas of the form:

$$\forall y \exists x Pyx \wedge \exists x_0 \forall y_1 y_2 \ldots \forall y_n G \tag{1}$$

where G is quantifier-free and contains only dyadic and monadic predicate letters and the identity sign $\equiv$. Moreover, if a dyadic predicate letter R occurs in G and R is different from P and different from $\equiv$, then we require that the following is a tautology:

$$G \rightarrow ((Py_1y_2 \rightarrow Ry_1y_2) \wedge (Ry_1y_2 \wedge Ry_2y_3 \rightarrow Ry_1y_3)). \tag{2}$$

The aim of this paper is to give an outline of a proof of the following theorem.

Theorem 1.

The decision problem for the class 5.7b defined above is solvable.

In this paper let L be a language with only dyadic and monadic predicate letters and the identity sign $\equiv$, without function symbols. Constant symbols are permitted.

Definition 2a.

A structure $\mathcal{A}$ for L with domain $|\mathcal{A}| = A$ is called skew-periodic with period p and distance d (here p and d are non-negative integers and $p \geq 1$) by virtue of f iff f is a mapping of $\omega = \{0,1,2,\dots\}$ onto $|A|$ such that the following three conditions are satisfied for all $j \geq d$.

(i) $Qf(j)$ iff $Qf(j+p)$

(ii) Suppose also $i \geq d$, then

$Rf(i)f(j)$ iff $Rf(i+p)f(j+p)$

(iii) Suppose either $i \leq j-d$ or $j+d+p \leq i$, then

$Rf(i)f(j)$ iff $Rf(i)f(j+p)$

and

$Rf(j)f(i)$ iff $Rf(j+p)f(i)$

Here Q is an interpretation of a monadic predicate letter in L and R is an interpretation of a dyadic predicate letter in L.

Definition 2b.

A structure A for L is called skew-periodic iff there exist positive integers p and d and a mapping of ω onto $|A|$ such that A is skew-periodic with distance d and period p by virtue of f.

Given a formula ϕ in the class 5.7b; then let L_ϕ be the language where the non-logical symbols consist of all predicate symbols in ϕ, and let $L_\phi(\omega)$ be the language consisting of L_ϕ and in addition the constant symbols $\{\bar{i} \mid i \in \omega\}$.

Definition 3.

A is a normal numerical structure for ϕ iff A is a structure for $L_\phi(\omega)$ and the following conditions are satisfied.

(i) $|A| = \omega$

(ii) If $i \in \omega$ then $\bar{i}^A = i$ (that is $\bar{i}$ is interpreted as i in A)

(iii) If $i, j \in \omega$ and $j = i+1$ then $A \models P\bar{i}\bar{j}$.

(Here P is the dyadic predicate letter explicitly mentioned in the definition of the class 5.7b. $A \models P\bar{i}\bar{j}$ means $P^A ij$ is true in A, where P^A is the interpretation of P in A.)

Definition 4.

$\mathbb{A}$ is an S-structure for ϕ iff $\mathbb{A}$ is a normal numerical structure for ϕ and $\mathbb{A} \models P\bar{i}\bar{j}$ iff $i+1=j$ for all $i>0$, $i\in\omega, j\in\omega$.

Definition 5.

$\mathbb{A}$ is a normal numerical model for ϕ in the class 5.7b iff $\mathbb{A}$ is a normal numerical structure for ϕ and

$$\mathbb{A} \models \forall y \exists x Pyx \wedge \forall y_1 \forall y_2 \ldots \forall y_n G(x_0/\bar{0})$$

where ϕ is of the form (1).

Here $G(x_0/\bar{0})$ means the result of replacing x_0 by $\bar{0}$ in G. Note also that when $\mathbb{A}$ is a normal numerical structure for ϕ then $\mathbb{A} \models \forall y \exists x Pyx$.

Definition 6.

Let ϕ be in the class 5.7b. Then $\mathbb{A}$ is an S-model for ϕ iff $\mathbb{A}$ is a normal numerical model for ϕ and $\mathbb{A}$ is an S-structure for ϕ.

Definition 7.

Let ϕ be in the class 5.7b. Then $\mathbb{A}$ is an F-structure for ϕ iff there exist integers r and k, $r<k$, such that $|\mathbb{A}| = \{0,1,2\ldots,k\}$ and the following conditions are satisfied

(i) $\mathbb{A}$ is a structure for $L_\phi(\{0,1,2,\ldots,k\})$
(ii) $0 \leq i \leq k$ implies $\bar{i}^{\mathbb{A}} = i$
(iii) $\mathbb{A} \models P\bar{0}\bar{1}$ if $k>0$, otherwise $\mathbb{A} \models P\bar{0}\bar{0}$
(iv) $i>0$ implies $\mathbb{A} \models P\bar{i}\bar{j}$ iff either $2 \leq i+1 = j \leq k$ or both $i=k$ and $j=r$.

Definition 8.

Let ϕ be in the class 5.7b. Then $\mathbb{A}$ is an F-model for ϕ iff $\mathbb{A}$ is an F-structure for ϕ and

$$\mathbb{A} \models \forall y \exists x Pyx \wedge \forall y_1 \forall y_2 \ldots \forall y_n G(x_0/\bar{0})$$

where ϕ is of the form (1)

Note that an F-model $\mathbb{A}$ for ϕ is also skew-periodic. We simply choose $d=r$ and $p=k-r+1$. If $i \leq k$ let $f(i) = i$ otherwise let $f(i) = f(i-p)$. Then f is a mapping of ω onto $|\mathbb{A}|$ and is skew-periodic by virtue of this function.

Definition 9.

Let ϕ be in the class 5.7b. Then $\mathbb{A}$ is a standard skew-periodic structure for ϕ iff either $\mathbb{A}$ is an F-structure for ϕ or the following two conditions are satisfied:

(i) $\mathbb{A}$ is a normal numerical structure for ϕ

(ii) $\mathbb{A}$ is a skew periodic structure by virue of the identity function I (i.e. $I(i)=i$ for all $i \in \omega$)

Definition 10.

Let ϕ be in the class 5.7a. Then $\mathbb{A}$ is a standard skew-periodic model for ϕ iff either $\mathbb{A}$ is standard-skew periodic structure for ϕ and $\mathbb{A}$ is a normal numerical model for ϕ or $\mathbb{A}$ is an F-model for ϕ.

Definition 11.

Let ϕ be in the class 5.7b. Then $\mathbb{A}$ is a skew-periodic S-structure for ϕ iff A is an S-structure for ϕ and A is a standard skew-periodic structure for ϕ.

Definition 12.

Let ϕ be in the class 5.7b. Then A is a skew-periodic S-model for ϕ iff A is a skew-periodic S-structure for ϕ and A is a standard skew-periodic model for ϕ.

3. Outline of a proof of Theorem 1.

We shall outline a proof of Theorem 1 by starting a sequence of lemmas, sometimes without proof, and sometimes with the basic steps only.

Given the formula ϕ where ϕ is of the form

$$\forall y\exists xPyx \wedge \exists x_0\forall y_1\forall y_2\ldots\forall y_n G \tag{1}$$

in the class 5.7b; we shall define the formulas $\phi_{r,k}$, ϕ'_m, ϕ''_r as follows: Let $r \leqslant k$, then $\phi_{r,k}$ is

$$\forall y\exists xPyx \wedge \exists x_0\exists x_1\ldots\exists x_k\forall y_1\forall y_2\ldots\forall y_n$$

$$(G \wedge Px_kx_r \wedge \bigwedge_{i=1}^{k} Px_{i-1}x_i)$$

ϕ'_m is

$$\forall y\exists xPyx \wedge \exists x_0\exists x_1\ldots\exists x_m\forall y_1\forall y_2\ldots\forall y_n$$

$$(G \wedge (\bigwedge_{i=1}^{m} Px_{i-1}x_i) \wedge \bigwedge_{i=1}^{m} ((\bigwedge_{j=0}^{i} \neg Px_ix_j) \quad \bigwedge_{j=i+2}^{m} \neg Px_ix_j))$$

Let $r > n$. Then ϕ''_r is

$$\forall y\exists xPyx \wedge \exists x_0\forall y_1\forall y_2\ldots\forall y_r$$

$$(G \wedge ((\bigwedge_{i=2}^{r} Py_{i-1}y_i) \rightarrow \bigvee_{i=3}^{r} \quad \bigvee_{j=1}^{i-2} Py_jy_i))$$

(Here $\bigwedge_{i=1}^{m} \phi_i$ means $\phi_1 \wedge \phi_2 \wedge\ldots\wedge \phi_m$ and $\bigvee_{i=1}^{m} \phi_i$ means $\phi_1 \vee \phi_2 \vee\ldots\vee\phi_m$.)

Lemma 1.

Suppose ϕ is in the class 5.7b. Then ϕ is consistent iff at least one of the following conditions is satisfied.

(i) There exist integers r,k where $r \leq k$ such that $\phi_{r,k}$ is consistent.

(ii) ϕ'_m is consistent for all integers $m > 0$.

(iii) There exists an integer $r \geq n$ such that ϕ''_r is consistent.

Lemma 2.

Suppose $\phi_{r,k}$ is consistent for some r and k where $r \leq k$. Then ϕ has an F-model.

Lemma 3.

Suppose ϕ'_m is consistent for all $m > 0$. Then ϕ has a skew-periodic S-model.

Lemma 4.

Suppose ϕ''_r is consistent for some $r \geq n$. Then ϕ has a standard skew-periodic model.

We shall not in this paper give any details of the proof of Lemmas 1 and 2. But we shall outline proofs of Lemmas 3 and 4.

Outline of a proof of Lemma 3.

Suppose ϕ'_m is consistent for all $m > 0$. By the compactness theorem we can then show that ϕ'_m and hence ϕ has an S-model $\mathfrak{A}$. $\mathfrak{A}$ need not be skew-periodic. We shall construct a skew-periodic S-model $\mathfrak{B}$ for ϕ. In order to define $\mathfrak{B}$ we need certain new sets and concepts.

Let $U_s = \{z_0, z_1, z_2, \ldots, z_s\}$ be a finite set of new variables. Let $V_0(\phi, s) = \{\psi \mid \psi$ is a signed atomic formula containing a predicate letter of ϕ and variables among $U_s\}$. Note that the set $V_0(\phi, s)$ is finite.

We shall use the letter h to denote mappings of U_s into $\{\bar{i} \mid i \in \omega\}$. (We do not require that h is one-to-one). Suppose $\psi \in V_0(\phi, s)$. Then we shall define $h(\psi)$ to be the result of replacing each z_j in ψ by $h(z_j)$. The profile of $\mathcal{A}$ relative to h and ϕ is defined as follows:

$$\mathrm{Prf}(\mathcal{A}, h, \phi) = \{\psi \mid \psi \in V_0(\phi, s) \wedge \mathcal{A} \models h(\psi)\}.$$

Since $\mathrm{Prf}(\mathcal{A}, h, \phi) \subseteq V_0(\phi, s)$ it follows that $\mathrm{Prf}(\mathcal{A}, h, \phi)$ is a finite set.

Let

$$H'_k = \{h \mid h \text{ maps } U_n \text{ into } \{\bar{0}, \bar{1}, \bar{2}, \ldots, \bar{k}\},\ h(z_0) = \bar{0} \text{ and } h(z_n) = \bar{k}\}$$

$$H''_k = \{h \mid h \text{ maps } U_{n+1} \text{ into } \{\bar{0}, \bar{1}, \ldots, \bar{k}, \overline{k+1}\},\ h(z_0) = \bar{0},\ h(z_n) = \bar{k} \text{ and } h(z_{n+1}) = \overline{k+1}\}$$

Let $V_1(\phi, s) = \{X \mid X \subseteq V_0(\phi, x)\}$ and $V_2(\phi, s) = \{Y \mid Y \subseteq V_1(\phi, s)\}$
Since $V_0(\phi, s)$ is finite it follows that $V_1(\phi, s)$ and $V_2(\phi, s)$ are finite.

For each $k > 0$, we define sets of profiles as follows.

$$\Gamma'_k = \{\mathrm{Prf}(\mathcal{A}, h, \phi) \mid h \in H'_k\}$$

$$\Gamma''_k = \{\mathrm{Prf}(\mathcal{A}, h, \phi) \mid h \in H''_k\}$$

and let

$$\Delta' = \{\Gamma'_k \mid k \in \omega\},\ \Delta'' = \{\Gamma''_k \mid k \in \omega\}$$

Then $\Gamma'_k \subseteq V_1(\phi, n), \Gamma'_k \in V_2(\phi, n), \Gamma''_k \subseteq V_1(\phi, n+1), \Gamma''_k \in V_2(\phi, n+1)$, $\Delta' \subseteq V_2(\phi, n)$ and $\Delta'' \subseteq V_2(\phi, n+1)$. Hence Γ'_k, Γ''_k, Δ' and Δ'' are all finite sets.

Let $X \in V_1(\phi, n+1)$ we then define $\sigma(X) = X \cap V_0(\phi, n)$ and

$$\rho(X) = \{\psi(z_{n+1}/z_n) \mid \psi \in X \text{ and } z_n \text{ does not occur in } \psi\}.$$

Moreover let $Y \in V_2(\phi, n+1)$, we then define

$$\sigma(Y) = \{\sigma(X) \mid X \in Y\}$$

$$\rho(Y) = \{\rho(X) \mid X \in Y\}$$

The following lemma is easily proved.

Lemma 5.

$$\sigma(\Gamma''_k) = \Gamma'_k \quad \text{and} \quad \rho(\Gamma''_k) = \Gamma'_{k+1}.$$

Given $Y \in \Delta'$, we define $\tau(Y)$ as follows: Let k be the first integer such that $Y = \Gamma'_k$, then let $\tau(Y) = \Gamma''_k$. Hence we have $\sigma(\tau(Y)) = Y$.

Then we define infinite sequences $Y'_0, Y'_1, Y'_2, \ldots$ and Y''_0, Y''_1, Y''_2 as follows.

$$Y'_0 = \Gamma'_0,\ Y''_0 = \Gamma''_0$$

$$Y'_{i+1} = \rho(Y''_i) \qquad i = 0,1,2,\ldots$$

$$Y''_{i+1} = \tau(Y'_{i+1}) \qquad i = 0,1,2,\ldots$$

Then there exsist numbers d and p such that $Y'_{i+p} = Y'_i$ and $Y''_{i+p} = Y''_i$ for all $i > d$.

Lemma 6.

There exists a unique S-structure $\mathcal{B}$ such that the following conditions are satisfied.

(i) $Y'_k = \{\mathrm{Prf}(\mathcal{B},h,\phi) \mid h \in H'_k\}$
(ii) $Y''_k = \{\mathrm{Prf}(\mathcal{B},h,\phi) \mid h \in H''_k\}$
(iii) For each dyadic predicate letter R occuring in ϕ and R different from P and $\equiv$ $\mathcal{B}$ satisfies the following condition:

$$\mathcal{B} \models \forall y_1 \forall y_2 \forall y_3((Py_1y_2 \rightarrow Ry_1y_2) \wedge (Ry_1y_2 \wedge Ry_2y_3 \rightarrow Ry_1y_3)).$$

Next we may prove the following lemma:

Lemma 7.

$\mathcal{B}$ is a skew-periodic S-model for ϕ.

This completes the outline of the proof of the Lemma 3.

Outline of a proof of Lemma 4.

Suppose ϕ_r'' is consistent for some $R > n$. Note that if ϕ is in the class 5.7b then the corresponding ϕ_r'' is in the class 5.7b. Since $\phi_r'' \to \phi$ it follows that ϕ has a finite model if ϕ''_r has a finite model. Hence we may without loss of generality suppose that ϕ_r'' is not satisfiable in a finite domain. Then ϕ_r'' has a normal numerical model $\mathfrak{A}$. We shall now define a one-to-one mapping J of ω into ω such that the following conditions are satisfied.

(i) $J(0) = 0$ and $\mathfrak{A} \models J(j)\ J(j+1)$ for all $j \in \omega$.

(ii) $\mathfrak{A} \restriction$ Range (J) is skew-periodic.

The following lemma is easily proved.

Lemma 8.

Suppose $a,b \in \omega$ and $a < b$. Then there exists an integer $s < r$ and a chain of integers $a_0, a_1, \ldots, a_s$ such that $a_0 = a$, $a_s = b$, $\mathfrak{A} \models P\bar{a}_j\bar{a}_{j+1}$ and $a_j < a_{j+1}$ for all $j = 0,1,\ldots,s-1$.

We may without loss of generality suppose that ϕ_r'' contains the identity sign $\equiv$. Let F be mapping of $\omega\times\omega\times\{0,1,2,\ldots,r\}$ into ω such that the following conditions are satisfied.

(i) If $a < b$ then $F(a,b,0) = a$

(ii) If $a < b$ then $F(a,b,r) = b$

(iii) If $0 < j < r-1$, $a < b$ and $F(a,b,j) = b$, then $F(a,b,j) < F(a,b,j+1)$

(iv) If $a < b$ and $\sigma < j < r-1$ then $F(a,b,j) < F(a,b,j+1)$ and $\mathfrak{A} \models (P\ \overline{R(a,b,j)}\ \overline{F(a,b,j+1)} \vee (\overline{F(a,b,j)} \equiv \bar{b}))$

(v) If $b < a$ then $F(a,b,j) = F(b,a,j)$.

Let $s = 5r$ and let $g(a,b,c,d,e,z_j)$ be defined as follows.

$$g(a,b,c,d,e,z_j) = \begin{cases} F(0,a,j) & \text{if } 0 \leq j \leq r \\ F(a,b,j-r) & \text{if } r < j \leq 2r \\ F(b,c,j-2r) & \text{if } 2r < j \leq 3r \\ F(c,d,j-3r) & \text{if } 3r < j \leq 4r \\ F(d,e,j-4r) & \text{if } 4r < j \leq 5r \end{cases}$$

We shall now use a combinatorial result due to Ramsey (1928). Ramsey's theorem is also stated and used in Dreben and Goldfarb 1979 pp.211-212 and stated an proved in Drake 1974 pp. 70-72 and pp. 204-206.)
If S is a set and $m > 0$, we let $S^{(m)}$ be the class of m-membered subsets of S.

Ramsey's Partition Theorem A.

For every infinite set S, every $m > 0$ and every function K on $S^{(m)}$ with finite range, there is an infinite $T \subseteq S$ such that if $t_1 \in T^{(m)}$ and $t_2 \in T^{(m)}$ then $K(t_1) = K(t_2)$.

Let $S = \{\bar{i} \mid i \in \omega\}$. Let $t \in S^{(5)}$. Then we define h_t as follows. Choose $a,b,c,d,e \in \omega$ such that $t = \{a,b,c,d,e\}$ and $a < b < c < d < e$. Then let $h_t(z_j) = g(a,b,c,d,e,z_j)$. Then it follows that for each $t \in \omega^{(5)}$ h_t is a mapping of U_s into $\{\bar{j} \mid j \in \omega\}$. (Recall $U_s = \{z_0, z_1, z_2, \ldots, z_s\}$ where $s=5r$.) Let $K(t) = \mathrm{Prf}(\mathbb{A}, h_t, \phi''_r)$. Then K is a mapping of $S^{(5)}$ into $V_1(\phi'', s)$, which is finite. In accordance with Ramsey's partition theorem, we can choose an infinite $T \subseteq S$ such that if $t_1, t_2 \in T^{(5)}$, then $K(t_1) = K(t_2)$. Let J_0 be a mapping of ω into ω such that $J_0(0) = 0$, $J_0(i+1) > J(i)$ for all $i \in \omega$ and $T = \{J_0(i) \mid i \in \omega$ and $i > 0\}$. Let

$$A' = \{F(J_0(i), J_0(i+1), j) \mid i \in \omega \text{ and } 0 \leq j \leq r\}.$$

Choose J to be a mapping of ω into ω such that $A' \equiv \{J(i) \mid i \in \omega\}$ and $J(i) < J(i+1)$ for all $i \in \omega$. Then it follows for all $i \in \omega$ that $\mathbb{A} \models P\overline{J(i)}\ \overline{J(i+1)}$. Let $\mathbb{A}' = \mathbb{A} \upharpoonright A'$. (Here $\mathbb{A} \upharpoonright A'$ is the result of restricting the structure $\mathbb{A}$ to A'. Then $|\mathbb{A} \upharpoonright A'| = |\mathbb{A}| \cap A' = A'$ since $A' \subseteq |\mathbb{A}|$. Moreover if $i,j \in A'$ then $\mathbb{A}' \models R\bar{i}\bar{j}$ iff $\mathbb{A} = R\bar{i}\bar{j}$ and $\mathbb{A}' \models Q\bar{i}$ iff $\mathbb{A} \models Q\bar{i}$). Let $\mathbb{B}$ be a monadic numerical structure for ϕ''_r such that when Q is a normal predicate in

ϕ_r'', then $\mathcal{B} \models Q\bar{i}$ iff $A' \models Q\, J(i)$ for all $i \in \omega$, and if R is a dyadic predicate in ϕ_r'', then $\mathcal{B} \models R\bar{i}\bar{j}$ iff $A' \models R\, \overline{J(i)}\ \overline{J(j)}$. Since $t_1, t_2 \in T^{(5)}$ implies $K(t_1) = K(t_2)$, it is easy to prove that $\mathcal{B}$ is a standard skew-periodic model for ϕ. This finishes the outline of the proof of Lemma 4.

From lemmas 1,2,3 and 4 follows immediately the following lemma.

Lemma 9.

Let ϕ be in the class 5.7b. Then ϕ is consistent iff ϕ has a standard skew-periodic model.

Definition 13.

Let ϕ be in the class 5.7b. Then we shall say that $\mathcal{B}$ is a finite base for standard periodic structure for ϕ with distance d and period p iff there is a number $r > d + 3p$ such that the following conditions are satisfied.

(i) $|\mathcal{B}| = \{0,1,2,\dots,r\}$

(ii) If $d \leq j \leq r-p$ then $\mathcal{B} \models Q\bar{j} \leftrightarrow Q\overline{j+p}$ for all monadic predicates Q which occur in ϕ.

(iii) If $d \leq j \leq r-p$ and $d \leq i \leq r-p$, then

$$\mathcal{B} \models R\bar{i}\bar{j} \leftrightarrow R\overline{i+p}\ \overline{j+p}$$

for all dyadic predicated R which occur in ϕ.

(iv) Suppose $d \leq j \leq r-p$ and either $0 \leq i \leq j-d$ or $j+d+p \leq i \leq r$, then

$$\mathcal{B} \models (R\bar{i}\bar{j} \leftrightarrow R\bar{i}\ \overline{j+p}) \wedge (R\bar{j}\bar{i} \leftrightarrow R\overline{j+p}\ \bar{i})$$

for all dyadic predicated R which occur in ϕ.

It is now possible to prove the following lemma.

Lemma 10.

Let ϕ be in the class 5.7b and let ϕ be of the form

$$\forall y\, \exists x Pyx \wedge \exists x_0 \forall y_1 \ldots \forall y_n G.$$

Then ϕ is consistent iff there exist numbers d,p and $r = 3(p+d+1)(n+1)$ and a finite base $\mathcal{B}$ for a standard periodic structure with distance d and period p such that $|\mathcal{B}| = \{0,1,2,\ldots,r\}$ and

$$\mathcal{B} \models \forall y_1 \forall y_2 \ldots \forall y_n\; G(x_0/\,\bar{0}).$$

By using Lemma 10 we then have the following decision procedure for satisfiability of the class 5.7b. At the same time we make a test for $r = 1,2,3\ldots$ to see whether there exists a proof of the inconsistency of ϕ of length r or there exists a finite base $\mathcal{B}$ for a standard periodic structure with period p and distance d such that $r = (d+p+1)(n+1)$ and $|\mathcal{B}| = \{0,1,2,\ldots r\}$ and such that $\mathcal{B} \models \forall y_1 \forall_2 \ldots \forall y_n\; G(x_0/\bar{0})$. In accordance with Lemma 10 this procedure will terminate for some r. This finishes the outline of the proof of Theorem 1.

References

Ackermann, Wilhelm

1928 "Über die Erfüllbarkeit gewisser Zählaüsdrucke," Math. Ann. V. 100 pp. 638-649,

1936. "Beiträge zum Entscheidungsproblem der mathematischen Logik". Math. Ann.V. 112 pp. 414-432.

Drake, Frank R,

1974 Set theory. An introduction to large cardinals. North-Holland. Amsterdam.

Denton, John

1963 Applications of the Herbrand theorem. Ph.D. thesis, Harvard University, Cambridge, Mass.

Dreben, Burten and John Denton

1963 "Three solvable cases," Notices Amer.Math.Soc. V.10 p 590.

Dreben, Burton and Goldfarb, Warren D.,
1979. The decision problem. Solvable classes of quantificational formulas. Addison-Wesley Reading, Massachusetts

Kostyrko, V.F.
1962 "On a case of the decision problem in the restricted predicate calculus". Uspekhi.Mat.Nauk V.17 (106) p.213.

Ramsey, F.P.,
1930 "On a problem of formal logic", Proc. London Math. Soc (Ser.2) V.30 pp. 264-286.

REDUCTIONS FOR THE SATISFIABILITY WITH A SIMPLE INTERPRETATION OF THE PREDICATE VARIABLE

MICHAEL DEUTSCH
UNIVERSITÄT BREMEN

In 1964 Kostyrko proved the following

Theorem: For every formula α of the predicate logic one can find a formula β of $\forall\exists\ldots\exists\forall(0,1)$ with
α is satisfiable
$\Longleftrightarrow$ β is satisfiable,
α is satisfiable in a finite model
$\Longleftrightarrow$ β is satisfiable in a finite model.

A fórmula β belongs to $\Pi(0,1)$, if β is a prenex formula of the first order predicate logic with the prefix Π without identity, without function variables und with only one predicate variable E, which is binary.

Gurevich proved a similar result with $\beta \in \forall\exists\forall\exists\ldots\exists(0,1)$.

One might ask, whether it is possible to find simple models for β, if it is satisfiable or satisfiable in a finite model. Kostyrko's and Gurevich's proofs do not supply simple models. In 1981 I published the following

Lemma: For every formula α of the theory of types one can find a formula β of $\forall\exists\ldots\exists\forall(0,1)$ with
α is satisfiable
$\Longleftrightarrow$ β is satisfiable on N with interpretation of E by $\subseteq^{\circ}$,
α is satisfiable in a finite model
$\Longleftrightarrow$ β is satisfiable on a segment $\{u \mid u \leq v\}$ of N with interpretation of E by $\subseteq^{\circ}$.

For all natural numbers x,y we define

$x \in^{\circ} y \Longleftrightarrow$ y has the digital 1 in the xth position of the binary representation (starting counting with position No. 0 from the right),

$x \subseteq^{\circ} y \Longleftrightarrow x \in^{\circ} y \vee x=y$.

For the satisfiability we allow nonstandardsemantics.

This Lemma can be strengthened to get a refining of Kostyrko's theorem, which I will publish soon. In this paper we want to change the prefix of β at the cost of a more complicated interpretation of the predicate variable E. The prefixes $\forall\exists\ldots\exists$ and $\exists\forall\exists\ldots\exists$, which are used in this paper, are unattainable in the classical theory of reduction types. This means that the structure (N;I(E)) used herein resp. (U;I(E)) for subsets U of N can not be defined by axioms with the prefix $\exists\ldots\exists\forall\exists\ldots\exists$. The more complicated interpretation of E is necessary to recognize special properties of natural numbers x by checking whether xI(E)x is true or not. The results which are presented here are applications of the standard form of enumerable predicates on "Quotiententermmengen" published in [8].

Definition: Let N_2 be the smallest subset of N, which contains 0 and 1 and with all x,y for $x \neq y$ also 2^x+2^y. For $x,y \in N$ we define

$$x \in_2^o y \iff \begin{cases} x \in^o y & \text{for } x \neq y \\ x \in N_2 \setminus \{1\} & \text{for } x=y. \end{cases}$$

Theorem 1: For every formula α of the theory of types one can find a formula β of $\exists\forall\exists\ldots\exists$ (0,1) with

1.) α is satisfiable

$\iff$ β is satisfiable on N with interpretation of E by $\in_2^o$

$\iff$ β is satisfiable on N or a finite subset of N with interpretation of E by $\in_2^o$,

2.) α is satisfiable in a finite model

$\iff$ β is satisfiable on a finite subset of N with interpretation of E by $\in_2^o$.

3.) Let be $\beta \equiv \exists u \forall r \exists s_1 \ldots s_\mu \beta_o$, where β_o does not contain quantifiers, and let be R_o the corresponding quantifierfree expression containing as prime formulas only $u \in_2^o v$. Then we can give boundaries for the $\exists$-quantifiers, if α is satisfiable:

$$\alpha \text{ is satisfiable} \iff \forall c \underset{[s_1 \leq 2^c]\ [s_2,\ldots,s_\mu \leq c]}{\exists s_1 \ldots\ldots\ldots\ldots s_\mu u} (R_o u c s_1 \ldots s_\mu \wedge u=0)$$

The remarks in square brackets may be added or omitted. Moreover we can write

$$u=0 \iff \underset{[v,w \leq 3]}{\exists vw} (u \in^o v \wedge \neg v \in_2^o v \wedge u \in^o w \wedge v \in^o w \wedge w \in_2^o w) ,$$

where $a \in^o b$ is a abbreviation for $a \in_2^o b \wedge \neg b \in_2^o a$.

We will now start to prove the theorem. The predicate $\in^o$ on N is the projection of the predicate $\in$ on the following domain $\mathfrak{M}$.

Definition: Let $\mathfrak{M}$ be the smallest set, which contains $\emptyset$ and with all x,y also $\{x\}$ and $x \cup y$. The function $\mathfrak{n}: \mathfrak{M} \mapsto N$ with

$$\begin{aligned}
\mathfrak{n}(\emptyset) &= 0 \\
\mathfrak{n}(\{x\}) &= 2^{\mathfrak{n}(x)} \\
\mathfrak{n}(x \cup y) &= \mathfrak{n}(x) + \mathfrak{n}(y) \quad \text{for } x \cap y = \emptyset
\end{aligned}$$

is a one-one mapping of $\mathfrak{M}$ onto N, and for all $x,y \in \mathfrak{M}$ we have

$$x \in y \iff \mathfrak{n}(x) \in^{\circ} \mathfrak{n}(y) .$$

The subset N_2 of N corresponds to the following subset $\mathfrak{M}_2$ of $\mathfrak{M}$.

Definition: Let $\mathfrak{M}_2$ be the smallest subset of $\mathfrak{M}$, which contains $\emptyset$ und $\{\emptyset\}$ and with all x,y for $x \neq y$ also $\{x,y\}$. For $x,y \in \mathfrak{M}$ we define

$$x \in_2^* y \iff \begin{cases} x \in y & \text{for } x \neq y \\ x \in \mathfrak{M}_2 \wedge x \neq \{\emptyset\} & \text{for } x = y \end{cases} .$$

Obviously $\in_2^*$ corresponds to $\in_2^{\circ}$, if we use $\mathfrak{n}$ as mapping. One can easily verify the following equivalences for all $x,y \in \mathfrak{M}$:

$$\begin{aligned}
x \in y &\iff x \in_2^* y \wedge \neg\, y \in_2^* x \\
x = \{\emptyset\} &\iff \neg\, x \in_2^* x \wedge \exists u (x \in u \wedge u \in_2^* u) \\
x = \emptyset &\iff \exists y\, (y = \{\emptyset\} \wedge x \in y) \\
x \neq y &\iff \exists u\, (u \in x \wedge u \notin y \vee u \in y \wedge u \notin x) \\
x \in \mathfrak{M}_2 &\iff x = \{\emptyset\} \vee x \in_2^* x \\
x \in \mathfrak{M} \setminus \mathfrak{M}_2 &\iff \neg\, x \in_2^* x \wedge \exists y\, (y = \{\emptyset\} \wedge x \neq y)
\end{aligned}$$

For $x,y \in \mathfrak{M}_2$ we have additionaly

$$x = y \iff x = \{\emptyset\} \wedge y = \{\emptyset\} \vee x \in_2^* y \wedge y \in_2^* x .$$

But for all $x,y \in \mathfrak{M}$ we can not define $x=y$ without using $\forall$-quantifiers. The subset $\mathfrak{M}_2$ of $\mathfrak{M}$ is a "Quotiententermmenge" (compare [8]) with the atomic element $\emptyset$ and the binary successor function

$$n_o(x,y) = \begin{cases} \{x,y\} & \text{for } x \neq y \\ \{\emptyset\} & \text{otherwise.} \end{cases}$$

For the application of the results of [8] we need a one-one mapping of $\mathfrak{M}_2 \times \mathfrak{M}_2$ in $\mathfrak{M}_2$. We cannot use the ordinary pair $\{\{x\},\{x,y\}\}$, because this might be in $\mathfrak{M} \setminus \mathfrak{M}_2$. For $x,y \in \mathfrak{M}_2$ we define $f_o(x) := n_o(x,\emptyset) = \{x,\emptyset\}$ and

$$\begin{aligned}
[x,y] &:= \{\{x,\emptyset\},\{y,\{\emptyset\}\}\} \quad \text{for } x \neq \emptyset \wedge y \neq \{\emptyset\} , \\
\langle x,y\rangle &:= [f_o \circ f_o(x),\, f_o \circ f_o(y)] .
\end{aligned}$$

The function $\mathfrak{p} := \lambda xy\ \langle x,y\rangle$ is a one-one mapping of $\mathfrak{M}_2 \times \mathfrak{M}_2$ in $\mathfrak{M}_2$. Let Π be the set of all $\langle x,y\rangle$ for $x,y \in \mathfrak{M}_2$ and $\mathfrak{p}_1$, $\mathfrak{p}_2$ any functions on $\mathfrak{M}_2$ with

$$\mathfrak{p}_1(\langle x,y\rangle) = x \wedge \mathfrak{p}_2(\langle x,y\rangle) = y \quad \text{for all } x,y \in \mathfrak{M}_2 .$$

Let $\mathcal{R}_1$ be the set of predicates on $\mathfrak{M}_2$, which are formed by $\wedge$ and $\vee$ out of the atomic predicates $x \neq \emptyset$ and $x \notin \Pi$ and equations. The functions on both sides of the equations are formed out of $\lambda\emptyset$, n_o, $\mathfrak{p}$, $\mathfrak{p}_1$, $\mathfrak{p}_2$ and identity functions $\lambda x_1 \ldots x_\varrho\, x_i$ $(1 \leq i \leq \varrho)$ by substitution. According to [8] one can find for every ϱ-ary enumerable predicate P on $\mathfrak{M}_2$ a $(\varrho+2)$-ary predicate $R \in \mathcal{R}_1$ with

$$Px_1 \ldots x_\varrho \iff \underset{[L\emptyset r]}{\exists r}\ \underset{Lsr}{\forall s}\ Rrsx_1 \ldots x_\varrho \qquad (1).$$

Lsr means: s is a member of the smallest subset of $\mathfrak{M}_2$, which contains r and with any $\mathfrak{p}(u,v)$ also u.

The proposition $L\emptyset r$ in square brackets may be added or not. Moreover the functions $\mathfrak{p}_1$ and $\mathfrak{p}_2$ are used only on Π and all terms or subterms t used on the righthand side of (1) have the property Vtr where Vxy means: x is a member of the smallest subset of $\mathfrak{M}_2$, which contains y und with any $\mathfrak{p}(u,v)$ also u and v. Obviously we have on $\mathfrak{M}_2$

$$z = n_o(x,y) \iff x \neq y \wedge x \in z \wedge y \in z \vee x = y \wedge z = \{\emptyset\},$$

$$z = \mathfrak{p}_1(x) \wedge x \in \Pi \iff \exists y\ x = \langle z,y \rangle,$$

$$z = \mathfrak{p}_2(x) \wedge x \in \Pi \iff \exists y\ x = \langle y,z \rangle,$$

$$|z| < 2 \iff z = \emptyset \vee z = \{\emptyset\},$$

$$\neg \exists u\ z = f_o \circ f_o(u) \iff |z| < 2 \vee \exists z_1 z_2 (z_1 \neq z_2 \wedge z_1 \in z \wedge z_2 \in z \wedge \neg(z_1 = \emptyset \wedge \emptyset \in z_2 \vee z_2 = \emptyset \wedge \emptyset \in z_1)),$$

$$\begin{aligned} z \notin \Pi \iff & |z| < 2 \vee \exists xy\ (x \neq y \wedge x \in z \wedge y \in z \wedge (|x| < 2 \vee |y| < 2)) \\ & \vee \exists x x_1 x_2 y y_1 y_2\ (x \neq y \wedge x_1 \neq x_2 \wedge y_1 \neq y_2 \wedge x \in z \wedge y \in z \wedge x_1 \in x \\ & \quad \wedge x_2 \in x \wedge y_1 \in y \wedge y_2 \in y \wedge \neg(\emptyset \in x \wedge \{\emptyset\} \in y \vee \{\emptyset\} \in x \\ & \quad \wedge \emptyset \in y)) \\ & \vee \exists x x_1 y y_1\ (x \neq y \wedge x_1 \neq \emptyset \wedge y_1 \neq \{\emptyset\} \wedge x \in z \wedge y \in z \wedge x_1 \in x \\ & \quad \wedge \emptyset \in x \wedge y_1 \in y \wedge \{\emptyset\} \in y \wedge (\neg \exists u\ x_1 = f_o \circ f_o(u) \\ & \qquad \vee \neg \exists v\ y_1 = f_o \circ f_o(v))) \end{aligned}$$

We have now

Lemma 1: For every ϱ-ary $(\varrho \geq 1)$ recursive enumerable predicate P on $\mathfrak{M}_2$ one can find a $\tau \in N$ and a $(\varrho+\tau+2)$-ary quantifierfree expression R_1 containing as prime formulas only $u \in_2^* v$ with

$$Px_1 \ldots x_\varrho \iff \underset{\substack{[L\emptyset r] \\ [Vx_i r] \\ 1 \leq i \leq \varrho}}{\exists r}\ \underset{Lsr}{\forall s}\ \underset{\substack{[Vt_\nu r] \\ 1 \leq \nu \leq \varrho}}{\exists t_1 \ldots t_\tau}\ R_1 r s t_1 \ldots t_\tau x_1 \ldots x_\varrho .$$

Any remark in square brackets may be added or omitted. The quantifiers are interpreted on $\mathfrak{M}_2$.

We restrict now ourselves to the representation of recursive enumerable predicates on the following model $\tilde{N} \subseteq \mathfrak{M}_2$ of the set of natural numbers:

$$0 := \{\emptyset, \{\emptyset\}\}$$
$$n+1 := \{n, \emptyset\}$$

Let $\mathfrak{g}$ be a Gödelnumbering of the formulas of the theory of types onto $\tilde{N}$, and let $\bar{\alpha}$ be the Gödelnumber of any formula α. The one-place predicate P_1 on $\tilde{N}$ with

$P_1 \bar{\alpha} \iff \alpha$ is satisfiable in a finite model

is obviously recursive enumerable. In the following Lemma we use one $\exists$-quantifier of the weak second order logic. All quantifiers are interpreted on $\mathfrak{M}$.

Lemma 2: For every ϱ-ary ($\varrho \geq 1$) recursive enumerable predicate P on $\tilde{N}$ one can find a $\tau \in N$ and a $(\varrho+\tau+1)$-ary quantifierfree expression $S_{\mathfrak{M}}$ containing as prime formulas only $u \in_2^* v$ on $\mathfrak{M}$ with

$$Px_1 \ldots x_\varrho \iff \underset{\substack{M \neq \emptyset \\ [|M| \geq 3]}}{\exists M} \ \underset{Mp}{\forall p} \ \underset{Ms_1,\ldots,Ms_\tau}{\exists s_1 \ldots s_\tau} \ S_{\mathfrak{M}}\, p s_1 \ldots s_\tau x_1 \ldots x_\varrho \ .$$

The remark $[|M| \geq 3]$ may be added or omitted.

Proof: We start with the representation of $Px_1 \ldots x_\varrho$ according to Lemma 1. For $x,y \in \mathfrak{M}$ we define

$x \lhd y \iff x$ belongs to the smallest subset of $\mathfrak{M}$, which contains the elements of y and with each w also the elements of w,

$x \unlhd y \iff x \lhd y \vee x=y$.

One can easily verify (compare the proof of Satz 1 in [8]), that we can add in the representation of $Px_1 \ldots x_\varrho$ in Lemma 1 the remark

$$[\forall s\, (s \unlhd r \wedge s \in \Pi \rightarrow \mathfrak{p}_2(s) \neq \{\emptyset\}],$$

in case that P is a recursive enumerable predicate on $\tilde{N}$ only. Therefore we have in this case

$$Px_1 \ldots x_\varrho \iff \underset{M \neq \emptyset}{\exists M} \ \underset{Mp}{\forall p} \ \underset{Mq_1,\ldots,Mw}{\exists q_1 q_2 q' q'' q_{21} q_{21} t_1 \ldots t_\tau r q w}$$
$$(p \in \mathfrak{M}_2 \wedge q' = \langle\langle r,r\rangle, \{\emptyset\}\rangle$$
$$(p \notin \Pi \vee p = \langle q,w\rangle \wedge (w \neq \{\emptyset\} \vee q \notin \Pi)$$
$$\vee\ p = \langle\langle q_1,q_2\rangle, \{\emptyset\}\rangle \wedge R_1 q_1 q_2 t_1 \ldots t_\tau x_1 \ldots x_\varrho$$
$$\wedge (q_2 = \emptyset \vee q_2 = \langle q_{21}, q_{22}\rangle \wedge q'' = \langle\langle q_1, q_{21}\rangle, \{\emptyset\}\rangle)))$$

About "$\Leftarrow$": Obviously we have on $\mathfrak{M}_2$ $\underset{Lsr}{\exists r \forall s} \exists t_1 \ldots t_\tau\, R_1 r s t_1 \ldots t_\tau x_1 \ldots x_\varrho$.

About "$\Longrightarrow$": As M we can choose the smallest set $A \in \mathfrak{M}$ with
$\forall x\, (x \triangleleft A \rightarrow x \in A)$ and $\{\langle\langle r,s\rangle, \{\emptyset\}\rangle \mid Lsr\} \subseteq A$.

With Lemma 2 we get of course

Lemma 3: For every formula α of the theory of types one can find a formula β_I of $\forall\exists\ldots\exists\ (0,1)$ with
α is satisfiable in a finite model
$\Longleftrightarrow$ β_I is satisfiable on a finite non empty subset U of N with interpretation of E by $\in_2^{\circ}$.

For getting a similar reduction for the satisfiability of α we need a representation of the coenumerable predicates on $\mathfrak{M}_2$, that means, of the complements of the recursive enumerable predicates on $\mathfrak{M}_2$. Starting with equivalence (1) we have in the weak second order logic

$$Px_1\ldots x_\varrho \Longleftrightarrow \underset{M\neq\emptyset}{\exists M}\ \underset{Mq}{\forall q}\ (q=\langle\mathfrak{p}_1(q),\mathfrak{p}_2(q)\rangle \wedge R\,\mathfrak{p}_1(q)\mathfrak{p}_2(q)x_1\ldots x_\varrho$$
$$\wedge M\langle\mathfrak{p}_1(q),\mathfrak{p}_1(q)\rangle \wedge (\mathfrak{p}_2(q)=\emptyset$$
$$\vee \mathfrak{p}_2(q)=\langle\mathfrak{p}_1\circ\mathfrak{p}_2(q),\mathfrak{p}_2\circ\mathfrak{p}_2(q)\rangle$$
$$\wedge M\langle\mathfrak{p}_1(q),\mathfrak{p}_1\circ\mathfrak{p}_2(q)\rangle)).$$

Instead of using the weak second order logic on $\mathfrak{M}_2$ we can use the first order logic on $\mathfrak{M}$:

$$Px_1\ldots x_\varrho \Longleftrightarrow \underset{m\neq\emptyset}{\exists m}\ \underset{q\in m}{\forall q}\ (q\in\mathfrak{M}_2 \wedge q=\langle\mathfrak{p}_1(q),\mathfrak{p}_2(q)\rangle$$
$$\wedge R\,\mathfrak{p}_1(q)\mathfrak{p}_2(q)x_1\ldots x_\varrho \wedge \langle\mathfrak{p}_1(q),\mathfrak{p}_1(q)\rangle\in m$$
$$\wedge(\mathfrak{p}_2(q)=\emptyset \vee \mathfrak{p}_2(q)=\langle\mathfrak{p}_1\circ\mathfrak{p}_2(q),\mathfrak{p}_2\circ\mathfrak{p}_2(q)\rangle$$
$$\wedge\langle\mathfrak{p}_1(q),\mathfrak{p}_1\circ\mathfrak{p}_2(q)\rangle\in m)).$$

For every coenumerable predicate $Qx_1\ldots x_\varrho$ on $\mathfrak{M}_2$ we have the representation

$$Qx_1\ldots x_\varrho \Longleftrightarrow \underset{m\neq\emptyset}{\forall m}\ \underset{q\in m}{\exists q}\ (q\notin\mathfrak{M}_2 \vee q\notin\Pi \vee \neg R\,\mathfrak{p}_1(q)\mathfrak{p}_2(q)x_1\ldots x_\varrho \qquad (2)$$
$$\vee\langle\mathfrak{p}_1(q),\mathfrak{p}_2(q)\rangle\notin m \vee \mathfrak{p}_2(q)\neq\emptyset$$
$$\wedge(\langle\mathfrak{p}_1(q),\mathfrak{p}_1\circ\mathfrak{p}_2(q)\rangle\notin m \vee \mathfrak{p}_2(q)\notin\Pi)).$$

Here $\neg R\,\mathfrak{p}_1(q)\mathfrak{p}_2(q)x_1\ldots x_\varrho$ is formed by $\wedge$ and $\vee$ out of the atomic predicates $x=\emptyset$, $x\in\Pi$ and negated equations. The functions on both sides of the negated equations are formed out of $\lambda\emptyset$, n_0, $\mathfrak{p}$, $\mathfrak{p}_1$, $\mathfrak{p}_2$ and identity functions by substitution. With the same equivalences which we used in the proof of Lemma 1 we get now

Lemma 4: For every ϱ-ary ($\varrho\geq 1$) coenumerable predicate Q on $\mathfrak{M}_2$ one can find a $\delta\in N$ and a $(\varrho+\delta+1)$-ary quantifierfree expression R_2

containing as atomic predicates only $u \in^*_2 v$ on $\mathfrak{M}$ with

$$Qx_1 \dots x_\varsigma \Longleftrightarrow \forall m \exists s_1 \dots s_\delta \; R_2 m s_1 \dots s_\delta x_1 \dots x_\varsigma \quad (3).$$

The quantifiers are interpreted on $\mathfrak{M}$.

Because the set of satisfiable formulas corresponds to a coenumerable predicate on $\widetilde{N} \subseteq \mathfrak{M}_2$ we have:
For every formula α of the theory of types one can find a $\delta \in N$ and a $(\delta+1)$-ary quantifierfree expression R_3 containing as atomic predicates only $u \in^\circ_2 v$ on N with

$$\alpha \text{ is satisfiable} \Longleftrightarrow \underset{\in N}{\forall m} \underset{\in N}{\exists} s_1 \dots s_\delta \; R_3 m s_1 \dots s_\delta \quad (4).$$

Now we want to find boundaries for the $\exists$-quantifiers used in (4). The representation (2) of $Qx_1 \dots x_\varsigma$ shows, that the $\exists$-quantifiers in (3) stand for elements x with $x \triangleleft m$ or for elements y, which are obtained from such an x by finitely many applications of the functions $n_o, \mathfrak{p}, \mathfrak{p}_1, \mathfrak{p}_2$. More $\exists$-quantifiers are used for describing the Gödelnumber $\bar{\alpha} \in \widetilde{N}$ of the formula α. Because of

$$z = n_o(x,y) \Longrightarrow \mathfrak{a}(z) \leq 2^{\mathfrak{a}(x)} + 2^{\mathfrak{a}(y)}$$

we have an $r \in N$ such that we can add for

$$f_o(x) := x$$
$$f_{n+1}(x) := 2^{f_n(x)}$$

under the $\exists$-quantifiers of (4) the boundaries $[s_1,\dots,s_\delta \leq f_r(m)]$.
Obviously we have

$$\begin{aligned}
&\forall m \underset{[s_1,\dots,s_\delta \leq f_r(m)]}{\exists s_1 \dots s_\delta} R_3 m s_1 \dots s_\delta \\
\Longleftrightarrow\ &\forall m \; (m = 0 \\
&\quad \vee \underset{[m_1 \leq m]}{\exists m_1} \; (m_1 \in^\circ m \wedge m_1 = 0) \\
&\quad \vee \underset{[m_1, m_2 \leq m]}{\exists m_1 m_2} \; (m_1 \in^\circ m_2 \wedge m_2 \in^\circ m \wedge m_1 = 0) \\
&\quad \vee \\
&\quad \vdots \\
&\quad \vee \underset{[m_1,\dots,m_{r-1} \leq m]}{\exists m_1 \dots m_{r-1}} \; (m_1 \in^\circ m_2 \wedge m_2 \in^\circ m_3 \wedge \dots \wedge m_{r-1} \in^\circ m \wedge m_1 = 0) \\
&\quad \vee \underset{[m_1,\dots,m_r,s_1,\dots,s_\delta \leq m]}{\exists m_1 \dots m_r s_1 \dots s_\delta} \; (m_1 \in^\circ m_2 \wedge \dots \wedge m_{r-1} \in^\circ m_r \wedge m_r \in^\circ m \\
&\qquad\qquad \wedge R_3 m_1 s_1 \dots s_\delta))
\end{aligned}$$

Because of

$$x=0 \Longleftrightarrow \underset{[y,z \leq 3]}{\exists yz} \; (\neg x \in^\circ_2 x \wedge x \in^\circ y \wedge \neg y \in^\circ_2 y \wedge x \in^\circ z \wedge y \in^\circ z \wedge z \in^\circ_2 z)$$

we have

Lemma 5: For every formula α of the theory of types one can find a $\delta \in N$ and a $(\delta+1)$-ary quantifierfree expression T containing as atomic predicates only $u \in_2^\circ v$ with

$$\alpha \text{ is satisfiable} \iff \forall m \underset{[b_1,\ldots,b_\delta \leq \max(3,m)]}{\exists b_1 \ldots b_\delta} \qquad Tmb_1 \ldots b_\delta \quad .$$

The quantifiers are interpreted on N and their boundaries in square brackets may be added or omitted.

Proof of Theorem 1: Let be α a formula of the theory of types and $\beta_T(m,b_1,\ldots,b_\delta)$ the quantifierfree formula of the predicate logic corresponding to $Tmb_1\ldots b_\delta$ and containing E as the only predicate variable. For α we choose a formula $\beta_I \in \forall\exists\ldots\exists(0,1)$ having the properties of Lemma 3, $\beta_I \equiv \forall c \exists s_1 \ldots s_\tau \ \beta_S(c,s_1,\ldots,s_\tau)$ with quantifierfree β_S. As β we choose

$$\exists u \forall c \exists b_1 \ldots b_\delta \, s_1 \ldots s_\tau \, yw \, ((\beta_T(c,b_1,\ldots,b_\delta) \wedge c\hat{E}y \vee w\hat{E}u)$$
$$\wedge (c\hat{E}u \longrightarrow \beta_S(c,s_1,\ldots,s_\tau) \wedge s_1\hat{E}u \wedge \ldots \wedge s_\tau\hat{E}u)).$$

In fact we have proved more than the first Theorem.

Theorem 2: Let P be a ρ-ary ($\rho \geq 1$) recursive enumerable predicate on N and Q be a ρ-ary coenumerable predicate on N with $\forall x_1 \ldots x_\rho \ (Px_1\ldots x_\rho \longrightarrow Qx_1 \ldots x_\rho)$. Then one can find for P and Q and arbitrary $x_1,\ldots,x_\rho \in N$ a formula $\beta \in \exists\forall\exists\ldots\exists(0,1)$ with

1) $Px_1\ldots x_\rho \iff \beta$ is satisfiable on a finite subset of N with interpretation of E by $\in_2^\circ$,
2) $Qx_1\ldots x_\rho \iff \beta$ is satisfiable on N with interpretation of E by $\in_2^\circ$.
3) Let $\beta \equiv \exists u \forall r \exists s_1 \ldots s_\mu \ \beta_0$, where β_0 does not contain quantifiers, and let be R_0 the corresponding quantifierfree expression containing as atomic predicates only $u \in_2^\circ v$ on N. Then we have

$$Qx_1\ldots x_\rho \iff \underset{[u=0]}{\exists u} \ \forall c \underset{[s_1 \leq 2^c]\,[s_2,\ldots,s_\mu \leq c]}{\exists s_1 \ldots s_\mu} \qquad R_0ucs_1\ldots s_\mu \quad .$$

The third Theorem doesn't mention the rather complicated predicate $\in_2^\circ$ any longer but uses the more complicated (but still very interesting) prefix $\exists\forall\forall\exists\ldots\exists$ instead of $\exists\forall\exists\ldots\exists$ and at exactly one place in the formulas of the reduction class the sign of equality.

To remember this, we write $\exists\forall\forall\exists\ldots\exists^{=}(0,1)$ instead of $\exists\forall\forall\exists\ldots\exists(0,1)$. For any interpretation I on N we define:

$I(E) \approx \epsilon^{\circ} \Longleftrightarrow$ For all $x,y \in N$ with $x \neq y$ we have $x\ I(E)\ y$ if and only if $x \epsilon^{\circ} y$.

Theorem 3: For every formula α of the theory of types one can find a formula α' of $\exists\forall\forall\exists\ldots\exists^{=}(0,1)$ with

1) α is satisfiable
$\Longleftrightarrow$ α' is satisfiable on N by an interpretation I(E) with $I(E) \approx \epsilon^{\circ}$,

2) α is satisfiable in a finite model
$\Longleftrightarrow$ α' is satisfiable on a segment $\{u \mid u \leq v\}$ of N by an interpretation I(E) with $I(E) \approx \epsilon^{\circ}$.

3) Let be $\alpha' \equiv \exists m \forall cz \exists s_1 \ldots s_\mu\ \alpha_o'$, where α_o' does not contain quantifiers, and let be R_0' the corresponding quantifier-free expression containing as prime formulas only $u \in_3^{\circ} v$ with for all $u,v \in N$:

$$u \in_3^{\circ} v \Longleftrightarrow \begin{cases} u \epsilon^{\circ} v & \text{for } u \neq v \\ u=0 \vee u \text{ has the digital 1 in exactly two positions} & \\ \text{of the binary representation} & \text{for } u=v. \end{cases}$$

Then we can give boundaries for the $\exists$-quantifiers, if α is satisfiable:

α is satisfiable

$$\Longleftrightarrow \quad \forall c\ \underset{[s_1 \leq 2^c]\ [s_2,\ldots,s_\mu \leq c]}{\exists s_1 \ldots s_\mu} \qquad R_0 9 c c s_1 \ldots s_\mu$$

The remarks in square brackets may be added or omitted.

Theorem 3 may be compared with the Lemma mentioned on the first page. As it is possible to get a refining of Kostyrko's result by using this Lemma it might also be possible to use Theorem 3 for answering the famous question whether $\exists^n\forall\forall\exists\ldots\exists^{=}(0,1)$ is a conservative reduction type or not. (But compare the note on the last page!)

We start now to prove Theorem 3. The predicate $\in_3^{\circ}$ on N corresponds to the following two-place predicate on $\mathfrak{M}$:

$$x \in_3^{*} y \Longleftrightarrow \begin{cases} x \in y & \text{for } x \neq y \\ x = \emptyset \vee |x| = 2 & \text{for } x = y\ . \end{cases}$$

Using the model $\tilde{N}$ of the set of natural numbers and the pair function $\mathfrak{p} := \lambda xy\ \{\{f_o \circ f_o(x), \emptyset\}, \{f_o \circ f_o(y), \{\emptyset\}\}\}$ on $\mathfrak{M}$ we can still prove without using the domain $\mathfrak{M}_2$ (compare [8] and the proofs of Lemma 2 and Lemma 3)

Lemma 6: For every ϱ-ary ($\varrho \geq 1$) recursive enumerable predicate P on $\widetilde{N}$ one can find a natural number τ and a ($\varrho + \tau + 1$)-ary predicate S on $\mathfrak{M}$ with

$$Px_1 \ldots x_\varrho \iff \underset{\substack{p \neq \emptyset \\ [\emptyset \in p]}}{\exists p} \ \underset{q \in p}{\forall q} \ \underset{s_1, \ldots, s_\tau \in p}{\exists s_1 \ldots s_\tau} \quad Sqs_1 \ldots s_\tau x_1 \ldots x_\varrho \ ,$$

where S is a quantifierfree expression containing as prime formulas only $u \in_3^* v$, $u = \emptyset$ and $u = \{\emptyset\}$. All quantifiers are interpreted on $\mathfrak{M}$.

Remark: The prime formulas $u = \emptyset$ and $u = \{\emptyset\}$ are used also for describing the Gödelnumbers $\bar{\alpha}$ of the formulas α with a $\exists \ldots \exists$-prefix, for example

$$x = \{\emptyset, \{\emptyset\}\} \iff \exists uv \, (u = \emptyset \wedge v = \{\emptyset\} \wedge u \in x \wedge v \in x \wedge x \in_3^* x) .$$

Just as Lemma 5 we can also prove

Lemma 7: For every formula α of the theory of types one can find a $\delta \in N$ and a ($\delta + 1$)-ary quantifierfree expression T containing as prime formulas only $u \in_3^\circ v$, $u=0$ and $u=1$ with

$$\alpha \text{ is satisfiable} \iff \forall m \underset{[b_1, \ldots, b_\delta \leq m]}{\exists b_1 \ldots b_\delta} \quad Tmb_1 \ldots b_\delta .$$

Proof of Theorem 3: We have to find an axiom $\gamma \in \exists\forall^2\exists\ldots\exists^{=}(0,1)$, which is true on N resp. on any segment of N for the interpretation of E by $\in_3^\circ$ and has the following property for any formula β of the predicate logic with E as the only free variable:

($\beta \wedge \gamma$) is satisfiable on N resp. on a segment of N by an interpretaton I(E) with $I(E) \approx \in^\circ$ $\Longrightarrow$

($\beta \wedge \gamma$) is satisfiable on N resp. on a segment of N with the interpretation of E by $\in_3^\circ$

We use as abbreviations:

$x\hat{E}y \iff xEy \wedge \neg\, yEx$

$x \cong y \iff xEy \wedge yEx$

$x \neq y \iff \exists u \, (u\hat{E}x \wedge \neg\, u\hat{E}y \vee u\hat{E}y \wedge \neg\, u\hat{E}x)$

("$x \neq y$" is regarded as an abbreviation, because we want to use the equality sign only once in α'.)

$|c| \geq 2 \iff \exists c_1 c_2 \, (c_1\hat{E}c \wedge c_2\hat{E}c \wedge c_1 \neq c_2)$

$|c| \geq 3 \iff \exists c_1 c_2 c_3 \, (c_1\hat{E}c \wedge c_2\hat{E}c \wedge c_3\hat{E}c \wedge c_1 \neq c_2 \wedge c_2 \neq c_3 \wedge c_1 \neq c_3)$

$z \nsubseteq c \Longleftrightarrow \exists u\,(u\hat{E}z \wedge \neg u\hat{E}c)$

As the axiom γ we choose

$$\exists m\,(\forall z \exists m_1 m_2 m_{21} m_{22} (mEm \wedge m_2 E m_2 \wedge m_1 \hat{E} m \wedge m_2 \hat{E} m \wedge m_{21} \hat{E} m_2 \wedge m_{22} \hat{E} m_2 \wedge m_1 \neq m_2$$

$$(1) \quad \wedge m_{21} \hat{E} m_{22} \wedge (|m_1| \geq 3 \wedge m_{21} \hat{E} m_1 \vee m_1 \cong m_{21})$$

$$\wedge \neg z \hat{E} m_{21} \wedge (\neg z \hat{E} m_{22} \vee z \cong m_{21}))$$

(2) $\{ \wedge \forall c\,(\exists x\, x\hat{E}c \vee cEc)$

(3) $\{ \wedge \forall c\,(\forall z\, \neg z\hat{E}c \vee |c| \geq 2 \vee \neg cEc)$

$$(4) \quad \wedge \forall c\,(\forall z\,(z \nsubseteq c \vee z = c \vee \exists u_1 u_2 (zEu_1 \wedge z\hat{E}u_2 \wedge u_1 \neq u_2 \wedge u_1 \hat{E} m \wedge u_2 \hat{E} m)) \vee |c| \geq 3 \vee cEc)$$

$$(5) \quad \wedge \forall c\,(\forall z\,(z \nsubseteq c \vee z \cong c \vee \neg zEz \vee \exists u_1 u_2 (zEu_1 \wedge z\hat{E}u_2 \wedge u_1 \neq u_2 \wedge u_1 \hat{E} m \wedge u_2 \hat{E} m)) \vee \neg cEc))\,.$$

Obviously we have $\mathrm{Mod}_{\mathfrak{m}} I\,\gamma$ for $I(E) = \in_3^*$, if we choose $m = \{m_1, \{\emptyset, \{\emptyset\}\}\}$ for any m_1 with $m_1 = \emptyset \vee |m_1| \geq 3 \wedge \emptyset \in m_1$.

Let now be $\mathrm{Mod}_{\mathfrak{m}} I\,\gamma$ and $I(E) \approx \in$.

Because of (2) we have $\emptyset I(E) \emptyset$.

Because of (3) we have $\neg cI(E)c$ for $|c| = 1$.

Because of (4) we have $cI(E)c$ for all but finite many c with $|c| = 2$.

For any exception c with $|c| = 2 \wedge \neg cI(E)c$ we have for all subsets $z \subseteq c$ with $|z| \leq 1$ and two different elements u_1, u_2 of m

$z \in u_1 \wedge z \notin u_2$.

Because of (5) we have $\neg cI(E)c$ for all but finite many c with $|c| \geq 3$.

For any exception c with $|c| \geq 3 \wedge cI(E)c$ we have for all subsets $z \subseteq c$ with $|z| = 2$ and two different elements u_1, u_2 of m

$\neg zI(E)z \vee z \in u_1 \wedge z \notin u_2$.

a) $|m| = 2$.

Proof: Otherwise there would be a subset $\{z_1, z_2\} \subseteq m$ with $z_1 \neq z_2$ such that $\{z_1, z_2\}$, $\{z_1\}$ and m have the same rank. This is a contradiction because it must be either $\{z_1, z_2\} \in u$ or $\{z_1\} \in u$ for some $u \in m$.

b) $m = \{m_1, m_2\}$ for m_1, m_2 with $(|m_1| \geq 3 \vee m_1 \in m_2) \wedge |m_2| = 2$.

Proof: Because of (1) we have $|m_2| \geq 2$ for some $m_2 \in m$. For $|m_2| \geq 3$ we get a subset $\{z_1, z_2\} \subseteq m_2$ with $z_1 \neq z_2$ such that $\{z_1, z_2\}$, $\{z_1\}$ and m_2 have the same rank. This is a contradiction because we have either $\{z_1, z_2\} \notin m_2$ or $\{z_1\} \notin m_2$. For the other element $m_1 \in m$ we have because of (1): $|m_1| \geq 3 \vee m_1 \in m_2$.

c) $m = \{m_1, \{\emptyset, \{\emptyset\}\}\}$ for some m_1 with $|m_1| \geq 3 \wedge \emptyset \in m_1 \vee m_1 = \emptyset$.

This is a consequence of (1).

d) For all c with $|c|=2$ we have $cI(E)c$.

Proof: Not for both subsets $z\subseteq c$ with $|z|=1$ can $z\subseteq\{\emptyset,\{\emptyset\}\}$ be possible.

e) For all c with $|c|\geq 3$ we have $\neg cI(E)c$.

Proof: Not for all subsets $z\subseteq c$ with $|z|=2$ it is possible that $z\subseteq\{\emptyset,\{\emptyset\}\}$.

Now we have for any I on $\mathfrak{M}$ with $I(E)\approx\in$ and $\mathrm{Mod}_{\mathfrak{M}}\, I\ \gamma$ obviously $I(E)=\in^*_3$ and $m=\{m_1,\{\emptyset,\{\emptyset\}\}\}$ for some m_1 with $|m_1|\geq 3\wedge\emptyset\in m_1\vee m_1=\emptyset$.

Remark: In γ we use "=" only once for getting (4) true for $I(E)=\in^*_3$ and any c with $|c|=1$. Here we can not substitute "$\cong$" for "=". Otherwise we can use "=" as an abbreviation:

$x=y \Longleftrightarrow \exists uu_1\ (uEu\wedge u_1\hat{E}u\wedge x\hat{E}u\wedge y\hat{E}u\wedge x\neq u_1\wedge y\neq u_1)$.

We continue to prove Theorem 3. Because of Lemma 6 we have for every formula α of the theory of types a $\tau\in N$ and a $(\tau+3)$-ary quantifierfree expression $\tilde{S}$ on $\mathfrak{M}$ containing as prime formulas only $u\in^*_3 v$ with

α is satisfiable in a finite model

$$\Longleftrightarrow \underset{p\neq\emptyset}{\exists p}\ \underset{q\in p}{\forall q}\ \underset{s_1,\ldots,s_\tau,u,v\in p}{\exists s_1\ldots s_\tau uv}\qquad (\tilde{S}uvqs_1\ldots s_\tau\wedge u=\emptyset\wedge v=\{\emptyset\}).$$

Because of Lemma 7 we have for a $\delta\in N$ and a $(\delta+3)$-ary quantifierfree expression $\tilde{T}$ on N containing as prime formulas only $u\in^\circ_3 v$ with

α is satisfiable

$$\Longleftrightarrow \forall m\ \underset{[b_1,\ldots,b_\delta,u,v\leq\max(m,1)]}{\exists b_1\ldots b_\delta uv}\qquad (\tilde{T}uvmb_1\ldots b_\delta\wedge u=0\wedge v=1).$$

Let be $\beta_{\tilde{S}}(u,v,q,s_1,\ldots,s_\tau)$ and $\beta_{\tilde{T}}(u,v,m,b_1,\ldots,b_\delta)$ the to $\tilde{S}$ and $\tilde{T}$ corresponding quantifierfree formulas of the predicate logic with E as the only predicate variable and $\gamma\equiv\exists m\ \gamma_o(m)$. The we can choose as α' the formula

$$\begin{aligned}\exists m\ (\gamma_o(m)\wedge\forall c\,\exists b_1\ldots b_\delta ys_1\ldots s_\tau m_1m_2m_{21}m_{22}w\ (&m_1\hat{E}m\wedge m_2\hat{E}m\wedge m_1\neq m_2\\ &\wedge m_2Em_2\wedge m_{21}\hat{E}m_2\wedge m_{22}\hat{E}m_2\wedge m_{21}\hat{E}m_{22}\\ &\wedge(\beta_{\tilde{T}}(m_{21},m_{22},c,b_1,\ldots,b_\delta)\wedge c\hat{E}y\vee w\hat{E}m_1)\\ &\wedge(c\hat{E}m_1\longrightarrow\beta_{\tilde{S}}(m_{21},m_{22},c,s_1,\ldots,s_\tau)\wedge\underset{1\leq i\leq\tau}{\forall i}\quad s_i\hat{E}m_1))).\end{aligned}$$

Of course we have proved also

<u>Theorem 4</u>: Let P be a ϱ-ary ($\varrho\geq 1$) recursive enumerable predicate on N and Q be a ϱ-ary coenumerable predicate on N with $\forall x_1\ldots x_\varrho\ (Px_1\ldots x_\varrho\longrightarrow Qx_1\ldots x_\varrho)$. Then one can find for P and Q and

arbitrary $x_1,\dots,x_\varsigma \in N$ a formula $\beta \in \exists\forall^2\exists\dots\exists^=(0,1)$ with

1) $Px_1\dots x_\varsigma \Longleftrightarrow \beta$ is satisfiable on a segment of N by an interpretation I(E) with $I(E) \approx \in^\circ$,

2) $Qx_1\dots x_\varsigma \Longleftrightarrow \beta$ is satisfiable on N by an interpretation I(E) with $I(E) \approx \in^\circ$.

3) Let be $\beta \equiv \exists m \forall cz \exists s_1\dots s_\mu\ \beta_o$, where β_o does not contain quantifiers, and let R be the to β_o corresponding quantifierfree expression containing as prime formulas only $u \in_3^\circ v$. Then we have

$$Qx_1\dots x_\varsigma = \underset{[s_1\leq 2^c][s_2,\dots,s_\mu\leq c]}{\forall c \exists s_1\dots s_\mu} \quad R9ccs_1\dots s_\mu$$

Similar results to Theorem 3 and 4 can be formulated for the two remaining classical reduction types (without the equality sign) which have in the prefix behind the $\forall$-quantifiers an unlimited number of $\exists$-quantifiers.

<u>Theorem 5</u>: For every formula α of the theory of types one can find a formula $\beta_1 \in \forall\exists\forall\exists\dots\exists(0,1)$ and a formula $\beta_2 \in \forall^3\exists\dots\exists(0,1)$ with for $i \in \{1,2\}$

1) α is satisfiable
$\Longleftrightarrow \beta_i$ is satisfiable on N by an interpretation I(E) with $I(E) \approx \in^\circ$,

2) α is satisfiable in a finite model
$\Longleftrightarrow \beta_i$ is satisfiable on a segment of N by an interpretation I(E) with $I(E) \approx \in^\circ$.

3) Let be $\beta_1 \equiv \forall c \exists d \forall z \exists s_1\dots s_\mu\ \beta_{01}(d,c,z,s_1,\dots,s_\mu)$ where β_{01} does not contain quantifiers, and let be $R_1dczs_1\dots s_\mu$ the to $\beta_{01}(d,c,z,s_1,\dots,s_\mu)$ corresponding quantifierfree expression containing as prime formula only $u \in_3^\circ v$. Then we have
α is satisfiable

$$\Longleftrightarrow \underset{[s_1,\dots,s_\mu\leq z]}{\forall z \exists s_1\dots s_\mu} \quad R_1 01zs_1\dots s_\mu .$$

4) Let be $\beta_2 \equiv \forall cuz \exists s_1\dots s_\mu\ \beta_{02}(c,u,z,s_1,\dots,s_\mu)$ where β_{02} does not contain quantifiers, and let be $R_2uczs_1\dots s_\mu$ the to $\beta_{02}(c,u,z,s_1,\dots,s_\mu)$ corresponding quantifierfree expression containing as prime formula only $u \in_3^\circ v$. Then we have
α is satisfiable

$$\Longleftrightarrow \underset{[s_1=1][s_2,\dots,s_\mu\leq z]}{\forall z \exists s_1\dots s_\mu} \quad R_2 01zs_1\dots s_\mu .$$

The remarks in square brackets may be added or omitted.

Instead of Theorem 5 we will prove the stronger result

Theorem 6: Let be P a ϱ-ary ($\varrho \geq 1$) recursive enumerable predicate on N and Q be a ϱ-ary coenumerable predicate on N with $\forall x_1 \dots x_\varrho \, (Px_1 \dots x_\varrho \longrightarrow Qx_1 \dots x_\varrho)$.
Then one can find for P and Q and arbitrary $x_1, \dots, x_\varrho \in N$ a formula $\beta_1 \in \forall\exists\forall\exists \dots \exists(0,1)$ and a formula $\beta_2 \in \forall^3 \exists \dots \exists(0,1)$ with for $i \in \{1,2\}$

1) $Px_1 \dots x_\varrho \Longleftrightarrow \beta_i$ is satisfiable on a segment of N by an interpretation I(E) with $I(E) \approx \in^\circ$,
2) $Qx_1 \dots x_\varrho \Longleftrightarrow \beta_i$ is satisfiable on N by an interpretation I(E) with $I(E) \approx \in^\circ$.
3) Let be R_1 and R_2 defined as in Theorem 5. Then we have for $i \in \{1,2\}$

$$Qx_1 \dots x_\varrho \Longleftrightarrow \underset{[s_1=1]\,[s_2,\dots,s_\mu \leq z]}{\forall z \exists s_1 \dots s_\mu} \quad R_i 0 1 s_1 \dots s_\mu .$$

Proof of Theorem 6: Compare the proof of Theorem 3. We will first define β_1. As axiom γ we choose now

$$(1) \quad \forall c\, (\exists x\, x\hat{E}c \vee cEc)$$
$$(2) \quad \wedge \forall c\, (\forall z\, \neg z\hat{E}c \vee |c| \geq 2 \vee \neg cEc)$$
$$(3) \begin{cases} \wedge \forall c \exists d\, ((d\hat{E}c \vee \forall z\, \neg z\hat{E}c) \\ \qquad \wedge \forall z\, (z \notin c \vee zEz \vee d\hat{E}z \vee |c| \geq 3 \vee cEc) \\ \qquad \wedge \forall z\, (z \notin c \vee \neg zEz \vee \neg cEc \vee \neg d\hat{E}z \vee z \cong c)) \end{cases}$$

Obviously we have $\mathrm{Mod}_{\mathfrak{m}} I\, \gamma$ for $I(E) = \in_3^*$. Let now be $\mathrm{Mod}_{\mathfrak{m}} I\, \gamma$ and $I(E) \approx \in$.
Because of (1) we have $\emptyset I(E) \emptyset$.
Because of (2) we have $\neg\, cI(E)c$ for $|c|=1$.
Because of (3) we have $cI(E)c$ for $|c|=2$ and $\neg\, cI(E)c$ for $|c| \geq 3$.
Thus we have for $\mathrm{Mod}_{\mathfrak{m}} I\, \gamma$ and $I(E) \approx \in$ now $I(E) = \in_3^*$. Because of Lemma 6 we have

$$Px_1 \dots x_\varrho \Longleftrightarrow \underset{p \neq 0\;\; q \in^\circ p\;\; s_1,\dots,s_\tau,u,v \in^\circ p}{\exists p\, \forall q \exists s_1 \dots s_\tau uv} \quad (\tilde{S}uvqs_1 \dots s_\tau \wedge u=0 \wedge v=1)$$

where $\tilde{S}$ is a $(\tau+3)$-ary quantifierfree expression on N containing as prime formula only $a \in_3^\circ b$.
Like Lemma 7 we get for any $x_1, \dots, x_\varrho \in N$

$$Qx_1 \dots x_\varrho \Longleftrightarrow \underset{[b_1,\dots,b_\delta,u,v \leq \max(m,1)]}{\forall m \exists b_1 \dots b_\delta uv} \quad (\tilde{T}uvmb_1 \dots b_\delta \wedge u=0 \wedge v=1)$$

where $\tilde{T}$ is a $(\delta+3)$-ary quantifierfree expression on N containing as atomic formula only $a \in_3^\circ b$. Let be $\beta_{\tilde{S}}(u,v,q,s_1,\dots,s_\tau)$ and $\beta_{\tilde{T}}(u,v,m,b_1,\dots,b_\delta)$ the to $\tilde{S}uvqs_1 \dots s_\tau$ and $\tilde{T}uvmb_1 \dots b_\delta$ corresponding quantifierfree formulas of the predicate logic with E as the only predicate variable and $\gamma \equiv \forall c \exists d\ \gamma_0(c,d)$. Then we define

$$\beta_1 \equiv \forall c \exists d\ (\gamma_0(c,d) \wedge (c \neq 1 \vee \forall z \exists b_1 \ldots b_\delta\ \beta_{\tilde{T}}(d,c,z,b_1,\ldots,b_\delta))$$
$$\wedge (c \neq 0 \vee d \neq 0 \wedge dEd \wedge \forall z \exists s_1 \ldots s_\tau m_1 m_2 y\ ((z\hat{E}y \vee m_2\hat{E}m_1)$$
$$\wedge (\neg z\hat{E}m_2 \vee z \cong c)$$
$$\wedge (\neg z\hat{E}m_1 \vee \beta_{\tilde{S}}(c,m_2,z,s_1,\ldots,s_\tau) \wedge \forall_{1 \leq i \leq \tau}\ s_i\hat{E}m_1)$$
$$\wedge m_1\hat{E}d \wedge m_2\hat{E}d \wedge c\hat{E}m_2)))$$

In this proof we use as abbreviations

$u \neq 0$ for $\exists v\ v\hat{E}u$,

$u \neq 1$ for $uEu \vee \exists v\ (v\hat{E}u \wedge v \neq 0)$.

Obviously β_1 has the properties of Theorem 6.

For defining β_2 we choose now as axiom γ the following formula of $\forall^3 \exists \ldots \exists\ (0,1)$:

(1) $\forall c\ (\exists x\ x\hat{E}c \vee cEc)$

(2) $\wedge \forall c\ (\forall u\ \neg u\hat{E}c \vee |c| \geq 2 \vee \neg cEc)$

(3) $\wedge \forall cu(u \notin c \vee uEu \vee \forall z(z\hat{E}c \leftrightarrow z\hat{E}u) \vee |c| \geq 3 \vee cEc)$

(4) $\wedge \forall cu(u \notin c \vee \neg uEu \vee \neg cEc \vee \forall z\ \neg z\hat{E}u \vee u \cong c)$

Obviously we have $Mod_{\mathfrak{M}} I\ \gamma$ for $I(E) = \in_3^*$. Let now be $Mod_{\mathfrak{M}} I\ \gamma$ and $I(E) \approx \in$. Because of (1) we have $\phi\ I(E) \phi$, because of (2) $\neg cI(E)c$ for $|c|=1$, because of (3) $cI(E)c$ for $|c|=2$, and because of (4) $\neg cI(E)c$ for $|c| \geq 3$. Thus we have $I(E) = \in_3^*$ for $Mod_{\mathfrak{M}} I\ \gamma$ and $I(E) \approx \in$.

With the same $\tilde{S}$, $\tilde{T}$, $\beta_{\tilde{S}}$, $\beta_{\tilde{T}}$ like above and the new axiom γ we define

$$\beta_2 \equiv \gamma \wedge \forall cuz \exists c_1 u_1\ (c \neq 1 \vee \underline{c}_1\hat{E}c \wedge \beta_{\tilde{T}}(c_1,c,z,b_1,\ldots,b_\delta)$$
$$\wedge (|u| \geq 2 \vee \exists y\ u\hat{E}y \vee u_1\hat{E}u \wedge u_1 \neq 0$$
$$\wedge (\neg z\hat{E}u_1 \vee \beta_{\tilde{S}}(c_1,c,z,s_1,\ldots,s_\tau) \wedge \forall_{1 \leq i \leq \tau}\ s_i\hat{E}u_1))))$$

Obviously β_2 has the properties of Theorem 6.

The Lemma mentioned on the first page is the fist step for proving the following theorem.

<u>Definition</u>:

1) For $x,y \in N$ we define

$$x \in^{\circ\circ} y \Longleftrightarrow y \neq 0 \wedge (x \in^{\circ} y \vee x=y).$$

2) On any finite subset U of N with $0 \in U$ and the largest element u_{max} we define

$$x \in^{\infty} y \Longleftrightarrow y \neq 0 \wedge x \neq u_{max} \wedge (x \in^{\circ} y \vee x=y).$$

<u>Theorem 7</u>: For every formula α of the theory of types one can find a formula $\beta \in \forall \exists \ldots \exists \forall\ (0,1)$ with

1) α is satisfiable

$\Longleftrightarrow \beta$ is satisfiable

$\Longleftrightarrow \beta$ is satisfiable on N with interpretation of E by $\subseteq^{\infty}$.

2) α is satisfiable in a finite model

$\Longleftrightarrow \beta$ is satisfiable in a finite model

$\Longleftrightarrow \beta$ is satisfiable on a finite subset $U \subseteq N$ with $0 \in U$ and interpretation of E by $\subseteq^{\infty}$.

3) Let be $\beta \equiv \forall x \exists s_1 \ldots s_\mu \forall z \; \beta_0(x, s_1, \ldots, s_\mu, z)$ where β_0 does not contain quantifiers, and let be $R_0 x s_1 \ldots s_\mu z$ the to $\beta_0(x, s_1, \ldots, s_\mu, z)$ corresponding qunatifierfree expression on N containing as prime formula only $u \subseteq^{\infty} v$. Then we have

α is satisfiable

$$\Longleftrightarrow \forall x \exists s_1 \ldots s_\mu \forall z \quad R_0 x s_1 \ldots s_\mu z$$
$$[s_1 \leq 2^x]\,[s_2, \ldots, s_\mu \leq x]\,[z \leq x]$$

The remarks in square brackets may be added or omitted.

For the rather complicated proof of Theorem 7 compare [9].

Theorem 7 is a refining both of the Lemma and the well known result of Kostyrko. It is a completely new access for getting reduction types (via normal forms of recursive enumerable predicates) which offers a simple interpretation of the only predicate variable at the same time. I suppose that it is possible to get a similar refining of Theorem 5 (might be even of Theorem 3) and the corresponding results of the reduction theory. It could even be possible to get informations about the length of the disjunctions used in the prenex conjunctive normal form because of the following theorems.

<u>Theorem 8</u>: For every ϱ-ary ($\varrho \geq 1$) recursive enumerable predicate P on N one can find natural numbers τ, m and a ($\varrho + \tau + 3$)-ary quantifier-free expression R with $u \subseteq^{\circ} v$ as the only prime formula and

$$Px_1 \ldots x_\varrho \Longleftrightarrow \exists p \forall r \exists s_1 \ldots s_\tau \forall t\, Rprs_1 \ldots s_\tau t x_1 \ldots x_\varrho$$
$$[r \leq p][s_1, \ldots, s_\tau, t \leq f_m(\max(p, x_1, \ldots, x_\varrho))]$$

$$\text{for } f_m(x) = \underbrace{2^{\cdot^{\cdot^{2^{(2^x)}}}}}_{m \text{ times}} .$$

The expression $Rprs_1 \ldots s\, tx_1 \ldots x_\varrho$ can be chosen as a conjunction of disjunctions of the maximal length 3.

In every disjunction we use not more than one negated formula.

The remarks in square brackets may be added or omitted.

Proof of Theorem 8: We use Theorem 4 (a) in [8] which says:
For every ϱ-ary recursive enumerable predicate P on $\mathfrak{M}$ there is a $\tau \in N$ and a $(\varrho+\tau+2)$-ary function f, which is formed out of $\lambda \emptyset$, $\lambda x\, \{x\}$, $\lambda xy\, x \cup y$, and identity functions by substitution, with[1]

$$Px_1 \ldots x_\varrho \Longleftrightarrow \exists r \underset{s \in r}{\forall s} \exists t_0 \ldots t_\tau\ f(r,s,t_1,\ldots,t_\tau,\varphi) = \{t_0\}.$$

Thus we have

$$Px_1 \ldots x_\varrho \Longleftrightarrow \exists r \underset{[s \in r]}{\forall s} \exists uvt_0 \ldots t_\tau((\neg\, s \in r \vee u=v) \wedge u=f(r,s,t_1,\ldots,t_\tau,\varphi) \wedge v = \{t_0\})$$

and

$$u = v \Longleftrightarrow u \subseteq v \wedge v \subseteq u$$
$$u \in v \Longleftrightarrow u \subseteq v \wedge \neg\, v \subseteq u$$
$$u \notin v \Longleftrightarrow \neg\, u \subseteq v \vee v \subseteq u$$
$$z = \{x\} \Longleftrightarrow x \in z \wedge \forall t(t \notin z \vee t=x)$$
$$\begin{aligned} z=x\cup y \Longleftrightarrow\ & \forall t(t \notin x \vee t \subseteq z) \wedge \neg\, z \in x \\ & \wedge \forall t(t \notin y \vee t \subseteq z) \wedge \neg\, z \in y \\ & \wedge \exists a(\underset{[t \in z]}{\forall t}(\neg\, t \subseteq z \vee t \subseteq a \vee t \subseteq y) \wedge \underset{[t \in x \cup \{z\}]}{\forall t}(\neg\, t \subseteq a \vee t \subseteq x \vee z \subseteq t) && (1) \\ & \wedge(\neg\, x \in z \vee x \subseteq y) \wedge (\neg\, y \in z \vee y \subseteq x) && (2) \\ & \wedge(\exists u(u \subseteq x \wedge \neg\, u \subseteq y) \vee z=y) && (3) \end{aligned}$$

We have to prove $(1) \wedge (2) \wedge (3) \Longleftrightarrow z \subseteq x \cup y$.
Ad "$\Longrightarrow$": Because of (1) we have $z \subseteq x \cup y \cup \{x\} \cup \{y\}$, because of (1) $\wedge$ (2) thus $z \subseteq x \cup y \cup (\{x\} \cap \{y\})$. For $z=y$ we get obviously $z \subseteq x \cup y$. For $z \neq y$ we have $\exists u(u \subseteq x \wedge \neg\, u \subseteq y)$, that means $x \neq y$ and thus $z \subseteq x \cup y$.
Ad "$\Longleftarrow$": (1) and (2) are trivial with $a := x \cup \{z\}$. For proving (3) we assume $z \neq y$, that means $x \nsubseteq y$. We have to find an u with $u \subseteq x \wedge \neg\, u \subseteq y$. Because of $x \nsubseteq y$ there is an u_0 with $u_0 \in x \wedge \neg\, u_0 \in y$. For $u_0 \neq y$ we choose $u=u_0$. For $u_0=y$ we have $y \in x$ and choose $u=x$.

Theorem 9: For every ϱ-ary ($\varrho \geq 1$) coenumerable predicate Q on N one can find natural numbers δ, m and a $(\varrho+\delta+2)$-ary quantifier-free expression S with $u \subseteq^{\circ} v$ as the only atomic formula and

$$Qx_1 \ldots x_\varrho \Longleftrightarrow \forall r \exists s_1 \ldots s_\delta \underset{[s_1,\ldots,s_\delta,t \leq f_m(\max(r,x_1,\ldots,x_\varrho))]}{\forall t} Srs_1 \ldots s_\delta t x_1 \ldots x_\varrho.$$

The expression $Srs_1 \ldots s_\delta t x_1 \ldots x_\varrho$ can be chosen as a conjunction of disjunctions of the maximal length 3.
In any disjunction we use not more than one negated formula.
The remarks in square brackets may be added or omitted.

[1] We write "φ" for "$x_1,\ldots,x_\varrho$".

Proof of Theorem 9: We start with Theorem 7 in [8] (compare also the proof of the following Theorem 10) which says:
For every ϱ-ary enumerable predicate $Px_1\ldots x_\varrho$ we have the standard form

$$Px_1\ldots x_\varrho \Longleftrightarrow \exists r \underset{s\leq r}{\forall s}\; Rrsx_1\ldots x_\varrho ,$$

where $Rrsx_1\ldots x_\varrho$ is a disjunction of equations and the terms are composed out of $\lambda\emptyset$, $\lambda x\,\{x\}$, $\lambda\, xy\; x\cup y$, variables and functions $\mathfrak{p}_1$, $\mathfrak{p}_2$ with

$\mathfrak{p}_1(z) = x$ for $\exists y\; z = \{\{x\},\{x,y\}\}$

$\mathfrak{p}_1(z) = \emptyset$ otherwise,

$\mathfrak{p}_2(z) = y$ for $\exists x\; z = \{\{x\},\{x,y\}\}$

$\mathfrak{p}_2(z) = \emptyset$ otherwise.

For every ϱ-ary coenumerable predicate $Qx_1\ldots x_\varrho$ we have thus the standard form

$$Qx_1\ldots x_\varrho \Longleftrightarrow \forall r \exists s\; (s\leq r \wedge \neg Rrsx_1\ldots x_\varrho).$$

This leads to Theorem 9 because of the following equivalences. We use the abbreviations $\langle x,y\rangle := \{\{x\},\{x,y\}\}$ and $\Pi := \{z \mid \exists xy\; z=\langle x,y\rangle\}$.

$$x \neq y \Longleftrightarrow \exists z\; (z=\{y\} \wedge x\notin z)$$

$$x = \mathfrak{p}_1(z) \Longleftrightarrow \exists y\tilde{z}vw\; (v=\langle x,y\rangle \wedge w=\emptyset \wedge \tilde{z}\notin \Pi \wedge (z=v \vee z=\tilde{z} \wedge x=w)$$

$$y = \mathfrak{p}_2(z) \Longleftrightarrow \exists x\tilde{z}vw\; (v=\langle x,y\rangle \wedge w=\emptyset \wedge \tilde{z}\notin \Pi \wedge (z=v \vee z=\tilde{z} \wedge y=w)$$

$$|u|\geq 2 \Longleftrightarrow \exists u_1u_2(u_1\in u \wedge u_2\in u \wedge u_1\neq u_2)$$

$$|u|\geq 3 \Longleftrightarrow \exists u_1u_2u_3(u_1\in u \wedge u_2\in u \wedge u_3\in u \wedge u_1\neq u_2 \wedge u_1\neq u_3 \wedge u_2\neq u_3)$$

$$z\notin\Pi \Longleftrightarrow \exists u_2v_2v_3abc\; (|u_2|\geq 2 \wedge |v_2|\geq 2 \wedge |v_3|\leq 3 \wedge a\notin b \wedge (z=\emptyset \vee z=v_3 \vee z=\{u_2,v_2\} \vee z=\{\{a\},b\} \vee \emptyset\in z \vee v_3\in z))$$

$$u\in z \Longleftrightarrow \exists v\; z=\{u\}\cup v$$

$$q=x_1 \vee q=x_2 \vee q=x_3 \vee q=x_4$$

$$\Longleftrightarrow \exists u_1u_2u_3u_4v_1v_2v_3abw(\underset{1\leq i\leq 4}{\forall i}\; u_i=\{q,x_i\} \wedge v_1=\{u_1,u_2\} \wedge v_2=\{u_3,u_4\} \wedge w=\{v_1,v_2\} \wedge v_3\in w \wedge v_3=\langle a,b\rangle\;)$$

$$\exists i\; (1\leq i\leq 6 \wedge q=x_i) \Longleftrightarrow \exists u_1u_2u_3u_4u_5u_6v_1v_2v_3ww_1w_2w_3w_{31}\; (v_1=\{u_1,u_2\} \wedge v_2=\{u_3,u_4\} \wedge v_3=\{u_5,u_6\} \wedge \forall i\; (1\leq i\leq 6 \longrightarrow u_i=\{q,x_i\}) \wedge w_1=\{v_1,v_2\} \wedge w_2=\{v_2,v_3\} \wedge w=\{w_1,w_2\} \wedge w_3\in w \wedge w_{31}\in w_3 \wedge \exists ab\; w_{31}=\langle a,b\rangle\;)$$

Definition:

1) Let be $\mathfrak{p}$ the standard pair function $\lambda xy\;\;\{\{x\},\{x,y\}\}$ on $\mathfrak{M}$ and $\Pi := \{\mathfrak{p}(x,y) \mid x\in\mathfrak{M} \wedge y\in\mathfrak{M}\}$.

$$xA^*y :\Longleftrightarrow \begin{cases} x\in y & \text{for } x\neq y \\ x\in\Pi & \text{for } x=y\,. \end{cases}$$

2) Let be N^* the smallest subset of $\mathfrak{M}$ with $\emptyset\in N^*$ and $\{x\}\in N^*$ for

any $x \in N^*$.

3) For $x,y \in N$ we define $xA^\circ y \iff \mathfrak{a}^{-1}(x)A^* \mathfrak{a}^{-1}(y)$.

4) $N^\circ := \{\mathfrak{a}(x) | x \in N^*\}$.

Theorem 10: 1) For every ϱ-ary ($\varrho \geq 1$) recursive enumerable predicate P on N° one can find natural numbers τ,m and a ($\varrho+\tau+2$)-ary quantifierfree expression R with $uA^\circ v$ as the only type of atomic subformulas such that on N

$$Px_1\ldots x_\varrho \iff \underset{}{\exists p}\ \underset{[r\leq p]}{\forall r}\ \underset{[s_1,\ldots,s_\tau \leq f_m(\max(p,x_1,\ldots,x_\varrho))]}{\exists s_1\ldots s_\tau}\ Rprs_1\ldots s_\tau x_1\ldots x_\varrho \quad .$$

2) For every ϱ-ary coenumerable predicate Q on N° one can find natural numbers δ,m and a ($\varrho+\delta+2$)-ary quantifierfree expression S with $uA^\circ v$ as the only type of atomic subformulas such that on N

$$Qx_1\ldots x_\varrho \iff \underset{[p_0=0]}{\exists p_0}\ \forall r\ \underset{[s_1,\ldots,s_\delta \leq f_m(\max(p,x_1,\ldots,x_\varrho))]}{\exists s_1\ldots s_\delta}\ Sp_0rs_1\ldots s_\delta x_1\ldots x_\varrho \quad .$$

3) The expressions $Rprs_1\ldots s_\tau x_1\ldots x_\varrho$ and $Srs_1\ldots s_\delta x_1\ldots x_\varrho$ can be chosen as Krom- and Horn-formulas (conjunctive normal form with disjunctions of the maximal length 2; in any disjunction not more than one nonnegated formula).

The quantifiers are interpreted on N.

The remarks in square brackets may be added or omitted. This means that no additional predicate or constant (except the rather simple predicate A°) has to be used for the representation of the enumerable and coenumerable predicates.

Unlike the famous result of Matijasevic we allow only one binary standard predicate, but on the other hand one $\forall$-quantifier and a larger range for the quantifiers than the domain in which the enumerable predicates are represented.

Proof of Theorem 10: We prove first the corresponding standard form for the recursive enumerable predicates on $N^* \subseteq \mathfrak{M}$ and repeat some parts of the proof of [8] in order to show that we can eliminate the truth functions.

Step 1:

As atomic subformulas we allow only $y=\emptyset$ and $z=\{x,y\}$, as truth function only "$\wedge$", and any prefix consisting of any number of $\exists$-quantifiers and any number of the following special combination of a $\exists$-quantifier and a $\forall$-quantifier:

$$\exists p\ \underset{[L\emptyset p]}{\forall q}\ Lqp$$

The binary predicate L is defined by

$Lxy \Longleftrightarrow$ x belongs to the smallest subset of $\mathfrak{M}$ which contains y and with any $\mathfrak{p}(x,y)$ also x.

For $\mathfrak{p}(x,y)$ we write $\langle x,y\rangle$ and for $\langle x_0,\langle x_1,\dots,x_n\rangle\rangle$ also $\langle x_0,\dots,x_n\rangle$ $(n \geq 2)$.

We have to show that the graph of any primitiv recursive function on $N^* \subseteq \mathfrak{M}$ can be described in the mentioned way. This is possible because for ("$\mathfrak{x}$" is used as an abbreviation for "$x_1,\dots,x_\varrho$")

$$f(\mathfrak{x},0) = g(\mathfrak{x})$$

$$f(\mathfrak{x},\{y\}) = h(\mathfrak{x},y,f(\mathfrak{x},y))$$

we get

$$f(\mathfrak{x},y) = z \wedge x_1 \in N^* \wedge \dots \wedge x_\varrho \in N^* \wedge y \in N^* \wedge z \in N^*$$

$$\Longleftrightarrow \underset{[L\emptyset p]}{\exists p}\ \underset{Lqp}{\forall q}\ (\exists p_1\ p=\langle p_1,y,z\rangle \wedge (q=\emptyset \vee \exists u\ q=\langle u,\emptyset,g(\mathfrak{x})\rangle \vee \exists v_1v_2v_3\ q=\langle\langle v_1,v_2,v_3\rangle,\{v_2\},h(\mathfrak{x},v_2,v_3)\rangle))$$

$$\wedge\, x_1 \in N^* \wedge \dots \wedge x_\varrho \in N^* \wedge y \in N^* \wedge z \in N^*$$

and

$$x \in N^* \Longleftrightarrow \underset{[L\emptyset p]}{\exists p}\ \underset{Lqp}{\forall q}\ (\exists p_1\ p=\langle p_1,x\rangle \wedge (q=\emptyset \vee \exists v_1v_2\ q=\langle\langle v_1,v_2\rangle,\{v_2\}\rangle \vee q=\langle\emptyset,\emptyset\rangle))$$

The truth function "$\vee$" can be eliminated like in the proof of Theorem 9 with $v_3 \in w \Longleftrightarrow \exists z\ w=\{v_3,z\}$.

Step 2:

Now we can reach for any ϱ-ary recursive enumerable predicate P on $N^* \subseteq \mathfrak{M}$ the standard form

$$Px_1\dots x_\varrho \Longleftrightarrow \underset{[L\emptyset p]}{\exists p}\ \underset{Lqp}{\forall q}\ \exists r_1\dots r_\tau\ Rpqr_1\dots r_\tau x_1\dots x_\varrho\ ,$$

where $Rpqr_1\dots r_\tau x_1\dots x_\varrho$ is a conjunction of the atomic subformulas $z=\emptyset$ and $z=\{x,y\}$, because of the following two equivalences

a) $\underset{[L\emptyset r_2]}{\exists r_1 r_2}\ \underset{Lsr_2}{\forall s}\ Pr_1r_2s\mathfrak{x}$

$$\Longleftrightarrow \underset{[L\emptyset r]}{\exists r}\ \underset{Lsr}{\forall s}\ \exists r_1r_2s_1s_2\ (r=\langle r_2,r_1\rangle \wedge (s=\emptyset \vee s=\langle s_1,s_2\rangle) \wedge Pr_1r_2s_1\mathfrak{x})$$

b) $\underset{[L\emptyset x_0]}{\exists x_0}\ \underset{Lrx_0}{\forall r}\ \underset{[L\emptyset s]}{\exists s}\ \underset{Lts}{\forall t}\ Prstx_0\mathfrak{x}$

$$\Longleftrightarrow \underset{[L\emptyset p]}{\exists p}\ \underset{Lqp}{\forall q}\ \exists p_1p_2p_3q_1q_2q_3q_4u_1u_2\tilde{u}_2u_3u_4\tilde{u}_4 (p=\langle p_1,p_2,p_3,p_3\rangle \wedge Pq_2q_3q_4p_2\mathfrak{x}$$

$$\wedge (q=\emptyset \vee q=\langle q_1,q_2,q_3,q_4\rangle) \wedge (q=\emptyset$$

$$\vee\, q=\langle\langle u_1,q_2,q_3,u_4\rangle,q_2,q_3,\langle u_4,\tilde{u}_4\rangle\rangle$$

$$\vee\, q=\langle\langle u_1,u_2,u_3,u_3\rangle,\langle u_2,\tilde{u}_2\rangle,q_3,\emptyset\rangle$$

$$\vee\, q=\langle\emptyset,\emptyset,q_3,\emptyset\rangle))$$

Of course we could eliminate the last truth function "$\wedge$" because of $u=v \wedge a=b \Longleftrightarrow \langle u,a\rangle = \langle v,b\rangle$.

Step 3:

Because of

$$\exists p \underset{[L\phi p]\,Lqp}{\forall q}\ \tilde{R}pqx_1\ldots x_\varrho$$

$$\Longleftrightarrow \underset{r\neq\phi}{\exists r}\ \underset{sA^*r}{\forall s}\ \exists ps_1s_{11}(\langle p,p\rangle\in r\wedge\tilde{R}ps_1x_1\ldots x_\varrho\wedge(s=r\vee s=\langle s_1,p\rangle)\wedge\langle s_{11},p\rangle\in r\wedge(s_1=\phi\vee s_1=\langle s_{11},p\rangle))$$

and the following equivalences we get the standard form of Theorem 10 for $Px_1\ldots x_\varrho$.

$$x\in y\Longleftrightarrow xA^*y\wedge\neg yA^*x$$

$$x\notin y\Longleftrightarrow\neg xA^*y\vee yA^*x$$

$$u=\{x\}\Longleftrightarrow\exists vw\ (wA^*w\wedge u\in w\wedge v\in w\wedge x\in u\wedge x\in v\wedge u\in v)$$

Ad "$\Longrightarrow$": Take $w=\langle x,\{x\}\rangle\wedge u=\{x\}\wedge v=\{x,u\}$.

Ad "$\Longleftarrow$": The supposition $|u|\geq 2$ would lead to a contradiction because of $|v|\geq 2\wedge v\neq u$.

$$z=\{x,y\}\Longleftrightarrow x\in z\wedge y\in z\wedge\exists\tilde{z}vw_1w_2(z\in v\wedge\tilde{z}\in v\wedge vA^*v\wedge x\in\tilde{z}\wedge w_1=\{x\}\wedge(yA^*w_1\wedge w_1\in\tilde{z}\wedge\neg w_1\in z\vee\neg yA^*w_1))$$

Ad "$\Longrightarrow$": For $x\neq y$ choose $\tilde{z}=\{x\}$. For $x=y$ choose $\tilde{z}=\{x,\{x\}\}$.

Ad "$\Longleftarrow$": For $x\neq y$ we have $z=\{x,y\}$ because of $z\in v\in\Pi$. For $x=y$ we have $\tilde{z}=\{x,\{x\}\}$ because of $\tilde{z}\in v\in\Pi$, and therefore $z=\{x\}$.

$$\neg sA^*r\vee z=\{x,y\}\Longleftrightarrow\exists uv(u=\{x,y\}\wedge v=\{u\}\wedge(\neg sA^*r\vee zA^*v))$$

$$\underset{p\neq 0}{\exists pp_o}\forall s((\neg sA^*p\vee App_orx_1\ldots x_\varrho)\wedge p_o=\phi)$$

$$\Longleftrightarrow\exists\bar{p}\forall s\exists pp_ou_1u_2vw(\bar{p}A^\circ\bar{p}\wedge u_1\in\bar{p}\wedge u_2\in\bar{p}\wedge p_o\in u_1\wedge p_o\in u_2\wedge p\in u_2\wedge v=\{p\}\wedge\neg p_oA^*v\wedge\neg p\in u_1\wedge\neg sA^*p_o\wedge w\in p\wedge(\neg sA^*p\vee App_orx_1\ldots x_\varrho))$$

Now we want to prove the standard form for the coenumerable predicate

$$Qx_1\ldots x_\varrho\Longleftrightarrow\neg Px_1\ldots x_\varrho\ ,$$

where P is a recursive enumerable predicate on $N^*\subseteq\mathfrak{M}$.

Step 1:

Eliminating "$\wedge$" by $x=y\wedge u=v\Longleftrightarrow\langle x,u\rangle=\langle y,v\rangle$ we get for $Px_1\ldots x_\varrho$ the standard form

$$Px_1\ldots x_\varrho\Longleftrightarrow\exists p\underset{Lqp}{\forall q}\ \exists r_1\ldots r_n\ f(p,q,r_1,\ldots,r_n,\varphi)=g(p,q,r_1,\ldots,r_n,\varphi)$$

where the terms $f(\ldots)$ and $g(\ldots)$ are formed out of $\lambda\phi$, $\lambda xy\ \{x,y\}$ and variables. Let be on $\mathfrak{M}$ for all x,y,z

$\mathfrak{p}_1(\langle x,y\rangle)=x$

$\mathfrak{p}_1(z)=\phi$ for $z\notin\Pi$,

$\mathfrak{p}_2(\langle x,y\rangle)=y$

$\mathfrak{p}_2(z)=\phi$ for $z\notin\Pi$,

and $\mathfrak{t}_1=\mathfrak{p}_1$, $\mathfrak{t}_2=\mathfrak{p}_1\circ\mathfrak{p}_2$, $\mathfrak{t}_3=\mathfrak{p}_2\circ\mathfrak{p}_2$. Then we have for all x_1,x_2,x_3 $\mathfrak{t}_i(\langle x_1,x_2,x_3\rangle)=x_i$ $(1\leq i\leq 3)$ and

$\exists r \; \forall s \; \exists t \; Srst\varphi$
$[L\emptyset r] \, Lsr$

$\Longleftrightarrow \exists p \; \forall q \; (p = \langle \dagger_1(p), \dagger_2(p), \dagger_3(p) \rangle$
$[L\emptyset p] \, Lqp \; \wedge (q = \emptyset \vee q = \langle \dagger_1(q), \dagger_2(q), \dagger_3(q) \rangle \wedge S\dagger_2(p)\dagger_2(q)\dagger_3(q)\varphi \;)$
$\wedge (\dagger_2(q) = \emptyset \vee \dagger_2(q) = \langle \wp_1 \circ \dagger_2(q), \wp_2 \circ \dagger_2(q) \rangle$
$\wedge \dagger_1(q) = \langle \dagger_1 \circ \dagger_1(q), \wp_1 \circ \dagger_2(q), \dagger_3 \circ \dagger_1(q) \rangle)).$

Therefore we have

$Px_1 \ldots x_\varrho \Longleftrightarrow \exists p \; \forall q \; Tpqx_1 \ldots x_\varrho$
$[L\emptyset p] \, Lqp$

where $Tpqx_1 \ldots x_\varrho$ is a disjunction of equations. The terms are formed out of $\lambda \emptyset$, $\lambda xy \, \{x,y\}$, $\wp_1$, $\wp_2$ and variables.

Step 2:

Because of

$\exists r \; \forall s \; Trsx_1 \ldots x_\varrho$
$[L\emptyset r] \, Lsr$

$\Longleftrightarrow \exists p \; \forall q \; (q = p \vee q = \langle \dagger_1(q), \dagger_2(q), \dagger_3(q) \rangle \wedge T \dagger_1(q) \dagger_2(q) x_1 \ldots x_\varrho$
$p \neq \emptyset \;\; qA^*p \quad \wedge (\dagger_2(q) = \emptyset \vee \dagger_2(q) = \langle \wp_1 \circ \dagger_2(q), \wp_2 \circ \dagger_2(q) \rangle$
$\wedge \langle \dagger_1(q), \wp_1 \circ \dagger_2(q), \langle \dagger_3(q), \wp_2(\dagger_2(q)) \rangle \rangle A^*p)$
$\wedge (\dagger_1(q) = \dagger_2(q) \vee \dagger_3(q) = \langle \wp_1 \circ \dagger_3(q), \wp_2 \circ \dagger_3(q) \rangle$
$\wedge \langle \dagger_1(q), \langle \dagger_2(q), \wp_2 \circ \dagger_3(q) \rangle, \wp_1 \circ \dagger_3(q) \rangle A^*p))$

we get for $Qx_1 \ldots x_\varrho$ the standard form

$Qx_1 \ldots x_\varrho \Longleftrightarrow \forall p \, \exists q \; (p = \emptyset \vee (qA^*p \wedge Spqx_1 \ldots x_\varrho))$

where $Spqx_1 \ldots x_\varrho$ is a conjunction of negated equations and disjunctions of the kind $t_1 \neq t_2 \vee \neg \, t_3A^*p$. The terms are formed out of $\lambda \emptyset$, $\lambda xy \, \{x,y\}$, $\wp_1$, $\wp_2$ and variables. We have

$x = \wp_1(z) \Longleftrightarrow \exists y \;\; z = \langle x,y \rangle \vee \neg \, zA^*z \wedge x = \emptyset$
$\Longleftrightarrow \exists yuvww_1ab \;\; (u = \langle x,y \rangle \wedge v = \{u\} \wedge w_1 \in w \wedge b = \{w\} \wedge a = \{w, \emptyset\} \wedge x \in a$
$\wedge (zA^*v \vee \neg \, zA^*z \wedge \neg \, xA^*b))$

$y = \wp_2(z) \Longleftrightarrow \exists x \;\; z = \langle x,y \rangle \vee \neg \, zA^*z \wedge y = \emptyset$
$\Longleftrightarrow \exists xuvww_1ab \;\; (u = \langle x,y \rangle \wedge v = \{u\} \wedge w_1 \in w \wedge b = \{w\} \wedge a = \{w, \emptyset\} \wedge y \in a$
$\wedge (zA^*v \vee \neg \, zA^*z \wedge \neg \, yA^*b))$

$x \neq y \Longleftrightarrow \exists u \;\; (u = \{x\} \wedge \neg \, yA^*u)$

$t_1 \neq t_2 \vee \neg \, t_3A^*p \Longleftrightarrow \exists xyzu \;\; (x = t_1 \wedge y = t_2 \wedge z = t_3 \wedge u = \{x\} \wedge (\neg \, yA^*u \vee \neg \, zA^*p))$

For the proof of Theorem 10 it is sufficient to show

$Qx_1 \ldots x_\varrho \Longleftrightarrow \forall p \, \exists q \; ((p = \emptyset \vee qA^*p) \wedge Spqx_1 \ldots x_\varrho).$

We have to prove

$\exists q \; S\emptyset qx_1 \ldots x_\varrho$,

that means we have to find a q with

$\neg (\, q = \emptyset \vee q = \langle \dagger_1(q), \dagger_2(q), \dagger_3(q) \rangle \wedge \ldots).$

Choose any q with $q \neq \emptyset \wedge q \notin \Pi$.

Theorem 11: For every formula α of the theory of types one can find a Krom- and Horn-formula $\beta \in \forall\exists^{\infty}\forall(0,1)$ with

1) α is satisfiable
$\Longleftrightarrow \beta$ is satisfiable on N with interpretation of E by A°,

2) α is satisfiable in a finite model
$\Longleftrightarrow \beta$ is satisfiable on a segment of N with interpretation of E by A°.

3) Let be $\beta \equiv \forall r \exists s_1 \ldots s_\mu \forall t \ \beta_o(r, s_1, \ldots, s_\mu, t)$ with quantifierfree $\beta_o(r, s_1, \ldots, s_\mu, t)$ and let be $R_o r s_1 \ldots s_\mu t$ the corresponding quantifierfree expression on N containing as atomic subexpression only $uA^{\circ}v$. Then we have for a natural number n:

α is satisfiable

$\Longleftrightarrow \forall r \exists s_1 \ldots s_\mu \forall t \ R_o r s_1 \ldots s_\mu t$
$[s_1, \ldots, s_\mu, t \leq f_n(r)]$

All remarks in square brackets may be added or omitted.
The second $\forall$-quantifier occurs only at two places in β_o, once left of the predicate variable E (to express y=0) and once in a binary disjunction on the right of the variable E (to characterize the largest number 2^k of a segment of N).

Proof of Theorem 11: We choose a Gödelnumbering of the formulas of the theory of types on $N^* \subseteq \mathfrak{M}$. Let be $\hat{\alpha}$ the Gödelnumber of α. Then we have

α is satisfiable $\Longleftrightarrow Q\hat{\alpha}$

for a coenumerable predicate Q on $N^* \subseteq \mathfrak{M}$ and

α is satisfiable in a finite model $\Longleftrightarrow P\hat{\alpha}$

for a recursive enumerable predicate P on $N^* \subseteq \mathfrak{M}$. Let be $\mathfrak{n}$ the Gödelnumbering of $\mathfrak{M}$ onto N. With the proof of Theorem 10 we get on $\mathfrak{M}$

$$P\hat{\alpha} \Longleftrightarrow \exists p \forall r \ \exists s_1 \ldots s_\tau x \quad (x = \hat{\alpha} \wedge Rprs_1 \ldots s_\tau x)$$
$$p \neq \emptyset \ rA^* p \quad [\mathfrak{n}(s_1), \ldots, \mathfrak{n}(s_\tau), \mathfrak{n}(x) \leq f_m(\max(\mathfrak{n}(p), \mathfrak{n}(\hat{\alpha})))]$$
$$\Longleftrightarrow \exists p \forall r \ \exists s_1 \ldots s_k \ \tilde{R}prs_1 \ldots s_k$$
$$p \neq \emptyset \ rA^* p \quad [\mathfrak{n}(s_1), \ldots, \mathfrak{n}(s_k) \leq f_n(\mathfrak{n}(p))]$$

for a natural number n which depends on $\hat{\alpha}$. The expressions $Rprs_1 \ldots s_\tau x$ and $\tilde{R}prs_1 \ldots s_k$ are conjunctions of the atomic formulas $z = \emptyset$, $z = \{x, y\}$ and $u \in p$. Similarly we have on $\mathfrak{M}$

$$Q\hat{\alpha} \Longleftrightarrow \forall r \exists s_1 \ldots s_\delta xw \ (Srs_1 \ldots s_\delta xw \wedge x = \hat{\alpha} \wedge w = \emptyset)$$
$$\Longleftrightarrow \forall r \exists u_1 \ldots u_i w \ (\hat{S}ru_1 \ldots u_i w \wedge w = \emptyset)$$
$$[\mathfrak{n}(u_1), \ldots, \mathfrak{n}(u_i), \mathfrak{n}(w) \leq f_n(\mathfrak{n}(r))]$$

where $\hat{S}ru_1 \ldots u_i w$ is a Krom- and Horn-formula with uA^*v as the only type of atomic subformulas.

Let be B the expression

$\forall r \exists r_1 \ldots r_n s_1 \ldots s_k u_1 \ldots u_i w_o \ldots w_n yzw$

$$((r=\emptyset \vee r_1 \in r) \wedge \underset{1 \leq \nu \leq n-1}{\forall \nu} (r_\nu = \emptyset \vee r_{\nu+1} \in r_\nu) \wedge \hat{S} r_n u_1 \ldots u_i w \wedge \forall t \neg t A^* w$$

$$\wedge z = \{w\} \wedge \underset{0 \leq \nu \leq n-1}{\forall \nu} w_\nu = \{w_{\nu+1}\} \wedge (r \in y \vee \forall t \neg w_o A^* t \wedge \neg w_n A^* z)$$

$$\wedge (\neg r A^* w_n \vee \tilde{R} w_n r s_1 \ldots s_k))$$

with (for $\forall t \neg t A^* w$)

$r_\nu = \emptyset \vee r_{\nu+1} \in r_\nu$

$$\Longleftrightarrow \exists \tilde{w} \hat{z} \tilde{z} w_1 (w_1 \in \tilde{w} \wedge \hat{z} = \{w, \tilde{w}\} \wedge \tilde{z} = \{\tilde{w}\} \wedge r_\nu \in \hat{z} \wedge (\neg r_\nu A^* \tilde{z} \vee r_{\nu+1} \in r_\nu))$$

Let β be the corresponding formula of the predicate logic after we use in B the descriptions of $u=\{x\}$, $x \in y$, $\neg s A^* r \vee z = \{x,y\}$ mentioned in the proof of Theorem 10. Obviously β has the properties of Theorem 11.

<u>Theorem 12</u>: For every formula α of the theory of types one can find a Krom- and Horn-formula $\beta_1 \in \forall\exists\forall\exists^\infty(0,1)$, a Krom- and Horn-formula $\beta_2 \in \exists\forall^2\exists^\infty(0,1)$ and a Krom-formula $\beta_3 \in \forall^3\exists^\infty(0,1)$ with for $i \in \{1,2,3\}$

1) α is satisfiable

$\Longleftrightarrow \beta_i$ is satisfiable on N with interpretation of E by A^o.

2) α is satisfiable in a finite model

$\Longleftrightarrow \beta_i$ is satisfiable on a segment of N with interpretation of E by A^o.

3) Let be $\beta_1 \equiv \forall r \exists \bar{w} \forall t \exists s_1 \ldots s_\mu \; \beta_o(r,\bar{w},t,s_1,\ldots,s_\mu)$ with quantifierfree $\beta_o(r,\bar{w},t,s_1,\ldots,s_\mu)$ and let $R_o r \bar{w} t s_1 \ldots s_\mu$ the corresponding quantifierfree expression on N containing as atomic subexpression only $uA^o v$. Thenwe have for natural numbers k,n

α is satisfiable

$$\Longleftrightarrow \forall r \forall t \exists s_1 \ldots s_\mu \; R_o r k t s_1 \ldots s_\mu . \; [t \leq k][s_1,\ldots,s_\mu \leq f_n(r)]$$

4) Let be $\beta_3 \equiv \forall rst \exists s_1 \ldots s_\mu \; \beta_{o3}(r,s,t,s_1,\ldots,s_\mu)$ with quantifierfree $\beta_{o3}(r,s,t,s_1,\ldots,s_\mu)$ and let $R_{o3} rsts_1 \ldots s_\mu$ the corresponding quantifierfree expression on N containing as atomic subexpression only $uA^o v$. Then we have for a natural number n

α is satisfiable

$$\Longleftrightarrow \forall r \forall t \exists s_1 \ldots s_\mu \; R_{o3} r 0 t s_1 \ldots s_\mu . \; [t \leq r][s_1,\ldots,s_\mu \leq f_n(r)]$$

<u>Note</u>: In this theorem we don't give axioms for the standard predicate A^o like we give axioms for the predicate $\in^\infty$ in the proof of Theorem 7. This is why we don't need the equality sign for β_2 and can even reach a Krom- and Horn-formula, what is impossible in the

classical reduction theory for the prefix class $\exists^\infty \forall^\infty \exists^\infty$.

Proof of Theorem 12: Compare the proof of Theorem 11.

Ad $\beta \in \forall\exists\forall\exists^\infty(0,1)$: Let B be the expression

$$\forall r \exists \bar{w} \forall t \exists r_1 \dots r_n s_1 \dots s_k u_1 \dots u_i w_{-3} \dots w_n yzw$$
$$((\neg tA^* r \vee r_1 \in r) \wedge \bigwedge_{1 \le v \le n-1} (\neg tA^* r_v \vee r_{v+1} \in r_v) \wedge \bar{w} = \langle w_0, w \rangle$$
$$\wedge \bigwedge_{-3 \le v \le n-1} w_v = \{w_{v+1}\} \wedge \neg tA^* w \wedge z = \{w\} \wedge \hat{S} r_n u_1 \dots u_i w$$
$$\wedge (r \in y \vee \neg w_{-3} A^* t \wedge \neg w_n A^* z) \wedge (\neg rA^* w_n \vee \tilde{R} w_n r s_1 \dots s_k))$$

For $\beta \in \exists\forall^2\exists^\infty(0,1)$ just substitute $\exists\bar{w}\forall r \forall t$ for $\forall r \exists \bar{w} \forall t$.

Ad $\beta \in \forall^3\exists^\infty(0,1)$: Let B be the expression

$$\forall rst \exists r_1 \dots r_n s_0 \dots s_k u_1 \dots u_i y$$
$$((\neg tA^* r \vee r_1 \in r) \wedge \bigwedge_{1 \le v \le n-1} (\neg tA^* r_v \vee r_{v+1} \in r_v) \wedge \bigwedge_{0 \le v \le n-1} w_v = \{w_{v+1}\}$$
$$\wedge (s \ne \emptyset \vee s = s_0) \wedge w_n \ne \emptyset \wedge \hat{S} r_n u_1 \dots u_i s_0 \wedge (\neg rA^* w_n \vee \tilde{R} w_n r s_1 \dots s_k \tfrac{s_0}{\emptyset})$$
$$\wedge (|t| \ge 2 \vee \exists y\; t \in y \vee t = \{w_0\}))$$

For the last line we can write

$$\exists \tilde{w}_0((\exists y\; t \in y \vee \tilde{w}_0 \in t) \wedge (\exists t_1 t_2 \hat{t}_2 (t_1 \in t \wedge t_2 \in t \wedge \hat{t}_2 = \{t_2\} \wedge \neg t_1 A^* t_2)$$
$$\vee \exists \hat{w}_0 (\hat{w}_0 = \{w_0\} \wedge \tilde{w}_0 A^* \hat{w}_0))).$$

For $s \ne \emptyset \vee s = s_0$ we can write

$$\exists s_1\; s_1 \in s \vee \exists \hat{s}_0\; (\hat{s}_0 = \{s_0\} \wedge sA^* \hat{s}_0).$$

Of course the predicates $\in_2$, $\in_3$, A^* and any I(E) with $I(E) \approx \in$ are antisymmetric (i.e. $\forall xy\ (xRy \wedge yRx \longrightarrow x=y)$) and antitransitiv (i.e. $\forall xyz\ (xRy \wedge yRz \wedge xRz \longrightarrow x=y \vee y=z)$). So many complexity results of the reduction theory seem to have their origin in the theory of standard forms of enumerable and coenumerable predicates.

References

[1] Aanderaa, S.O.: On the decision problem for formulas in which all disjunctions are binary. Proc. of the Second Scand. Log. Symp. (1971), pp.1-18.

[2] Aanderaa, S.O., Börger, E., Gurevich, Yu.: Prefix classes of Krom formulae with identity. Archiv math. Logik 22 (1982), pp.43-49.

[3] Aanderaa, S.O., Lewis, H.: Prefix classes of Krom formulas. Journal of Symbolic Logic 38 (1973), pp.628-642.

[4] Börger, E.: Reduktionstypen in Krom- und Hornformeln. Diss. Münster 1971.

[5] Börger, E.: Eine entscheidbare Klasse von Kromformeln. Zeitschrift f. Math. Logik u. Grundl. d. M. 19 (1973),pp.117-12o.

[6] Börger, E.: Decision problems in predicate logic. Bericht Nr. 153/1983 der Abteilung Informatik der Universität Dortmund.

[7] Deutsch, M.: Zur Reduktionstheorie des Entscheidungsproblems. Zeitschr. f. Math. Logik u. Grundlagen d. Math. 27 (1981), pp. 113-117.

[8] Deutsch, M.: Zur gödelisierungsfreien Darstellung der rekursiven und primitiv-rekursiven Funktionen und rekursiv aufzählbaren Prädikate über Quotiententermmengen. Zeitschr. f. Math. Logik u. Grundlagen d. Math. 26 (1980), pp. 1-32.

[9] Deutsch, M.: Eine Verschärfung eines Satzes von Kostyrko zur Reduktionstheorie. To appear in Zeitschr. f. Math. Logik u. Grundlagen d. Math.

[10] Goldfarb, W.D.: On the Gödel class with identity. The Journal of Symbolic Logic 46 (1981), pp. 354-364.

[11] Goldfarb, W.D., Gurevich, Yu., Shelah, S.: On the Gödel class with identity. To appear.

[12] Gurevich, Yu.: Ob effektivnom raspoznavanii vipolnimosti formul UIP. Algebra i Logika 5 (1966), pp. 25-55.

[13] Gurevich, Yu.: Formulas with one $\forall$. In Memory of A.I. Maltsev, 1973 (in Russian).

[14] Gurevich, Yu.: The decision problem for standard classes. The Journal of Symbolic Logic 41 (1976), pp. 460-464.

[15] Kostyrko, V.F.: Klass cvedeniya $\forall\exists^n\forall$. Algebra i Logika 3 (1964), pp. 45-65.

[16] Kostyrko, V.F.: Klass cvedeniya $\forall\exists\forall$. Kibernetika 2 (1966), pp. 15-19.

[17] Krom, M.R.: The decision problem for a class of first-order formulas in which all disjunctions are binary. Zeitschr. f. Math. Logik u. Grundlagen d. Math. 13 (1967), pp. 15-20.

[18] Krom, M.R.: The decision problem for formulas in prenex conjunctive normal form with binary disjunctions. The Journal of Symbolic Logic 35 (1970), pp.210-216.

[19] Lewis, H.R.: Krom formulas with one dyadic predicate letter. The Journal of Symbolic Logic 41 (1976), pp.341-362.

[20] Lewis, H.R.: Unsolvable classes of quantificational formulas. Reading (Mass.) 1979.

[21] Deutsch, M.: Zum Reduktionstyp $\forall^3\exists^\infty(0,1)$. To appear in Zeitschrift f. Math. Logik u. Grundlagen d. Mathematik.

Appendix

1.) Mr. Börger informed me that Mr. Goldfarb has recently proved, that $\exists^n \forall^2 \exists^\infty(0,1)$ is a conservative reduction class.

2.) In between I could prove an analogous result to theorem 7 for the prefix $\forall^3 \exists^\infty$:

Theorem 13: There is a binary predicate B on N with for all $x,y \in N$ with $x \neq y$

$xBy \Longleftrightarrow x \in^\circ y$

such that for every formula α of the theory of types one can find a formula $\beta \in \forall^3 \exists^\infty(0,1)$ with

a) α is satisfiable

$\Longleftrightarrow \beta$ is satisfiable

$\Longleftrightarrow \beta$ is satisfiable on N with interpretation of E by B.

b) α is satisfiable in a finite model

$\Longleftrightarrow \beta$ is satisfiable in a finite model

$\Longleftrightarrow \beta$ is satisfiable on a finite subset $U \subseteq N$ with interpetation of E by a binary predicate B^* on U which has the following property:

Let x_{max} be the largest element of U. Then we have for all x,y with $x \neq x_{max} \wedge y \neq x_{max}$

$xB^*y \Longleftrightarrow xBy$

The proof is similar to that one of Theorem 7 and will appear in [21]. The proof supplies also boundaries for the quantifiers like in 3.) of Theorem 7.

It seems to be possible to replace the reduction class $\forall^3 \exists^\infty(0,1)$ in Theorem 13 both by $\forall\exists\forall\exists^\infty(0,1)$ and by $\exists^n \forall^2 \exists^\infty(0,1)$!

I hope to be able to publish the proofs soon.

Furthermore one could ask for "simple" interpretations of the predicate variables in the remaining classical reduction types.

With the old methods of constructing a model of α out of any model of β in case we have

α is satisfiable (in a finite model)

$\Longleftrightarrow \beta$ is satisfiable (in a finite model)

we get a new way of finding models for α in case α is satisfiable (or satisfiable in a finite model) at all.

The Computational Complexity of the Unconstrained Limited Domino Problem (with Implications for Logical Decision Problems)

Martin Fürer

University of Zürich

ABSTRACT

The unconstrained domino or tiling problem is the following. Given a finite set T (of tiles), sets $H, V \subseteq T \times T$ and a cardinal $k \leq \omega$, does there exist a function $t: k \times k \rightarrow T$ such that

$(t(i,j), t(i+1,j)) \in H$ *and* $(t(i,j), t(i,j+1)) \in V$

for all $i, j < k$?

The unlimited domino problem (k infinite) has played an important role in the process of finding the undecidability proof for the $\forall\exists\forall$ prefix class of predicate calculus. The limited domino problem is similarly connected to some decidable prefix classes.

The limited domino problem is NP-complete, if k is given in unary. The same was conjectured for k given in binary, but we show an $\Omega(c^n)$ nondeterministic time lower bound (upper bound $O(d^n)$). As a consequence, the nondeterministic time complexity of the decision problem for the $\forall\exists \wedge \forall\forall$ class of predicate calculus lies between $\Omega(c^{n/\log n})$ and $O(d^{n/\log n})$.

1. Introduction

To be precise, we could make a distinction between the tiling problem, as described in the Abstract, and the domino problem. In the latter problem the types of tiles are identified by colors on the edges, and the relations H and V are encoded in the colors of adjacent edges. (T,T') ∈ H iff the right edge of T and the left edge of T' have the same color, and (T,T') ∈ V iff the lower edge of T and the upper edge of T' have the same color. It is not allowed to turn the dominoes.

Clearly the two versions of the tiling problem are equivalent. To transform the tiling problem into the domino problem, we use for all tiles T, T', T'' with (T,T') ∈ H and (T,T'') ∈ V a domino whose edges are colored as follows:

left: (T,0), upper: (T,1), right: (T',0), lower: (T'',1).

From now on we forget about the distinction and also call the tiling problem sometimes 'domino problem'.

We use the notation ω for the set of *non-negative integers*, and we identify each $k \in \omega$ with the set $\{j : j \in \omega$ and $j < k\}$. A *tiling* is a function t from $k \times k$ into the set of tiles T (fulfilling the adjacency conditions V and H). Intuitively, a chess board of length k is covered with unit square tiles.

In the origin constrained domino problem some tiles are forbidden to be placed on the origin. Then undecidability results (for k infinite) and lower bounds (for k finite) are easily obtained by encoding Turing machine computations into the pattern of the tiling. Horizontal rows encode configurations of a one-tape TM. Subsequent rows encode subsequent configurations. And the origin (0,0) marks the beginning of the start configuration. We need a blank tile encoding a blank tape cell, and all the neighboring tiles are allowed to be of the same type. So we see that removing the origin constraint allows us to tile the whole plane with the blank tile. Therefore the unconstrained domino problem needs a more subtle encoding of computations.

The domino problem has recently been used by van Emde Boas [13] and others as a main NP-complete problem, avoiding the reduction to SATISFIABILITY for showing EXACT COVER and KNAPSACK to be NP-complete.

Initially Wang [14] has introduced the domino problem as an intermediate step between TM computations and their description by formulas of predicate calculus. It has promoted the solution of the famous old problem of finding the largest set of quantifier prefixes, such that the satisfiability problem is decidable for all sentences of predicate calculus without equality (and without function symbols) whose prefix is in the set. The prefix classes are the classes of formulas in prenex form with a given sequence of quantifier types in the prefix. Or for a finer classification, conjunctions of formulas in prenex form are investigated.

It has turned out that the classical decidable prefix classes namely

$\exists \ldots \exists \forall \ldots \forall$	(Schönfinkel-Bernays class)
$\exists \ldots \exists \forall \exists \ldots \exists$	(Ackermann class)
$\exists \ldots \exists \forall \forall \exists \ldots \exists$	(Gödel class)

are the only ones. (For references see e.g. [3,4,9] and references thereof.) The last hurdle was the undecidability proof for the $\forall\exists\forall$ class, whereas the class $\exists\forall\exists\forall$ or in the finer classification even the $\exists \wedge \forall\exists\forall$ class was shown to be undecidable by Büchi [2]. Formulas of this type in functional form are built up from say a, x, f(x) and y. An origin constrained tiling can be described by such a formula. E.g.

$$R_j(a,a)$$

means that the origin is tiled by j, and

$$\neg(R_i(x,y) \wedge R_j(f(x),y))$$

means that tile j is not allowed to be to the right of tile i. Here and in the following, we always assume the x-axis to point downwards and the y-axis to point to the right.

We see that the first $\exists$-quantifier is used to express the origin constraint. And therefore an undecidability result for the unlimited origin constrained domino problem implies the undecidability of the class $\forall\exists\forall$.

It has turned out that the $\forall\exists\forall$ class has first been shown to be undecidable [6] via a diagonal constrained domino problem (only special tiles are allowed on points with coordinates (j,j)), which does not require the additional existential quantifier. Then finally Berger [1] showed the unlimited unbounded domino problem to be undecidable by a rather involved construction. (Simplified proofs are due to Robinson [11] and Aanderaa (see [9]).) The contents of a tape cell at a given time is not only encoded by one tile, but by all the tiles placed on the points of a regular grid. And a complicated system of tiled paths for information flow ensures all the points in the grid to encode the same symbol. The mesh size of the grid grows exponentially in the coordinates (t,s) ($t=$ time and $s=$ position on the tape). So a square of length k encodes only $\log k$ pieces of configurations of length $\log k$, and it does not seem possible to do better. Therefore, it is reasonable to conjecture that also in the limited case, exponentially big squares could encode only polynomially big computations, and then the complexity of the domino problem with k given in binary would still be in NP. The purpose of this paper is to show that this conjecture of Lewis [7] is wrong. With a limited tiling (i.e. finite $k \times k$ square) we can in fact have a compact encoding of computations.

As a Corollary to this, we obtain the result, that not only the $\exists\wedge\forall\forall\exists$ class, but even the $\forall\forall\exists$ class has an $\Omega(c^{n/\log n})$ lower bound. For a discussion of the complexity of the decidable prefix classes, the reader is referred to [3,4,8].

2. Counting with tiles

We call (T,V,H) with $V,H \subseteq T \times T$ a *tiling system*, to distinguish it from the instances of the *unary* or *binary tiling problems* (T,V,H, k) (unconstrained) and (T,T_0,V,H, k) (constrained; only the tiles of T_0 are allowed to be put on the origin (0,0)).

Definition. An *m-counting tiling system* is a tiling system (T,V,H) with the property: There are subsets T_V and T_H of T such that for every tiling of a $k \times l$ rectangle there is a displacement $(d_V,d_H) \in \omega \times \omega$ such that (x,y) is tiled with a tile of T_V iff $m \mid (x-d_V)$, and with a tile of T_H iff $m \mid (y-d_H)$. In these cases, we say that x belongs to a *distinguished column* or *row*.

Definition. An $f(n)$-counting tiling system (T,V,H) is *polytime constructable* iff there is a polynomial time bounded TM which on input n in binary, produces the output (T,V,H).

Lemma 1. There are polytime constructable $\lfloor \log\log n \rfloor$-counting tiling systems.

Proof: Let $m = \lfloor \log\log n \rfloor$. (All logarithms are to the base 2.)
We use tiles $T = m \times m$ with

$$((t_1,t_2),(t_3,t_4)) \in V \quad iff \quad t_1+1 \equiv t_3 \ mod \ m,$$

$$((t_1,t_2),(t_3,t_4)) \in H \quad iff \quad t_2+1 \equiv t_4 \ mod \ m,$$

$$T_V = \{(0,t) : t \in m\} \text{ and}$$

$$T_H = \{(t,0) : t \in m\}.$$

Clearly m and (T,V,H) can be computed by a polynomial time bounded TM. □

Remark. With this method, Garey, Johnson, Papadimitriou (see [5]) and Lewis [7] have shown the unary tiling problem (T,V,H,k) (k in unary) to be NP-complete.

Definition. A *marked binary number* is a binary number where exactly the rightmost 1 and all the subsequent 0's are underlined.

Lemma 2. If there exists a polytime constructable $f(n)$-counting tiling system (T,V,H), then there exists also a polytime constructable $2^{f(n)}$-counting tiling system (T',V',H').

Proof: We use the horizontal yardstick (of length $f(n)$) given by the tiling system (T,V,H) to write binary numbers of length $f(n)$ horizontally, and with their help, we count vertically from 0 to $f(n)-1$ (see Figure). We do the same with 'horizontally' and 'vertically' interchanged. For marked binary numbers it is locally testable, whether a binary number q written (right-adjusted) below p is equal to $p+1$.

2 Z	0 Z	1 Z	2 Z	0 Z	1 Z	2 Z	0 Z
2 $\underline{1}$	0 0	1 0	2 $\underline{1}$	0 0	1 0	2 $\underline{1}$	0 0
2 $\underline{0}$	0 0	1 $\underline{1}$	2 $\underline{0}$	0 0	1 $\underline{1}$	2 $\underline{0}$	0 0
2 $\underline{1}$	0 0	1 1	2 $\underline{1}$	0 0	1 1	2 $\underline{1}$	0 0
2 $\underline{0}$	0 $\underline{1}$	1 $\underline{0}$	2 $\underline{0}$	0 $\underline{1}$	1 $\underline{0}$	2 $\underline{0}$	0 $\underline{1}$
2 $\underline{1}$	0 1	1 0	2 $\underline{1}$	0 1	1 0	2 $\underline{1}$	0 1
2 $\underline{0}$	0 1	1 $\underline{1}$	2 $\underline{0}$	0 1	1 $\underline{1}$	2 $\underline{0}$	0 $\underline{0}$
2 $\underline{1}$	0 1	1 1	2 $\underline{1}$	0 1	1 1	2 $\underline{1}$	0 $\underline{1}$

Figure: A tiling of a 2^3-counting tiling system. Note that the last column may count wrong.

We use tiles with two tracks and require a tiling with track two removed to be still a tiling of (T,V,H). Furthermore, we require the binary numbers to start at the distinguished columns of the $f(n)$-counting. And the row containing 0 (distinguished by a special zero symbol Z) will be the distinguished row of the $2^{f(n)}$-counting.

So we define

$$T' = T \times \{0,\underline{0},1,\underline{1},Z\} - T_H \times \{\underline{0}\}$$

$$((t_1,t_2),(t_3,t_4)) \in H' \;\; iff \;\; ((t_1,t_3) \in H) \wedge (t_2 \in \{\underline{0},\underline{1}\} \leftrightarrow (t_3 \in T_H \vee t_4 = \underline{0})) \wedge (t_2 = Z \leftrightarrow t_4 = Z)$$

$$((t_1,t_2),(t_3,t_4)) \in V' \;\; iff \;\; ((t_1,t_3) \in V) \wedge (\, t_2 \in \{0,\underline{0},Z\} \leftrightarrow t_4 \in \{0,\underline{1}\})$$

This marks every $2^{f(n)}$-th row by tiles of $T'_H = T \times \{Z\}$. Clearly, with the help of a third track, the same can be done with 'horizontal' and 'vertical' interchanged, to mark every $2^{f(n)}$-th column. □

Lemma 3. There exists a polytime constructable $2^{2^{\lfloor \log\log n \rfloor}}$-counting tiling problem.

Proof: This follows from the Lemmas 1 and 2. □

3. The origin-constrained tiling problem

Proposition. The complexity of the origin-constrained limited $k \times k$ domino problem with k given in binary is nondeterministic exponential time.

In the main part of the proof, which can be found in Lewis [7], an origin constrained tiling system is constructed in such a way that the tilings of the $k \times k$ square encode the accepting computations (of length k) of a given nondeterministic TM. We do not repeat this part, because it involves standard techniques, as described e.g. in [9,10]. We just want to point out that there is an additional problem. In order to describe the start configuration corresponding to an input of length n, we need a tiling system of length $O(n \log n)$, because the names of the tiles have length $O(\log n)$. Therefore, in order to obtain a (really exponential) $\Omega(c^n)$ lower bound instead of $\Omega(c^{n/\log n})$, we need to describe computations with small inputs, i.e. computations whose complexity is $\omega(c^{n/\log n})$.

Theorem 1. The complexity of the origin-constrained limited $k \times k$ domino problem, with k given in binary, is bounded from above by $O(4^n \log^2 n)$ and from below by $\Omega((\sqrt{2}-\varepsilon)^n)$ nondeterministic time (for all $\varepsilon > 0$). This even holds for k restricted to integers of the form 2^{2^l} for $l \in \omega$.

Proof: First, the tiles are renamed, if they have names longer than $\log n$. Then, while a tiling is guessed line by line, two lists are built, namely the sequences of vertically and horizontally adjacent pairs in the tiling. Now these lists of length $O(2^{2n} \log n)$ are sorted lexicographically. This adds another factor $\log n$ to the TM time. Finally the sorted lists (compressed to length n by avoiding repetitions) are quickly checked for correctness.

Let us define the function g for example by

$$g(n) = 2^{\lfloor \log n + 2\log\log n \rfloor}.$$

Instead of the coefficient 2 of $\log\log n$, any other rational $c > 1$ would be fine. To obtain a lower bound, let us take a $(2^{g(n)}-2)$-bounded nondeterministic TM M accepting a language L which is not accepted by an $o((\sqrt{2}-\delta)^{g(n)})$-bounded nondeterministic TM for any $\delta > 0$. The nondeterministic Hierarchy Theorem [12] ensures the existence of such a language, because

$$(\sqrt{2}-\delta)^{g(n+1)} \leq (\sqrt{2}-\delta)^{2g(n)} = o(2^{g(n)}-2).$$

We assume M has a counter built in, such that all non-accepting computations are stopped after $2^{g(n)}-2$ steps, and in accepting states M continues the computation staying in accepting states (busy waiting). We construct an origin-constrained tiling system (T,V,H), such that exactly the accepting computations of length $2^{g(n)}-1$ ($2^{g(n)}-1$ steps corresponding to $2^{g(n)}$ configurations one below the other) correspond to tilings of a $2^{g(n)}\times 2^{g(n)}$ square. The number of tiles is $c+n$ for a constant c depending on M. The n tiles are used to describe the input. If $w=a_1...a_n$ is the input word and q_0 is the starting state, then we use the tiles

$$t_1=(1,(q_0,a_1)),\ t_2=(2,a_2),\ldots,t_n=(n,a_n),\ t_{n+1}=t_{n+2}=(0,b),$$

to describe the start configuration. We require the origin to be tiled with t_1, and t_{j+1} to be the only possible right neighbor of t_j (*for* $j=1,\ldots,n+1$). Now there is a constant c such that this tiling problem (T,$\{t_1\}$,V,H, k) ($k=2^{g(n)}$ in binary) of length

$$g(n)+O(n\log n)=g(n)+o(g(n))$$

cannot be solved in nondeterministic time $O((\sqrt{2}-\delta)^{g(n)})$ (otherwise we had a fast algorithm to accept L), and therefore not in time $O((\sqrt{2}-\varepsilon)^k)$ for all $\varepsilon>\delta$. So we have an $\omega((\sqrt{2}-\varepsilon)^k)$ lower bound for the origin-constrained tiling problem, even if we restrict k to the form $2^{2^{g(n)}}$.

□

4. The unconstrained tiling problem

Lemma 4. If there exists a polytime constructable m-counting tiling system, then there is a tiling system (T,V,H), such that there is a tiling of a $k\times k$ square iff $k\leq m$. Furthermore, for $k=m$, (0, 0) is distinguished, i.e. the tiles which can be used to tile (0, 0) cannot be used to tile any other square.

Proof: Remove $((t_1,t_2),(t_3,t_4))$ from V iff $t_3\in T_V$, and remove $((t_1,t_2),(t_3,t_4))$ from H iff $t_4\in T_H$. Then only the tiles of $T_V\cap T_H$ can be used to tile (0, 0), and these tiles cannot be used for any other square. □

Theorem 2. The complexity of the unconstrained limited $k\times k$ domino problem, with k given in binary, is bounded from above by $O(4^n\log^2 n)$ and from below by $\Omega((\sqrt{2}-\varepsilon)^n)$ nondeterministic time (for all $\varepsilon>0$).

Proof: The upper bound is clear. The lower bound is obtained by a transformation from the origin constrained tiling problem. Let (T,T_0,V,H, k) be any instance of the origin-constrained tiling problem with k of the form 2^{2^l}. Then by Lemma 3, there exists a polytime constructable k-counting tiling system. This is modified according to Lemma 4 to obtain the system (T',V',H'). Then (T",V",H", k) with

$$T''=T_0\times(T'_V\cap T'_H)\ \cup\ T\times(T'-T'_V\cap T'_H)$$

$$V''=V\times V'$$

$$H''=H\times H'$$

can be computed in polynomial time from (T, T_0,V,H, k), and the length of (T",V",H", k) is smaller than $1+\varepsilon$ times the length of (T,T_0,V,H, k) for all $\varepsilon>0$. □

5. Related Problems

As a corollary to Theorem 2, we obtain a sharp lower bound for the complexity of the $\forall\exists\wedge\forall\forall$ class of predicate calculus. Lewis [8] has shown the same lower bound for the prefix class $\exists\wedge\forall\exists\wedge\forall\forall$.

Theorem 3. The $\forall\exists\wedge\forall\forall$ class of the decision problem of predicate calculus, without equality and without function symbols, has a nondeterministic time complexity between $\Omega\ (c^{n/\log n})$ and $O(d^{n/\log n})$ for some constants $c,d > 1$.

Proof: We can just copy the proof of Lewis [8] for the prefix class with an additional existential quantifier, and observe that this quantifier is only used to describe the origin constraint. □

Theorem 4. The complexity of the unconstrained $\omega\times k$ tiling problem is nondeterministic exponential space.

Proof: If there are |T| tiles, then there is a tiling of the infinite stripe, iff there is a tiling of a $(|T|^k+1)\times k$ rectangle, which can be guessed and checked line by line keeping only the previous line and the line number stored on a tape. For the lower bound, we proceed as in the time bounded case, except that we count vertically not only up to $2^{g(n)}$, but to $2^{2^{g(n)}}$. □

Remark. It is tempting to think that the domino problem is already difficult for a constant number of of dominoes. Naturally this is not the case. For each constant, there is even a finite automaton which can solve this problem.

But one might reasonably ask, how many tiles are necessary for an m-counting tiling system. From Lemma 2, it is easy to see that $O(\log \cdots \log m)$ tiles are sufficient for every constant number of iterations of the logarithm. It is more involved to show that even about $2^{2^{\alpha(n)}}$ tiles are enough, for α the inverse of the Ackermann function.

6. References

[1] **Berger, R.** The Undecidability of the Domino Problem, *Memoirs of the American Mathematical Society 66,* 1966.

[2] **Büchi, J. R.** Turing Machines and the Entscheidungsproblem. *Mathematische Annalen 148* (1962), pp. 201-213.

[3] **Dreben, B., and W. D. Goldfarb.** *The Decision Problem: Solvable Classes of Quantificational Formulas.* Reading, Mass. Addison-Wesley, 1979.

[4] **Fürer, M.** Alternation and the Ackermann Case of the Decision Problem. *L'Enseignement Mathématique 27* (1981), pp. 137-162.

[5] **Garey, M. R. and D. S. Johnson.** *Computers and Intractability: A Guide to the Theory of NP-Completeness.* Freeman, San Francisco, 1979.

[6] **Kahr, A. S., E. F. Moore and Hao Wang.** Entscheidungsproblem Reduced to the $\forall\exists\forall$ Case. *Proc. Nat. Acad. Sci. USA 48,* (1962), pp. 365-377.

[7] **Lewis, Harry R.** Complexity of Solvable Cases of the Decision Problem for Predicate Calculus. *Proc. 19th Symp. on Foundations of Computer Science,* 1978, pp. 35-47.

[8] **Lewis, Harry R.** Complexity Results for Classes of Quantificational Formulas. *Journal of Computer and System Sciences 21,* Dec. 1980, pp. 317-353.

[9] **Lewis, Harry R.** *Unsolvable Classes of Quantificational Formulas.* Addison-Wesley, Advanced Book Program, Reading, Massachusetts, 1979.

[10] **Lewis, Harry R. and Chr. H. Papadimitriou.** *Elements of the Theory of Computation.* Prentice-Hall, 1981.

[11] **Robinson Raphael. M.** Undecidability and Nonperiodicity for Tilings of the Plane. *Inventiones Mathematicae 12,* (1971), pp. 177-209.

[12] **Seiferas, J. I., M. J. Fischer and A. R. Meyer.** Separating Nondeterministic Time Complexity Classes. *Journal of the Association for Computing Machinery 25,* (1978), pp. 146-167.

[13] **van Emde Boas, P.** Dominoes are Forever. *Report on the* 1st *GTI-workshop Paderborn, FRG,* Reihe Theoretische Informatik 13, L. Priese Ed., Univ. of Paderborn 1982, pp. 75-95.

[14] **Wang, Hao.** Proving Theorems by Pattern Recognition II. *The Bell System Technical Journal 40,* (1961), pp. 1-41.

Implicit definability of finite binary trees by sets of equations (+)

Jerzy Tiuryn
Institute of Mathematics
University of Warsaw

00-901 Warsaw, PKIN IX p
Poland

Introduction

Let T be the set of all finite binary trees which are pairwise nonisomorphic. Let $\mathcal{L}$= {mktree, nil} be a signature consisting of a binary function symbol mktree and a constant symbol nil. We make T into an $\mathcal{L}$-algebra $\mathcal{T}$ by intrepreting nil as the one-node tree and letting mktree (t_1,t_2) to be the tree with left child t_1 and right child t_2. $\mathcal{T}$ is a free algebra in the class of all algebras of signature $\mathcal{L}$.

The algebra $\mathcal{T}$ has very interesting computational properties. It has been proved by H.Friedman [1], and independently by M.Paterson and C.Hewitt [3] that there exists a function computable in $\mathcal{T}$ by a recursive procedure which cannot be computed by any flow-diagram. P.Urzyczyn [5] has shown that every total function computable in $\mathcal{T}$ by a flow-diagram can be computed by a loop-free program. From this result he derived that no universal formula of Regular Dynamic Logic(cf. [2] for necessary definitions) is capable of defining $\mathcal{T}$ uniquely.

The main result of this paper is a generalized and polished version of the crucial combinatorial fact used in [4] to prove that no logic of programs using bounded memory is capable of defining $\mathcal{T}$ uniquely up to an isomorphism. Therefore, every logic of programs which can uniquely define $\mathcal{T}$ is not reducible to any logic of programs with bounded memory. As stated in [4], this result solves some open problems in logics of programs. It also implies the results of [1], [3] and [5], mentioned above.

In this paper we study the power of implicit definitions in $\mathcal{T}$. An implicit definition is a set Γ of equations of the form $x_o = t$ such that the left side of any equation in Γ is the same variable x_o, and the maximal number of different variables which may occur in a term t is bounded. The least $k \in N$ such every term in Γ has at most k different

(+) This work was supported in part by the National Science Foundation, Grant No. MC 5 8010707, and by a grant to the MIT Laboratory for Computer Science by the IBM Corporation.

variables is called the width of Γ(cf. 1.4 for details). Given an algebra $\underline{\underline{A}}$ of signature $\mathcal{L}$, an implicit definition Γ defines a set $\Gamma(\underline{\underline{A}}) = \{a \in A : \Gamma(a/x_o)$ is consistent in $\underline{\underline{A}}\}$. Implicit definitions are equivalent in $\underline{\underline{A}}$ iff they define the same set in $\underline{\underline{A}}$.

The main result of this paper is the following theorem.

Theorem

There exists a primitive recursive function $Q: N \to N$ such that for every $k \in N$ and for every implicit definition Γ of width k, if $\Gamma(\mathcal{T}) \neq \emptyset$, then there is a subset $\Delta \subset \Gamma$ such that
(i) card $(\Delta) \leq Q(k)$
(ii) Γ and Δ are equivalent in $\mathcal{T}$.

In fact we prove a somewhat stronger result. We show that condition (ii) holds in a certain class of $\mathcal{L}$-algebras which is defined in 1.1o.

Section 1 contains all necessary definitions and preliminary results. In section 2 we prove the main theorem.

1. Preliminary definitions and results

1.1 We fix throughout the paper the signature $\mathcal{L} = \{$mktree,nil$\}$. Let X be a countable set of individual variables. We will denote variables by x,y,z,u,v,w.

By a term over $\mathcal{L}$ we mean either a variable, or nil, or an expression of the form mktree (t_1, t_2), where t_1 and t_2 are terms. Terms will be denoted by r,s,t.

Individuals, i.e. elements of the carrier of an $\mathcal{L}$-algebra, will be denoted by a,b,c,d. In the above notations we allow for sub- or superscripts. We will also be using vector notation: $\vec{x}$ denotes a vector of variables, and $|\vec{x}|$ its length. Thus if $|\vec{x}| = k$, then $\vec{x} = (x_1, \ldots, x_k)$. The same applies to vectors of terms and to vectors of individuals.

For a term t we write $t(x_1, \ldots, x_k)$ to indicate that all variables occurring in t are among $\{x_1, \ldots, x_k\}$. If $t_1, \ldots, t_k$ are terms, then $t(t_1/x_1, \ldots, t_k/x_k)$ denotes the term obtained from t by simultaneously substituting t_i for all occurences of x_i in t, for every $1 \leq i \leq k$. The same notation for substitution applies to equations and sets of equations. We may also substitute individuals rather than terms.

1.2 An equation $t_1 = t_2$ is consistent in an $\mathcal{L}$-algebra $\underline{\underline{A}}$ iff it has a solution in $\underline{\underline{A}}$, i.e. if there exists a valuation val $\in A^X$ such that $(\underline{\underline{A}}, \text{val}) \models t_1 = t_2$ holds.

Let γ be an equation and let $a \in A$. $\gamma(a/x)$ is consistent in $\underline{\underline{A}}$ iff there exists a valuation $v \in A^X$ such that $v(x) = a$ and $(\underline{\underline{A}}, v) \models \gamma$ holds.
A set of equations Γ is consistent in $\underline{\underline{A}}$ iff Γ has a solution in $\underline{\underline{A}}$.

1.3 Let Δ and Γ be sets of equations over $\mathcal{L}$, and let $\vec{x}$ be a vector of variables, with $|\vec{x}| = k$ for some $k \in N$. Δ and Γ are said to be $\vec{x}$-equivalent over an $\mathcal{L}$-algebra $\underline{\underline{A}}$ iff for every $\vec{a} \in A^k$, $\Delta(a_1/x_1,\ldots,a_k/x_k)$ is consistent in $\underline{\underline{A}}$ iff $\Gamma(a_1/x_1,\ldots,a_k/x_k)$ is consistent in $\underline{\underline{A}}$.

1.4 A set of equations $\Gamma = \{x_o = t_i : i \in I\}$, where $I \subseteq N$ is an implicit definition provided there exists $k \in N$ such that in every term t_i occur at most k different variables. The least k satisfying the above property is called the width of Γ.

An implicit definition Γ defines in an $\mathcal{L}$-algebra $\underline{\underline{A}}$ a set $\Gamma(\underline{\underline{A}}) = \{a \in A: \Gamma(a/x_o)$ is consistent in $\underline{\underline{A}}\}$.

Implicit definitions Γ and Δ are said to be equivalent over $\underline{\underline{A}}$ iff $\Gamma(\underline{\underline{A}}) = \Delta(\underline{\underline{A}})$.

1.5 Let $\{0,1\}^*$ denote the set of all finite words $\{0,1\}$. A finite binary tree is a partial function $a:\{0,1\}^* \to \{mktree, nil\}$ which satisfies the following conditions.

1.5.1 The domain of a, dom(a), is finite and nonempty.

1.5.2 For every $\pi \in dom(a)$, if $a(\pi) = mktree$, then $\pi 0, \pi 1 \in dom(a)$, else $\pi 0, \pi 1 \notin dom(a)$.

1.5.3 For every $\pi \in dom(a)$, if π_1 is an initial subword of π, i.e. if $\pi = \pi_1 \pi_2$ for some $\pi_2 \in \{0,1\}^*$, then $\pi_1 \in dom(a)$.

1.6 There is a one-to-one correspondence between terms and finite binary trees over the extended language of $\{mktree, nil\} \cup X$, the latter can be defined as partial functions $a:\{0,1\}^* \to \{mktree, nil\} \cup X$ which satisfy 1.5.1-1.5.3. We will be often using this intuitive representation of terms.

In the context of trees, finite words over {0,1} will be called paths. A path π is contained in a tree a, if $\pi \in dom(a)$. We say that a path π leads to (an occurrence of) a symbol s in a tree a, if $\pi \in dom(a)$ and $a(\pi) = s$.

A subtree of a tree a determined by a path π is a tree $a\restriction\pi$ such that $dom(a\restriction\pi) = \{\rho \in \{0,1\}^* : \pi\rho \in dom(a)\}$, and $a\restriction\pi(\rho) = a(\pi\rho)$, for all $\rho \in dom(a\restriction\pi)$.

According to 1.6 we may use phrases like "a path contained in a tree t," "π leads in t to an occurrence of a variable x", or "a subterm of t determined by a path π''.

A depth of a tree a is the maximum of lengths of all paths contained in a. For $\pi, \rho \in \{0,1\}^*$ we define $\pi \leq \rho$ iff $\rho = \pi\pi_o$ for some $\pi_o \in \{0,1\}^*$.

1.8 Let T denote the set of all finite binary trees. We interpret in T the constant symbol nil as the unique one-element tree nil, defined by $nil(\Lambda) = nil$, where Λ is the empty word. We also define a binary operation mktree: $T^2 \to T$ as follows. For $a_1, a_2 \in T$, $dom(mktree(a_1,a_2)) = \{0\pi : \pi \in dom(a_1)\} \cup \{1\pi : \pi \in dom(a_2)\}$ and we set $mktree(a_1,a_2)(0\pi) = a_1(\pi)$, for $\pi \in dom(a_1)$ and $mktree(a_1,a_2)(1\pi) = a_2(\pi)$, for $\pi \in dom(a_2)$.

The above interpretation of mktree and nil makes T into an $\mathcal{L}$-algebra $\mathcal{T} = \langle T, mktree, nil \rangle$.

Obviously, $\mathcal{T}$ is an initial algebra in the class of all algebras of similarity type $\mathcal{L}$.

1.9 Let $\underline{\underline{A}}$ be an $\mathcal{L}$-algebra. An element $a \in A$ will be called an atom iff for every term t having at least two different occurences of the same variable, the equation $a = t$ is inconsistent in $\underline{\underline{A}}$.

It should be obvious that the only atom in is nil.

1.1o An $\mathcal{L}$-algebra $\underline{\underline{A}}$ will be called a term-like algebra iff $\underline{\underline{A}}$ satisfies the following conditions.

1.1o.1 $mktree(x_1,x_2) = mktree(y_1,y_2) \to (x_1 = y_1 \wedge x_2 = y_2)$

1.1o.2 $x \neq nil \Leftrightarrow \exists x_1 \exists x_2\ mktree(x_1,x_2) = x$

1.1o.3 $t \neq x$, for every term t not being a variable, and every variable x which occurs in t.

1.1o.4 $\underline{\underline{A}}$ is generated by its atoms, i.e. for every $a \in A$ there exist a term $t(x_1,\dots,x_k)$, and atoms $a_1,\dots,a_k \in A$ such that

$$a = t(a_1/x_1,\dots,a_k/x_k) \text{ holds in } \underline{\underline{A}}.$$

1.11 $\mathcal{T}$ obviously is a term-like algebra, and every term-like algebra contains an isomorphic copy of $\mathcal{T}$ as a subalgebra. In every term-like algebra $\underline{\underline{A}}$, nil is an atom. Atoms other than nil will be called non-standard atoms. Below we give an example of a term-like algebra, non-isomorphic to $\mathcal{T}$.

1.12 Let $BIN = \{bin_\pi : \pi \in \{0,1\}^*\}$ be a set of constant symbols disjoint from $\mathcal{L} \cup X$. Let T[BIN] be the set of all partial functions $a: \{0,1\}^* \to \mathcal{L} \cup BIN$ which satisfy 1.5.1-1.5.3. Similarly as in 1.8 we make T[BIN] into an $\mathcal{L}$-algebra. Denote this algebra by $\mathcal{T}$[BIN].

Let $\mathcal{T}^\infty$ be a quotient algebra $\mathcal{T}[BIN]/\sim$, where $\sim$ is a congruence generated by the pairs $(bin_\pi, mktree(bin_{\pi o}, bin_{\pi 1}))$, for $\pi \in \{0,1\}^*$.

The reader should think of bin_Λ as a full infinite binary tree. Then bin_π is a name of a subtree of bin_Λ determined by the path π. The equation $bin_\pi = mktree(bin_{\pi o}, bin_{\pi 1})$ in $\mathfrak{T}^\infty$ expresses the property that $bin_{\pi o}$ is the left subtree of bin_π, and $bin_{\pi 1}$ is the right subtree of bin_π. The elements $bin_\pi \in T^\infty$, for $\pi \in \{0,1\}^*$ will be called nonstandard atoms. $\mathfrak{T}^\infty$ is obviously a non-isomorphic extension of $\mathfrak{T}$.

1.13 Let t be a term. Register complexity of t, denoted by r.c.(t), is the least number of program variables, used in a sequence of elementary assignments to compute t.

Elementary assignments are assignments of either of the following three forms:

$v \leftarrow c$, where c is either nil or an individual variable;

$v \leftarrow mktree(v', v'')$;

$v \leftarrow v'$.

The variables v, v', v'' in the above definition are program variables i.e. registers.

Every term has finite register complexity. Since finite binary trees can be thought as variable-free terms, every $a \in T$ has finite register complexity. Nonstandard atoms can be thought as objects of infinite register complexity.

1.14 For every $n \in N$ we define a term r_n with 2^n different variables $x_1, \dots, x_{2^n}$, as follows:

$r_o = x_1$, $r_{n+1} = mktree(r_n, r_n(x_{2^n+1}/x_1, \dots, x_{2^{n+1}}/x_{2^n}))$.

For every $i \in N$ let $\underline{\underline{i}} \in T$ be the tree $r_i(nil/x_1, \dots, nil/x_{2^i})$. It is easy to check that for every $i \in N$, r.c.$(\underline{\underline{i}}) = 1$.

However, there are trees of arbitrary large register complexity. The next result is implicitly contained in [1] and in [3].

1.15 Proposition

Let $n \in N$ and let $t = r_n(t_1/x_1, \dots, t_{2^n}/x_{2^n})$, where $t_1, \dots, t_{2^n}$ are pairwise different terms. Then r.c.$(t) \geq n+1$. If for $1 \leq i \leq 2^n$, $t_i = \underline{\underline{i}}$, then r.c.$(t) = n+1$.

The next result follows directly from the definition of register complexity.

1.16 Proposition

Let t and r be terms such that r.c.$(t) \leq n$ and r.c.$(r) \leq n$ for some $n \in N$. Then

(i) every subterm of t has register complexity at most n.

(ii) $r.c.(t(r/x)) \leq n+1$ for every variable x.

1.17 In the sequel of the paper we will be using the following two formal rules for transforming equations.

1.17.1 R-rule

$$\frac{x = t,\ x = t'}{y = r},$$

provided there is a path π leading in t to an occurrence of y, and r is a subterm of t' determinded by π.

1.17.2 S-rule

$$\frac{x = t,\ y = r}{x = t'},$$

where t' is obtained from t by replacing a certain occurence of y in t by r.

Obviously R- and S-rules are sound in every term-like algebra.

$\Gamma \vdash \gamma$ means that γ is derivable from Γ by a finite number of applications of R,S-rules.

1.18 Proposition

Let $\underline{\underline{A}}$ be a term-like algebra. For every $a \in A$ and for every $k \in N$ there exists $m \in N$ such that for every term t with at most k different variables, if $a = t$ is consistent in $\underline{\underline{A}}$, then depth of t is at most m.

Proof. We first prove by induction on k that for every atom $a \in A$ and for every term t with k different variables, if $a = t$ is consistent in $\underline{\underline{A}}$, then depth of t is $\leq k$.

Next, let $a \in A$ be an arbitrary element. Let $|a|$ be the minimal depth of a term $t(x_1,\ldots,x_p)$, for some $p \in N$, such that $a = t(e_1/x_1,\ldots,e_p/x_p)$ holds in $\underline{\underline{A}}$ for some atoms $e_1,\ldots,e_p \in A$. Then it is easy to prove that for every $k \in N$, $m = |a|+k$ satisfies 1.18.

1.19 Corollary

Let $\underline{\underline{A}}$ be a term-like algebra and let Γ be an implicit definition consistent in $\underline{\underline{A}}$. There exists $m \in N$ such that depth of every term in Γ is at most m.

Proof: Take $m = \min\{|a| : a \in \Gamma(A)\}+\text{width}(\Gamma)$.

1.2o Register complexity of an implicit definition

Γ, $r.c.(\Gamma)$, is the least $n \in N$ such that for every equation $x_o = t \in \Gamma$, $r.c.(t) \leq n$. If there is no such n, then we set $r.c.(\Gamma) = \infty$.

It follows from 1.19 that every implicit definition which is consistent in a certain term-like algebra has finite register complexity.

1.21 Proposition

Let $\underline{\underline{A}}$ be a term-like algebra, and let Γ be an implicit definition consistent in $\underline{\underline{A}}$.
There exists a term s such that
(i) $\Gamma \vdash x_o = s$
(ii) Γ and $\{x_o = s\}$ are equivalent in all term-like algebras.

Moreover if in Γ there are at most m+1 different variables then r.c.(s) $\leq$ m+r.c.(Γ).

Proof: Let $\Gamma^* \supset \Gamma$ be the least set closed under R-rule. Let V be the set of all variables which occur in Γ. Define an equivalence relation $\sim$ in V as follows.
For $u,w \in V$, $u \sim w$ iff $u = w \in \Gamma^*$.

From every equivalence class $[w]_\sim$ of the relation $\sim$ choose one variable $x_{[w]} \in [w]$. Observe that since Γ is consistent in $\underline{\underline{A}}$, $[x_o] = \{x_o\}$. Let Δ be the set of all equations which are obtained from Γ by replacing every variable w by $x_{[w]}$. It follows from 1.19 that Δ is finite. Moreover $\Gamma \vdash \Delta$ and Γ and Δ are equivalent in all term-like algebras.

Let $\Delta^* \supset \Delta$ be the least set closed under R-,S-rules. Since Δ is finite and consistent in $\underline{\underline{A}}$, it follows from 1.1o.3 that Δ^* is finite as well. Let $x_o = s \in \Delta^*$ be such an equation that for every variable w which occurs in s, if there is a term t such that $w = t \in \Delta^*$, then t is w.

Obviously (i) holds for s. We prove (ii). Let $\underline{\underline{B}}$ be a term-like algebra and let $b \in B$ be such that $b = s$ is consistent in $\underline{\underline{B}}$. We shall prove that $\Delta(b/x_o)$ is consistent in $\underline{\underline{B}}$. It will follow then that $\Gamma(b/x_o)$ is consistent in $\underline{\underline{B}}$. The implication in opposite direction is obvious.

Let W be the set of all variables which occur in equations of Δ. For $w_1, w_2 \in W$ we set $w_1 \leq w_2$ iff there exists an equation $w_2 = t \in \Delta^*$ such that w_1 occurs in t. Since Δ^* is consistent in $\underline{\underline{A}}$, and it is closed under R,S-rules and equations of the form $w_1 = w_2$ do not belong to Δ^*, unless w_1 and w_2 are identical - it follows that $\leq$ is a partial order in W. If $w \in W$ is a minimal element in $\langle W, \leq \rangle$, then either

1.21.1 w occurs in s, or

1.21.2 for some variable-free term t, $w = t \in \Delta^*$.

We define a valuation val:W $\to$B such that $(\underline{\underline{B}}, \text{val}) \models \Delta^*$. Let $w \in W$ be a minimal element in $\langle W, \leq \rangle$. If (1.21.1) holds, then let val(w) be the "w-component" of the unique solution of b = s in $\underline{\underline{B}}$. If (1.21.2) holds, then we set val(w) to be the value of t in $\underline{\underline{B}}$.

Let $w \in W$ be a minimal element in the set of those variables for which val has not been defined yet. Since w is not a minimal element in $\langle W, \leq \rangle$ there is an equation $w = t \in \Delta^*$ such that t is not a variable. All variables in t are less than w, and val is defined for them. Let val(w) be the value of t in $\underline{\underline{B}}$ determined by values of all variables in t. This completes the definition of val.

Let $w \in W$ be a minimal element such that for some equation $w = t \in \Delta^*$, $(\underline{\underline{B}}, val) \models w \neq t$.

Obviously w cannot be a minimal element in $\langle W, \leq \rangle$. Let $w = t_o \in \Delta^*$ be the equation which has been used to determine val(w). Since $w = t$ and $w = t_o$ are consistent in $\underline{\underline{A}}$, it follows that there is an equation $w' = r$ such that $w = t$, $w = t_o$ $w' = r$, and $(\underline{\underline{B}}, val) \models w' \neq r$. Since w' is less than w, this contradicts minimality of w.

In this way we have proved that $(\underline{\underline{B}}, val) \models \Delta^*$. Since $(\underline{\underline{B}}, val) \models b = s$, it follows that $val(x_o) = b$. Therefore $\Delta^*(b/x_o)$ is consistent in B, and $\Delta(\mathbf{b}/x_o)$ is consistent in $\underline{\underline{B}}$ too.

In order to finish the proof of 1.21 observe that we may get the term s in the following way. Let $x_o = t_o \in \Delta$ be any equation. Assume that $x_o = t_o, \ldots, x_o = t_{n-1}$ has been defined for some $n \geq 1$.

Let W_n be the set of all variables which occur in t_{n-1}. Let $w \in W_n$ be a variable, maximal with respect to $\leq$, such that there exist a term t which is not a variable, and $w = t \in \Delta^*$. The equation $x_o = t_n$ is then $x_o = t_{n-1}(t/w)$. If there is no such w, then t_{n-1} is the sought term s. Choosing w maximal guarantees that w will never occur again in later equations. Therefore we make at most m substitutions until we reach s. The second part of 1.21 follows now from 1.16.

<u>1.22 Corollary</u>

If an implicit definition Γ is consistent in a term-like structure, then it is consistent in all term-like structures.

<u>1.23 Corollary</u>

If implicit definitions Γ and Δ are equivalent in a term-like structure, then they are equivalent in all term-like structures.

<u>1.24</u> The term s of 1.21 will be called a <u>general term</u> of Γ. It follows from 1.21 that if s_1 and s_2 are general terms of Γ_1, then there is a permutation $f: X \to X$ of the set of all variables such that s_2 may be obtained from s_1 by simultaneously replacing every occurence of a variable w in s_1 by f(w). The same relationship holds between any two general terms of any two implicit definitions equivalent in all term-

like algebras.

2. The main result

It follows from 1.19 and 1.23 that for every implicit definition Γ which is consistent in a term-like algebra there exists a finite subset $\Delta \subset \Gamma$ such that Δ and Γ are equivalent in all term-like algebras. The cardinality of Δ depends on Γ. We shall show that there is an upper bound on the cardinality of Δ which depends only on the width of Γ.

The following result is a special case of the main result. It will be used in the proof of 2.3.

2.1 Proposition

Let $k \in N$ and let Γ be an implicit definition having all its variables among $\{x_o,\ldots,x_k\}$. Assume that Γ is consistent in a term-like algebra. There exists $\Delta \subset \Gamma$ such that

(i) card $(\Delta) \leq (k+1)!$

(ii) Δ and Γ are $\vec{x}$-equivalent in all term-like algebras, where $\vec{x} = (x_o,\ldots,x_k)$.

Proof: is by induction on k. For $k = 0$, if Γ is consistent, then $\text{card}(\Gamma) \leq 1$ and we set $\Delta = \Gamma$.

Let $k > 0$ and assume that Γ contains equations of the form $x_o = t$, where t is a term with all its variables in $\{x_1,\ldots,x_k\}$. Let $\gamma_o \in \Gamma$ be any equation such that the right side of γ_o is not x_o. Let Σ be the set of all equations σ such that for some $\gamma \in \Gamma, \sigma$ is obtainable from $\{\gamma_o,\gamma\}$ by an application of R-rule.

For $1 \leq i \leq k$ let $\Sigma_i = \{\sigma \in \Sigma : \text{left side of } \sigma \text{ is } x_i\}$. By induction assumption for every $1 \leq i \leq k$ there exists $\Sigma_i^* \subset \Sigma_i$ such that $\text{card}(\Sigma_i^*) \leq k!$, and Σ_i^* and Σ_i are $(x_1,\ldots,x_k)$-equivalent in all term-like algebras.

For every $\sigma \in \{\Sigma_i^* : 1 \leq i \leq k\}$ let $\gamma_\sigma \in \Gamma$ be an equation such that R-rule applied to γ_o and γ_σ yields σ. It is easy to prove that $\Delta = \{\gamma_o\} \cup \{\gamma_\sigma : \sigma \in \{\Sigma_i^* : 1 \leq i \leq k\}\}$ satisfies (i) and (ii) of 2.1.

2.2 Corollary

Let $k \in N$ and let Γ be a set of equations whose variables are among $\{x_o,\ldots,x_k\}$. If Γ is consistent in a term-like algebra then there exists $\Delta \subset \Gamma$ such that

(i) $\text{card}(\Delta) \leq (k+2)!$

(ii) Δ and Γ are $\vec{x}$-equivalent over all term-like algebras.

Now we are ready to prove the main result of this paper. Let $Q: N \to N$ be a function defined as follows $Q(0) = 1, Q(m+1) = ((m+1)^2 Q(m)+2)!$

2.3 Theorem

Let $m \in N$ and let Γ be an implicit definition of width m. If Γ is consistent in a term-like algebra, then there exists $\Delta \subset \Gamma$ such that
(i) $\mathrm{card}(\Delta) \leq Q(m)$
(ii) Δ and Γ are equivalent in all term-like algebras.

Proof.: Let $\Gamma = \{x_o = t_i : i \in I\}$, where $I \subset N$, and let t_i have all its variables in $\{z_1^i, \ldots, z_m^i\}$, for all $i \in I$. Since the width of Γ is m we may assume that:

2.3.1 there exists $i \in I$ such that in t_i occur m different variables.

The proof of Theorem 2.3 is by induction on m. For $m = 0$, $\mathrm{card}(\Gamma) \leq 1$ and we set $\Delta = \Gamma$.

For $i \in I$, let γ_i denote the equation $x_o = t_i$. Assume now $m > 0$ and 2.3 holds for all $k < m$. In Γ we define a binary relation $[$ as follows, for $i,j \in I$, define $\gamma_i [\gamma_j$ iff there exist $1 \leq p \leq m$ and a path π which leads in t_i to an occurence of z_p^i such that π determines in t_j a subtree of depth ≥ 1 which contains all variables $z_1^j, \ldots, z_m^j$.

Observe that if $m \geq 2$, then any tree containing m different variables has depth ≥ 1. We claim that

2.3.2 $\exists i \in I \forall j \in I \quad \gamma_i [\gamma_j$

If 2.3.2 does not hold, then we obtain an infinite chain: $\gamma_{i1} [\gamma_{i2} [\cdots$ It follows from the definition of $[$ that in this case we would derive from Γ equations $x_o = s_n$ with unbounded depth of s_n. By 1.22 this contradicts consistency of Γ in $\mathcal{T}$. Changing enumeration of equations in Γ we may assume that γ_o is the equation whose existence asserts 2.3.2.

If t_o is a term without variables, then obviously $\Delta = \{\gamma_o\}$ is the sought subset of Γ. If t_o is a term being just a variable, then either $m = 1$ and by 2.3.2 Γ is the set of equations of the form $x_o = y$, or $m \geq 2$ and by 2.3.2 we get a contradiction with 2.3.1. In the former case we set $\Delta = \{\gamma_o\}$. Therefore we may assume that

2.3.3 depth of t_o is at least 1.

Let Σ be the set of all equations obtained from Γ by R-rule with one of the premises γ_o.

By 2.3.2, for every $1 \leq p \leq m$ if $z_p^o = r$ belongs to Σ, then r has at most m-1 different variables. For $1 \leq p \leq m$ let $\Sigma_p = \{\sigma \in \Sigma : \text{left side of } \sigma \text{ is } z_p^o\}$. By induction assumption for every $1 \leq p \leq m$ we can find $\Sigma_p^* \subset \Sigma_p$ such that

2.3.4 $\mathrm{card}(\Sigma_p^*) \leq Q(m-1)$, and

2.3.5 Σ_p^* and Σ_p are equivalent in all term-like algebras.

Let $i \in I$ and let z_p^i $(1 \leq p \leq m)$ be a variable which occurs in t_i. Let π be a path which leads in t_i to z_p^i. We have the following two possibilities:

(I). π is contained in t_o. In this case we associate with the pair (i, π) the equation $z_p^i = r(\vec{z}^o)$, where r is a subterm of t_o determined by π.

(II). π is not contained in t_o. Let $\pi_1 < \pi$ be the longest initial subpath of π which is contained in t_o. By 2.3.3 π_1 is nonempty. By consistency of Γ in $\mathcal{T}$, π_1 leads in t_o to a variable z_n^o for some $1 \leq n \leq m$. Let π_2 be such that $\pi = \pi_1 \pi_2$, and let $s(\vec{z}^i)$ be a subterm of t_i determined by π_2. It follows that π_2 is contained in s and the equation $z_n^o = s(\vec{z}^i)$ belongs to Σ_n. By 2.3.5 and 1.21 there exists an equation $z_n^o = r^*(\vec{v})$ which is derivable from Σ_n^* by R/S-rules such that π_2 is contained in $r^*(\vec{v})$.

Let $r(\vec{v})$ be a subterm of r^* determined by π_2. We associate with the pair (i, π) the equation $z_p^i = r(\vec{v})$.

Let Γ^* be the set of all equations associated with pairs (i, π) such that $i \in I$ and π leads in t_i to a variable. Let $\Sigma^* = \bigcup \{\Sigma_p^* : 1 \leq p \leq m\}$. We claim that

2.3.6 Γ and $\{\gamma_o\} \cup \Gamma^* \cup \Sigma^*$ are x_o-equivalent over $\mathcal{T}$.

Proof Since all equations in $\{\gamma_o\} \cup \Gamma^* \cup \Sigma^*$ are derivable from Γ by R/S-rules, for every $a \in T$, if $\Gamma(a/x_o)$ is consistent in $\mathcal{T}$, then $(\{\gamma_o\} \cup \Gamma^* \cup \Sigma^*)(a/x_o)$ is consistent in $\mathcal{T}$.

Assume that $(\{\gamma_o\} \cup \Gamma^* \cup \Sigma^*)(a/x_o)$ is consistent in $\mathcal{T}$ and $\Gamma(a/x_o)$ is inconsistent in $\mathcal{T}$. The latter part of the assumption implies the following two possibilities.

(A) There exists $i \in I$ and a path π contained in t_i which leads in t_i to a symbol which is not a variable, and π leads in a to a different symbol.

(B) There exist $i, j \in I$ and paths π_1, π_2 such that π_1 leads in t_i to a variable z, π_2 leads in t_j to the same variable z, and subtrees determined in a by π_1 and π_2 are different.

We will prove that neither (A) nor (B) is possible.

Suppose (A) holds. Let $\pi_1 < \pi$ be the longest initial subpath of π which is contained in t_o. Since Γ is consistent in $\mathcal{T}$, and $\gamma_o(a/x_o)$ is consistent in $\mathcal{T}$ it follows that π_1 leads in t_o to a variable z_p for some $1 \leq p \leq m$. Let r be a subterm of t_i determined by π_1. We have $z_p^o = r(\vec{z}^i)$

belongs to Σ_p^o. Let b be a subtree of a determined by π_1. It follows that $\Sigma_p(b/z_p^o)$ is inconsistent in $\mathcal{T}$.

On the other hand, since $(\{\gamma_o\} \cup \Sigma^*)(a/x_o)$ is consistent in $\mathcal{T}$, we have $\Sigma_p^*(b/z_p^o)$ is consistent in $\mathcal{T}$ as well. This contradicts 2.3.5.

Suppose (B) holds. Let $z = r_1(\vec{v})$ and $z = r_2(\vec{v})$ be equations associated with (i,π_1) and (j,π_2) respectively. It follows from the construction of these equations given in (I.) and (II.) and from (B) that $(\{\gamma_o\} \cup \Sigma^* \cup \{z = r_1(\vec{v}),\ z = r_2(\vec{v})\})(a/x_o)$ is inconsistent in $\mathcal{T}$. Since the above two equations belong to Γ^* we obtain inconsistency of $(\{\gamma_o\} \cup \Gamma^* \cup \Sigma^*)(a/x_o)$ in $\mathcal{T}$.

This completes the proof of 2.3.6.

Let $\Phi = \{r(\vec{v}) = s(\vec{v})$: for some variable $z, z = r(\vec{v})$ and $z = s(\vec{v})$ belong to $\Gamma^*\}$. Using the same argument as in the proof of 2.3.6 we obtain

2.3.7 Γ and $\{\gamma_o\} \cup \Phi \cup \Sigma^*$ are x_o-equivalent over $\mathcal{T}$.

Observe that by 2.3.4 there are at most $m\cdot(m-1)\cdot Q(m-1)$ different variables occuring in equations of Φ. We have $m(m-1)Q(m-1) \leq m^2Q(m-1)$. By corollary 2.2 there exists $\Phi^* \subset \Phi$ such that

2.3.8 $\operatorname{card}(\Phi^*) \leq (m^2\cdot Q(m-1)+1)!$, and

2.3.9 Φ^* and Φ are $\vec{v}$-equivalent in all term-like algebras, where $\vec{v}$ is a vector of all variables occuring in equations of Φ.

Observe that for any equation $\sigma \in \Sigma$ there exists $\gamma_\sigma \in \Gamma$ such that γ_o and γ_σ yield σ by R-rule. Let $\Delta_1 = \{\gamma_\sigma : \sigma \in \Sigma^*\} \cup \{\gamma_o\}$.

Notice also that for every equation $\sigma \in \Gamma^*$ there exists $\gamma_\sigma \in \Gamma$ such that $\Delta_1 \cup \{\gamma_\sigma\}$ yields σ by R/S-rules.

In order to get equations in Φ^* from equations in Γ^* we need a new rule.

T-rule

$$\frac{z = t(\vec{v}),\ z = r(\vec{x})}{t(\vec{v}) = r(\vec{x})}$$

It follows that for any equation $\varphi \in \Phi^*$ there exists a two element set $\Gamma_\varphi \subset \Gamma$ such that $\Delta_1 \cup \Gamma_\varphi$ yields φ by R/S/T-rules.

Let $\Delta_2 = \bigcup\{\Gamma_\varphi : \varphi \in \Phi^*\}$. We set $\Delta = \Delta_1 \cup \Delta_2$.

Since applying R/S/T-rules we can derive from Δ all equations of $\{\gamma_o\} \cup \Phi^* \cup \Sigma^*$, by 2.3.7 and 2.3.9 we obtain that Δ and Γ are equivalent over $\mathcal{T}$. By 1.23 we conclude that Δ and Γ are equivalent in all term-like algebras. Moreover we have $\operatorname{card}(\Delta) \leq 1+m\cdot Q(m-1)+2\cdot(m^2Q(m-1)+1)!$ $\leq (m^2\cdot Q(m-1)+2)!$ for all $m \geq 1$. This completes the proof of 2.3.

2.4 Corollary

Let Γ be an implicit definition of width m. If Γ is consistent in a term-like algebra, and s is a general term of Γ, then $r.c.(s) \leq r.s.(\Gamma)+m\cdot Q(m)$.

Proof follows from 2.3 and 1.21.

The upper bound given by Theorem 2.3 is a nonelementary recursive function. We leave this as an open problem: whether there is an elementary recursive upper bound in 2.3.

References

[1] Friedman, H. Algorithmic Procedures, Generalized Turing Algorithms, and Elementary Recursion Theory. In Gandy and Yates (eds.) Logic Colloquium '69, North-Holland Publ. Co.,Amsterdam 1977,pp.361-39o.

[2] Harel, D. First order Dynamic Logic. Lecture Notes in Computer Science 68, Springer Verlag, Berlin 1979.

[3] Paterson, M.S. and Hewitt, C.E., Comparative Schematology. Record of Project MAC Conference on: Concurrent Systems and Parallel Computation, ACM, New York 197o, pp. 119-128.

[4] Tiuryn,J. Unbounded ProgramMemory, Adds to the Expressive Power of First-Order Dynamic Logic (Extended Abstract). Proc. of the 22nd Annual Symposium on Foundations of Computer Sience, Nashville, Tenn. 1981, pp. 335-339.

[5] Urzyczyn, P. The Unwind Property in Certain Algebras. Information and Control, 5o. No. 2.,1981, pp.91-1o9.

Egon Börger

Spektralproblem and Completeness of Logical Decision Problems

Lehrstuhl Informatik II
Universität
D-4600 Dortmund

Abstract. We develop a simple general technique for **minimal logical implementation of Turing machine programs.** This implementation provides for completeness of logical decision problems which via the implementation correspond to limited or unlimited halting problems.

We exemplify our method by a new, simple proof for the automata-theoretical characterization of spectra of arbitrary order n as nondeterministically in n-fold exponential time Turing recognizable sets. We show that via our implementation technique the **underlying completeness phenomenon** is literally the same as in the Church/Turing theorem on the Σ_1-completeness of Hilbert's Entscheidungsproblem, as in Trachtenbrot's theorem on the Σ_1-completeness of the finite satisfiability problem for first order logic, but also as in Cook's theorem on the NP-completeness of the decision problem for propositional logic and in many well known lower complexity bound results for logical decision problems.

The Spektralproblem is in the title and in the center of our discussion because it has been officially formulated 33 years ago by the founder of the Institut für mathematische Logik und Grundlagenforschung of the University of Münster where our symposium takes place today. For this reason we include also a summary of the **history of the Spektralproblem** from 1951 to today, which at some points touches also the history of that Institut.

INTRODUCTION

In 1950 Boris TRACHTENBROT complemented CHURCH's (1936) and TURING's (1937) unsolvability result for Hilbert's Entscheidungsproblem by the theorem that **HILBERT's Entscheidungsproblem** is unsolvable also **for finite satisfiability,** i.e. that there is no algorithm which decides for any first order formula whether it is satisfiable by a model with a finite domain. Trachtenbrot's proof is based on a construction showing that precisely the graphs of partial recursive functions f have a spectral representation, i.e. admit a first order formula F with =, without function symbols and with monadic predicate symbols P_i such that for all $x_1,\dots,x_{n+1}$:

$f(x_1,\dots,x_n) = x_{n+1}$ iff

there is a finite model satisfying F where $P_1,\dots,P_{n+1}$ are interpreted as sets of cardinalities $x_1,\dots,x_{n+1}$ resp.

This last result depends on and yields **normal forms for recursively enumerable predicates** and has been strengthened later on in various ways (see Trachtenbrot 1953); Deutsch (1975) for example obtains from the Davis-Putnam-Robinson (1961) exponential diophantine normal form for r.e. predicates closed prenex formulae F with prefix $\wedge\vee \dots \vee\wedge$, only one occurrence of = and besides the P_i only one, a binary, predicate with a fixed interpretation; for stronger reduction results see Deutsch's contribution to these Proceedings.

At the Institut für mathematische Logik und Grundlagenforschung of the University of Münster where this symposium takes place today, 33 years ago Trachtenbrot's proof has led Gisbert HASENJAEGER - the senior organizer of our symposium here -, Werner MARKWALD and Heinrich SCHOLZ to the question what are the sets of cardinalities of all finite models - called spectrum (F) - of first order formulae F (read: finitely axiomatizable logical theories with = but without function symbols since functions can be represented by their graphs).

The official formulation of this problem to characterize the spectra for first order formulae (including = but without function symbols) has been given in Scholz (1952) from where the problem has taken the name "**Scholz's Spektralproblem**".

The same **problem applies to finitely axiomatizable classes of finite structures:** a finitely (first order) axiomatizable projective class of finite type in the sense of Tarski is a class of precisely those finite structures - i.e. with finite domain and finitely many finite relations over that domain - which are models of a formula

$$\bigvee_{P_1} \dots \bigvee_{P_r} F$$

with a first order formula F without free individual variables containing besides the P_i some free occurring predicate variables $R_1,\dots,R_s$. In analogy to spectra these classes have been called also $R_1,\dots,R_s$-spectrum of $\bigvee_{P_1} \dots \bigvee_{P_r} F$ or simply generalized (first order) spectra.

In the same way the **Spektralproblem** can and has been considered **for type theory formulae** of any positive order n asking for a characterization for every n of

$SPECTRA_n := \{spectrum(F) | F \text{ logical formula of order } n\}$

where only finite models over the naturals are considered to avoid set theoretical difficulties connected with the problem of having standard or nonstandard semantics.

The main answer to Scholz's Spektralproblem is the following theorem which has been discovered and rediscovered various times - in different forms - in the course of the last thirty years, as will be briefly sketched in the next section:

CHARACTERIZATION THEOREM FOR SPECTRA. **The spectra of order n are precisely the sets of positive natural numbers which are acceptable by nondeterministic Turing machines in n-fold exponential time** , symbolically:

$Spectra_n = NTIME(f_n)$-sets of positive integers

with $f_n(x) = 2^{2^{\cdot^{\cdot^{\cdot^{2^x}}}}}$ n-times and the class NTIME(f) of sets which are accepted by a nondeterministic Turing machine in time $f(c \cdot |x|)$ for a constant c and length $|x|$ of inputs x.

As immediate **corollary** one obtains from well-known properties of subelementary time complexity classes among others the following HIERARCHY THEOREM FOR SPECTRA. **The classes $Spectra_n$ of n-th order spectra form a strict subelementary hierarchy containing (strictly?) the class of all E_2-sets of positive integers and being contained in and exhausting the class of all E_3-sets of positive integers in the Grzegorczyk (1953) hierarchy.**

The special case n = 1 of the characterization theorem has become especially famous and has attracted attention of computer scientists since it expresses that the **first order spectra are precisely the NEXPTIME-sets of positive integers** (resp. modulo unary instead of binary encoding the NP-sets of positive integers) or by rephrasing the construction that the **generalized first order spectra coincide with the NP-sets of positive integers.**

The AIM OF THIS PAPER is to give a new proof for the nontrivial direction of the characterization theorem for spectra which shows that essentially **this characterization is only a particular case of a general completeness phenomenon** (in the sense of recursion and complexity theory) other instances of which are the Church-Turing theorem on the unsolvability of Hilbert's Entscheidungsproblem, Trachtenbrot's version of the latter for finite satisfiability, Cook's (1970) theorem on NP-completeness of the decision problem of propositional logic and many other well known lower complexity bound results for logical or mathematical theories (see for ex. Lewis 1980, Fürer 1981, Denenberg & Lewis 1984, Young 1983). Our proof succeeds in extracting as core of those unsolvability or lower complexity bound theorems a strikingly simple, natural and

minimal logical implementation of (Turing-)machine programs. This construction **reveals an intimate structural relation between the (computational) complexity of programs and the logical complexity of their implementing formulae:** unbounded computations correspond to infinite satisfiability problems, bounded computations to finite satisfiability problems (of propositional logic or higher order logic over finite domains or structures), deterministic programs to Horn formulae etc.

We hope that our construction besides from exhibiting explicitly and in a technically very simple way the fundamental and uniform reason which underlies many famous undecidability and lower complexity bound results, will help to obtain finer complexity theoretical characterizations of such decision problems w.r.t. well established criteria of logical complexity like quantifier structure, propositional structure, structure of atomic formulae. For spectra in particular, interesting contributions to such a refined analysis can be found in the papers by E. Dahlhaus, E. Grandjean and B. Scarpellini in the Proceedings of this symposium.

HISTORY OF METHODS

In 20 years, from 1951 to 1971, the essence of the above stated solution of the Spektralproblem has been (almost completely) discovered three times, based on different approaches. The troubled succession of these discoveries is a particularly well documented facette of the history of the **slow growth of the machine oriented computational approach to complexity** of functions out of the classical number theoretical approach of the 50'-ies and illuminates at the same time the **difficulty which the concept of nondeterministic computation encountered to be understood** in its importance for theoretical investigations into complexity phenomena for algorithms.

This symposium takes place at the Institut für mathematische Logik und Grundlagenforschung of the University of Münster where its founder, Heinrich Scholz, now 33 years ago wrote the official formulation of the Spektralproblem. The senior organizer of our symposium today at that time was also member of this institute and as I mentioned above was involved in the elaboration of the problem; one of the solutions of the problem has been found by Dieter Rödding, the second organizer of this symposium and today one of the directors of the Logic Institute in Münster, in collaboration with Helmut Schwichtenberg who is present at this symposium and who at that time was assistant in this institute; through their "solution" I learnt the problem in 1972, after my doctoral examination in Münster, and here I want to explain my recent discovery that and how by pushing Turing's (1937) approach of logical description of machine computations to its extreme the Spektralproblem turns out to be - together with many other fundamental and famous decision problems - a particular and conceptually simple subcase of Hilbert's Entscheidungsproblem.

Therefore the story of the Spektralproblem tells also some episodes of the life of three academic generations in the institute where 33 years after the formulation of the problem we are together today. I take this as an occasion to tell - and indeed in some more detail as would have been done under other circumstances - the history of methods of attack to the problem.

Already the first investigations into the Spektralproblem showed that the latter has to do with complexity theoretical problems at lower levels of the Grzegorczyk hierarchy $(E_n)_{0\underline{<}n}$ of primitive recursive functions: ASSER (1955) proved that **the class of first-order spectra is properly contained in the class E_3 of Kalmar--Csillag elementary sets** of positive integers by constructing an elementary enumeration function for the characteristic functions of first-order spectra. This construction is based on a direct number theoretical description, for arbitrarily given formulae F in Skolem normal form, of models for F over n = {0,1,...,n-1} which uses essentially finite sums and products (corresponding to logical (finite!) quantifications) and an elementary encoding of k-ary relations over n.

ANDRZEJ MOSTOWSKI independently obtained similar results: moreover in Mostowski (1956) we find essentially a first order description of the graphs of certain bounded number theoretical functions which can be obtained from constant functions, projections, successor (bounded by the second argument) and predecessor by substitution and bounded primitive recursion; since this axiomatization works also for simultaneous bounded primitive recursion it shows that **every Grzegorczyk E_2-set is a first-order spectrum.** Mostowski's construction consists in a straightforward logical description (read: over n = {0,...,n-1}) of how function values $\leq$ n for arguments $\leq$ n are computed using the allowed defining schemata; this formalization contains already all ingredients which by our implementation technique to be explained in the next section turn out to be necessary and sufficient for a machine oriented proof of the full characterization theorem for spectra: encode natural numbers up to n-read: the data of the computation process to be encoded into logical formulae - by **describing with respect to a linear ordering** (read: over n = {0,1,...,n}) the first element (**"zero"**) **and** the next element (if there is any) to a given one (**"successor"**); based on such an encoding it is obvious how to paraphrase logically the computation of function values through the given defining schemata.

In his doctoral dissertation at Princeton, **Bennett** (1962) on a suggestion by Abraham Robinson considered the Spektralproblem for the natural extension to predicate logic of arbitrary order n (many sorted theory of types) and studied especially the connection between higher order spectra and the constructive arithmetic resp. rudimentary relations defined at that time by Smullyan (1961). Investigating further and extending the sharp analysis of complexity phenomena at low levels of Grzegorczyk's hierarchy which had been obtained by Smullyan, Bennett was able to sharpen Asser's and Mostowski's earlier methods and results: from a characterization of the class of n-th order spectra in terms of Smullyan's constructive arithmetic resp. rudimentary relations Bennett obtains above all that these **classes** $\mathbf{Spectra}_n$ **form a strict hierarchy** exhausting all E_3-sets of positive integers; other interesting complexity theoretical features come into light like the **reductive characterization of spectra of arbitrarily high order** n+1 as inverse images under $\lambda x.f_n(x^c)$ of first order spectra for some constant c, or the corresponding characterization of spectra of weak order n+1 - i.e. spectra of formulae of order n+1 without predicate symbols of order n+1 - as inverse images under $\lambda x.f_n(x^c)$ of the constructive arithmetic sets for some constant c.

Interestingly enough Bennett mentions Ritchie's 1960 Princeton Doctoral Dissertation, but he does not make use of the **computational approach to complexity** appearing in Ritchie's dissertation. Bennett obtained and formulated his results - as has been generally the case for all investigations into the complexity of primitive recursive functions up to the beginning of the 60-ies - in classical terms of (complexity of defining schemes for) number theoretical functions or relations, not of (resource complexity of) machine programs computing functions or relations. This may be the reason why Bennett's Ph.D.Thesis - apart from not having been published - has been forgotten for ten years and why his results have become known to the scientifique community only after an independent rediscovery which was based on a computational approach to complexity of primitive recursive functions.

The computational approach to complexity has been developed starting in the 60-ies, strongly motivated by new experiences and interests coming from computer science. At the Institut für mathematische Logik und Grundlagenforschung here in Münster, G. HASENJAEGER had worked already in the 50-ies at physical realizations of small universal Turing machine models. As we have seen from the demonstration and explanations of various such models "in actu" at one of the evenings of this symposium, Hasenjaeger's machine building activities in the 50-ies have created a strong tradition in the Logic Institute in Münster taken care of later by Dieter Rödding: since the 50-ies up to today, using the rapidly changing technological means at disposition, the conceptual **logical experiments with automata** are **accompanied** in this institute **by hardware realizations** of and tests on machine models implementing the concepts under study. From Hasenjaeger's and Rödding's comments in occasion of the above mentioned demonstration of some of their machine models we have learnt several examples of fruitful reciprocal feedbacks between theoretical and practical problems and results especially in connection with the invention of small universal machine models (see also the contributions by Hasenjaeger and Christodoulakis to this symposium.) I want to add one more example which is of historical interest for the computational solution to the Spektralproblem.

Influenced by the idea of nonerasing machines from Wang's (1957) variant to Turing's theory of computing machines - to which by the way also Walter Oberschelp, at that time working at the Logic Institute in Münster and today participating in our symposium, contributed with an interesting investigation into the forms of number representations on Turing tapes which allow the elimination of erasing operations, see Oberschelp 1958 - Hasenjaeger and Rödding between 1958 and 1960 built a small 3-tape programmable (universal) Turing machine model - known under the name "WANG" - where the following theoretical question grew out in connection with hardware problems for the implementation of a "simple" memory device for numbers: is it possible to encode arbitrary (finite sequences of) numbers on a 0-1-tape of a universal Turing machine using only a prefixed finite number of occurrences of the unique nonblank symbol 1? **Rödding** in 1959/60 found a positive solution of this problem which went into the construction of the new **"register memory"** of the universal machine WANG; conceptually the essential idea was to represent numbers n as distance of a unique occurrence of the nonblank symbol to a distinguished field ("zero position") on the otherwise empty (one way infinite) tape. This idea to **use the Turing tape as "number register"** and to conceive a multi-tape Turing machine as "register machine" capable of direct number operations (addition, subtraction and zero test) led to positive solutions of many other problems about the computational power of various combinatorial systems (like the universality of Post's TAG systems or of 2-register machines etc.) which Rödding has reported in 1960/61 in colloquium talks at the Logic Institute in Münster and which nowadays are used in many introductory courses and books due to the independent discovery (and publication) of the register machine model and the related results by **Minsky** (1961) and **Shepherdson & Sturgis** (1963).

Subsequently Rödding pursued this machine oriented approach to recursive function theory by an **investigation of structural and time complexities of register machine programs** computing functions or relations, which found its way into the 1967 Leeds lectures, see Rödding (1968). In particular one finds there the strict Rödding hierarchy (rephrased) DSPACE(f_n) of functions starting from E_2 and contained in and exhausting E_3; this hierarchy has been formulated and proved there in terms of n-fold exponential time bounded register machines, but by Christen's (1974) doctoral dissertation we can rephrase it as done above in terms of the more widely used Turing machine concept.

During the summer term 1968 Rödding gave a special course in logic where among others after the presentation of some classical Surany (1959)-like methods for obtaining reduction classes - i.e. r.e. complete decision problems - of first order logic, **Rödding showed how to adapt to register machines Büchi's** (1962) **important improvement of Turing's** (1937) **logical description technique of Turing machine halting problems**, simplifying thereby Büchi's proof for the reduction class of all prenex formulae with prefix EAEA (see the mimeographed lecture notes Rödding 1968a written up by Helmut Schwichtenberg, at that time assistant at the Institut für mathematische Logik und Grundlagenforschung).

In the course of the following winter term 1968/69, these two lines of research, constituted by a) the discovery of a hierarchy of complexity classes characterized by bounds on the computation time of register machine programs and b) the discovery of a technically simple and direct way to describe register machine computations (halting problems) by logical formulae, came together and led RÖDDING and SCHWICHTENBERG in a natural way to a **logical description of f_n(p)-limited computations of register machine programs** for polynomials p **by finite satisfiability problems of type formulae of order n+1.** Rödding and Schwichtenberg draw only the conclusion that all DSPACE(f_n)-sets of positive integers are spectra of order n+1; comparison with our construction in the next section will show that an appropriate modification of this Rödding/Schwichtenberg technique applied to nondeterministic computations is already sufficient for a proof of the whole characterization theorem for spectra.

Another interesting way of lifting the Rödding/Schwichtenberg result from the inclusion of the class of DSPACE(f_n)-sets of positive integers in Spectra_{n+1} to the corresponding inclusion for NTIME(f_{n+1}) has been found by **Christen** (1974) in his doctoral dissertation written under his supervisor E. Specker who has opened our symposium with his lecture: Christen develops very carefully the analogue of **Rödding's (1968) computation time classification** of subelementary functions **for wordfunctions** using a register machine model from Shepherdson & Sturgis (1963) which works directly on words over a finite alphabet (instead of numbers as in Rödding 1968) and in a normalized but nondeterministic way resembling Post's (1943) normal canonical systems (and by the way also a simple (nondeterministic) form of Wang's non-erasing Turing machines as we have been able to show in Börger 1982). Using also ideas from Cleave (1963) and Ritchie (1965) Christen adapts to wordfunctions Rödding's inductive characterizations of these time complexity classes, essentially by a **refined Kleene normal form theorem defining a T-predicate of low complexity.** Natural Gödelization then permits the above mentioned characterization of Rödding's classes as classes DSPACE(f_n). The improved T-predicate (read: E_2-description of register machine computations) allows an improved existential representation of NTIME(f_n)-acceptable sets with linear time computable matrix and f_n-bounded quantifiers, and this representation is then reduced to the finite satisfiability problem of appropriate type formulae of order n. Together with the analogue for word register machines of Cook's (1971) theorem for Turing machines that NSPACE(f_n) $\subseteq$ DTIME(f_{n+1}), these constructions yield the extension of the Rödding/Schwichtenberg result to the full characterization theorem for spectra.

Rödding and Schwichtenberg applied their axiomatization technique also to what they called bounded exponential arithmetical predicates (read: Smullyan's (1961) rudimentary predicates) and rediscovered thereby and by an appropriate bounded exponential arithmetical Gödelization of type formulae of weak order n Bennett's reductive characterization of spectra of weak higher order and the inclusion of Spectra$_{n+1}$ in the class of all DSPACE(f_{n+1})-sets of positive integers. Schwichtenberg reported these results - obtained without knowledge about Bennett's work and by a completely different, namely a machine oriented method - in 1969 at the Logic Spring Meeting in Oberwolfach (see Schwichtenberg 1969) and again in two colloquium talks at the Institut für mathematische Logik und Grundlagenforschung of the University of Münster on may 9 and 16, 1969; the full version has been received for publication on february 26, 1971 and has appeared in Rödding & Schwichtenberg (1972).

In february 1972, N.D. JONES and A. SELMAN announced in the AMS Notices 19,2 under no. * 72T-E28 the case n = 1 of the characterization theorem for spectra; they presented it to the spring's STOC conference in may 1972 and published it again in Jones & Selman (1974). As Alan Selman who has been asked to deliver to this symposium lectures on the Spektralproblem and its connections to complexity and finite model theory has written to me on april 5, 1983, he obtained the result with Jones between spring and fall 1971 and spoke about it privately to participants of a Computational Complexity Conference at New York University in October 1971 when also the extended abstract for the spring's 1972 STOC had been prepared, he "became aware of rumours that others had obtained our results independently and/or earlier than us" but (also up to the time of this symposium) has not seen the Rödding & Schwichtenberg 1972-paper which as a matter of fact is not cited in the 1974-publication. March 2, 1972, is the publication date of the preprint version "Turing machines and the spectra of first-order formulas with equality" written at the Computer Science Department of Carnegie-Mellon University, no. CMU-CS-72-113, pp. 11, which appeared with minor changes in the Proc. Fourth ACM Symposium on Theory of Computing, 1972, pp. 157-167, and again in JSL 1974.

Jones & Selman (1974) give a first order description of halting problems of what they call spectrum automata by which they simulate arbitrary nondeterministic $f_1 \cdot c$-bounded Turing machine computations for any constant c. Due to the concept of spectrum automaton their construction is technically much more involved, less natural and more restricted than the simple and general logical register machine description in Rödding & Schwichtenberg (1972), but the Jones/Selman paper has the merit of having **connected** such **logical formalizations of (bounded) computations with** the already more than 10 years old **notion of nondeterministic (Turing) machines** the importance of which became more and more clear after Cook's discovery in 1971 of the class NP, the fundamental theorem of the NP-completeness of the decision problem for propositional logic and the problem whether P = NP.

In particular Jones and Selman noticed that **Asser's complement problem is connected to the P = NP-problem.** Asser's (1955) complement problem asks whether the complement of every first order spectrum is again a first order problem; the characterization of spectra as NEXPTIME-sets of positive integers implies that a negative answer to Asser's complement problem yields $P \neq NP$. Remember that a function is f_1-bounded in the length of the binary representation of a natural number n iff it is bounded by a polynomial in the length n of the unary representation of n, and that the encoding from unary to binary can be done in polynomial time. Therefore the characterization of Spectra_1 can be read also as: Spectra_1 = class of all NP-sets of positive integers modulo unary input encoding.

There is no consideration of nondeterministic computations in Rödding & Schwichtenberg (1972) and this prevented them from applying their simple but general logical machine description to nondeterministic Turing instead of deterministic register machines what would have yielded the full (automata theoretical!) characterization theorem for spectra instead of their formulation of lower and upper bounds $\text{DSPACE}(f_n)$-resp. $\text{DSPACE}(f_{n+1})$-sets of positive integers for Spectra_{n+1}.

As Alan Selman has told us during his lectures at this symposium, he learned the Spektralproblem in a logic course given by William Craig at Berkeley. FAGIN's Doctoral Dissertation written in 1973 in the Department of Mathematics at the University of California in Berkeley took up again the Spektralproblem and by considering its **generalization to finitely first order axiomatizable projective classes of finite type** in the sense of Tarski opened the way to many very illuminating connections of the (generalized) Spektralproblem to complexity theory (see Fagin 1974) and to interesting model theoretical techniques and results for finite structures (see Lynch 1982, also for further references of this fruitful line of research.) Fagin's characterization of generalized first order spectra as class of all NP-sets of positive integers is again a particular subcase of our translationschema of machines into logical formulae to be explained in the next section.

Why has it taken almost half a century since Turing's (or less directly from Church's) unsolvability proof for first order logic to understand that Hilbert's Entscheidungsproblem, Trachtenbrot's problem, Scholz's Spektralproblem, but also Cook's more recent complexity result for the decision problem of propositional logic and many other now already classical lower complexity bound results for logical theories are nothing else than subcases of just one completeness phenomenon? And - what is still more surprising - of a completeness phenomenon which underlies all those many axiomatizations in the literature not only by vaguely felt and sometimes expressed "analogy" but which can be stated explicitely and be derived very easily in terms of a natural logical definition of the meaning of programs to be explained in the next section where the main concern is parsimony, that means to impose on the models only those conditions which are directly involved and unavoidable in **any** description of the effect of single program instructions on the given data. Comparing in the light of my construction - and that means post festam - the myriad of preceding individual constructions in the relevant literature one can observe that axiomatizations of computations were greatly influenced by a kind of **fear that the simulating model could be richer than the encoded computation** and therefore possibly allow some unexpected and uncontrollable detour which would then destroy the correctness of the simulation: in fact many descriptions are even categorical (i.e. describe the simulating model uniquely up to isomorphism), and always one finds formulations of a lot of side conditions which are not really necessary for a correct implementation of the given computation. It was the psychologically motivated statement of such in re unnecessary (mostly global frame) conditions which hindered the discovery of the simple and uniform structure of all those reductions.

This observation is confirmed by my **own personal experience** with the invention of new encoding techniques in connection with small logical reduction classes or with degree preserving simulations between various models of computations; for the latter see the unifying general approach (and its history) in Börger 1979, for the former see the well documented stepwise and slow progress made in understanding the essence of Turing's (1936) first order description of machine problems in Suranyi 1959, Büchi 1962, Kostyrko 1964, 1966, Gurevich 1966, Aanderaa 1971 and Börger 1971, Lewis 1979, Börger 1983.

To cite some other clear **witnesses in the literature** for my above stated observation: look at the history of Hilbert's and Bernays' (1939: pg. 191) conjecture that not every non contradictory formula of the restricted lower predicate calculus (i.e. without = or functions) admits also a recursive model (see Kreisel 1953, Mostowski 1953, 1955, Rabin 1958, Aanderaa 1971, Börger 1975); the development of the problem whether every reduction class w.r.t. Hilbert's Entscheidungsproblem can be strengthened also w.r.t. Trachtenbrot's concern about finite models (see Gurevich 1976, Aanderaa & Börger & Lewis 1982); the slowly growing consciousness about the intimate connections between Gödel's incompleteness theorem technique and techniques for establishing lower complexity bounds for logical decision problems (see Machtey & Young 1978, 1981, Young 1983. cf. also Germano's (1976) very weak sufficient condition for an r.e. consistent theory to be incomplete - namely to admit term representations for addition and multiplication using =, 0 and successor! - and the new machine oriented proof discovered by Jones & Matijasevich (1984) for the Davis-Putman-Robinson 1961 exponential diophantine representation of r.e. sets.)

IMPLEMENTATION METHOD AND REDUCTION PROOFS

The nontrivial part of the characterization theorem for spectra is the claim that every set of positive integers which is accepted by a nondeterministic Turing machine in time $f_n(c \cdot |x|)$ for a constant c is spectrum of a type formula of order n. We view this claim as a **completeness statement**: every f_n-time-bounded halting problem of a nondeterministic Turing machine can be reduced to a corresponding logical decision problem, here the satisfiability problem for type formulae of order n over finite domains with given cardinality. In this way the (proof for the) characterization theorem for spectra becomes a special case of (the construction for) the following general theorem which applies mutatis mutandis to different cases of machines and limited or unlimited halting problems and corresponding notions of logical formulae and decision problems:

THEOREM ON MINIMAL LOGICAL IMPLEMENTATION OF TURING MACHINES:

(i) To every deterministic or nondeterministic Turing machine program M and every M-configuration C we associate a **configuration formula** $\bar{C}$ and a **program formula** Π_M which describes the meaning of M on C in the sense of the following **Simulation Property**: Let A be any model for Π_M. Let C be any M-configuration and t an arbitrary natural number. If $\bar{C}$ is true in A, then also $\bar{D}$ is true in A for at least one of the configurations D which M may reach after precisely t computation steps started in C.

(ii) To every instance of a given M-halting problem H we associate an **input formula** α and a **stop formula** ω with a corresponding logical decision problem D which describes H in the sense of the following **Reduction Property:** Any instance of the M-halting problem H has a positive answer iff the decision problem D for the corresponding formula $\Pi_M \wedge \alpha \wedge \omega$ has a positive answer.

PROOF, part (i) (IMPLEMENTATIONSCHEME). It is useful for an intuitive understanding of the minimality of the following logical implementation Π_M of M to **view the Turing machine as a Central Processor Unit with a Memory:** the memory is the (finite but potentially infinite) Turing tape of numbered tape cells, the Central Processor Unit consists of the finite automaton which is defined by the program and an accumulator for the content of precisely one memory cell per time interval, which we identify with the tape cell where the reading head is positioned. A (logical) description of M-computations needs therefore a description of the state of CPU and Memory at time t through statements (formulae) with the following (intended) meaning:

$CPU_t\ (i,u)$	iff the CPU at time t is in state i and is processing the u-th tape cell
$Tape_t\ (u, j)$	iff the tape at time t contains in the u-th cell the symbol j
$Memory_t\ (v, j)$	iff at time t the v-th tape cell contains the symbol j and the CPU is processing another tape cell

Let M be given by a program of instructions (i, j, k, $\pm 1, \ell$) for states i,ℓ and tape symbols j, k of M with the meaning: if in state i symbol j is scanned under the reading head, then M may print symbol k, move the reading head one position to the right (+1) resp. to the left (-1) and go to the successive state ℓ.

A (logical form of a) definition of the meaning of these M-instructions for transforming given configurations at time t into successive configurations at time t' is expressed by Π_M which we determine as (finite or infinite) conjunction over all parameters of the following implications (Stt' denotes: t' is successor of t):

$$CPU_t(i,\ u) \wedge Tape_t(u,\ j) \wedge Stt' \wedge Suu' \to \underset{(i,j,k,+1,\ell)\varepsilon M}{\overset{\text{disjunction}}{\bigvee}} CPU_{t'}(\ell,\ u') \wedge Tape_{t'}(u,\ k)$$

$$\ldots\ldots\ldots u' \ldots\ldots\ldots u' \ldots\ldots\ldots\ldots\ldots \to \underset{(i,j,k,-1,\ell)\varepsilon M}{\overset{\text{disjunction}}{\bigvee}} \ldots\ldots\ldots u \ldots\ldots\ldots\ldots u' \ldots$$

$$Memory_t(v,j) \wedge Stt' \quad \to \quad Tape_{t'}(v,j)$$

NB. For easier formulation but without loss of generality we have assumed that M contains for any instruction entry (i,j) either only right-movement or only left-movement instructions. NB. Logically we may formulate infinite conjunctions over the time or memory position parameters t,u,v by universal quautifications. Define $\bar{C}$ in the obvious way as conjunction of the one positive CPU_o- and the (possibly infinitely many) positive $Tape_o$-statements which describe C. From the definition it is obvious that Π_M and $\bar{C}$ satisfy the simulation property. (If your intuition needs help from a formal proof, try an induction on the computation length t: for t=0 take D=C. For t+1 one knows by induction hypothesis that $\bar{E}$ is true in the given model for at least one M-configuration E reached at time t from C; truth of $\bar{E}$ assures truth of a CPU_t- and the corresponding $Tape_t$- and $Memory_t$-premisses in Π_M so that from the truth of Π_M one obtains the truth of one possible successive configuration D of E defined by the $CPU_{t'}$- and $Tape_{t'}$ -statements in the conclusion.)

<u>Remark</u>. Note that in Π_M and in $\bar{C}$ there is no formulation of assumptions about unique state, unique reading head positon or uniqueness of cell content. Π_M expresses only the local changements of data effected by a CPU-operation and that all other memory positions which are not under actual control of the CPU remain unchanged. Accordingly $\bar{C}$ consists only of positive statements expressing a) through CPU_o (i_o, u_o) the initial state i_o and the initial reading head position u_o and b) through statements $Tape_o$ (v,j) that initially the content of the v-th cell is j.

For this reason Π_M is a Horn formula if M is deterministic (and if CPU_t, $Tape_t$ and $Memory_t$ are defined without use of negation).

For the derivation of the simulation property there is no need for further specification of CPU_t-, $Tape_t$- and $Memory_t$-statements in more explicit "logical" terms. This is needed however when we want to reduce particular (bounded or unbounded) halting problems to particular logical decision problems.

PROOF, part (ii) (SPECIFICATION FOR PARTICULAR HALTING PROBLEMS). By part (i) we need only to specify in more detail the statements $CPU_t(i,u)$, $Tape_t(u,j)$, $Memory_t(v,j)$ and to add accordingly to Π_M the start condition α (time parameter t=0) and the intended stop condition ω (time parameter $t=\ell$ (last step if any)). This specification amounts in our case to a representation of data by logical terms - here: natural numbers 0,t' for time moments and positions of tape cells - and to a definition of the CPU_t-, $Tape_t$- and $Memory_t$-statements and consequently of α and ω by logical formulae $I_i(t)$, $P(t,u)$ and $T_j(t,u)$ with the following intended meaning:

$I_i(t)$ iff at time t the CPU is in state i (instruction label)
$P(t,u)$ iff .... is processing the u-th memory cell
$T_j(t,v)$ iff ...... the content of the v-th tape cell is j

For the sake of illustration we fill in now the last details for four characteristic cases of machine halting problems H and corresponding logical decision problems D, ending up with the characterization theorem for spectra. The reader who has got already the idea may skip the rest of this paper; the reader who is still sceptical is encouraged to try to write down the rest of this paper by himself.

In order to simplify our notation but without loss of generality we assume for the following examples that the machines M are equipped with a tape which is infinite in one direction, say to the right, with cells numbered 0,1,2,.... Let us agree on 0 as initial state, cell 0 as initial position of the reading head and 1 as (accepting) halting state; we assume that in a halting state the machine computation becomes constant, say because no program contains an instruction with a label which is a halting state.

EXAMPLE 1. **Unlimited TURING machine halting problem and Hilbert's Entscheidungsproblem for first order predicate logic: Σ_1-completeness.**

In the case of (even restricted) first order logic, by Büchi's (1962) important observation natural numbers can be explicitely expressed in Π_M as terms 0,t' built up from an individual constant 0 by a monadic function symbol '. For reasons of uniformity we leave nevertheless in Π_M the explicit formulation

$\bigwedge_{tt'}\bigvee$Stt' of the successor relation; the program part of Π_M then has universally quantified variables t,t',u,u',v where by Skolem's normal form theorem we can restrict our attention to models of Π_M over the naturals $\{0,0',0'',\dots\}$ where Stu means u=t'.

We specify therefore with predicate symbols I_i,P,T :

$$CPU_t(i,u) :\equiv I_i(t) \wedge P(t,u)$$
$$Tape_t(u,j) :\equiv T_j(t,u)$$
$$Memory_t(v,j) :\equiv Tape_t(v,j) \wedge P(t,u) \wedge u \neq v$$

Let us consider the Σ_1-complete **halting problem** if M eventually stops in state 1 when started in state 0 at the left end of the empty tape. Then we can define (take 0 as blank symbol):

$$\alpha :\equiv I_0(0) \wedge P(0,0) \wedge \bigwedge_u T_0(0,u) \qquad \omega :\equiv \bigwedge_t \text{not } I_1(t)$$

The **reduction property** in this case reads:
M does not reach state 1 when started in state 0 on the empty tape

iff $\Pi_M \wedge \alpha \wedge \omega$ is satisfiable

The only-if-direction is assured by the above indicated intended meaning of I_t,P,T_j,S which yields a model over the naturals for $\Pi_M \wedge \alpha$, and also for ω if M does not reach state 1.

The if-direction is a direct consequence of the simulation property of part (i): if M - which in this case is deterministic - would reach state 1 at some time moment t, then in the given model in particular $I_1(t)$ would be true which contradicts ω.

Corollary. Hilbert's Entscheidungsproblem is Σ_1-complete

Remark. The above proof yields by easy modifications as a by-product for machines over a binary alphabet already Büchi's (1962) reduction class of closed prenex formulae with prefix EAEA and besides monadic only 3 -binary-predicate symbols. Sharper results have been obtained by more elaborate machine encodings into logical formulae but follow the same proof pattern as the one above (see Börger 1984 for a survey).

EXAMPLE 2. **Polynomical time bounded TURING machine halting problem and Entscheidungsproblem for propositional logic: NP-completeness.**

We specify our construction for the reduction property:

M accepts $q\epsilon\{0,1\}^n$ in ℓ steps iff $\Pi_{M,\ell}\wedge\alpha_{M,n}(q)\wedge\omega_{M,\ell}$ is satisfiable where $\Pi_M\wedge\alpha\wedge\omega$ is polynomial time bound computable from n, and ℓ is a polynomial in the length n of the input q.

Π_M is as in example 1 with $I_i(t)$, $P(t,u)$, $T_j(t,v)$ to be read as propositional variables for $t,u,v\leq\ell$ and with $u\neq v$, Stt', Suu' deleted from the formula but added as external conditons $u\neq v, t'=t+1, u'=u+1$ on the parameters. Since we are considering here a possibly nonempty input we have to replace the input conjunct in α of example 1 by the conjunction of the following formulae for $0\leq j<n, n\leq k<\ell$ (with blank symbol 2 and new propositional variables x_j which are substituted by $q_j\epsilon\{0,1\}$ to obtain $\alpha(q)$):

$$T_0(0,j)\leftrightarrow\neg x_j \quad T_1(0,j)\leftrightarrow x_j \quad T_2(0,k)$$

Since we are considering acceptance in state 1 instead of nonhalting in state 1 we replace the "no halt in state 1"-statement ω of example 1 by the "no non-accepting state in the last step"-formula:

$$\omega :\equiv \bigwedge_{i\neq 1} \text{not } I_i(\ell)$$

Corollary. The Satisfiability problem for propositional logic is NP-complete.
Remark. Since the decision problem for Horn formulae is in P and every $\Pi_M\wedge\alpha(q)\wedge\omega$ is (trivially equivalent to) a Horn formula for deterministic M, it is natural to consider for any Boolean function $f^{(n)}$ such defining pseudo-Horn formulae of minimal length: by the above given proof of Cook's theorem this complexity measure has been intimately connected to the P=NP-problem and later on been shown to be polynomially equivalent to network and Turing machine complexity, see Aanderaa & Börger 1979, 1981.

EXAMPLE 3. **Double unlimited TURING machine halting problem and Hilbert/Trachtenbrot's Entscheidungs- and finite satisfiability problem: Σ_1-completeness.**

Consider the **extension of the** halting **problem** of example 1 to machines M which started in state 0 at the left end of the empty tape either reach the halting state 1 or the halting state 2 or go into an infinite computation which does not become periodic. The corresponding so called **conservative reduction property** is the conjunction of a) and b):

a) the if-direction of the reduction property of example 1 for the halting state 1
b) if M started in state 0 on the empty tape eventually halts in state 2 then $\Pi_M\wedge\alpha\wedge\omega$ has a finite model

To prove b) one only has to cut off at ℓ the model constructed for a) over the naturals, where ℓ-1 is the length of the computation terminating in state 1 (set $\ell':=\ell$ in the model).
Since the sets of machines M which started in state 0 on the empty tape eventually halt in state 1 resp. 2 are recursively unseparable, by a) and b) the sets of contradictory resp. finitely satisfiable formulae are recursively unseparable too and in particular both non recursive.
By Gurevich's (1976) general recursion theoretical argument these two reductions a) and b) already suffice for assuring that all $\Pi_M \wedge \alpha \wedge \omega$ form a conservative **reduction class,** i.e. such that one can associate effectively to any first order formula F a formula F which is satifiable resp. finitely satisfiable iff F is satisfiable resp. finitely satisfiable.
One can show this also directly by a slight but natural modification of our formulae Π_M: one obtains also the only-if-direction in the reduction properties a) and b) if one adds to Π_M some conjuncts which **express the special feature of M's double halting problem,** namely that if M does not reach state 2 but also not state 1 then its computation will become infinite and not periodic.
Since a nonperiodic computation will encounter arbitrarily big reading head position numbers, we can assure the existence of such big elements in the model by adding to (the successoraxiomatization of) Π_M the axiomatization of a relativized "greater-than-relation" G over naturals:

$P(u,t) \to G(t',t)$	"working positions have a greater successor"
$G(t,u) \to G(t',u)$	"transitivity"
not $G(t,t)$	irreflexivity

Obviously the reduction property a) and its inverse implication are both true as before with G to be interpreted as > over the naturals; the same works in b) because any $t \leq \ell-1$ has a "greater" successor t' in the model, but since by definition ℓ is never reading head position it does not need any more a greater successor and therefore allows $\ell':=\ell$. The crucial inverse implication of b) is now also correct: if M does not reach state 2, it may reach state 1 or not. In the first case by a) our formula has no model at all; in the second case it may have - and in fact by a) does have - models but no finite one because in any model the arbitrarily big (infinitely many) reading head position numbers occuring in the computation impose the existence of new "greater" elements on the model.

Corollary. Contradictory, finitely satisfiable and only infinitely satisfiable first order formulae are recursively inseparable.
Remark. For a more sophisticated conservative reduction technique see Aanderaa & Börger & Lewis 1982. By the above technique for describing the double halting problem of machines M which enumerate recursively unseparable sets one gets (with minor rewritings) Π_M as simple example for an axiom of an essentially

undecidable and incomplete first order theory - due to the recursively unseparable theoremhood and logical falsehood - or for a (E_n-) satisfiable formula without (E_{n-1}-) recursive models (see Aanderaa 1971; Börger 1984).

EXAMPLE 4. **n-fold exponential time bounded Turing machine halting problems and Trachtenbrot's finite satisfiability problem for n-th order formulae: characterization of spectra.**

We only need to reformulate and to adapt to finite satisfiability the reduction property stated in our proof of Cook's theorem with $f_n(p)$- time bounds instead of polynomial bounds p: since a function is f_1-bounded in the length of the binary representation of a natural number q iff it is bounded by a polynomial in (the length of the unary representation 1^q of) q and since encoding from unary to binary ("keep dividing") can be done in polynomial time, NTIME (f_{n+1}) $\subseteq$ Spectra $_{n+1}$ for $0 \leq n$ follows from the **Reduction Property:** For all $0 \leq n, 1 \leq s, 2 \leq q$ and arbitrary M:

M accepts q in $f_n(q^s)$ steps iff $\Pi_{M,s} \wedge \alpha_s \wedge \omega_{M,s}$ is satisfiable over q={0,1,...,q-1} where $\Pi_M \wedge \alpha \wedge \omega$ are formulae of order n+1.

In order to enable use of our Π_M for the description, over a domain with q elements, of the M-computation with input q but of length $f_n(q^s)$ we have to find an encoding of the $f_n(q^s)$-many time and tape cell numbers into a structure over the domain q: starting from the s-ary cartesian product over q we provide by n succesive power set constructions $f_n(q^s)$ "objects" of a type p_n of order n, impose on these objects a linear order K with first-element-predicate Z (for "zero") and successor relation S, say by Mostowski's (1956) formula $Ord_K(Z,S)$, take the latter for our Π_M of example 1 as zero/successor axiomatization part and use the objects t,u,v to express their unique order number $|t|, |u|, |v| \leq f_n(q^s)$ with respect to K.

To be more explicit remember that **types** are inductively defined by: o is a type (of domain elements) of order $|o|=0$; with $t_1, \ldots t_s$ also $(t_1, \ldots, t_s)$ is a type (of s-ary predicates) of order $|(t_1, \ldots, t_s)| := \max\{|t_i| \mid 1 \leq i \leq s\}+1$. In particular the **power set types** over q^s are defined by $p_{i+1} := (p_i)$ such that over q={0,1,...,q-1} there are exactly $f_n(q^s)$ objects of type p_n where $f_o(x)=x$ and by objects of type p_o we understand s-tuples of elements of the domain.

For any type t we assume to have countably many variables $x^t, y^t, \ldots$ (for objects) of type t from which **type formulae** are inductively defined in the usual way by saying that every

$$x^{(t_1,\ldots,t_s)}\, y^{t_1} \ldots y^{t_s} \quad \text{and } x^t = y^t$$

is a type formula (an atomic formula) and with F, G also $\neg F$, $(F \wedge G)$, $\bigwedge_{x^t} F$. F is said to be of order n if all bounded variables in F have types of order $\leq n-1$ and all free occuring variables in F have types of order $\leq n$.
Therefore read in Π_M t,t',u,u',v as (bounded) variables of type p_n and I_i, Z resp. P, T_j, K, S as (free occuring variables of type (p_n) resp. $(p_n\, p_n)$.
The "in the last moment no nonaccepting state"-formula ω reads now

$$\bigwedge_t (\, (\text{not} \bigvee_u Ktu) \rightarrow \bigwedge_{i \neq 1} \text{not } I_i(t)\,)$$

The formula α for the initial configuration is slightly more complicated for the encoding of the initial tape content and consists of the conjunction of the following formulae:

- **embedding** formula stating that the given domain of objects of type 0 (read: $q=\{0,1,\ldots,q-1\}$ for some q) is embedded into an initial segment of the K-ordering of p_n-objects by an injective function (with graph) F. Symbolically (let x,y be variables of type 0):

  $\bigwedge_{xu} \bigvee Fxu$ {existence} $\wedge$ $\bigwedge\bigwedge\bigwedge_{xuv}$ $(Fxu \wedge Fxv \rightarrow u=v)$ {uniqueness}

  $\wedge \bigwedge\bigwedge\bigwedge_{xyu}$ $(Fxu \wedge Fyu \rightarrow x=y)$ {injectivity}

  $\wedge \bigwedge\bigwedge\bigwedge\bigwedge\bigwedge_{uvwxy}$ $(Kuv \wedge Kvw \wedge Fxu \wedge Fyw \rightarrow \bigvee_z Fzv)$ {range is a segment}

  $\wedge \bigwedge_u$ $(Zu \rightarrow \bigvee_x Fxu)$ {embedding starts with the first element}

- $\bigwedge_t (Zt \rightarrow I_o t \wedge Ptt)$ {state 0 and reading head position 0 at time 0}

- $\bigwedge_t (Zt \rightarrow \bigwedge_u (\bigvee_x Fxu \rightarrow T_1 tu) \wedge \bigwedge_u (\, (\text{not} \bigvee_x Fxu) \rightarrow T_o tu)\,)$

  {at time 0 every tape cell (with number) in the range of F has content 1 and every other cell content 0}

  Corollary: The characterization theorem for spectra.

  Remark: Fagin's (1974) characterization theorem for generalized first order spectra (=NP-sets of positive integers) follows also as subcase of the above construction with only 2 modifications in α: a new embedding formula expressing that the domain of type-0-objects is embedded in an order preserving way with respect to an ordering of type-0-objects with Zero-predicate {0} and Successor-predicate $\lambda x,y.y=x+1$; another initial tape description expressing that M starts with the encoding of the (new) monadic input predicate I. All predicate symbols except I have to be existentially bounded.

References

Aanderaa, S.O. (1971): On the decision problem for formulas in which all disjunctions are binary. Proc. of the Second Scand. Log. Symp., pp. 1-18.

Aanderaa, S.O., Börger, E. (1979): The Horn complexity of Boolean functions and Cook's problem. in: F.V. Jensen, B.H. Mayoh, K.K. Moller (Ed.): Proceedings from 5th Scandinavian Logic Symposium, Aalborg University Press, pp. 231-256.

Aanderaa, S.O., Börger, E. (1981): The equivalence of Horn and network complexity for Boolean functions. Acta Informatica 15, 303-307.

Aanderaa, S.O., Börger, E., Lewis, H.R. (1982): Conservative reduction classes of Krom formulas. The Journ. of Symb. Logic 47, 110-129.

Asser, G. (1955): Das Repräsentantenproblem im Prädikatenkalkül der ersten Stufe mit Identität. Zeitschrift für mathematische Logik und Grundlagen der Mathematik 1, pp. 252-263.

Bennett, J. (1962): On spectra, Doctoral Dissertation, Princeton University, N.J.

Börger, E. (1971): Reduktionstypen in Krom- und Hornformeln. Dissertation. Münster, see Beitrag zur Reduktion des Entscheidungsproblems auf Klassen von Hornformeln mit kurzen Alternationen. in: Archiv für math. Logik und Grundlagenforschung 16 (1974), 67-84.

Börger, E. (1975): On the construction of simple first-order formulae without recursive models. Proc. Coloquio sobra logica simbolica, Madrid, 9-24.

Börger, E. (1979): A new general approach to the theory of the many-one equivalence of decision problems for algorithmic systems, in: Zeitschrift für math. Logik und Grundlagen der Mathematik 25, pp. 135-162.

Börger, E. (1982): Undecidability versus degree complexity of decision problems for formal grammars. in: Proceedings of GTI (Ed. L. Priese), Paderborn 1982, pp. 44-55.

Börger, E. (1984): Decision problems in predicate logic. in: G. Lolli (Ed.): Proc. Logic Colloquium Florence 1982, North-Holland Pu.Co.

Büchi, J.R. (1962): Turing machines and the Entscheidungsproblem, Mathematische Annalen 148, pp. 201-213.

Christen, C.-A. (1974): Spektren und Klassen elementarer Funktionen. Dissertation ETH Zürich.

Church, Alonzo (1936): A note on the Entscheidungsproblem. Journal of Symbol. Logic 1, pp. 40-41; Correction ibid, pp. 101-102.

Cleave, J.P. (1963): A Hierarchy of Primitive Recursive Functions. in: Zeitschr. für math. Logik und Grundlagen der Math. 9, 331-345.

Cook, S.A. (1970): The complexity of theorem proving procedures. Conf. Rec. of 3rd ACM Symp. on Theory of Computing, 151-158.

Cook, S. (1971a): Characterization of push-down machines in terms of time bounded computers. in: Journal of the Association for Computing Machinery 18, 4-18.

Davis, M. & Putnam, H. & Robinson, J. (1961): The decision problem for exponential diophantine equations. in: Annals of Math. 74, 425-436.

Denenberg, L. & Lewis, H.R.: The complexity of the satisfiability problem for Krom formulas. Proc. Logic Colloquium Aachen 1983 (to appear).

Deutsch, Michael (1975): Zur Theorie der spektralen Darstellung von Prädikaten durch Ausdrücke der Prädikatenlogik I. Stufe. in: Archiv math. Logik 17, pp. 9-16.

Fagin, Ronald (1974): Generalized First-Order Spectra and Polynomial-Time Recognizable Sets. in: R. Karp (Ed.): Complexity of Computation, SIAM - AMS Proceedings Vol. 7, pp. 43-73.

Fürer, M. (1981): Alternation and the Ackermann case of the decision problem, in: L'Enseignement mathématique XXVII, 1-2, pp. 137-162.

Germano, G. (1976): An arithmetical reconstruction of the liar's antinomy using addition and multiplication. in: Notre Dame J. of Formal Logic XVII, 457-461.

Grzegorczyk, A. (1953): Some classes of recursive functions. Rozprawy Matematyczne IV, Warzsaw.

Gurevich, Yuri (1966): Ob effektivnom raspoznavanii vipolnimosti formul UIP. Algebra i Logika 5, 25-55.

Gurevich, Yuri (1976): Semi-conservative reduction. in: Archiv für mathematische Logik 18, 23-25.

Hilbert, D. & Bernays, P. (1939): Grundlagen der Mathematik, Vol. II. Berlin

Jones, J.P & Matijasevich, Yu. (1984): Register machine proof of the theorem on exponential diophantine representation of enumerable sets. in: The Journal of Symbolic Logic.

Jones, N.G. & Selman, A.L. (1974): Turing machines and the spectra of first-order formulas, Journal of Symbolic Logic 39, pp. 139-150.

Kostryko, V.F. (1964): Klass cvedeniya $\forall\exists^n\forall$. Algebra i Logika 3, pp. 45-65.

Kostyrko. V.F. (1966): Klass cvedeniya $\forall\exists\forall$. Kibernetika 2, pp. 17-22; English translation, Cybernetics 2, pp. 15-19.

Kreisel, G. (1953): Note on arithmetic models for consistent formulae of the predicate calculus II. in: Proc. XI-th Int. Congr. Philosophy, vol. 14, pp. 39-49.

Lewis, H.R. (1979): Unsolvable classes of quantificational formulas. Reading (Mass.)

Lewis, H.R. (1980): Complexity results for classes of quantificational formulas. JCSS.

Lynch, (1982): Complexity Classes and Theories of Finite Models. in: Math. Systems Theory 15, 127-144.

Machtey, M. & Young, P. (1978): An Introduction to the General Theory of Algorithmus. North-Holland Pu. Co., Amsterdam.

Machtey, M. & Young, P. (1981): Remarks on recursion versus diagonalization and exponentially difficult problems. in: J. Computer and System Sciences 22, 442-453.

Minsky, M.L. (1961): Recursive unsolvability of Post's problem of 'Tag' and other topics in theory of Turing machines, Ann. of Math. 74, 437-455.

Mostowski, A. (1953): On a system of axioms which has no recursively enumerable model. in: Fundamenta Mathematicae 40, 56-61.

Mostowski, A. (1955): A formula with no recursively enumerable model. in: Fundamenta Mathematicae 43, 125-140.

Mostowski, A. (1956): Concerning a problem of H. Scholz, Zeitschrift für mathematische Logik und Grundlagen der Mathematik, 2, pp. 210-214.

Oberschelp, W. (1958): Varianten von Turingmaschinen. in: Archiv für math. Logik 4, 53-62.

Post, E. (1943): Formal Reductions of the General Combinatorial Decision Problem. in: American Journal of Math. 65, 197-268.

Rabin, MO. (1958): On recursively enumerable and arithmetic models of set theory. in: The Journal of Symbolic Logic 23, pp. 408-416.

Ritchie, R.W. (1960): Classes of Recursive Functions of Predictable Complexity. Doctoral Dissertation at Princeton University.
Ritchie, R.W. (1963): Classes of Predictably Computable Functions. in: Trans. Amer. Math. Soc. 106, 139-173.

Ritchie, R.W. (1965): Classes of Recursive Functions Based on Ackermann's Function. in: Pacific J. Math. 15, 1027-1044.

Rödding, D. (1968): Klassen rekursiver Funktionen. in: Proc. of the Summer School in Logic, Leeds 1967, Springer Lecture Notes in Math. vol. 70, 159-222.

Rödding, D. (1968a): Höhere Prädikatenlogik: Interpolationstheorem, Reduktionstypen. Lecture Notes (Mimeographed), Institut für math. Logik und Grundlagenforschung of the University of Münster.

Scholz, Heinrich (1952): Ein ungelöstes Problem in der symbolischen Logik. in: The Journal of Symbolic Logic 17, pp. 160.

Schwichtenberg, H. (1969): Bemerkungen zum Spektralproblem. in: Tagungsberichte des Math. Forschungsinstituts Oberwolfach, 8, pg. 12.

Shepherdson, J.C. & Sturgis, H.E. (1963): Computability of Recursive Functions. in: Journal of the Association for Computing Machinery 10, 217-255.

Smullyan, R. (1961): Theory of Formal Systems. in: Annals of Mathematics Studies vol. 47, Princeton University Press, Princeton/New Jersey.

Specker, E. & Strassen, V. (1976): Komplexität von Entscheidungsproblemen. Springer Lecture Notes in Computer Science Vol. 43.

Suranyi, J. (1959): Reduktionstheorie des Entscheidungsproblems im Prädikatenkalkül der ersten Stufe. Verlag der Ungarischen Akademie der Wissenschaften, Budapest.

Trachtenbrot, B.A. (1950): Impossibility of an algorithm for the decision problem in finite classes. in: Dokl. Akad. Nauk SSS R 70, 1950, pp. 569-572. English translation in: AMS Transl. Ser. 2 Vol. 23 (1963, pp. 1-5.

Turing, Alan M. (1937): On computable numbers, with an application to the Entscheidungsproblem. Proceedings of the London Mathematical Society, 2nd series 42, pp. 230-265; correction ibid., 43, pp. 544-546.

Wang, H. (1957): A variant to Turing's theory of computing machines. in: Journal of the Association for Computing Machinery 4, 63-92.

Young, P. (1983): Gödel Theorems, Exponential Difficulty and Undecidability of Arithmetic Theories: An Exposition. in: Proc. AMS Summer School on Recursion Theory, Ithaca 1982.

Rödding, D., Schwichtenberg, H. (1972): Bemerkungen zum Spektralproblem. Zeitschrift für math. Logik und Grundlagen der Mathematik, 18, pp. 1-12

REDUCTION TO NP-COMPLETE PROBLEMS BY INTERPRETATIONS

Elias Dahlhaus
Fachbereich Mathematik
Technische Universität
Berlin
D-1000 Berlin 12

1) Introduction

Fagin(4) has shown that every class of finite models, which is closed by isomorphisms, is in NP iff it is the class of finite models, which fulfill a certain formula with existential quantification on predicates (shortly Σ_1^1-formula). Aim of this paper is to define and to realize a strong notion of logical NP-completeness, which could give a new combinatorial aspect of the P=NP-problem. We show that the well known NP-complete sets SAT, CLIQUE and Directed Hamilton Cycle are complete also in this new proposed sense, but not for example 3-SAT.

2) Fundamental Definitions

2.1. Mod L is the class of all L-structures for any set L of pairs(r,n) with relation symbols r and arity n. $\mathrm{Mod}_f L$ is the class of all finite L-structures.

2.2. For any L-structure M, $|M|$ is the set of individuals of M.

2.3. $I: \mathrm{Mod}\ L_1 \to \mathrm{Mod}\ L_2$ is <u>interpretative</u> iff $|I(M)|$ is a finite disjoint union of finite powers of $|M|$ and every L_2-relation on $|I(M)|$ is definable by a quantifierfree L_1-formula, formally:

i) there exist n and n_i, $i<n$, s.t. for each $M \in \mathrm{Mod}\ L_1$:
$|I(M)| = \bigcup_{i<n} \{i\} \times |M|^{n_i}$

ii) for each relational symbol r of L_2 with arity k and each $i_1 .. i_k < n$ there is a quantifierfree formula H(dependent on $r, i_1 .. i_k$), s.t. for each $M \in \mathrm{Mod}\ L_1$ and each n_{i_j}-tuple $\bar{x}_j \in |M|^{n_{i_j}}$:
$r^{I(M)}((i_1,\bar{x}_1)..(i_k,\bar{x}_k))$ iff $M \models H(\bar{x}_1 ... \bar{x}_k)$.

This notion of reduction is in some sense an intensification of elementary reductions in sense of Lovász and Gács(9) and the first order translations in sense of Immerman(6), compare also interpretations in Tarski(12) and Enderton(3).

2.4. $\underline{M} \subset Mod_f L$ is <u>interpretative complete</u> in NP: iff for each isomorphism closed $\underline{M}' \subset Mod_f L_1$ in NP there is an interpretative I: Mod $L_1 \to$ Mod L_2, s.t. $\underline{M}' = I^{-1}[M]$ and M is in NP.

2.5. A formula H in prenex conjunctive or disjunctive normal form is <u>monotonic</u> in the relational symbol r iff r appears in H only non-negated.

3) Description of some NP-complete problems as model problems

3.1. SAT is the class of all finite $M := (|M|, V^M, C^M, P^M, N^M)$, such that $M \models \exists V' \subset V \forall c \in C \exists v \in V((v \in V' \wedge P(v,c)) \vee (v \notin V' \wedge N(v,c)))$.
V is a relational symbol for the variables, C a relational symbol for the clauses, P a relational symbol for positive and N for negative occurence of a variable in a clause. Note that V′ plays the role of a truth assignment of the variables occuring in the clauses.

3.2. CLIQUE is the class of all finite $M = (|M|, E^M, N^M)$, s.t. $M \models \exists F(dom(F) = N \wedge \forall i,j\ (N(i) \wedge N(j) \wedge i \neq j \to (E(F(i), f(j)) \wedge F(i) \neq F(j)))$.

3.3. DHC (Directed Hamilton Cycle) is an isomorphism closed class of finite models with one 2-ary relational symbol E.

4) The Theorems

4.1. SAT, CLIQUE and DHC are interpretative complete in NP.

Moreover:

4.2. Each isomorphism closed NP-problem, defined by a Σ_1^1-formula which is monotonic in r can be reduced to SAT, s.t. nonlogical symbols appear only in the N, P defining formulas of the interpretative reduction and this formulas are monotonic in r.

4.3. Each isomorphism closed NP-problem, whose Σ_1^1-formula is monotonic in r, can be reduced to DHC, s.t. the formulas of the interpretative reduction are monotonic in r.

5) Proofs of the Theorems

5.1. Completeness of SAT:

5.1.1. By Skolemisation one can find for each Σ_1^1-formula an equivalent Σ_1^1-formula of the form:

$$\exists G_1..G_n \exists F_1..F_m \bigwedge_{i=1}^{m} \bar{x}_i \; y_i F(\bar{x}_i, y_i) \wedge \forall y_1..y_r H \qquad (1*)$$

where H is a quantifierfree formula with the free variables $y_1..y_r$, which is in conjunctive normal form.

5.1.2. Let $\underline{M}$ be a class of finite models being in NP and defined by a formula as in (1*), s.t. G_i has arity k_i, F_i has arity l_{i+1}, $H=\bigwedge_{i=1}^{p} C_i$, and $C_i = \bigvee A_i \vee \bigvee B_i$,
where A_i resp. B_i is a finite set of positive or negated prime-formulae with fixed resp. variable relational symbols.

5.1.3. Construction of the interpretative reduction to SAT:
For any L-structure M define: $I(M)=(|I(M)|, V_M, C_M, N_M)$, where $|I(M)| = V_M \cup C_M$,
V_M describes the replacement of the variable predicates, therefore set: $V_M := \bigcup_{i=1}^{n} \{i\} \times |M|^{k_i} \cup \bigcup_{i=1}^{m} \{n+i\} \times |M|^{l_i+1}$.
For each F_i and each $\bar{x} \in |M|^{l_i}$ there is a clause which means (i) $\exists y F(\bar{x},y)$ and for each C_i and $y_1..y_r$ there is a clause which controls that (ii) $C_i(y_1..y_r) = \bigvee A_i(y_1..y_r) \vee \bigvee B_i(y_1..y_r)$ is fullfilled.
Set $C_M := \bigcup_{i=1}^{m} \{n+m+i\} \times |M|^{l_i} \cup \bigcup_{i=1}^{p} \{n+2m+i\} \times |M|^{r}$.
Note that V_M and C_M depend only on the arity of the relation symbols, n, m, and p.
P_M resp. N_M represent positive resp. negative occurences of variables (read: elements of the quantified relations F_i, G_j) in clauses.
Set therefore:
to fulfill (i): $((n+i,\bar{x},y),(n+m+i,\bar{x})) \in P_M$ for each $\bar{x} \in |M|^{l_i}$ and $y \in |M|$.
to fulfill (ii): $(v,(n+2m+i,y_1..y_r)) \in P_M \cap N_M$ if $A_i(y_1..y_r)$ is true in M, where v is any variable, (the clause $c:=(n+2m+i,y_1..y_r)$ contains all variables positively and negatively, and is therefore true for each truthassignment, if $A_i(y_1..y_r)$ is true in M. Moreover monotonicity is preserved.)
$((i,y_{f1}..y_{fk_i}),(n+2m+j,y_1..y_r)) \in P_M$ (resp. N_M), if $G_i(y_{f1}..y_{fk_i})$ (resp. $\neg G_i(y_{f1}..y_{fk_i})$) is in B_j,
$((n+i,y_{f1}..y_{fl_i+1}),(n+2m+j,y_1..y_r)) \in P_M$ (resp. N_M), if $F_i(y_{f1}..y_{fl_i+1})$ (resp. $\neg F_i(y_{f1}..y_{fl_i+1})$) is in B_j.
There are no other pairs in N_M or P_M. It is easily checked that I is an interpretative reduction of $\underline{M}$ to SAT and that monotonicity

is preserved.

5.2.Completeness of CLIQUE:Karp's(8)reduction from SAT to CLIQUE is interpretative,and interpretativity of functions is preserved under composition.

5.3.Completeness of DHC:We define an interpretative reduction of SAT to DHC.For brevity we write directed edges in both directions as undirected edges:$\{x,y\}\in E$:iff$(x,y),(y,x)\in E$.

For each finite model(M,V,C,P,N)with the relational symbols of SAT construct the following digraph(K,E):=I(M,V,C,P,N),where

$K:=\{1,2,3\}xM\cup\{4,5,6,7,8,9\}xM^2$,

for all v,w,k,l the following are the edges of E:

$\{2\}$xM and $\{3\}$xM form each a clique,

$\{(2,v),(3,w)\}\in E$ iff $v\neq w$,

$\{(2,v),(4,v,k)\},\{(3,v),(9,v,k)\}\in E$,

(4,v,k)...(9,v,k) form an undirected chain of edges,

$\{(4,v,k),(9,v,l)\}\in E$ iff $k\neq l$,

$(6,v,k)\rightarrow(1,k)\rightarrow(7,v,k)\in E$ if P(v,k)

" $\leftarrow$ " $\leftarrow$ " $\in E$ if N(v,k) } and $k\in C$

" $\leftrightarrow$ " $\leftrightarrow$ " $\in E$ if $k\notin C$

With this definition following properties hold:

5.3.1.Each Hamilton cycle going $(6,v,k)\rightarrow(1,k)$

i)comes from (5,v,k),otherwise (5,v,k) would be isolated,

ii)goes$(1,k)\rightarrow(7,v,k)$,otherwise (7,v,k) would be isolated.

Vice versa for $(7,v,k)\rightarrow(1,k)$ coming from (8,v,k)and going to(6,v,k)

5.3.2.From 5.3.1. follows that Hamilton cycles must go the paths:

$(4,v,k)\rightarrow(5,v,k)\rightarrow(6,v,k)(\rightarrow(1,k))\rightarrow(7,v,k)\rightarrow(8,v,k)\rightarrow(9,v,k)$(positive direction) or vice versa(negative direction).

5.3.3.For each Hamilton cycle going a path$(2,v)\rightarrow(4,v,l)\rightarrow\cdots\rightarrow(3,v)$ or $(3,v)\rightarrow(9,v,k)\rightarrow\cdots\rightarrow(2,v)$in between $\cdots$passes all of the points (4,v,k)..(9,v,k) for all k in the positive resp. negative direction (this is true,because the only edges,by which the third component in a triple point can be changed,are $(4,v,k)\rightarrow(9,v,l)$ or vice versa).

5.3.4. By the foregoing for each v either all (v,k) have positive direction or all (v,k) have negative direction.Therefore we can say,v has positive or negative direction respectively.

5.3.5.By the definition of SAT the following equivalence is true:

(M,V,C,P,N) SAT iff there is a V' V and an $F:M\rightarrow V$,s.t. for each c C $(F(c)\in V'$ and $P(F(c),c))$ or $((F(c)\notin V'$ and $N(F(c),c))$(2*).

If (2) is true then there is a Hamilton Cycle in I(M,V,C,P,N),which is in positive direction for v if $v\in V'$resp. negative direction if $v\notin V'$, and goes $(6,F(c),c)\rightarrow(1,c)\rightarrow(7,F(c),c)$ resp. vice versa.

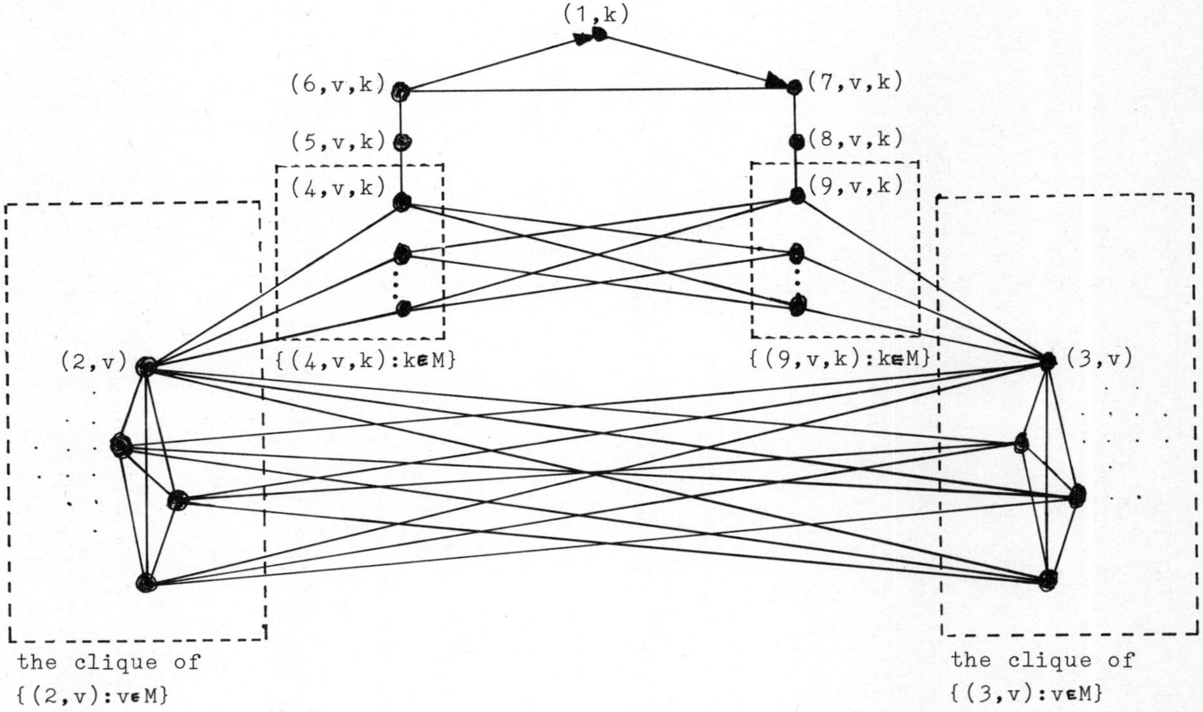

Figure 1: Interpretative reduction from SAT to DHC, the case, that the variable v appears positively in the clause k.

Conversely if there is a Hamilton cycle in I(M,V,C,P,N),define V':={v:v has positive direction} and F(c):=v :iff the Hamilton cycle goes $(6,v,c)\rightarrow(1,c)\rightarrow(7,v,c)$ or vice versa.This F and V' fulfill (2*).

5.3.6.The monotonicity properties are preserved,because P and N are used only positively.

6)<u>A counterexample</u>

For suitable encoding as model problem,3-SAT is such counterexample: For each boolean formula with exactly three literals per clause,t, we construct the model $M_t:=(M,V,K,P_1,P_2,P_3,G)$,where V:=set of variables of t,K:=set of clauses of t,the set of individuals of M_t,M:=K V,$P_i(k)$:iff i^{th} literal of k is positiv,and $G(k,v_1,v_2,v_3)$:iff for i=1,2,3,:v_i is the variable of the ith literal of k.

<u>3-SAT</u>:=$\{(M,V,K,P_1,P_2,P_3,G)$:M finite,$(M,V,K,P_1..P_3,G)\models\exists V'\subset V\forall v_1..v_3\in V$ $\forall k\in K(G(k,v_1..v_3)\rightarrow\bigvee_{i=1}^{3}((P_i(k)\wedge v_i\in V')\ (\neg P_i(k)\wedge v_i\notin V'))$.

As easily seen,$t\in$3-SAT iff $M_t\in$<u>3-SAT</u>.Therefore <u>3-SAT</u> is NP-complete, but interpretative completeness of <u>3-SAT</u> would implie,that every isomorphism closed NP-problem has a defining Σ_1^1-formula,whose first-order part is universal,which would implie the subobject closure of NP-problems,which is absurd.

7)<u>Some consequences for spectra</u>

This chapter gives some consequences of the interpretative completeness of CLIQUE to spectra.A spectrum is,as well known,the set of the cardinalities of the finite models of a certain first-order formula (see Scholz(11)).Jones and Selman(7)characterized spectra as tally languages which are in NP(compare also Rödding/Schwichtenberg(10)). That means,spectra can be described as classes of finite models of the empty set of nonlogical symbols,fulfilling a certain Σ_1^1-formula. For brevity n and $\{i:0\leq i<n\}$ are identified,if n is a natural number. W.l.o.g. the finite domains are natural numbers.

7.1. If we reduct a spectrum to CLIQUE by an interpretative map I, then $I(n)=(V_n,E_n,N_n)$, where (V_n,E_n) is a graph and has a clique of power $|N_n|$ iff n is in the spectrum above.

Moreover:

7.1.1. Define $p(n):=|N_n|$. Then p is a polynom.

7.1.2. $V_n=\dot{\bigcup}_{i=1}^{k}\{i\}\times n^{p_i}$ for each natural number n, and for each $i,j\in\{1..k\}$ there is a quantifierfree formula (without nonlogical symbols) H_{ij}, s.t. $\{(i,\bar{x}),(j,\bar{y})\}\in E_n$ iff $n\models H_{ij}(\bar{x},\bar{y})$.

7.2. Let $(n)^k$ be the set of all k-tuples of n, whose components are pairwise different.

$(V_n,E_n)_{n=0}^{\infty}$ is called a <u>system of equality graphs</u> iff

i) there is a $k\in\mathbb{N}$ and there are natural numbers $p_1..p_k$, s.t. for each $n\in\mathbb{N}$: $V_n=\dot{\bigcup}_{i=1}^{k}\{i\}\times(n)^{p_i}$, and

ii) E_n is defined by quantifierfree formulas as in 7.1..

i=1..k is called a <u>level</u> and p_i is the <u>dimension</u> of the level i.

7.3. By contracting equal components to one component it follows from 7.1.:

$S\subset\mathbb{N}$ is a spectrum iff there is a system of equality graphs $(V_n,E_n)_0^{\infty}$ and a polynom p, s.t. $n\in S$ iff (V_n,E_n) has a p(n)-clique.

7.4. <u>Remark</u>: Let $(G_n)_{n=0}^{\infty}$ be a system of equality graphs with the levels 1..k, the dimensions p_i, i=1..k, and the edge defining formulas H_{ij}, $i,j\in\{1..k\}$. W.l.o.g. we can assume that each H_{ij} is in complete disjunctive normal form, $H_{ij}=\bigvee_{t=1}^{m_{ij}}\bigwedge B_t$, and for each $l=1..p_i$, $l'=1..p_j$ $x_l=y_{l'}$ or $x_l\neq y_{l'}$ is in B_t. The relation $\{(l,l'):x_l=y_{l'}$ is in $B_t\}$ is a partial injection and is called the <u>equality color</u> of B_t. Vice versa equality colors define systems of equality graphs. For combinatorial aspects define an equality color to be <u>good</u> iff only equal components are joint.

7.5. Now systems of equality graphs with only one level, so called <u>homogeneous</u> systems of equality graphs, are of interest.

We can formulate following

<u>THEOREM</u>:

$S\subset\mathbb{N}$ is a spectrum iff there is a homogeneous system of equality graphs $(\bar{G}_n)_{n=0}^{\infty}$, a natural number k, and a polynom <u>p</u>, s.t. for almost all natural numbers n: $n\in S$ iff $\bar{G}_{n+k}$ has a <u>p</u>(n)-clique.

<u>Proof</u>:

1st step: Let $(G_n)_{n=0}^{\infty}$ be a system of equality graphs and <u>p</u> be a polynom, which describe the spectrum S as in 7.3.. Let $(G_n)_{n=0}^{\infty}$ have a number of levels of k and dimensions p_i, i=1..k.

Let p be the maximal dimension, $V_n^1:=\{1..k\}\times(n)^p$, and

$E_n^1:=\{\{(i,x_1..x_p),(j,y_1..y_p)\}:(i,x_1..x_{p_i}),(j,y_1..y_{p_j})\ G_n\text{-adjacent}\}$.

Then $G_n^1:=(V_n^1,E_n^1)$ has a $\underline{p}(n)$-clique iff G_n has a $\underline{p}(n)$-clique for almost all natural numbers n. Each level of $(G_n^1)_{n=0}^{\infty}$ has the same dimension.

2nd step: Set $V_n^2:=\{1..k\}\times(n)^{p+2}$ and $\{(i,\bar{x},x_1,x_2),(j,\bar{y},y_1,y_2)\}\in E_n^2$ iff $\{(i,\bar{x}),(j,\bar{y})\}\in E_n^1$, $x_1=y_1$, and ($x_2=y_2$ iff $i=j$).

Then $(V_{n+k+1}^2,E_{n+k+1}^2)$ has a $\underline{p}(n)$-clique iff G_n^1 has a $\underline{p}(n)$-clique. Moreover $(i,\bar{x}),(j,\bar{y})$ adjacent in $G_n^2:=(V_n^2,E_n^2)$ implies $\bar{x}$ and $\bar{y}$ are equal in at least one component and unequal in at least one component.

3rd step: Set $\bar{V}_n:=(n)^{k(p+2)}$ and $\{(\bar{x}_1..\bar{x}_n),(\bar{y}_1..\bar{y}_n)\}\in\bar{E}_n$ iff for some $i\leq j\leq k$: $\{(i,\bar{x}_i),(j,y_j)\}\in E_n^2$ and $\bar{x}_l=\bar{y}_l$ for $l<i$ and $l>j$ and $\bar{x}_{l+1}=y_l$ for $i\leq l<j$. Call the $\bar{x}_i$ resp. $\bar{y}_i$ <u>bytes</u>.

Then the adjacency of $\bar{x}$ and $\bar{y}$ in $(\bar{V}_n,\bar{E}_n)$ implies that only one pair of bytes in $\bar{x}$ resp. $\bar{y}$ is equal in at least one component and unequal in at least one component, because $\bar{x}$ and $\bar{y}$ have each pairwise unequal components.

Conclusion: $(\bar{V}_{n+k+1+(p+2)(k-1)},\bar{E}_{n+k+1+(p+2)(k-1)})$ has a $\underline{p}(n)$-clique iff G_{n+k}^2 has a $\underline{p}(n)$-clique.

Therefore the THEOREM is proved.

8) Some final remarks

With the auxiliary predicate of the total order Immerman(6) proved the NL-completeness of GAP via first order translations. Lovász and Gács(9) have shown the NP-completeness of SAT via "elementary reductions". Another possible combinatorial interesting form of reductions are the one-way log tape reductions of Hartmanis/Immerman/Mahaney(5). The THEOREMs of this paper are from (2).

References

(1) Asser — "Das Representantenproblem im Prädikatenkalkül der ersten Stufe mit Identität", ZML 1, p.252-263, 1955.

(2) Dahlhaus — "Kombinatorische und logische Eigenschaften von Reduktionen auf einige vollständige Probleme in NP und NL", Doctoral Dissertation, TU Berlin, 1982.

(3)Enderton: "A Mathematical Introduction to Logic",Academic Press,1972.

(4)Fagin: "Generalized First Order Spectra and Polynomial Time Recognizable Sets",Karp(ed.)"Complexity of Computations";SIAM-AMS Proc. 7,p.27-41.

(5)Hartmanis Immerman Mahaney: "One-Way Log Tape Reductions",IEEE-FOCS 19, p.65-72,1978.

(6)Immerman: "Languages which capture Complexity Classes", 15th ACM-STOC,p.347-354,1983.

(7)Jones,Selman: "Turing Machines and Spectra of First Order Formulas",JSL 39,p.139-150,1974.

(8)Karp: "Reducibility among Combinatorial Problems", Miller/Thatcher(eds.)"Complexity of Computations", Plenum Press,New York,1972.

(9)Lovász,Gács: "Some Remarks on Generalized Spectra",ZML 23, p.547-554,1977.

(10)Rödding, Schwichtenberg: "Bemerkungen zum Spektralproblem",ZML 18,p.1-12, 1972.

(11)Scholz: "Ein ungelöstes Problem der symbolischen Logik", JSL 17,p.160,1952.

(12)Tarski: "Undecidable Theories",North Holland Publishing Company,1953.

ZML:Zeitschrift für Mathematische Logik und Grundlagen der Mathematik,
JSL:Journal of Symbolic Logic
ACM-STOC:Annual Symp. on Theory of Computing of the ACM
IEEE-FOCS:IEEE Symp. on Foundations of Computer Science

UNIVERSAL QUANTIFIERS AND TIME COMPLEXITY OF RANDOM ACCESS MACHINES

E. Grandjean
Université Lyon 1, 43 Bd du 11 Novembre 1918
69622 VILLEURBANNE CEDEX / FRANCE

Abstract. Let $Sp(d\forall)$ denote the class of spectra of first-order sentences with d universal quantifiers. Let NRAM(T(n)) denote the class of sets (of positive integers) accepted by Nondeterministic Random Access Machines (with successor as the only arithmetical operation), in time O(T(n)) where n is the input integer. We prove

$$Sp(d\forall) = NRAM(n^d) \text{ for } d \geq 2.$$

A similar result holds for generalized spectra.

1. Introduction.

This paper states an equivalence between the two following complexity measures, the first one in automata theory, the second one in logic.

A successor RAM is a Random Access Machine (cf [1,3,14,17]) where the only arithmetical operations are incrementation and decrementation by one. It is shown in [17] that successor RAMs where we assume the uniform cost measure (i.e., the time cost of each instruction is one) are time equivalent to Storage Modification Machines (SMMs). Successor RAMs are compared with various computers in [14,17,19]. A nondeterministic version of successor RAM, denoted NRAM, is defined in [15]: it can guess any integer in one step. Let NRAM(T(n)) denote the class of sets (of positive integers) accepted by NRAMs (with uniform cost measure) in time cT(n), where c is a constant integer and n is the input integer.

We will consider <u>first-order</u> <u>formulas</u> with equality and possibly relation and function symbols (of arity $r \geq 0$). Let φ be a sentence, i.e. a first-order formula where each variable is quantified. The <u>spectrum</u> of φ, denoted Sp(φ), is the set of cardinalities of the finite structures which satisfy sentence φ. It is proved in [10] and [4] that for unary notation of integers, the spectra (resp. generalized spectra) are exactly the sets of positive integers (resp. sets of graphs) in NP. Let Sp($d\forall$) denote the class of spectra of sentences with d quantifiers (or, equivalently, with d universal quantifiers).

Let NTIME(T(n)) be the class similar to NRAM(T(n)) for Turing machines. Using the hierarchy result of classes NTIME(T(n)) and a correspondence between classes Sp($d\forall$) and NTIME(T(n)), Pudlák [16] proved that Sp($d\forall$) $\subsetneq$ Sp($d+1\ \forall$). An immediate consequence of our main result, that is Sp($d\forall$) = NRAM(n^d), and of the inclusions NTIME(T log T) $\subseteq$ NRAM(T) $\subseteq$ NTIME($T(\log T)^2$) stated in [15], is the following:

NTIME($n^d \log n$) $\subseteq$ Sp($d\forall$) $\subseteq$ NTIME($n^d(\log n)^2$),

for each integer $d \geq 2$. It slightly improves a result of [6]. Further, we get a result of pure logic:

Sp($d\forall$) = Sp($d\forall$, arity d), for $d \geq 2$.

(This last class is defined by the added requirement that the arity of each relation or function symbol of the sentences is at most d.) Note that we do not know how to prove this result by using only logical tools. It is interesting to compare our results with the following result of J. Lynch [13]:

NTIME(n^d) $\subseteq$ Sp(arity d, without function symbols), for $d \geq 2$.

We are convinced that from such sharp connections between automata theory and logic (more precisely, finite-model theory), one obtains results of pure logic which cannot be proved without automata theoretic arguments. The reason is that computers and their computations seem to be much more flexible concepts than sentences and finite structures which satisfy them. For example, Pudlák's hierarchy theorem above-mentioned is a logical version of the nondeterministic time hierarchy whose proof uses in an essential manner a padding argument and a recursion theorem.

In section 2, we present normalization lemmas for the proof of our main result which is given in section 3. Corollaries are presented in section 4.

2. Normalization of NRAMs and sentences.

Definition. A *simple NRAM* is a NRAM with 2 accumulators, denoted a,b, and the following set of instructions of easy meaning ($\langle a \rangle$ denotes the register pointed to by accumulator a):

(1) Read(a)	(7) b := a
(2) a := 0	(8) $\langle a \rangle$:= b
(3) a := a+1	(9) Go to inst p or q
(4) a := a-1	(10) If a=b then go to inst p else go to inst q
(5) Guess(a)	(11) Accept
(6) a := $\langle a \rangle$	(12) Reject

Remarks. Instruction Read(a) means that the contents of accumulator a are replaced by the input. The guessed integer of instruction Guess(a) is arbitrarily large but we can require that it be smaller than T (for a T time-bounded computation), since the time cost of each instruction is one.

Lemma 1. Each NRAM (as defined in [15], for example) is simulated by a simple NRAM in linear time.

Proof. Almost trivial (see [17]).□

We will use multidimensional NRAMs (i.e., NRAMs in which tuples of integers are addresses of registers) and will permit $<$-comparison (that is, instruction (10) with $a<b$ in place of $a=b$). The following lemma is implicitly proved in [15].

Lemma 2 (Monien [15]). Each multidimensional NRAM with $<$-comparison is simulated in linear time by an ordinary NRAM (i.e., a one-dimensional NRAM without $<$-comparison).

We will use a notion of spectrum with linear order.

Definition. Let $[m]$ denote the set of integers $\{1,2,\ldots m\}$. Let $\mathcal{T} = \{V_1,V_2,\ldots V_p\}$ be a (first-order) type, i.e., a finite set of relation or function symbols, and let φ be a sentence of type $\mathcal{T}\cup\{<\}$. The spectrum of φ, denoted $Sp_<(\varphi)$, is the set of cardinalities n of structures $\mathcal{S}$ of the form $([n], <, V_1,\ldots V_p)$ such that:

(i) $<$ is the natural order of the domain $[n]$;

(ii) $\mathcal{S}$ satisfies φ.

Let $Sp_<(d\,\forall)$ denote the class of spectra of sentences with d universal quantifiers.

Definition. An elementary formula of type $\mathcal{T}$ is a formula of one of the following forms:

$v_1 = v_2$; $v_1 < v_2$; $R(\bar{v}_r)$; $\neg R(\bar{v}_r)$; $F(\bar{v}_r) = v_{r+1}$,

where each v_i is a variable ($\bar{v}_r$ denotes the r-tuple $v_1,\ldots v_r$) and R,F are respectively relation and function symbols (of arity r) of $\mathcal{T}$.

<u>Lemma 3</u>. <u>If</u> $S \in Sp_<(d\forall)$, <u>then there is a type</u> $\mathcal{T}$ <u>and a sentence</u>

$$\varphi = \forall \bar{x}_d \ \exists \bar{y}_k \ \bigvee_{i \in [c]} \psi_i(\bar{x}_d, \bar{y}_k),$$

<u>such that</u>:

(i) $S = Sp_<(\varphi)$;

(ii) <u>each</u> ψ_i <u>is a conjunction of elementary formulas of type</u> $\mathcal{T}$.

<u>Proof</u>. It essentially uses Skolem functions (see [6]).□

Let us regard elements of $[n]$ as constant symbols.

<u>Definition</u>. An $[n]$-<u>formula of type</u> $\mathcal{T}$ is a formula of type $\mathcal{T} \cup [n] \cup \{<\}$ constructed by replacing each variable by a constant of $[n]$ in an elementary formula of type $\mathcal{T}$.

The algorithm of theorem 5 below is based on the following

<u>Lemma 4</u>. <u>Let sentence</u> φ <u>be as in lemma 3. Then we have for each integer</u> $n > 0$,

(∗) $n \in Sp_<(\varphi) \leftrightarrow$ there are functions $g : [n]^d \longrightarrow [n]^k$ and $h : [n]^d \longrightarrow [c]$ and a structure on domain $[n]$, which satisfies sentence $\varphi_{g,h}$

where $\varphi_{g,h}$ is the following sentence of type $\mathcal{T} \cup [n] \cup \{<\}$

$$\bigwedge_{\bar{a}_d \in [n]^d} \psi_{h(\bar{a}_d)}(\bar{a}_d, g(\bar{a}_d)).$$

<u>Proof</u>. Easy (see [6]).□

3. <u>Proof of the main result</u>.

<u>Theorem 5</u>. <u>For each integer</u> $d \geq 1$, $Sp_<(d\forall) \subseteq NRAM(n^d)$.

<u>Proof</u>.

Let $S = Sp_<(\varphi)$ where sentence φ is as in lemma 3. From equivalence

($\ast$) of lemma 4, we obtain the following nondeterministic algorithm for checking if $n \in S$.

(1) For each $\bar{a}_d \in [n]^d$, guess $\bar{b}_k = g(\bar{a}_d) \in [n]^k$ and $i = h(\bar{a}_d) \in [c]$.

(2) Check if $\varphi_{g,h}$ is satisfiable by a structure on domain $[n]$.

By lemma 2, it is sufficient to exhibit a multidimensional NRAM with $<$-comparison, denoted M, which accepts S in the required time bound. Exploiting the structure of such a machine, it is not necessary to perform the whole part (1) of the above algorithm before starting part (2) but we can check if $\varphi_{g,h}$ is satisfiable at the time when we guess the values $g(\bar{a}_d)$ and $h(\bar{a}_d)$.

M has d+k+1 accumulators, $a_1,\dots a_d$, $b_1,\dots b_k$ and n, respectively corresponding to variables $x_1,\dots x_d$, $y_1,\dots y_k$ and to the input. For each function, F, of arity r, occuring in φ, M has a r-dimensional sequence of registers. In this sequence, let $\langle i_1,\dots i_r\rangle_F$ denote the register whose address is $(i_1,\dots i_r)$. (Note that each relation symbol of φ can be assimilated to its characteristic function.)

Let $\bar{a}_d$ and $\bar{b}_k$ (resp. $\bar{1}^d$ and $\bar{n}^d$) respectively denote the tuples $(a_1,\dots a_d)$ and $(b_1,\dots b_k)$ (resp. the d-tuples $(1,\dots 1)$ and $(n,\dots n)$).

Program of M {in PASCAL-like style}

```
begin Read(n);
  for ā_d := 1̄^d to n̄^d do
  begin Guess(b̄_k); If b̄_k ∉ [n]^k then Reject
                    else go to inst 1 or 2... or c;
    ......
    inst i: {Comment: i ∈ [c]. Let us describe inst i on an example.
             Assume that ψ_i is x_1 < y_2 ∧ F(x_3,y_1) = x_1.
             inst i: inst i,1: If a_1 < b_2 then go to inst i,2
                                 else Reject;
                     inst i,2: If <a_3,b_1>_F ≠ a_1 and 0 then Reject
                                 else <a_3,b_1>_F := a_1; go to inst c+1}
    ......
    inst c+1: {Comment: ā_d becomes the next d-tuple.}
  end; Accept
end.
```

The loop controlled by $\bar{a}_d$ is executed n^d times. So, M accepts the set S in time $c_1 n^d$, for a constant c_1. □

Remark. An algorithm comparable to the algorithm above is presented in [11] (proposition 3.1) in a different context.

Notation. Let $\tilde{m}$ denote the set of integers $\{0,1,\dots m-1\}$.

Theorem 6. For each integer $d \geq 1$,

$$\mathrm{NRAM}(n^d) \subseteq \mathrm{Sp}_{<}(d\forall).$$

Proof.

It looks like the proof of inclusion $\mathrm{NTIME}(n^d) \subseteq \mathrm{Sp}_{<}(d\forall)$ in [6]. Let M be a simple NRAM which accepts a set of positive integers, S, in time

n^d, for an integer $d \geq 2$. Let inst 0, inst 1,..., inst k-1 be the sequence of instructions of M. We want to encode an accepting computation in a structure of domain $\tilde{n}$. We can require that all integers encountered in it be smaller than n^d. The ith instant (resp. i^{th} integer) can be encoded by the ith d-tuple of $\tilde{n}^d$ in lexicographical order. The type of our structure includes the list of relation and function symbols of figure 1. Each of them has arity d (i.e., has d arguments $v_1,...v_d$; let $\bar{v}$ abbreviate $\bar{v}_d = v_1,...v_d$) and each function symbol has d values.

Symbol	Intuitive meaning
relation I_i, $i<k$	$I_i(\bar{t})$ holds iff the current instruction is inst i
function A	$A(\bar{t})$ is the contents of accumulator a
function $\langle A\rangle$	$\langle A\rangle(\bar{t})$ is the integer read in register $\langle a\rangle$
function $\langle A\rangle'$	$\langle A\rangle'(\bar{t})$ is the integer written in register $\langle a\rangle$
function B	$B(\bar{t})$ is the contents of accumulator b

Figure 1

The argument $\bar{t}$ of the above relation and function symbols intuitively means: "at the instant encoded by $\bar{t}$".

The conjunction φ_0 of sentences ($\forall x\ \exists y\ \ F_{suc}(y) = x$) and ($\forall x \neq 1$) $[x<F_{suc}(x) \wedge F_{suc}(1) = 0]$ defines (on the domain $\tilde{n}$) the constants 0, 1= n-1 and the function F_{suc} such that $F_{suc}(e)$ = e+1 modulo

n, for each $e \in \tilde{n}$. It is easy to define constants 1,2,... by using 0 and F_{suc} and, for r-tuples $\bar{x}_r = (x_1,\ldots x_r)$ and $\bar{y}_r = (y_1,\ldots y_r)$, to define the expression $\bar{x}_r+1$ and the formula $\bar{x}_r < \bar{y}_r$ as abbreviations. The d-tuple $(\bar{0}^{d-2}, 1, 0)$ will be denoted Input. Let $\bar{0}$ and $\bar{1}$ abbreviate the d-tuples $\bar{0}^d$ and $\bar{1}^d$.

For purpose of describing an accepting computation of M, let us give some sentences of the type above.

$$\varphi_1 = [A(\bar{0}) = \bar{0} \wedge \langle A\rangle(\bar{0}) = \bar{0} \wedge B(\bar{0}) = \bar{0} \wedge I_0(\bar{0})].$$

φ_1 describes the initial conditions. The following sentence

$\varphi_2 = \forall \bar{t} \ \bigvee\!\!\!\bigvee_{i<k} I_i(\bar{t})$ (where $\bigvee\!\!\!\bigvee$ is the exclusive disjunction) means that exactly one instruction of the program is current at each instant. Let (Same a)$(\bar{t})$, (Same$\langle a\rangle$)$(\bar{t})$ and (Same b)$(\bar{t})$ abbreviate the formulas $A(\bar{t}+1) = A(\bar{t})$, $\langle A\rangle'(\bar{t}) = \langle A\rangle(\bar{t})$ and $B(\bar{t}+1) = B(\bar{t})$, respectively. Define the formula (Same a,$\langle a\rangle$,b)$(\bar{t})$, as the conjunction of these three formulas. Similarly, define (Same $\langle a\rangle$,b)$(\bar{t})$... Figure 2 associates to each instruction a formula which describes its action.

instruction inst i	formula $\psi_i(\bar{t})$
Read(a)	$A(\bar{t}+1) = \text{Input} \wedge (\text{Same } \langle a \rangle, b)(\bar{t}) \wedge I_{i+1}(\bar{t}+1)$
a := a+1 (similarly for a:= a-1 and a:= 0)	$A(\bar{t}+1) = A(\bar{t})+1 \wedge (\text{Same } \langle a \rangle, b)(\bar{t}) \wedge I_{i+1}(\bar{t}+1)$
Guess(a)	$\exists \bar{y}\ A(\bar{t}+1) = \bar{y} \wedge (\text{Same } \langle a \rangle, b)(\bar{t}) \wedge I_{i+1}(\bar{t}+1)$
a:= ⟨a⟩	$A(\bar{t}+1) = \langle A \rangle(\bar{t}) \wedge (\text{Same } \langle a \rangle, b)(\bar{t}) \wedge I_{i+1}(\bar{t}+1)$
b:= a	$B(\bar{t}+1) = A(\bar{t}) \wedge (\text{Same } a, \langle a \rangle)(\bar{t}) \wedge I_{i+1}(\bar{t}+1)$
⟨a⟩ := b	$\langle A \rangle'(\bar{t}) = B(\bar{t}) \wedge (\text{Same } a, b)(\bar{t}) \wedge I_{i+1}(\bar{t}+1)$
Go to inst p or q	$(\text{Same } a, \langle a \rangle, b)(\bar{t}) \wedge [I_p(\bar{t}+1) \vee I_q(\bar{t}+1)]$
If a=b then go to inst p else go to inst q	$(\text{Same } a, \langle a \rangle, b)(\bar{t}) \wedge [A(\bar{t}) = B(\bar{t}) \rightarrow I_p(\bar{t}+1)] \wedge [A(\bar{t}) \neq B(\bar{t}) \rightarrow I_q(\bar{t}+1)]$
Accept (or Reject)	$I_i(\bar{t}+1)$

Figure 2

Let φ_3 be the sentence $(\forall \bar{t} \neq \bar{1}) \bigwedge_{i<k} [I_i(\bar{t}) \longrightarrow \psi_i(\bar{t})]$.

We want to express such things as the following: if t' is the last time before instant t when register ⟨a⟩ is the same register as at intant t, then the integer read in ⟨a⟩ at instant t is the integer written at instant t' in the same register. For this purpose, we will define a new function symbol, N, which has arity d and 2d values. Let us be more explicit. Define the "current registers" as the couples (a(t), t) where a(t) is the contents of accumulator a at instant t, $t < n^d$. Let c.r.(j) denote the "current register" of rank j ($0 \leqslant j < n^d$) in lexicographical ordering. The intuitive meaning of the expression $N(\bar{x})$ is c.r.(j) where j is the integer encoded by $\bar{x}$ ($0 \leqslant j < n^d$). Clearly, function N is defined by the following sentences:

$$\varphi_4 = \forall \bar{t}\ \exists \bar{x}\ \ (A(\bar{t}), \bar{t}) = N(\bar{x})$$

$$\varphi_5 = (\forall \bar{x} \neq \bar{1})\ \ N(\bar{x}) < N(\bar{x}+1).$$

The following sentence, φ_6, expresses what is the integer read in register ⟨a⟩ at each noninitial instant according as the register has been pointed to before or has not.

$$\varphi_6 = (\forall \bar{t} \neq \bar{0})\ \exists \bar{t}'\ \exists \bar{x}$$
$$[N(\bar{x}) = (A(\bar{t}'), \bar{t}') \wedge N(\bar{x}+1) = (A(\bar{t}), \bar{t})$$
$$\wedge (A(\bar{t}) = A(\bar{t}') \rightarrow \langle A\rangle(\bar{t}) = \langle A\rangle'(\bar{t}'))$$
$$\wedge (A(\bar{t}) \neq A(\bar{t}') \rightarrow \langle A\rangle(\bar{t}) = \bar{0})\ ]$$

Let φ_7 be the sentence $\bigvee I_i(\bar{1})$ where the disjunction extends over all $i < k$ such that inst i = Accept.

Clearly the sentence $\varphi = \bigwedge_{i \leqslant 7} \varphi_i$ is satisfied by a structure on domain $\tilde{n}$ iff M accepts n. That is, $n \in Sp_{<}(\varphi)$ iff $n \in S$. Moreover it is easy to put sentence φ in an equivalent form with only d universal quantifiers.

The general case when S is accepted by a NRAM in time cn^d, for some integers $c,d \geqslant 1$, is similar but moreover it needs an easy manipulation of quantifiers and connectives which is left to the reader.□

4. Corollaries.

Corollary 7. For each integer $d \geq 1$,

$$Sp_<(d\forall) = Sp_<(d\forall, \text{arity } d).$$

Proof. It is sufficient to note that in the proof of inclusion $NRAM(n^d) \subseteq Sp_<(d\forall)$, each relation and function symbol of the sentence φ has arity d.□

From the fact (proved in [6]) that a linear order can be defined by an existential second-order formula of arity 2 with only 2 universal quantifiers, we obtain the

Corollary 8. For each integer $d \geq 2$,

$$NRAM(n^d) = Sp(d\forall) = Sp(d\forall, \text{arity } d).$$

Definition (cf [4]). The generalized spectrum of a (first-order) sentence φ of type $\{<, R, V_1, \ldots V_p\}$ where $<$ and R are specified binary relation symbols, is the set of structures (or graphs) $\mathcal{G}$ such that:

(i) $\mathcal{G}$ is a structure $([m], <, R)$ where $<$ is the natural order of the set of vertices $[m]$;

(ii) $\mathcal{G}$ satisfies the (existential second-order) formula

$$\exists V_1 \ldots \exists V_p \ \varphi.$$

Let $GenSp_<(d\forall)$ be the class of generalized spectra of sentences with d universal quantifiers. Let $NRAM(m^d)$ denote the class of sets of graphs accepted by NRAMs in time cm^d, for a constant c. (Note that $[m]$ is the set of vertices of the input graph.)

Corollary 9. For each integer $d \geq 2$,

$$GenSp_<(d\forall) = NRAM(m^d).$$

Proof. Similar to the corresponding proof for spectra. □

Remark. It follows from corollary 8 that $Sp(d\forall, \text{arity } k) = Sp(d\forall)$

for $k \geq d \geq 2$. It is an open and hard problem to characterize classes Sp(d $\forall$, arity k) (in case $k < d$) and Sp(arity k) in terms of computational complexity.

References

[1] A.V. Aho, J.E. Hopcroft and J.D. Ullman, The design and analysis of computer algorithms, Addison-Wesley, Reading (1974).

[2] S.A. Cook, A hierarchy for nondeterministic time complexity, J. Comput. Systems Sci. 7 (1973), pp. 343-353.

[3] S.A. Cook and R.A. Reckhow, Time bounded Random Access Machines, J. Comput. Systems Sci. 7 (1973), pp. 354-375.

[4] R. Fagin, Generalized first-order spectra and polynomial-time recognizable sets, Complexity of computations, R.M. Karp ed., Amer. Math. Soc., Providence (1974), pp. 43-73.

[5] R. Fagin, A spectrum hierarchy, Z. Math. Logik. Grundlag. Math. 21 (1975), pp. 123-134.

[6] E. Grandjean, The spectra of first-order sentences and computational complexity, to appear in SIAM J. Comput.

[7] N. Immerman, Number of quantifiers is better than number of tape cells, J. Comput. Systems Sci. 22 (1981), pp. 384-406.

[8] N. Immerman, Upper and lower bounds for first-order expressibility, J. Comput. Systems Sci. 25, 1 (1982).

[9] N. Immerman, Languages which capture complexity classes, 15th ACM SIGACT Symp. (1983), pp. 1-16.

[10] N.D. Jones and A.L. Selman, Turing machines and the spectra of first-order formulas with equality, J. Symb. Logic 39 (1974), pp. 139-150.

[11] H.R. Lewis, Complexity results for classes of quantificational formulas, J. Comput. Systems Sci. 21 (1980), pp. 317-353.

[12] L. Lovász and P. Gács, Some remarks on generalized spectra, Z. Math. Logik. Grundlag. Math. 23 (1977), pp. 547-553.

[13] J.F. Lynch, Complexity classes and theories of finite models, Math. Systems Theory 15 (1982), pp. 127-144.

[14] B. Monien, Characterizations of time-bounded computations by limited primitive recursion, 2nd Int. Colloq. Automata Languages Programming, Springer, Berlin (1974), pp. 280-293.

[15] B. Monien, About the derivation languages of grammars and machines, 4th Int. Colloq. Automata Languages Programming (1977), pp. 337-351.

[16] P. Pudlák, The observational predicate calculus and complexity of computations, Comment. Math. Univ. Carolin. 16 (1975), pp. 395-398.

[17] A. Schönhage, Storage Modification Machines, SIAM J. Comput. 9 (1980), pp. 490-508.

[18] J.I. Seiferas, M.J. Fisher and A.R. Meyer, Separating nondeterministic time complexity classes, J. Assoc. Comput. Mach. 25 (1978), pp. 146-167.

[19] R. Weicker, Turing machines with associative memory access, 2nd Int. Colloq. Automata Languages Programming (1974), pp. 458-472.

Second order spectra

Bruno Scarpellini at University of Basel

Introduction

In what follows we discuss second order spectra, in particular complete second order spectra. To start with, we briefly sumarize known facts about first order spectra. Let F be sentence of first order predicate calculus with equality. The spectrum Sp(F) is the set of natural numbers n such that F has model of cardinality n. H. Scholz suggested in [13] to classify spectra from a numbertheoretic point of view. A number of results have been obtained by different authors; we refer to [1],[2],[4],[7],[8], [10],[11],[13] for an incomplete list of relevant papers. In [8], N. Jones and A. Selman proved that for a set S of strings of the form 1^n , $\{ n \mid 1^n \in S \}$ is the spectrum of a first order sentence F iff there is a nondeterministic Turing machine, working in polynomial time, which accepts S . This may be considered as a satisfactory solution of Scholz's problem. In connection with this solution a still open question of G. Asser ([1]) regained actuality: is the complement of a spectrum again a spectrum? By the result of Jones - Selman a negative result would immediately yield a proof of Cooks hypothesis $P \neq NP$ while a positive answer would in any case constitute a breakthrough in complexity theory. Once the notion of spectrum for first order sentences has been introduced it is clear that one can extend this notion to formulas involving higher type predicates and quantifiers, thus giving rise to higher order spectra. Investigations on higher order spectra have been performed by Rödding - Schwichtenberg in [11] and extended by C.A. Christen in his thesis [4].

Complete second order spectra

Among higher order spectra, second order spectra distinguish themselves in that they are related to alternating machines in the same way as first order spectra are related to nondeterministic Turing machines. Let F be a sentence of second order predicate calculus with equality. The spectrum Sp(F) is the set of natural numbers n such that F is true in the universe $D_n = \{ 1, \dots ,n \}$, with individual variables ranging over D_n and k - place predicate variables ranging over the set of all k - place relations with domain D_n . A second order sentence is said to have standard form if it has the form $(Q_0)(Q_1)\dots(Q_k)(I)F'$ where :

(1) each Q_j denotes a string E^* of second order existential quantifiers or a string $\forall^*$ of second order universal quantifiers,

(2) if Q_j is E^* (or $\forall^*$) then Q_{j+1} is $\forall^*$ (or E^* respectively),

(3) I is a string of individual quantifiers,

(4) F' is quantifierfree.

We say that F belongs to the prefix class $(Q_0)(Q_1)\ldots(Q_k)$. It is not difficult to show that for every second order formula there is an equivalent standard sentence. Thus we may henceforth restrict our attention to standard sentences.

Theorem 1 :

Let $F=(Q_0)(Q_1)\ldots(Q_k)(I)F'$ be a standard sentence with $Q_0 = E^*$. Then there is a k-alternating machine T, whose initial state is existential and which works in polynomial time, such that $n \in Sp(F)$ iff 1^n is accepted by T. Conversely, there is for every k-alternating machine T, whose initial state is existential and which works in polynomial time, a standard sentence $F=(Q_0)(Q_1)\ldots(Q_k)(I)F'$ with $Q_0 = E^*$ such that 1^n is accepted by T iff $n \in Sp(F)$. Likewise with $\forall^*$ and "universal" in place of E^* and "existential". ∎

Theorem 1 is known and contained in the literature. One half is contained in [9], where the connection between the polynomial hierarchy and alternating machines is discussed. The other half is contained in [14] (remarks pg. 7-8) where the connection between the polynomial hierarchy and second order spectra is discussed. Below we will describe a direct proof of the second part of the theorem ; its details are then used for further considerations. Prior to this we will focus our attention on our main topic, namely complete spectra.

Definition 1 :

A set $M \subseteq \{1\}^*$ is said to be complete for the prefix class $(Q_0)(Q_1)\ldots(Q_k)$ if there is a mapping $h: \{1\}^* \to \{1\}^*$ such that :

(1) h is 1-1,

(2) the mapping $w \to h(w)$ is polynomial,

(3) $n \in Sp(F)$ iff $h(1^n) \in M$.

The existence of complete first order spectra of various kinds has been proved by R. Fagin in [7] and it has been pointed out by L. Stockmeyer in [14] (remarks pg. 7-8) that this method can be extended in order to obtain complete second order spectra for every prefix type. Our main result is

Theorem 2 :

For all integers $N \geq 1$, $k \geq 0$ and every prefix type $(Q_0)(Q_1)\ldots(Q_k)$ there is a standard sentence F_{Nk} of this prefix type such that $\{1^n \mid n \in Sp(F_{Nk})\}$ is complete with respect to the prefix type $(Q_0)(Q_1)\ldots(Q_k)$ and such that :

(1) there is a k-alternating multitape machine T_{Nk} whose initial state is existential if Q_0 is E^* and universal if Q_0 is $\forall^*$, which accepts the set S, where $S=\{1^n \mid n \in Sp(F_{Nk})\}$ in time $\leq Bn$, with B a constant depending on N, k,

(2) there is a deterministic machine T^*_{Nk} which accepts the set S in time $\leq 2^{d\sqrt[N]{n}}$, with d a constant depending on N, k. ∎

The proof of theorem 2 which is based on a rather involved padding argument and on certain particular alternating machines cannot be reproduced here because of its length. However, a full proof, containing all the details, is given in [12]. Here we stress some points of interest concerning theorem 2. Clause (2) states that there are highly nondeterministic languages, polynomially complete with respect to a certain family of languages, whose deterministic decision problem requires deterministic time $\leq 2^{O(\sqrt[N]{n})}$, that is, asymptotically less then the usual $2^{\lambda n}$. This is a technical fact with seems to be new and which reveals some interest of its own. A second point of interest lies in the possibility that clause (2) is already optimal in the following sense. It is not excluded that for some prefix type $(Q_0)(Q_1)\ldots(Q_k)$ there is a complete spectrum which for some N has $2^{O(\sqrt[N]{n})}$ as deterministic lower bound. In this case it is easy to show that no numbertheoretic function $g(n)$ exists such that:

(a) $\lim g(n)(\sqrt[N]{n})^{-1} = 0$ for all N,

(b) there is a complete spectrum of prefix type $(Q_0)(Q_1)\ldots(Q_k)$ whose deterministic time complexity is $\leq 2^{g(n)}$.

The most interesting aspect of clause (2) is the following. As far as the time complexity of complete sets is concerned there is practically no difference between the different prefix classes. For every prefix type $(Q_0)\ldots(Q_k)$ there is a list $S_0, S_1, \ldots$ of complete spectra of decreasing deterministic time complexity such that if $(Q_0')\ldots(Q_k')$ is any other prefix type with corresponding list $S_0', S_1', \ldots$ there is for every spectrum S_j a spectrum S_i' of asymptotically smaller deterministic time complexity and conversely. Now it is known ([14]) that if there are two different prefix types which yield different spectra classes, then $N \neq NP$ follows, thus generalizing Assers conjecture. The above situation shows that in order to differentiate between different prefix classes one should avoid deterministic time complexity arguments and look instead into other directions.

Clause (1) of theorem 2 is perhaps less important than clause (2). However it admits an interesting corollary, namely

<u>Corollary :</u>

The formulas F_{Nk} in theorem 2 can be chosen so as to contain only one and two place predicate variables. ■

This corollary is mentioned in [12] but not proved there. In the sequel we will give a proof of the second half of theorem 1, and based on this proof, a proof of the corollary.

Alternating machines

We start with a brief description of alternating machines, as far as we need it for the sequel. For details we refer to the basic articles [3], [9]. For simplicity we restrict us to single tape machines, but everything extends without change to multitape machines.

An alternating single tape machine is based on its alphabet, its states and its five tuples which describe the behaviour of the machine. Configurations and successor configurations are introduced in the usual way. For an input word w the computation tree T of M on w is the tree whose nodes are labeled with configurations such that :
(1) the root is labeled with the input configuration,
(2) if the node α is labeled with configuration D then the descendants of α correspond to the successor configurations of D.
We say that T is t-bounded if every path of T has length $\leq t$; the machine M is $t(n)$-bounded if the computation tree T on input word w is $t(|w|)$-bounded. The set of states of M is split into two disjoint subsets :
(a) the existential or $\vee$-states,
(b) the universal or $\wedge$-states.
A configuration is a $\vee$-configuration (a $\wedge$-configuration) if its state is existential (universal); the configuration is accepting if its state is accepting. We assume that if there are $\vee$-states at all, there is exactly one accepting $\vee$-state; likewise with $\wedge$-states. With every node of a finite computation tree we associate a truth value t (truth) or f (false) as follows :
(a) if a node labeled with configuration D has no descendants then its value is t iff D is accepting,
(b) a nonterminating node labeled with a $\vee$-configuration D has value t iff at least one descendant has value t,
(c) a nonterminating node labeled with a $\wedge$-configuration D has value t iff all descendants have value t.
The computation tree is accepting if its root has value t. We call T k-alternating if along every path at most k changes (alternations) from $\vee$-states to $\wedge$-states or conversely occur. A machine M is k-alternating if every computation tree is k-alternating. $\mathrm{Acc}(M)$ is the set of words w such that the computation tree of M on w is bounded and accepting. As in the deterministic case it is routine to show that k-alternating multitape machine M, working in time $t(n)$, can be transformed into a single tape machine M', whose tape is one sided, which works in time $\leq ct(n)^4$ (for some c) and which is k-alternating, such that $\mathrm{Acc}(M) \cap \{0,1\}^* = \mathrm{Acc}(M) \cap \{0,1\}^*$.

Description of alternating machines by second order formulas

(A) Our first aim is to prove the second half of theorem 1 by directly formalizing k-alternating machines by second order formulas. We thereby use the methods described in [5], [8]. Thus let M be a fixed k-alternating machine with alphabet $a_o, \dots, a_\alpha$ ($a_o = b = \text{blank}$, $a_1 = 1$, $a_2 = 0$) and states $q_o, \dots, q_\beta$. For simplicity we assume that

M has a single, one sided tape and that M works within time $< n^s$ for some fixed integer s. The general case where M has several, twosided tapes and works within time $< An^s$ (for some $A > 1$) can be handled either by reducing it to previous simpler one or or else by extending the method of Jones - Selman in such a way as to cover the general case. The second way will be appropiate for the proof of the corollary of theorem 2. The set $q_o, \ldots, q_\beta$ is split into two subsets $q_o, \ldots, q_m$ and $q_{m+1}, \ldots, q_\beta$ with $q_o, \ldots, q_m$ existential and $q_{m+1}, \ldots, q_\beta$ universal if the initial state q_o is existential, and with $q_o, \ldots, q_m$ universal and $q_{m+1}, \ldots, q_\beta$ existential otherwise. q_m, q_β are the accepting states.

(B) In order to formalize the behaviour of M we prefer to use the total order $<$ of individuals as a predicate constant instead of equality $=$. Thus if we interpret our formulas in a domain D_n of n individuals we assume D_n to be $\{1,\ldots,n\}$ and interpret $<$ as the total order on integers, restricted to D_n. With the aid of $<$ we can define the lexicographical ordering $<_s$ of s-tuples of individuals in a familiar way. We use x, y etc. as short for s-tuples $x_1, \ldots, x_s$, $y_1, \ldots, y_s$ of variables. As in [12] we introduce three types of predicate variables with describe the behaviour of M, namely:

(1) $L(x,y)$, which states that the reading head is at time x at place y,

(2) $S_j(x)$, $(j = 0,\ldots,\beta)$, which states that at time x the machine is in state q_j,

(3) $B_j(x,y)$, $(j = 0,\ldots,\alpha)$, which states that at the time x the tape contains at place y the symbol a_j.

A collection L, S_j, B_i $(j = 0,\ldots,\beta$, $i = 0,\ldots,\alpha)$ of predicate variables of the form (1), (2), (3) is called a "computation path" and denoted by P. By $P(x)$ we denote symbolically the configuration of P at time point x. By P' we denote any collection L', S'_j, B'_i $(j = 0,\ldots,\beta$, $i = 0,\ldots,\alpha)$ having the form (1), (2), (3). We also use S and B as short for the collections S_i, $i = 0,\ldots,\beta$ and B_j, $j = 0,\ldots,\alpha$ respectively. By proceding as in [12] we find a closed first order formula $H(<,L,S,B)$ (or more briefly $H(<,P)$) with the property:

(*) $H(<,P)$ is true in the domain $D_n = \{1,\ldots,n\}$ for some choice of the predicates L, S_j, B_i if and only if $P = (L,S,B)$ describes a computation path of the machine M of time length $< n^s$.

In order to formalize the behaviour of M we need a series of first order formulas which express properties of computation paths P, P' which are tacitly assumed to satisfy $H(<,P)$, $H(<,P')$. This formulas are introduced by the clauses below.

(I) $(P/x/P')$ expresses that the two computations paths P, P' coincide for all time points smaller or equal to x,

(II) $(P,\vee)_x^y$ expresses that for all time points z with $x \leq_s z <_s y$ the state of the configuration $P(z)$ is existential,

(III) $(P,\wedge)_x^y$ is likewise, with "universal" in place of "existential",

(IV) $(P,\vee)_x^\infty$ expresses that for all time points z with $x \leq_s z$ the state of the configuration $P(z)$ is existential,

(V) $(P, \wedge)^{\infty}_{x}$ is likewise, with "universal" in place of "existential",

(VI) $(P, \vee)_x$ expresses that the state of configuration $P(x)$ is existential,

(VII) $(P, \wedge)_x$ is likewise, with "universal" in place of "existential",

(VIII) $Acc(P)$ expresses that the state of the last configuration of P is accepting.

With the aid of formulas (I) - (VIII) we introduce second order formulas $G(h, \wedge, P, x)$ and $G(h, \vee, P, x)$ whose semantical properties are as follows. Assume $H(<,P)$ to be true in D_n for certain choice of L, S, B. Let $\xi = (\xi_1, \ldots, \xi_s)$ be an s-tuple of integers in D_n such that the following holds :

(A) the subtree of the computation tree of M, which has its root at the configuration $P(\xi)$, is h-alternating. Then the following holds :

(B) $G(h, \vee, P, \xi)$ is true iff the state of the configuration $P(\xi)$ is $\vee$ and if the truth value of $P(\xi)$ is t,

(B*) $G(h, \wedge, P, \xi)$ is true iff the state of $P(\xi)$ is $\wedge$ and the truth value of $P(\xi)$ is t.

We now proceed to the inductive definition of the G's, such that (B), (B*) hold whenever (A) is satisfied. For $G(0, \vee, P, x)$ we take the formula :

$$(P,\vee)_x \wedge H(<,P) \wedge (EP')(H(<,P') \wedge (P/x/P') \wedge (P',\vee)^{\infty}_{x} \wedge Acc(P')) .$$

For $G(o, \wedge, P, x)$ we take the formula :

$$(P,\wedge)_x \wedge H(<,P) \wedge (\forall P')(H(<,P') \wedge (P',\wedge)^{\infty}_{x} \wedge (P/x/P') \longrightarrow Acc(P')) .$$

We now assume that $G(h-1, \vee, P, x)$ and $G(h-1, \wedge, P, x)$ with the desired properties have already been defined. For $G(h, \vee, P, x)$ we take the formula :

$$H(<,P) \wedge (P,\vee)_x \wedge \{ (EP')(H(<,P') \wedge (P/x/P') \wedge (P',\vee)_x \wedge Acc(P')) . \vee .$$
$$(EP')(Ey)(H(<,P') \wedge (P/x/P') \wedge x <_s y \wedge (P',\wedge)_y \wedge (P',\vee)^{y}_{x} \wedge G(h-1,\wedge,P',y))\} .$$

For $G(h, \wedge, P, x)$ we take the formula :

$$H(<,P) \wedge (P,\wedge)_x \wedge (\forall P') \{ H(<,P') \wedge (P/x/P') \longrightarrow$$
$$((P',\wedge)^{\infty}_{x} \wedge Acc(P') \vee (Ey)(x <_s y \wedge (P',\wedge)^{y}_{x} \wedge (P',\vee)_y \wedge G(h-1,\vee,P',y))\}$$

We omit the easy verification that $G(h, \vee, P, x)$, $G(h, \wedge, P, x)$ satisfy (B), (B*) under assumption (A).

<u>Lemma 1 :</u>

(A) By moving quantifiers to the left, $G(h, \vee, P, x)$ can be transformed into a formula $G'(h, \vee, P, x)$ in prenex normal form with prefix structure $(Q_0)(Q_0')(Q_1)(Q_1')\ldots(Q_h)(Q_h')(I)$ where : (1) Q_j and Q_j' are either both E^* or both $\forall^*$, with Q_j second order quantifiers, Q_j' first order quantifiers, and with Q_0, Q_0' of type E^*,

(2) (I) denotes a series of first order quantifiers.

(B) Likewith with $G(h, \wedge, P, x)$, but with Q_0', Q_0 of type $\forall^*$. ■

The proof is by a straightforward induction with respect to h, based on the inductive definition of the formulas G. From lemma 1 we infer :

<u>Lemma 2 :</u>

(A) For h, $0 \leq h \leq k$ there is a standard formula $(Q_0)(Q_1)\ldots(Q_h)G^*(h, \vee, P, x)$ of prefix type $(Q_0)(Q_1)\dots(Q_h)$, with G^* quantifierfree and $Q_0 = E^*$, which is logically equivalent to $G(h,\vee,P,x)$.

(B) Likewise with $\wedge$ and $\forall^*$ in place of $\vee$ and E^* respectively.

Proof : According to R.Büchi (lemma 5.4 in [6]) we have the following identities which hold in every nonempty domain :

$$(Ex)(\forall Y)F \longleftrightarrow (EZ)(\forall Y)(\forall x)((Z(x) \longrightarrow F) \wedge (Es)Z(s)) \quad , \qquad (1)$$

$$(\forall x)(EY)F \longleftrightarrow (\forall Z)(EY)Ex)((Es)G(s) \longrightarrow (G(x) \wedge F)) \quad . \qquad (2)$$

The identities (1), (2), applied to the formulas $G'(h,\vee,P,x)$ and $G'(h,\wedge,P,x)$ of lemma 2 permit us to move successively all individual quantifiers to the right. After a number of steps and a final passage to prenex normal form we end up with equivalent standard formulas $(Q_0)(Q_1)\dots(Q_h)(I)G^*(h,\vee,P,x)$ and $(Q_0)(Q_1)\dots(Q_h)(I)G^*(h,\wedge,P,x)$ respectively whereby $Q_0 = E^*$ in the first case and $Q_0 = \forall^*$ in the second case. ∎

In a last step, we have to take the input into consideration. Let R be a new one place predicate variable. Then it is easy to construct a first order sentence $I(R,P,<)$, involving only R,L,S,B and $<$, whose properties are as follows. Let $H(<,P)$ be true in D_n and R a one place predicate over D_n. Let ξ_R be a 0,1 - string of length n such that $\xi_R(j) = 1$ iff $R(j)$ is true ($j = 1,\dots,n$). Then $I(R,P,<)$ is true if and only if the first (i.e. initial) configuration of the computation path P has the following structure :

(1) the tape inscript coded by $B_j(1,\dots,1,n_1,\dots,n_s)$, $n_1,\dots,n_s \in D_n$, $j = 0,\dots,\alpha$ starts with ξ_R followed by a string of blanks, i.e. is $\xi_R b^\gamma$, $\gamma = n^s - n - 1$,

(2) the state of the configuration is q_0 , i.e. $S_0(1,\dots,1)$ holds,

(3) the reading head is in its initial position, i.e. $L(1,\dots,1,1,\dots,1)$ is true.

By combining the formula $I(R,P,<)$ with the formula from lemma 2 we obtain

Theorem 3 :

Let M be a k - alternating, one sided, single tape machine, working in time $< n^s$, whose initial state is existential. There is a standard formula $G_M(R)$ of prefix type $(Q_0)(Q_1)\dots(Q_k)$, $Q_0 = E^*$, whose only free variable is the one place predicate variable R , all whose variables have at most 2s argument, which has the following property : (*) given a one place predicate R over the domain D_n , the formula $G_M(R)$ is true in D_n iff M accepts the 0,1 - string ξ_R , where $\xi_R(j) = 1$ iff $R(j)$ is true. Likewise with $\wedge$ and $\forall^*$ in place of $\vee$ and E^*.

Proof : Consider eg. the case where the initial state of M is universal. Let $N(<,x)$ be a first order formula expressing that $x\ (=(x_1,\dots,x_s))$ is the smallest element with respect to the lexicographical ordering $<_s$. Consider the formula

$$(\forall P)(\forall x)(H(<,P) \wedge I(R,P,<) \wedge N(<,x) \longrightarrow (Q_0)(Q_1)\dots(Q_k)(I)G^*(k,\wedge,P,x)) \quad ,$$

where the righthandside of the implication is the formula provided by lemma 2. This formula, denoted by $G'_M(R)$, is easily seen to have the semantical property (*) stated in the theorem. We now move the second order quantifiers $(Q_0)(Q_1)\dots(Q_k)$ to the left.

Afterwards we move the first order quantifiers $(\forall x)$ to the right proceeding as in the proof of lemma 2. After a final passage to prenex normal form we obtain a formula $G_M(R)$ which has the desired properties of the theorem. The case where the initial state is existential is handled similarly. ■

(C) In order to prove the corollary to theorem 2 it is necessary to modify the formal frame used in the proof of theorem 3. Below we describe the modifications which are necessary in order to convert the proof of theorem 3 into a proof of the corollary to theorem 2.

To start with, let M be a k-alternating multitape machine which works in time $\leq A\cdot n$, with A an integer >1. Let $a_0,\dots,a_\alpha$ be the alphabet and $q_0,\dots,q_\beta$ the states of M with splitting $\{q_0,\dots,q_m\}$, $\{q_{m+1},\dots,q_\beta\}$ into, say, existential and universal states; M has p tapes, numbered from 1 to p. We introduce three types of predicate symbols:

(a) for every $i\in\{0,\dots,\beta\}$, $t\in\{0,\dots,A\}$ a one place predicate symbol $S(t,i/x)$,

(b) for every $\tau\in\{-A,\dots,A\}$, $t\in\{0,\dots,A\}$, $k\in\{1,\dots,p\}$ a two place predicate symbol $L(\tau,t,k/x,y)$,

(c) for every $\tau\in\{-A,\dots,A\}$, $t\in\{0,\dots,A\}$, $j\in\{0,\dots,\alpha\}$, $k\in\{1,\dots,p\}$ a two place predicate symbol $B(\tau,t,k,j/x,y)$. Let $D_n=\{1,\dots,n\}$ be given; points of time are now represented by ordered pairs (t,ξ), $t\in\{0,\dots,A\}$, $\xi\in D_n$, which are ordered lexicographically; thus $(0,1)$ represents time zero. Similarly we assume the squares of each tape to be labeled by ordered pairs (τ,η), $\tau\in\{-A,\dots,A\}$, $\eta\in D_n$, which are ordered lexicographically; thus $(0,1)$ is the initial square, (A,n) the rightmost square and $(-A,n)$ the leftmost square. The interpretation of these symbols is as follows:

(1) $S(t,i/\xi)$ states that at time (t,ξ) the machine is in state q_i,

(2) $L(\tau,t,k/\xi,\eta)$ $(\xi,\eta\in D_n)$ states that at time (t,ξ) the reading head of tape k is on the square with label (τ,η),

(3) $B(\tau,t,k,j/\xi,\eta)$ states that at time (t,ξ) the square with label (τ,η) on tape k contains the symbol a_j. With this interpretation in mind it is easy to write down formulas whose conjunction states that the predicates $S(t,i/x)$, $L(\tau,t,k/x,y)$ and $B(\tau,t,k,j/x,y)$ describe a computation path of the machine M. Below the indices t,τ,k,i,j run through $\{0,\dots,A\}$, $\{-A,\dots,A\}$, $\{1,\dots,p\}$, $\{0,\dots,\beta\}$ and $\{0,\dots,\alpha\}$ respectively.

(I) First we state that at every time point (t,x) the machine is in exactly one state:

$$\bigwedge_t (\forall x)\Big(\bigvee_i S(t,i/x)\wedge \bigwedge_{i\neq i'}(\neg S(t,i/x)\vee \neg S(t,i'/x))\Big) .$$

(II) Next we state that at every time (t,x) there is on every square of every tape at least one a_j: $\bigwedge_t\bigwedge_\tau\bigwedge_k(\forall x)(\forall y)\big(\bigvee_j B(\tau,t,k,j/x,y)\big)$.

(III) We have to state that at every time (t,x) there is on every square of every tape at most one a_j:

$$\bigwedge_t\bigwedge_\tau\bigwedge_k\bigwedge_{j\neq j'}(\forall x)(\forall y)(\neg B(\tau,t,k,j/x,y)\vee \neg B(\tau,t,k,j'/x,y)) .$$

(IV) We express that at each time (t,x) the reading head of each tape is on at least one square : $\bigwedge_t \bigwedge_k (\forall x)(Ey)(\bigvee_\tau L(\tau,t,k/x,y))$.

(V) We express that at each time (t,x) the reading head of each tape is on at most one square by the conjunction of the following two formulas :

$\bigwedge_t \bigwedge_k \bigwedge_{\tau\neq\tau'} (\forall x)(\forall y)(\forall y')(\neg L(\tau,t,k/x,y) \vee \neg L(\tau,t,k/x,y'))$,

$\bigwedge_t \bigwedge_k \bigwedge_\tau (\forall x)(\forall y)(\forall y')(y\neq y' \longrightarrow (\neg L(\tau,t,k/x,y) \vee \neg L(\tau,t,k/x,y'))$.

(VI) Next we need a formula which formalizes the following. Let (t',y) be the time point which immediately succeeds (t,x). Assume that at time (t,x) we have :

(1) the reading head of tape k is not at place (τ,z) ,

(2) place (τ,z) of tape k is occupied by the symbol a_j . Then place (τ,z) of tape k is occupied by the symbol a_j at time (t',y). In order to obtain such a formula we need three first order formulas s(x,y), min(x), max(x) , expressing that y is the immediate successor of x , that x is minimal, respectively maximal with respect to < . Our formula is the conjunction of the two formulas :

$\bigwedge_k \bigwedge_\tau \bigwedge_t \bigwedge_j (\forall z)(\forall x)(\forall y)(\mathrm{succ}(x,y) \wedge \neg L(\tau,t,k/x,z) \wedge B(\tau,t,k,j/x,z).$

$\longrightarrow B(\tau,t,k,j/y,z))$,

$\bigwedge_k \bigwedge_\tau \bigwedge_{t<A} \bigwedge_j (\forall z)(\forall x)(\forall y)(\max(x) \wedge \min(y) \wedge \neg L(\tau,t,k/x,z) \wedge B(\tau,t,k,j/x,z).$

$\longrightarrow B(\tau,t+1,k,j/y,z))$.

(VII) Finally we need a formula which describes the relationship between a configuration and its successor configurations on the basis of the set of 3p+2 - tuples which governs the dynamics of the machine M . It is clear that it is only a matter of routine to write down such a formula, which is simply a straightforward extension of formula (23) on pg. 118 in [5]. However, since the formula in question would occupy several lines we refrain from writing it down and refer to it simply as formula (VII).

To sum up, let H(< ,P) be the conjunction of the formula I,...,VII , with P denoting symbolically the set S(t,i/x), L(τ,t,k/x,y), B(τ,t,k,j,/x,y) of predicate symbols. If H(< ,P) is true in D_n for some choice of P , then P describes a computation path of the machine M of time length $\leq (A+1)n - 1$, in the sense of the interpretation of P explained at the beginning. Conversely, every such computation path can be described by a suitable interpretation of P . With the formula H(< ,P) at hand we now proceed in exactly the same way as in the proof of theorem 3 , with the difference that at the end of the procedure we replace the one place variable R(x) by the formula x = x . The result is a closed formula F in standard form, of prefix type $(Q_0)(Q_1)\ldots(Q_k)$, with $Q_0 = E^*$ or $= \forall^*$ according to wether q_0 is existential or universal, such that:

(*) F is true in D_n iff M accepts 1^n ,

(**) F contains only one or two place predicate variables. The corollary to theorem 2 now follows by combining (*), (**) with clause (1) of theorem 2.

<u>Remark</u> In [12], the following slightly weaker form of theorem 2 is proved :

Theorem 2* :

There is a integer > 0 with the following properties. For every $N \geq 1$, $k \geq 0$ and every prefix type $(Q_0)(Q_1)\ldots(Q_k)$ there is a standard sentence F_{Nk} of this prefix type such that $S = \{1^n \mid n \in Sp(F_{Nk})\}$ is complete with respect to $(Q_0)(Q_1)\ldots(Q_k)$, such that (1) of theorem 2 holds and such that there is a deterministic machine T^*_{Nk} which accepts S in time $\leq 2^{d \sqrt[N]{n^{\varepsilon}}}$. ■

Our theorem 2 then follows by taking for F_{Nk} the formula $F_{\varepsilon N,k}$ of theorem 2*.

References

1. Asser, G. : Das Representantenproblem im Prädikatenkalkül erster Stufe mit Identität. Zeitschrift für math. Logik u. Grundl. d. Math., 1 (1955) 252-263.
2. Bennet, J. : On spectra. Doctoral Dissertation, Princton Univ.,Princeton, 1962.
3. Chandra, A. and Stockmeyer, L. : Alternation. 17 th. IEEE - Focs (1976) 98 - 108.
4. Christen, C.A. : Spektren und Klassen elemtarer Funktionen. Dissertation 5330, ETH Zürich, 1974.
5. Christen, C.A. : Spektralproblem und Komplexitätstheorie. Lecture Notes in Computer Science 43 (1976) 102 - 126.
6. Eglot, C. : Decision problems of finite automa design and related arithmetics. Trans AMS 98 (1961) 21 - 51, 103 (1962).
7. Fagin, R. : Generalized first - order spectra and polynomial time recognizable sets. SIAM - AMS Proc. 7 (1974) 43 - 73.
8. Jones, N.D. and Selman, A.L. : Turing machines and the spectra of first order formulas. J. Symbolic Logic 39 (1974) 139 - 150.
9. Kozen, D. : On parallelism in Turing machines. 17 th . IEEE - Focs (1976) 89 - 97.
10. Mostowski, A. : Concerning a problem of H. Scholz. Zeitschrift f. math. Logik u. Grundl. d. Math. 2 (1956) 210 - 214.
11. Rödding, D. und Schwichtenberg, H. : Bemerkungen zum Spektralproblem. Zeitschrift f. math. Logik u. Grundl. d. Math. 18 (1972) 1 - 12.
12. Scarpellini, B. : Complete second order spectra. To appear in Zeitschrift f. math. Logik u. Grundl. d. Math. .
13. Scholz, H. : Ein ungelöstes Problem in der symbolischen Logik. J. Symbolic Logic 17 (1952) 160.
14. Stockmeyer, L. : The polynomial - time hierarchy. Theor. Comp. Sci. 3 (1977) 1 - 22.

ON THE ARGUMENT COMPLEXITY OF MULTIPLY TRANSITIVE BOOLEAN FUNCTIONS

U. Hedtstück
Institut für Informatik
Universität Stuttgart
D-7000 Stuttgart 1

Introduction

We shall discuss the complexity of Boolean functions under the following aspect: Let $P:\{0,1\}^d \to \{0,1\}$ $(d \in \mathbb{N})$ be a Boolean function and $\mathfrak{X} \in \{0,1\}^d$. How many components x_i of $\mathfrak{X}$ does an optimal decision algorithm for P require in the worst case to decide whether $P(\mathfrak{X})=1$? This number is called the argument complexity of P denoted by C(P). Aanderaa and Rosenberg conjectured [4]: For nontrivial monotone properties P of graphs represented by their adjacency matrices, $C(P)=\Omega(d)$ (i.e., there exists some $c>0$ such that $C(P) \geq c \cdot d$ for all P,d under consideration). The Aanderaa-Rosenberg conjecture was proved by Rivest and Vuillemin [5]. They also stated the following more general conjecture: If P is invariant under the action of a transitive permutation group on the set $\{1,\dots,d\}$ and if $P(\mathbf{0}) \neq P(\mathbf{1})$, where $\mathbf{0}$ (respectively $\mathbf{1}$) represents the d-bit vector of zeroes (respectively ones), then C(P)=d. They were able to prove this for d a prime power (and some other cases), whereas the general case was disproved by Illies [2]. He constructed a counterexample for d=12 using the cyclic group C_{12} on $\{1,\dots,12\}$. In this paper we shall discuss the following question: Does the Rivest-Vuillemin conjecture hold if we assume multiply transitivity? We present a lower bound on the degree of transitivity depending on d by showing that k-transitivity with $k > \frac{d}{3}+1$ is sufficient for the Rivest-Vuillemin conjecture. In addition we shall make some investigations about the existence of a universal lower bound.

Definitions

Let S_d (respectively A_d) denote the symmetric (respectively alternating) group on $\{1,\dots,d\}$. Let's recall that a permutation group $G \leq S_d$ is said to be <u>k-transitive</u> $(1 \leq k \leq d)$ if, given any two ordered k-tuples $i_1,\dots,i_k$ and $j_1,\dots,j_k$ of distinct elements of $\{1,\dots,d\}$ (i.e., $i_\nu \neq i_\mu$, $j_\nu \neq j_\mu$ for $\nu \neq \mu$), there is some g in G such that $g(i_\nu)=j_\nu$ $(1 \leq \nu \leq k)$.

For k=1 we have the wellknown transitivity.

Let $P:\{0,1\}^d \to \{0,1\}$ be a Boolean function. For $x = x_1 \ldots x_d \in \{0,1\}^d$ and $\sigma \in S_d$ we define $\sigma(x) := x_{\sigma(1)} \ldots x_{\sigma(d)}$. $\Gamma(P)$ denotes the stabilizer of P: $\Gamma(P) := \{\sigma \in S_d \mid \forall x \in \{0,1\}^d \; P(x) = P(\sigma(x))\}$. Obviously, $\Gamma(P)$ is a group.

The Boolean function P is said to be k-transitive whenever $\Gamma(P)$ is itself k-transitive. Clearly, k-transitivity implies l-transitivity for $1 \leq l \leq k$. P is nontrivial, if it is onto. The weight $w(x)$ of a vector x is the number of ones in x. $\mathbf{0}$ denotes the vector 0...0, $\mathbf{1}$ denotes the vector 1...1.

Decision trees

Let T be a "decision tree" whose internal nodes are labelled with elements of $\{1,\ldots,d\}$ and whose leaves are labelled with one or zero. T may be viewed as an algorithm for the computation of a Boolean function $P:\{0,1\}^d \to \{0,1\}$ in the following sense: Let $x \in \{0,1\}^d$ be the input vector. If the root of T is labelled with i, then the algorithm determines the i-th component x_i of x. If $x_i = 0$, the algorithm continues with the component of x specified at the root of the left subtree of T, otherwise it proceeds to the right and so on. Thus for each input vector x there is a unique path from the root to a leaf. The label of this leaf specifies the value of P for x.

Let $c(T,x)$ denote the number of components of x which have to be determined using T for the computation of $P(x)$. We define $c(T) := \max\{c(T,x) \mid x \in \{0,1\}^d\}$ and $C(P) := \min\{c(T) \mid T$ is a decision tree for $P\}$. $C(P)$ is called the argument complexity of P. If $C(P)=d$, P is said to be exhaustive (some authors use the terms "elusive" or "evasive").

Results

Lemma: If the Boolean function $P:\{0,1\}^d \to \{0,1\}$ is k-transitive, then $P(x) = P(y)$ for any two vectors $x, y \in \{0,1\}^d$ with $w(x) = w(y) \leq k$ or with $w(x) = w(y) \geq d-k$.

Proof: Let r be such that $w(x)=w(y)=r\leq k$. We pick out the ones in x and y: Let $x_{i_1},\ldots,x_{i_r}$ and $y_{j_1},\ldots,y_{j_r}$ be all the components of x and y, respectively, with $x_{i_\nu}=1$ and $y_{j_\nu}=1$, $1\leq\nu\leq r$. Since P is k-transitive, i.e., $\Gamma(P)$ is k-transitive, and $r\leq k$, there is a $\sigma\in\Gamma(P)$ such that $\sigma(i_\nu)=j_\nu$, $1\leq\nu\leq r$. Hence $x_i=y_{\sigma(i)}$ and we obtain $P(x)=P(x_1\ldots x_d)=P(y_{\sigma(1)}\ldots y_{\sigma(d)})=P(\sigma(y))=P(y)$.

In the case $r\geq d-k$ we have to pick out the zeroes and then proceed analogously.

Remark: It is easy to see, that the Lemma holds even if we take a weaker form of k-transitivity where we consider k-element sets of indices instead of ordered k-tuples.

Theorem 1: Let $P:\{0,1\}^d\to\{0,1\}$, $d\geq 3$, be a nontrivial Boolean function. If $A_d\leq\Gamma(P)$, then $C(P)=d$.

Proof: We know, that A_d is (d-2)-transitive whenever $d\geq 3$. The Lemma implies, that $P(x)=P(y)$ for all $x,y\in\{0,1\}^d$ with $w(x)=w(y)$. Thus the value of P for a vector x only depends on the number of ones in x. Let us suppose, that $C(P)<d$. Then there exists a full decision tree for P with height d-1, that is a decision tree in which each internal node has exactly two sons and all paths have length d-1. Each leaf L in T represents exactly two vectors x_L and y_L with $w(x_L)=w(y_L)-1$ and $P(x_L)=P(y_L)$. Thus for each number i, $0\leq i\leq d-1$, there exists a leaf L(i) in T with $w(x_{L(i)})=i$. The neighbour $y_{L(i)}$ has weight $w(x_{L(i)})+1$ whereas $P(x_{L(i)})=P(y_{L(i)})$. Since the value of P for a vector x only depends on the weight of x, we can now show by induction on i, that all vectors with weight i have the same value under P as all vectors with weight $i+1$, $i=0,\ldots,d-1$. We have P is constant, which is a contradiction to the assumption that P is nontrivial.

Lower bounds

Theorem 2: If the Boolean function $P:\{0,1\}^d\to\{0,1\}$ is nontrivial and k-transitive, $k>\frac{d}{3}+1$, then $C(P)=d$.

Proof: From group theory we know, that if a permutation group $G \leq S_d$ is k-transitive, $k > \frac{d}{3}+1$, then $A_d \leq G$. See for example [1]. The result follows from Theorem 1. (The cases d=1,2 can be handled directly.)

Now we have a lower bound for k depending on d, such that k-transitivity is sufficient for the exhaustiveness of P. In the following we consider Boolean functions P with the more restrictive property $P(\mathbf{0}) \neq P(\mathbf{1})$, which is assumed in the Rivest-Vuillemin conjecture, and ask: Does there exist a least universal k_0 such that each k_0-transitive Boolean function P with $P(\mathbf{0}) \neq P(\mathbf{1})$ is exhaustive?

To exclude a natural number n for being such a universal lower bound k_0 it is sufficient to have an n-transitive Boolean function P with $P(\mathbf{0}) \neq P(\mathbf{1})$ and $C(P) < d$. As n-transitivity implies m-transitivity for $1 \leq m \leq n$, excluding n by a counterexample excludes automatically all m, $1 \leq m \leq n$. Thus we are looking for groups, which, being the stabilizer of a Boolean function P, are able to yield a counterexample. Now we shall show that some classes of groups are not suitable to yield counterexamples.

The Rivest-Vuillemin conjecture states that 1 is a candidate for k_0, whereas the counterexample of Illies [2] excludes 1. On the other hand, except the symmetric and alternating groups, the "trivial" groups, there is no 6-transitive group known. It is believed, that each 6-transitive group is symmetric or alternating. Assuming this (C_6), we can make the following applications of our Theorems:

Proposition 1 (C_6): There exists a least k_0, $2 \leq k_0 \leq 6$, such that if a Boolean function P with $P(\mathbf{0}) \neq P(\mathbf{1})$ is k_0-transitive, then P is exhaustive.

Remark: The Proposition remains valid, if we replace the condition "$P(\mathbf{0}) \neq P(\mathbf{1})$" by the condition "nontrivial".

Proposition 2 (C_6): If there exists a k-transitive Boolean function $P:\{0,1\}^d \rightarrow \{0,1\}$ with $2 \leq k \leq 6$ and $P(\mathbf{0}) \neq P(\mathbf{1})$, such that $C(P) < d$, then either k=2 or k=3 or $\Gamma(P)$ is one of the 5-transitive Mathieu groups M_{12} and M_{24}.

Proof: According to our assumption C_6, there are only two nontrivial 5-transitive groups, the Mathieu groups M_{12} and M_{24}, and two nontrivial 4-transitive groups, which are not 5-transitive, the Mathieu groups M_{11} and M_{23} [6]. The degrees of M_{11} and M_{23} are 11 and 23, respectively, which are primes. Rivest and Vuillemin proved their conjecture for d a prime power. Hence M_{11} and M_{23} cannot yield a counterexample.

It follows from Proposition 2, that a 4-transitive group, yielding a counterexample, must also be 5-transitive. In other words: If 5-transitivity suffices for the exhaustiveness of Boolean functions, then so does 4-transitivity. Since we are looking for a least lower bound, we have:

Corollary: $k_0 \neq 5$.

2-transitivity

Let us now consider those 2-transitive groups, which are not suitable to yield a counterexample for $k_0=2$. First we need a new definition:

Definition: A permutation group G on $\{1,\dots,d\}$ is sharply k-transitive, $1 \leq k \leq d$, if for any two ordered k-tuples of distinct elements of $\{1,\dots,d\}$ there is exactly one g in G which takes the first into the second.

Lemma: If $P:\{0,1\}^d \rightarrow \{0,1\}$ is a Boolean function with $P(\mathbf{0}) \neq P(\mathbf{1})$ and $\Gamma(P)$ is sharply 2-transitive, then $C(P)=d$.

Proof: We use a result from group theory: If G is a sharply 2-transitive permutation group of degree d, then d is a prime power [6]. Hence the result of Rivest and Vuillemin implies the Lemma.

Jordan has shown, that the only sharply 5-transitive group is the Mathieu group M_{12} and that the only sharply 4-transitive group is the Mathieu group M_{11} [3] . Now we can graphically demonstrate, for which groups we have shown, that they cannot yield a counterexample in our sense:

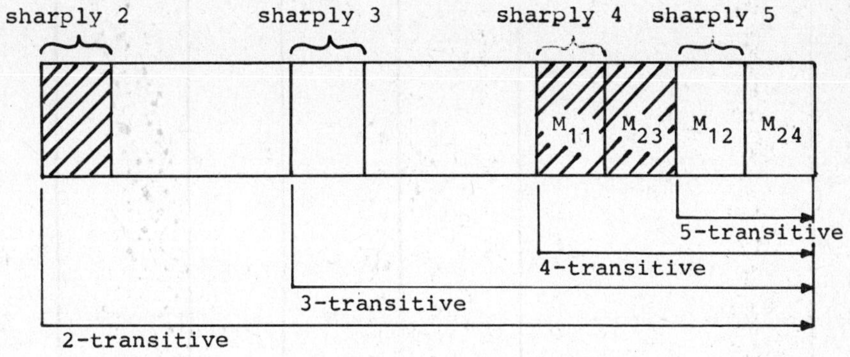

Conclusion

What we have done is just excluding some kinds of groups for being the stabilizer of a nonexhaustive Boolean function. The next step would be to investigate, if in the remaining set of groups there exists a group, which yields a counterexample. If there is no such group, then 2-transitivity of a Boolean function P with $P(\mathbf{0}) \neq P(\mathbf{1})$ would imply the exhaustiveness of P.

References

[1] R. D. Carmichael, Groups of finite order, Boston, Ginn&Co, 1937.

[2] N. Illies, A counterexample to the generalized Aanderaa-Rosenberg conjecture, Inform. Proc. Lett. 7(1978), no. 3, 154-155.

[3] C. Jordan, Sur la limite de transitivite des groupes non alternes, Bull. Soc. Math. France 1 (1873), 40-71.

[4] A. L. Rosenberg, On the time required to recognize properties of graphs: A problem, SIGACT News 5(1973), 15-16.

[5] R. L. Rivest, J. Vuillemin, On recognizing graph properties from adjacency matrices, Theor. Comp. Sci. 3(1976), 371-384.

[6] H. Wielandt, Finite permutation groups, New York - London, Academic Press, 1964.

THE VLSI COMPLEXITY OF BOOLEAN FUNCTIONS

M.R. Kramer and J. van Leeuwen

Department of Computer Science, University of Utrecht
P.O. Box 80.012, 3508 TA Utrecht, the Netherlands

Abstract. It is well-known that all Boolean functions of n variables can be computed by a logic circuit with $O(2^n/n)$ gates (Lupanov's theorem) and that there exist Boolean functions of n variables which require logic circuits of this size (Shannon's theorem). We present corresponding results for Boolean functions computed by VLSI circuits, using Thompson's model of a VLSI chip. We prove that all Boolean functions of n variables can be computed by a VLSI circuit of $O(2^n)$ area and period 1, and we prove that there exist Boolean functions of n variables for which every (convex) VLSI chip must have $\Omega(2^n)$ area.

Keywords and phrases: logic circuit, Boolean function, Lupanov's theorem, Shannon's theorem, VLSI, chip, area, period.

1. Introduction.

The requirements for VLSI circuits are not as relaxed as they are for logic circuits. For example, a VLSI circuit must be embedded in the plane and occupy only a small amount of area, counting what it takes both for gates and wires. Also there can be no unbounded fan-in or fan-out, as basic switching elements can handle only a small number of wires. In this paper we study the quantitative effect of these differences on the "circuit complexity" of arbitrary Boolean functions.

It is well-known that all Boolean functions of n variables can be computed by a logic circuit of $O(2^n/n)$ gates (Lupanov [2]) and that there exist Boolean functions of n variables for which any logic circuit actually requires $\Omega(2^n/n)$ gates (Shannon [6]). We shall prove that suitable analogs of these results hold when VLSI circuits are used, but with the $2^n/n$ replaced by 2^n in the bounding expressions. This suggests that there is a fundamental difference in complexity between traditional logic circuits and their counterparts in VLSI.

To prove upper- and lowerbounds on the VLSI complexity of Boolean functions we adopt Thompson's simplified model of a VLSI chip ([7]). Essentially a chip is a bounded (convex) region in the plane, with the plane divided into unit size cells as in the regular grid (see figure 1). Each cell may contain one basic switching element or (at most) two crossing wires. Wires connect processing elements, and are constraint to run in

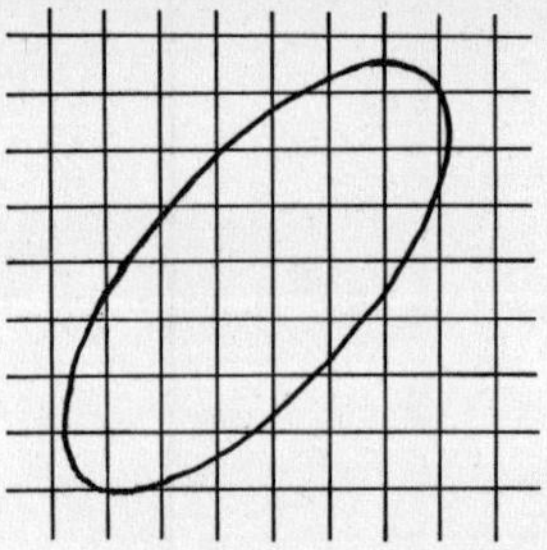

figure 1.

strictly horizontal or vertical directions. (Wires can bend from one direction into another in any cell, but can only cross and never run on top of one another.) The area of a chip is defined as the area of the grid that it occupies. We assume, like Thompson, that all switching elements act synchronously and that signals take unit time to propagate over a wire regardless its length. The period of a circuit is the minimum interval of time that must pass between consecutive sets of input data when the circuit is pipelined. We use the standard notations: A for area, T for time, and P for period.

The paper is organised as follows. In section 2 we recall some concepts from the theory of Boolean functions and provide a simple proof of Lupanov's theorem. It is included for having the opportunity to point out why the resulting construction of a logic circuit is not suited for VLSI. In section 3 we prove that all Boolean functions of n variables can be computed by a VLSI-circuit of $O(2^n)$ area and period 1. In section 4 we prove that there exist Boolean functions of n variables for which any (convex) VLSI circuit requires $\Omega(2^n)$ area. All proofs are elementary.

2. Preliminaries, and a classroom proof of Lupanov's theorem.

A minterm in $x_1,\ldots,x_n$ is any expression of the form $x_1^{\sigma_1}\ldots x_n^{\sigma_n}$ with $\sigma_i \in \{0,1\}$ $(1 \leq i \leq n)$, where $x_i^0 \equiv \bar{x}_i$ and $x_i^1 \equiv x_i$. Observe that $x_1^{\sigma_1}\ldots x_n^{\sigma_n} \equiv 1$ if and only if $x_i \equiv \sigma_i$ for all i. If $f(\sigma_1,\ldots,\sigma_n) = 1$, then $x_1^{\sigma_1}\ldots x_n^{\sigma_n}$ is called a minterm for f. It is well-known that every Boolean function f is the sum of its minterms. As there are only 2^n distinct minterm expressions in all, the resulting expression for f is finite. A Boolean function f is uniquely determined by the choice of its minterms. As there are 2^{2^n} distinct sets of minterms, there are 2^{2^n} distinct Boolean functions of n variables.

The representation by minterms can be rephrased in computational terms. Consider the lexicographic tree of minterms, modified such that the nodes "compute" the proper x_i or $\bar{x}_i$ and "and" it to the partial term values in x_1 to x_{i-1} $(1 \leq i \leq n)$. The resulting tree will be called the (general) mintree for n variables and is shown in figure 2(a). It is a circuit with $2^i + 2^{i-1}$ logic gates in the ith level and (hence) $\sum_1^n (2^i+2^{i-1}) =$ $= 3.(2^n-1)$ gates total. Every leaf of the mintree corresponds to a unique minterm in x_1 to x_n. Every Boolean function of n variables corresponds to a selection of leaves

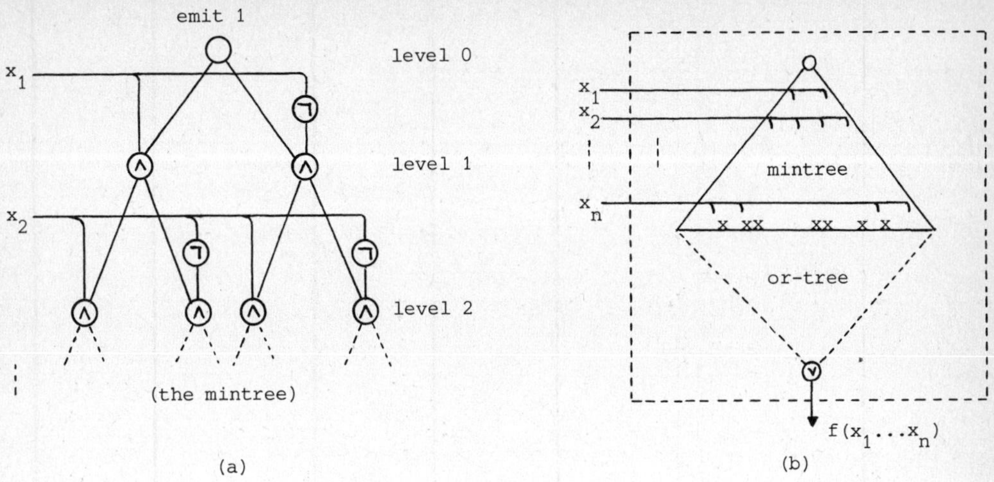

figure 2.

of the mintree. Let the selected leaves be marked (by x, see figure 2.b) and suppose another tree circuit is added that sums ("or-s") the computed values at the marked leaves. The resulting circuit is shown in figure 2(b) and clearly is a logic circuit of $O(2^n)$ gates for computing any Boolean function by the proper masking of components. We shall ignore the "or-" tree and concentrate on the mintree part. By optimizing it we shall be able to derive a weak form of Lupanov's theorem ([2]) in a simple way.

Let f be an arbitrary Boolean function in x_1 to x_n and let T be the mintree with marked leaves representing f. Cut T at level s, and consider the 2^s subcircuits (subtrees) attached to the nodes in this level. (The value of s will be fixed later, but

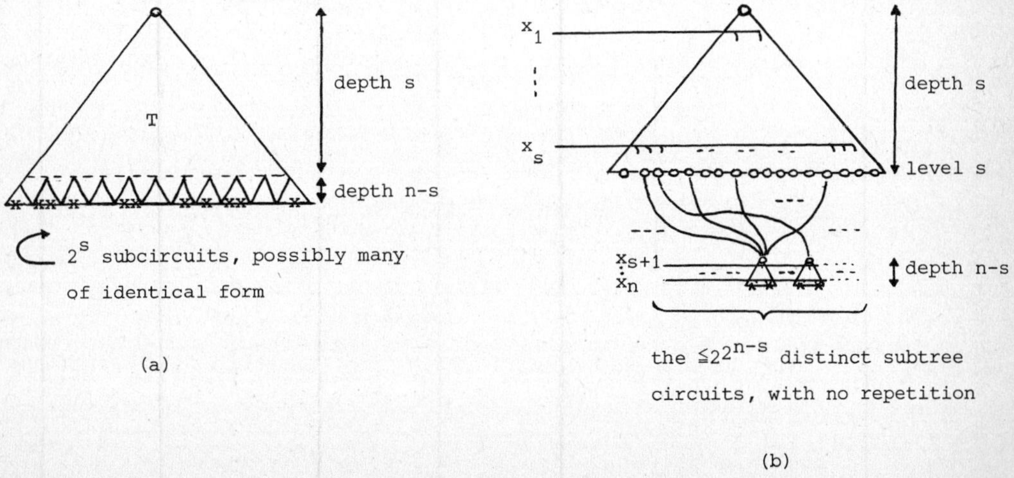

figure 3.

will be "almost equal" to n.) As circuits with input x_{s+1} to x_n the subtrees are identical, but they may differ through their marking. Because the subtrees are small but large in number, many must have the same marking and (hence) perform essentially the same computation. Thus, to save gates, we might as well split the subcircuits off from T and weed out all duplicates, and connect every intermediate node of level s to the proper and unique copy of the subcircuit that was originally attached to it. See figure 3(a)(b) for a pictorial illustration of this optimization of T. When phrased in terms of Boolean functions the construction is known as the "Lupanov decomposition" of f. Observe that every subcircuit of depth n-s has 2^{n-s} leaves, and thus there are $2^{2^{n-s}}$ distinct subcircuits if we take all possible markings into account.

Theorem 2.1 (Lupanov, 1958). Every Boolean function of n variables can be computed by a logic circuit of $O(2^n/n)$ gates, with no unlimited fan-in's.

Proof

Count the number of gates in the Lupanov decomposition of f's circuit. There are $3.(2^s-1)$ gates in the top part. Next there are $\leq 2^{2^{n-s}}$ small tree circuits with $3.(2^{n-s}-1)$ gates each. Suppose i_j nodes from the original level s are connected to the jth subtree circuit ($1 \leq j \leq 2^{2^{n-s}}$). To avoid the unlimited fan-in, we can combine them through an or-tree with i_j-1 gates. This accounts for an extra of $\sum_j (i_j-1) < \sum_j i_j = 2^s$ gates. Finally the results at the marked leaves must be summed in an or-tree to obtain $f(x_1 \ldots x_n)$ as output. Because we optimized the lower part of the mintree, there may actually be fewer marked nodes in the subcircuits altogether that still (correctly) represent f. Note that each of the $2^{2^{n-s}}$ subtrees can be paired to one with a complementary marking, to contribute a total of 2^{n-s} marked leaves. Thus there are at most $\frac{1}{2}.2^{n-s}.2^{2^{n-s}}$ marked leaves, and the or-tree summing for f needs to have no more gates than this. The total number of gates in the circuit is bounded by:

$$
\begin{aligned}
& 3.(2^s-1) + 3.(2^{n-s}-1).2^{2^{n-s}} + 2^s + \tfrac{1}{2}.2^{n-s}.2^{2^{n-s}} \leq \\
& \leq 4.2^s + 3\tfrac{1}{2}.2^{n-s}.2^{2^{n-s}} \leq \\
& \leq 4.(2^s + 2^{n-s}.2^{2^{n-s}}).
\end{aligned}
$$

Now choose s (integer) such that $2^{n-s} \leq n-\alpha\log n \leq 2^{n-s+1}$, for some α to be fixed later. It follows that $2^s \leq 2.2^n/n-\alpha\log n$ and $2^{n-s}.2^{2^{n-s}} \leq (n-\alpha\log n).2^n/n^\alpha \leq 2^n/n^{\alpha-1}$. Thus the number of gates in the circuit is bounded by

$$
\begin{aligned}
& 4.(2.\frac{2^n}{n-\alpha\log n} + 2^n/n^{\alpha-1}) = \\
& = 2^n/n.(8.\frac{n}{n-\alpha\log n} + 4/n^{\alpha-2}) = \\
& = O(2^n/n).
\end{aligned}
$$

, for any $\alpha \geq 2$. □

Note that we proved only a simplified version of Lupanov's theorem. The classical result states that by additional techniques one can bring the constant factor in the $O(2^n/n)$ term down to 1.

3. Area/time-efficient VLSI circuits for Boolean functions.

We now change our point of view to VLSI. We adopt all conventions as laid down in Thompson's model except that, for the moment, we will allow unlimited fan-outs by free "tapping" of a wire. The main issues are not gate counts, but area and time.

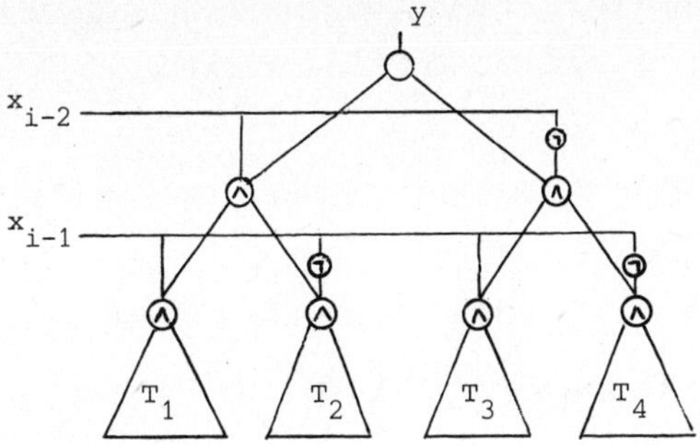

figure 4.

Trees are among the graphs that are easiest to lay out (see e.g. Thompson [7]) and only require "linear" area. The circuit lay-outs presented in this section derive from the common H-pattern design for (perfect) binary trees originally proposed by Mead and Rem [4].

Consider the representation of an arbitrary Boolean function f as suggested in fig. 2(b). For convenience we assume n even. The circuit of fig. 2(b) has a recursive structure that can be exploited to obtain an area-efficient lay-out. The essential idea is shown in figure 4. Build up the tree from the lowest level by adding two variables at the time, with the necessary logic to gate them into the four identical (but perhaps differently marked) subtrees that are needed on the variables x_i to x_n.

The construction suggests that a Lupanov-type optimization (see theorem 2.1) does not immediately lead to desirable effects for VLSI: (i) the many subtree circuits that Lupanov optimizes are all in the lowest part of the mintree and (hence) uniformly spread over the chip, and it would require much extra area and long wires to set the circuits apart for general access, (ii) by the same token it would lead to many more wire-crossings, which does not enhance the practicality for VLSI production and (iii) the unlimited fan-in (or: the or-trees that simulate it) is not equal at all places and (thus) creates severe problems of synchronization if the circuit is to be pipelined unless we "pretend" the or-trees to be all equal and complete. Redundancy has never been an issue in logic circuits, but it may be unavoidable in VLSI circuits!

We shall now rid ourselves of the assumption of unlimited fan-out. Rather than input the x_1 to x_n one after the other at separate levels of the mintree, we input them simultaneously in parallel at the top and distribute the signals down the tree while the computation of partial minterms in x_1 to x_i $(i \to n)$ is taking place. The idea is given in figure 5, where □-cells are used to represent the subcircuit that appends a next x_i to the computed terms at level i. Recursively continuing it leads

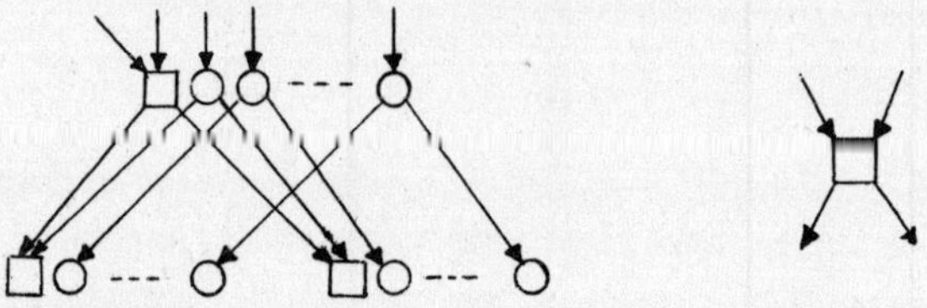

figure 5.

to another tree that computes all minterms at its leaves, the parallel mintree. Every Boolean function f of n variables can be represented by a marking of the leaves of the parallel mintree.

The parallel mintree has an area-efficient lay-out (see figure 6). The technique derives from the H-pattern construction and consists of adding two variables x_{i-1} and x_{i-2} at a time (for $i \to 1$) while combining the four recursively obtained sub-

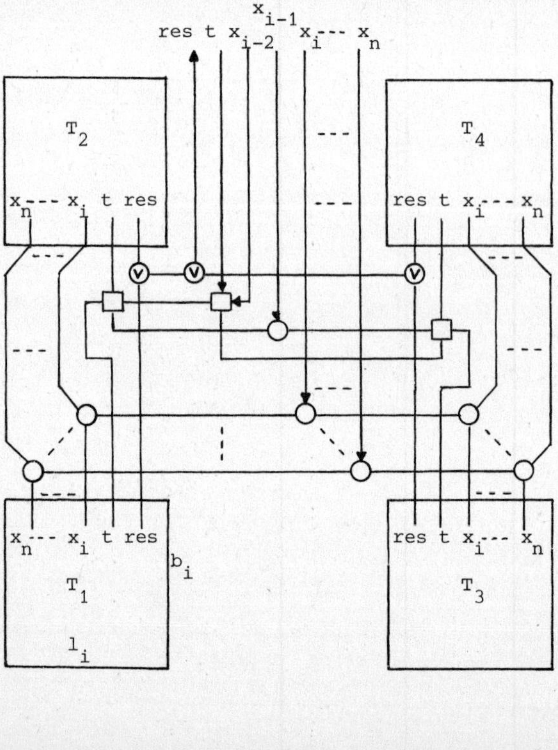

figure 6.

circuits in x_i to x_n. The □-cells in figure 6 actually are small circuits of some bounded size. The design uses a "t-" wire for accumulating partial minterms on the way down the tree, and a "res-" wire for summing the results from the marked leaves.

Theorem 3.1. Every Boolean function of n variables can be computed by a VLSI-circuit of $O(2^n)$ area, with a compute time of $O(n)$ and a period of $O(1)$.

Proof

Let $C=C_1$ be the resulting VLSI-circuit for f obtained by laying out the parallel mintree as suggested. Let the ith intermediate circuit C_i of the recursive construction (starting with i equal to n-1) fit in a rectangle of size $b_i \times l_i$. From figure 6 we conclude that $b_{i-2} = 2b_i+(n-1) + 9$ and $l_{i-2} = 2l_i+(n-i) + 8$. (The additive constants are really larger, because the detailed circuits for the □-cells require more than unit area.) It follows that $b_1 = O(2^{n/2}-n)$ and $l_1 = O(2^{n/2}-n)$ and (hence) that C occupies no more than $O(2^n)$ area.

The time bound of $O(n)$ for computations with C is obvious. The period of $O(1)$ follows because C can process an entire input tuple in parallel as a single wavefront going down, with the possibility of sequencing consecutive wavefronts at unit time distance. □

4. Optimal area/time bounds for Boolean VLSI-circuits.

For particular Boolean functions f of n variables the bounds of theorem 3.1. may not be best possible. (See e.g. Thompson [7] for many examples of essentially Boolean functions that require only $O(n)$ or $O(n^2)$ area, or $O(\log n)$ time.) The question is rather how tight the bounds are if the theorem is taken as a uniform statement for the entire class of Boolean functions of n variables. It is clear that the $O(1)$ bound on the period of circuits can not be improved. We show that none of the other bounds in theorem 3.1. can be improved either.

Theorem 4.1. There are Boolean functions of n variables such that any circuit to compute them requires $\Omega(n)$ time on some inputs.

(The result is a simple variant of the fact that there are Boolean functions that require circuit depth n, see Savage [5]. The proof is omitted.)

A natural question is whether the $O(2^n)$ bound on the area requirement for general Boolean functions can be improved.

Proposition 4.2. Suppose all Boolean functions of n variables can be computed by circuits that fit on a fixed chip of area A. Then necessarily $A = \Omega(2^n)$.

Proof

Recall that a chip was defined to be a connected region of the unit grid. Assume there is a fixed chip R of area A such that for all Boolean functions f of n variables there is a circuit to compute it that can be embedded in R. According to Thompson's model each cell of R can be occupied by at most one of the finitely many different constructs, say c altogether. Thus at most c^A different circuits can be embedded in R. On the other hand, the number of distinct Boolean functions of n variables is 2^{2^n}. It follows that $c^A \geq 2^{2^n}$, or $A = \Omega(2^n)$. □

The assumption that there be a "general chip" is too strong to measure the area requirement of Boolean functions. We shall now consider circuits fitted on their own, optimal area chips. To be realistic we require that the chips are convex. (This is a very common constraint, see e.g. Brent and Kung [1]). By the following two lemmas we shall be able to "mold" every circuit that fits on a convex chip of area A into a form that fits in a standard (iso-oriented) rectangle on the grid of area O(A).

Lemma 4.3. Every convex chip of area A can be enclosed by a (tilted) rectangular box of area 2A.

Proof

(The easy argument can be derived from e.g. Yaglom and Yaglom [8], solution to problem 120a.) Let R be a bounded convex region of area A (see figure 7). Let AB be the longest chord. The tangents to R at A and B are perpendicular to AB. Now "enclose" R by drawing the (two) lines of support that are parallel to AB. Let the

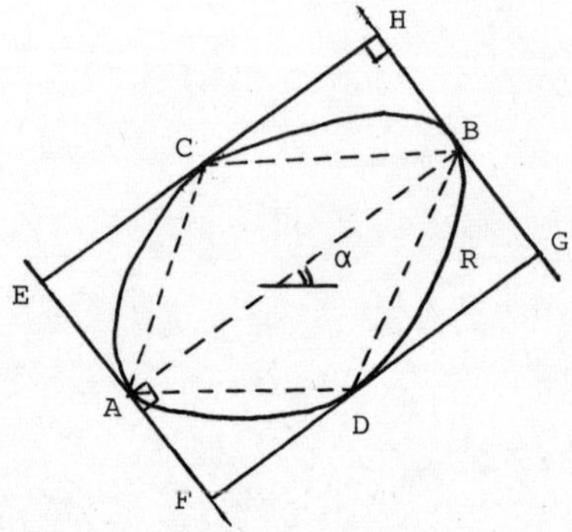

figure 7.

lines touch on R at C and D (see figure 7). In figure 7 is indicated that R is now enclosed by □EFGH, formed by the four intersecting tangents. By convexity □ADBC $\subseteq$ R and (hence) □ADBC has area $\leq$A. On the other hand, by geometric reasoning it follows immediately that □ABHE = 2.ΔABC and that □ABGF = 2.ΔABD and (hence) that □EFGH = = 2.□ADBC $\leq$ 2A. □

Lemma 4.4. Let circuit C be embedded on the unit grid within a rectangle of area A that is tilted by angle $\alpha (0 \leq \alpha \leq \pi/4)$. Then C can be embedded on the unit grid within an iso-oriented rectangle of area bounded by $(1 + 2\tan\alpha - \tan^2\alpha)A \leq 2A$.

Proof

Let C be embedded on the grid such that it is enclosed by a rectangle of size lxw ($l \geq w$ and $l.w=A$) whose longest side makes an angle α with the positive x-axis of the grid (see e.g. figure 7). It is no restriction to assume that $0 \leq \alpha \leq \pi/4$, for otherwise the argument holds for a properly reflected or rotated version of the problem. We also assume that no cell cut by the boundary of the rectangle is used by C.

Divide the rectangle (and correspondingly, the circuit) into parts A_1, A_2 and B as shown in figure 8. Parts A_1 and A_2 are "padded" to iso-oriented rectangular boxes

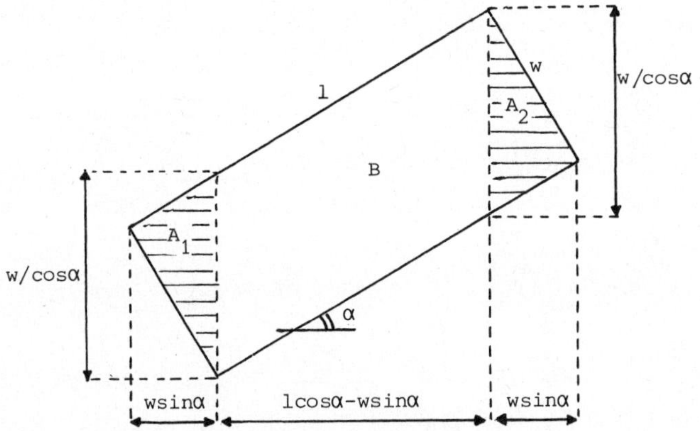

figure 8.

of size $w/\cos\alpha \times w\sin\alpha$ each. Part B consists of $l.\cos\alpha - w.\sin\alpha$ columns of $w/\cos\alpha$ cells each. Because $\alpha \leq \pi/4$, each column can differ by at most one cell at either end from the immediately preceding column (see figure 9). It means that we can bring B down to an iso-oriented form by shifting each of its columns downwards by at most one cell with respect to the immediately preceding column, resulting in a circuit as shown in figure 10. Figure 9 shows that this can be done within the constraints of Thompson's model, provided that we include a spacing of one extra column in between every pair

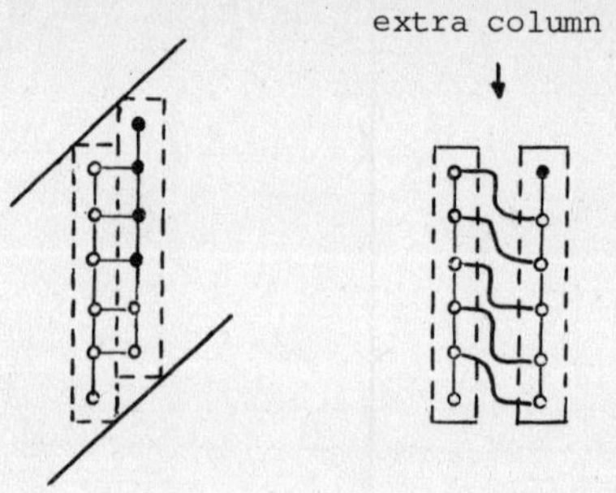

figure 9.

of columns. This explains that in figure 10 the length of the B-part is at most doubled. Thus circuit C appears embedded in an iso-oriented rectangle of width $w/\cos\alpha$ and length about $2.l.\cos\alpha$, and hence area $2.l.w = 2A$. By being a little more accurate, one can

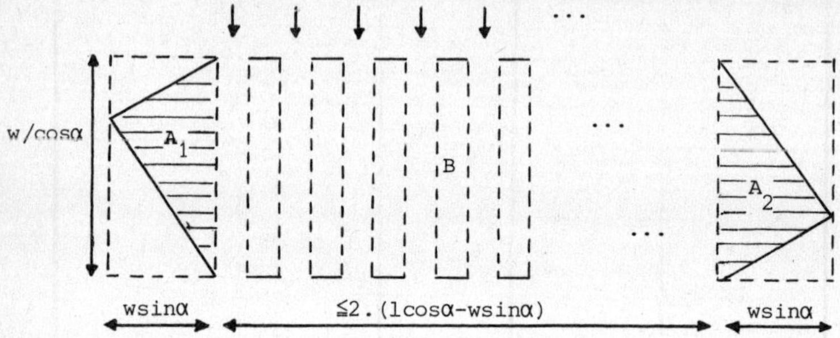

figure 10.

see that a down-shift over one cell (hence, the need for an extra column) is required only once in every $1/\tan\alpha$ columns. It follows that the length of α can remain at $(1+\tan\alpha)$ times the original. Hence C fits in an iso-oriented rectangle of size $(1+\tan\alpha) - l.\cos\alpha + (1-\tan\alpha).w.\sin\alpha$ by $w/\cos\alpha$, which has an area bounded by $(1+2\tan\alpha-\tan^2\alpha).A$. □

We conclude that the convexity assumption in VLSI-theory degenerates and does not go much beyond assuming that chips are rectangular! (In all fairness we should add that this only applies if one adopts the grid-model for the chip surface.)

<u>Corollary</u> 4.5. Every circuit that fits on a convex chip of area A can be embedded in an ordinary iso-oriented rectangle of area $\leq 4A$.

Theorem 4.6. There are Boolean functions of n variables such that every convex chip for them requires $\Omega(2^n)$ area.

Proof

Suppose that all Boolean functions of n variables can be computed by convex chips of area $\leqq A$. By corollary 4.5. we may as well assume that all circuits fit in an iso-oriented rectangle of area $\leqq 4A$. Now note that there are at most $\sqrt{4A} = 2.\sqrt{A}$ different (i.e., non-isomorphic) iso-oriented rectangles of area $\leqq 4A$, and that each of these rectangles can have at most c^{4A} different circuits. (The constant c is as in the proof of proposition 4.2.) It follows that necessarily $2\sqrt{A}.c^{4A} \geqq 2^{2^n}$, and (hence) that $A = \Omega(2^n)$. □

By an only slightly more sophisticated argument one can see that theorem 4.6. also holds without the convexity assumption. Note that every chip of area $\leqq A$ is fully described by its contour which, when "projected" onto the grid-lines, can be encoded into a string of length $\leqq 4A$ over the alphabet {L,R,U,D} (L for "one square left", etc.). It follows that there are at most $\sum_{1}^{4A} 4^1 \leq \frac{4}{3}.4^{4A} \leqq d^A$ (some d) chips of area $\leqq A$. On each chip at most c^A different circuits can be fitted. To accomodate all Boolean functions of n variables one must necessarily have $c^A.d^A = (cd)^A \geqq 2^{2^n}$, and (hence) that $A = \Omega(2^n)$. By a more detailed counting argument it follows that almost all Boolean functions of n variables require $\Omega(2^n)$ area.

5. References.

[1] Brent, R.P. and H.T. Kung, The area-time complexity of binary multiplication, J.ACM 28 (1981) 521-534.

[2] Lupanov, O.B., A method of circuit synthesis, Izv. V.U.Z. Radiofiz. 1 (1958) 120-140.

[3] Mead, C.A. and L.A. Conway, Introduction to VLSI systems, Addison-Wesley, Reading, Mass., 1980.

[4] Mead, C.A. and M. Rem, Cost and performance of VLSI computing structures, IEEE J. Solid State Circuits SC-14 (1979) 455-462.

[5] Savage, J.E., The complexity of computing, John Wiley & Sons, New York, N.Y., 1976.

[6] Shannon, C.E., The synthesis of two-terminal switching circuits, Bell Syst. Techn. J. 28 (1949) 59-98.

[7] Thompson, C.D., A complexity theory for VLSI, (Ph.D.Thesis), Techn. Rep. CMU-CS-80-140, Dept. of Computer Science, Carnegie-Mellon University, Pittsburgh, P.A., 1980.

[8] Yaglom, A.M. and I.M. Yaglom, Challenging mathematical problems with elementary solutions, Holden-Day, San Francisco, CA., 1967.

Fast parallel algorithms for finding all prime implicants for discrete functions

W. Oberschelp
Lehrstuhl für Angewandte Mathematik, insbesondere Informatik,
RWTH Aachen, 5100 Aachen, Germany

Abstract. We consider the problem of enumerating the prime implicants of a given discrete function as a basic task of circuit theory. First, we count PI's for random Boolean functions. Then we use the well known lattice differentiation as a tool for finding implicants. The concept of a peak admits to characterize prime implicants, at least those with no improper domains. The improper case can be reduced to a lower dimensional problem. Since the peak test is local, a parallel algorithm is available. The time and space complexity turns out to be low measured in the input size.

§ 1 Introduction: Enumerating all prime implicants as a basic task of circuit theory

The determination of all prime implicants for a given Boolean function $f(x_1,\dots x_n)$ is a basic technique asked for in Boolean logic. For small n, e.g. for $n=4$, the problem can be solved by visual inspection, e.g. by Karnaugh maps (cf. Klar). For arbitrary n the well known algorithms of Quine (1952) and McCluskey (1956) do the job in principle. It is interesting to find a good, possibly optimal design of switching network with input size n approximately 10 to 20 bits. This problem makes sense even in the time of VLSI, since in the logical design of big circuits optimal building blocks of medium size are the basis of good upper bounds.

We use a well known generalized concept of Boolean functions, viz. discrete functions over the base $\{+,\cdot\}$ with arbitrary finite range and domain S. Such a function

$$f(x_1,\dots,x_n) : S^n \to S$$

could be part of a parallel adder or multiplier or of a sorting network with e.g. $|S| = 2^{16}$ and $n=2$. In this way we have the chance to consider multiple output functions directly, whereas in the classical Boolean approach those switching functions have to be combined as a direct product of simple Boolean functions.

We mention one conceivable shortcoming of our optimization philosophy: Our approach amounts to minimize formula size with an additional

two level restriction, as it is considered in the usual sum of products concept with Boolean literals. Thus we are not interested in minimizing overall formula size in the sense of Neciporuk (cf. Savage), since we feel, that the run time for signals - which is measured by circuit depth - is the most important characteristics of a switching network. As a matter of fact, classical constructions like carry lookahead or carry bypass (cf. Spaniol) are working in the direction of making the number of levels small.
It is well known (cf. Davio et al., p. 472/73), that an optimal design for f can be obtained by solving two problems:

i) calculate all prime implicants of f

ii) select a minimal system of prime implicants, which "covers" all the entries of f.

The minimization in ii) may also take into account a nontrivial cost function for PI's. Curiously, this function decreases, if an implicant covers a bigger domain, since in this case the description of the PI tends to be simple. Thus finding short implicants is like winning in a lottery, and the proof of primality for an implicant means, that there is no further hope of winning more in that game. Even though problem ii) is conceptually simpler than i), it is NP-hard, since we are in a set covering situation (cf. Garey-Johnson); therefore we should forget the exact optimization of ii). The difficulties are obvious even for a Boolean example with 20 binary inputs. Using a result developed in section 2, we have to expect by $1{,}536 \cdot 10^6$ PI's for a randomly given Boolean function. We can therefore only offer heuristics for finding optimal selections of PI's.
Unfortunately, we don't know of generally good heuristics for the PI covering problem. The well known greedy-heuristic selects a cheap and short PI, which covers as many positive entries of f as possible and which dynamically updates the set of entries which still have to be covered. But in the case of the covering problem the greedy algorithm has a rather bad worst-case performance-factor log n, and one can only hope that the average performance is better.

We therefore restrict all our endeavours to solve problem i) exactly. There are many situations in circuit theory where an exact solution is asked for, e.g. the detection of hazards in circuits (cf. Davio et al. theorem 3.29 or Friedman, p. 246). Here exactness (i.e. completeness of the list of prime implicants) is of an importance, which could be compared to the safety standards which are required whenever the primality of a natural number is claimed.

§ 2 Counting prime implicants in the Boolean case

We take a short glance to the problem of counting rather than constructing prime implicants, and we restrict ourselves to the n-place Boolean case. Here we get some results for random situations. We start with some well known observations by Mileto-Putzolu (1964). Using an inclusion-exclusion argument, these authors made a global count of the occurences of prime implicants. They obtained for the total number $\Gamma_n(k,u)$ of occurences of prime implicants with length $n-k$ for a random function with u positive entries in the function table the

Lemma 1: $\Gamma_n(k,u) = \binom{n}{k} 2^{n-k} \sum_{i=0}^{n-k} (-1)^i \binom{n-k}{i} \binom{2^n-(i+1)2^k}{u-(i+1)2^k}$

Thus in the classical Karnaugh diagram case $n=4$ we have

u \ k	0	1	2	3	4	$\sum$	$\binom{2^n}{u}$
0							1
1	16					16	16
2	176	32				208	120
3	880	448				1328	560
4	2640	2816	24			5480	1820
5	5280	10496	288			16064	4368
6	7392	25792	1584			34768	8008
7	7392	43904	5280			56576	11440
8	5280	52864	11832	8		69984	12870
9	2640	45056	18624	64		66384	11440
10	880	26656	20832	224		48592	8008
11	176	10432	16320	448		27376	4368
12	16	2432	8544	560		11552	1820
13		256	2668	448		3392	560
14			384	224		608	120
15				64		64	16
16					1	1	1
$\sum$	32768	221184	86400	2040	1	342393	65536
$\frac{\Sigma}{2^{2^n}}$	0.5	3.375	1.318	0.031	0	5.2245	1

Note, that $k=0$ is the case of minterms, and in general k means the number of missing variables in an implicant; 2^k is the number of positive entries in the function table of an implicant.

The table shows, that for a random Boolean function with 4 variables we can expect a little more than five prime implicants, and in every second function we can expect a minterm occuring as a PI.

We don't know of a simple formula for the row sums. Meo (1964) has given a rough diagram for these numbers if $n=20$, which resulted from "some approximations for binomial coefficients." But we can give a simple formula for the column sums $S_k^{(n)}$. From column k we learn how many PI's of length n-k occur in total.

<u>Theorem 2</u>: $S_k^{(n)} = \binom{n}{k} 2^{2^n+n-(2^k+k)} \left(1 - \frac{1}{2^{2^k}}\right)^{n-k}$

for $0 \leqslant k \leqslant n$.

<u>Proof</u>: Summing up Lemma 1 over u and interchanging the order of summation we get

$$S_k^{(n)} = \sum_{u=0}^{2^n} \Gamma_n(k,u) = \binom{n}{k} 2^{n-k} \sum_{i=0}^{n-k} (-1)^i \binom{n-k}{i} \sum_{u=0}^{2^n} \binom{2^n-(i+1)2^k}{u-(i+1)2^k}$$

The sum over u can be started at $(i+1)2^k$. Changing the summation variable, we get for the innermost sum on the right side

$$\sum_{\rho=0}^{2^n-(i+1)2^k} \binom{2^n-(i+1)2^k}{\rho} = 2^{2^n-(i+1)2^k}$$

Therefore we get

$$S_n^{(k)} = \binom{n}{k} 2^{n-k+2^n-2^k} \sum_{i=0}^{n-k} (-1)^i \binom{n-k}{i} 2^{-i \cdot 2^k}$$

$$= \binom{n}{k} 2^{2^n+n-(2^k+k)} \left(1 - \frac{1}{2^{2^k}}\right)^{n-k} \qquad \text{q.e.d.}$$

If we devide $S_n^{(k)}$ by the total number 2^{2^n} of n-ary Boolean functions, we get the average number $\sigma_n^{(k)}$ of PI's of length n-k for Boolean functions. Summing up over k, we get as the average number of PI's $\sigma^{(n)} = \sigma_0^{(n)} + \sigma_1^{(n)} + \ldots$ with the special values

$$\sigma_0^{(n)} = \frac{1}{2}\ ; \quad \sigma_1^{(n)} = \frac{n}{4}\left(\frac{3}{2}\right)^{n-1}\ ; \quad \sigma_2^{(n)} = \frac{1}{16}\binom{n}{2}\left(\frac{15}{8}\right)^{n-2}$$

$$\sigma_3^{(n)} = \frac{1}{256}\binom{n}{3}\left(\frac{255}{128}\right)^{n-3}\ ; \quad \sigma_4^{(n)} = \frac{1}{65536}\binom{n}{4}\left(\frac{65535}{32768}\right)^{n-4}$$

Utilizing the general term

$$\sigma_k^{(n)} = \frac{1}{2^{2^k}}\binom{n}{k} 2^{n-k}\left(1 - \frac{1}{2^{2^k}}\right)^{n-k},$$

we expand $\left(1 - \frac{1}{2^{2^k}}\right)^{n-k} = 1 - \frac{n-k}{2^{2^k}} \pm \ldots$, and for the first approximation term we get, if $k \geqslant 4$ and $n \leqslant 50$, the estimate

$$\frac{n-k}{2^{2^k}} \leqslant 0{,}000702,$$

i.e. we have an error of less than 1 °/oo in this range, if we use the approaximation $\sigma_k^{(n)} = 2^{n-k-2^k}\binom{n}{k}$.

If $k \geq 5$ and $n \leq 10^6$, we get an even better estimate

$$\frac{n-k}{2^{2^k}} \leq 0{,}000233.$$

Furthermore, we find $\frac{\sigma_6^{(n)}}{\sigma_5^{(n)}} \approx \frac{n-6}{3 \cdot 2^{34}} \leq 0.000019$ for $n \leq 10^6$.

Thus, in any realistic range of the number n of variables, we can confine ourselves to the formulas extended up to $k = 5$, and there is an essentially unique k which gives the main term for approximating the number of PI's. This k is roughly approximated by $\log_2 \log_2 n$.
We illustrate this situation by giving an extract of the table of $\sigma_k^{(n)}$ (we write En short for 10^n):

k \ n	4	10	20	50	100	200	500
0	0,5	0,5	0,5	0,5	0,5	0,5	0,5
1	3,375	9,61 E1	1,11 E4	5,31 E9	6,78 E18	5,51 E36	9,26 E89
2	1,318	4,30 E2	9,74 E5	5,73 E14	1,76 E29	1,41 E57	7,02 E139
3	0,031	5,84 E1	5,46 E4	8,97 E15	6,85 E31	4,77 E62	4,73 E153
4	0	0,21	4,84 E3	2,47 E14	4,74 E30	3,93 E63	7,97 E153
5	-	1,88 E-6	0,12	1,74 E10	6,94 E26	2,96 E58	6,08 E150
Σ	5,2245	584,83	1,536 E6	9,785 E15	7,34 E31	4,41 E63	1,271 E154

In order to illustrate the simple calculability of these values, we give some more numbers:

$c_3^{(1000)} = 1{,}76$ E304; $c_4^{(1000)} = 4{,}17$ E305; $c_5^{(1000)} = 6{,}43$ E302.
$c_4^{(1000000)} = 9{,}29$ E301039; $c_5^{(1000000)} = 6{,}00$ E301046; $c_6^{(1000000)} =$ 1,17 E301042.

In Lemma 1, the <u>primality</u> of an implicant is taken care of by the sign factor $(-1)^i$. If we want to count arbitrary implicants, we have to take the positive sign in the last factor of Theorem 2. We therefore get the average implicant numbers

$$\tau_k^{(n)} = \frac{1}{2^{2^k}} \binom{n}{k} 2^{n-k} \left(1 + \frac{1}{2^{2^k}}\right)^{n-k}$$

For the relative proportion of prime implicants in the set of all implicants we have

$$q_k^{(n)} := \frac{\sigma_k^{(n)}}{\tau_k^{(n)}} = \left(\frac{1-2^{-2^k}}{1+2^{-2^k}}\right)^{n-k} = 1 - 2(n-k)2^{-2^k} \pm \ldots$$

For fixed k and $n \to \infty$ this quantity tends to zero. On the other hand even for rather small values of k and for rather big values of n

$q_k^{(n)}$ is close to 1. We give a table for some reciprocals of $q_k^{(n)}$:

k \ n	4	20	100	1000
0	81	3,49 E9	5,15 E47	1,32 E477
1	4,63	1,64 E4	9,18 E21	4,24 E221
2	1,28	9,52	2,12 E6	1,77 E54
3	1,008	1,14	2,13	2,41 E3
4	1	1,0005	1,0029	1,031

As further values, for $n = 1000000$ and $k = 4$ and $k = 5$ we note the entries 1,73 E13 and 1,00047 respectively.

The numbers show, that e.g. for $n = 1000$ approximately 97% of all the implicants with $k = 4$ missing variables are prime, but for $n = 10^6$ this proportion has decreased to 5.58 E-14. And for $n = 20$, a test for primality of an implicant g of a function f is only hopeful for failing if at most 3 variables are missing, i.e. if at most $2^3 = 8$ positive entries occur in the table of g, since in all other cases the probability, that g is prime, is practically 1. There is a somewhat critical remark, that surely the number of implicants is essentially bigger than the number of prime implicants, since each prime implicant with missing variables can be lengthened in many ways to an implicant. But relativized to the implicants of the <u>same size</u>, probability of an implicant to be prime is very close to 1 for $k \geq 4$ within the normal range of n. In spite of this descourageing fact, we nevertheless intend to develop primality tests in the following sections, since the non-primality of an implicant is a happy event which promises to find even simpler (prime) implicants. And we shall see, that for generalized discrete functions the situation is much more interesting than in the Boolean case. We intend for the future to publish combinatorial results also for the general case; in the present paper we simply get a transparent algorithmic book-keeping of the evolution of implicants towards primality.

§ 3 Lattice differentiation of discrete functions

The Boolean case can be generalized in at least two ways to arbitrary finite domains S instead of the Boolean set $S = \{0,1\}$. S can be considered alternatively

i) as a ring of residues mod $|S| =: s$. This ring is a field if s is prime

ii) as a lattice with max and min (in the sense of a fixed total ordering of S) as join and meet (the case $s = 2$ even yields a

Boolean algebra).

A detailed description of these generalizations is given in Davio et al. The well known Reed-Muller expansion of Boolean functions on the one hand, and of the minterm - maxterm approach on the other hand are the origins of these generalizations.

We are pursuing here the lattice approach for n-ary functions f from S^n to S. In order to normalize the total order $\leqslant$ given in S, we assume that $S = \{0,1,\ldots,n-1\}$. Then we can speak of the successor $x \oplus 1$ of an element x of S. In addition, we define 0 as the successor of n-1.

By an n-ary block (with domain D) we understand a function which is constant $\neq 0$ in a rectangular domain D and which is 0 else. D is the n-fold cartesian product of intervals of S. An interval is a connected part of the successor structure of S - thus an interval may extend over the element n-1 back to the beginning 0,1,... of S. If an interval ist equal to S, we call it improper. The introduction of a cyclic successor structure into S is a convenient - but by no means crucial - variant of the lattice generalization concept. An n-ary block g is an implicant of an n-ary function f, if $g \leqslant f$.

Definition: An implicant g of f is called a prime implicant, iff

i) g cannot be increased

ii) dom(g) cannot be extended.

Example for $n = 2$, $S = \{0,1,2,3,4\}$: (The empty entry in function tables always means the entry 0).

y \ x	0	1	2	3	4
0	2	2			
1	4			4	4
2		1	4	3	3
3		1	4	3	3
4	2	2	3	3	3

binary function f

y \ x	0	1	2	3	4
0					
1					
2					
3			3	3	3
4			3	3	3

implicant g of f

y \ x	0	1	2	3	4
0					
1					
2			3	3	3
3			3	3	3
4			3	3	3

prime implicant h of f

The problem of recognizing the prime implicants of f is the problem of recognizing maximal possible function values and maximal possible domains (shapes). In order to do this automatically, the concept of (partial) lattice derivation has been introduced (cf. Davio et al. and Thayse). We give the definition in the simplest case:

Definition: $\frac{\partial f}{\partial x} := \mathrm{Min}\,(f(x\oplus 1),\ f(x))$

This definition means for functions with more than one argument, that the arguments are kept identically at places, which are not touched by our notation. The definition extends analogously for higher derivatives. There are several obvious rules, e.g. that the order of derivation is irrelevant for the result.

In order to calculate $\frac{\partial f}{\partial x}$ for the argument x_o, we only need to know in the table of f the entry for x_o and for its right neighbour (in x-direction) $x_o \oplus 1$. Thus the discrete lattice derivative is a <u>local</u> operation, which can be calculated in one step by an ideal parallel computer for a whole function f.

From the table of f we can calculate all the tables of partial derivatives; we indicate the degree of differentiation by a degree-vector, i.e. we characterize the differentiation

$$\frac{\partial f}{\partial^{i_1}_{x_1} \partial^{i_2}_{x_2} \cdots \partial^{i_n}_{x_n}}$$ by the d-vector $(i_1, i_2, \ldots, i_n)$.

The function f itself is, of course, characterized by the d-vector $(0,0,\ldots,0)$.

We give a list of all the derivation tables for our example. Note, that we only have to go to the (n-1)st derivative in each direction, since obviously the n-th and higher derivatives don't change anything in the tables. Therefore the (n-1)st derivations in one direction is maximal; each table containing n-1 in its d-vector at least once is called <u>marginal</u>. The significance of the circles around some entries in the given tables is explained in section 4.

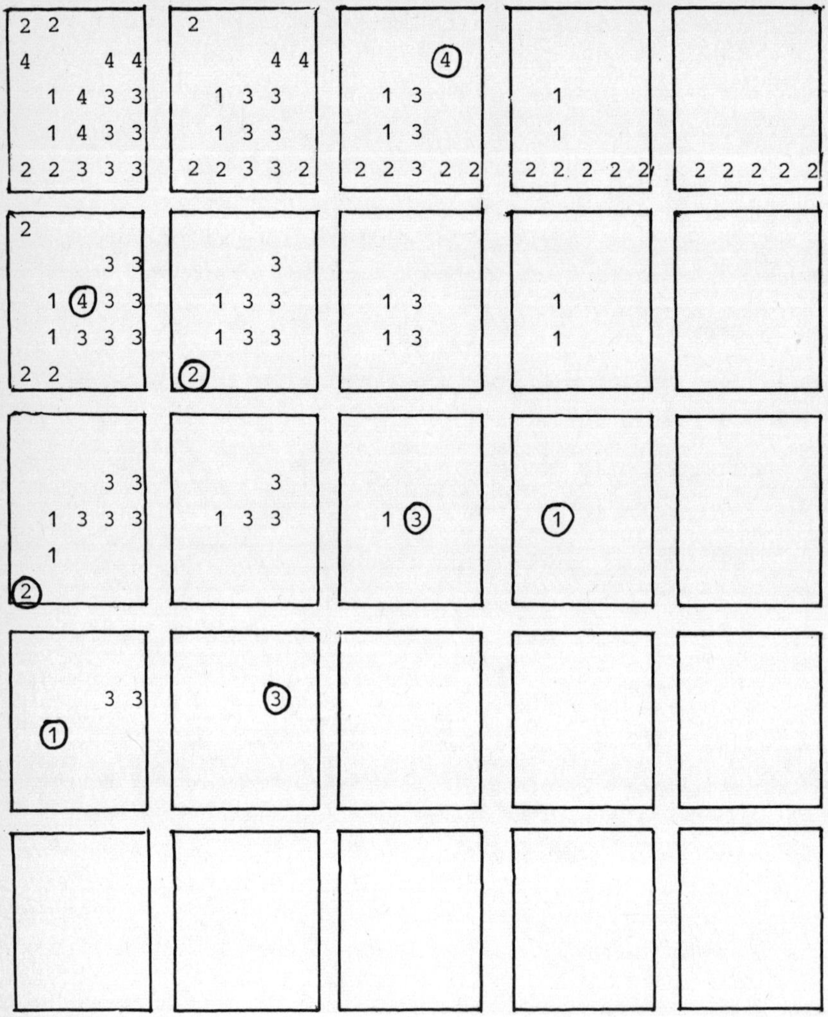

§ 4 Peaks and prime implicants: The proper case

It has been noticed (cf. Davio et al. p. 156, Theorem 3.8(a)), that any positive entry a in a derivation table of f with vector $(i_1, i_2, \ldots, i_n)$ for the argument $\mathfrak{x}_o = (x_1^{(o)}, \ldots, x_n^{(o)})$ indicates, that the function

$$g(\mathfrak{x}) := \begin{cases} a, & \text{if } x_j \in \{x_j^{(o)}, x_j^{(o)} \oplus 1, \ldots, x_j^{(o)} \oplus i_j\} \text{ for } j = 1, \ldots, n \\ 0 & \text{else} \end{cases}$$

is an implicant of f. Thus, in our example, the fact that $\left(\frac{\mathfrak{D} f}{\mathfrak{D} x^2 \mathfrak{D} y}\right)_{\mathfrak{x}=(2,3)} = 3$, indicates, that g is an implicant of f.

The central fact for the following is the observation, that prime implicants can be recognized from a derivation table by a simple local test. Moreover: Each prime implicant leaves one single trace in a certain derivation table (cf. also Oberschelp-Remlinger, chap. 4) in form of a peak. We have to distinguish two different siutations, ordinary peaks and improper peaks; we first consider the ordinary case:

Definition: A peak in a table is a strong relative maximum, i.e. an entry which is bigger than all neighbour-entries.

Note: Neighbourhood in a table means to be neighbour in rectilinear direction, i.e. to be a neighbour with respect to exactly one coordinate.

Theorem 3: The (ordinary) peaks and the improper peaks in the derivation tables of f characterize the prime implicants of f in a one-one manner. The size of their domains is described by the d-vector of the table, the function value is given by the value at the peak.

Example: Since

$$\left(\frac{\partial f}{\partial x^2 \partial y^2}\right)_{\mathfrak{r}=(2,2)} = 3, \text{ and since this is a peak in}$$

the (2,2)-derivation table, the function

$$h(x,y) := \begin{cases} 3, \text{ if } x \in \{2,3,4\} \text{ and } y \in \{2,3,4\} \\ 0 \text{ else} \end{cases}$$

is a prime implicant of f. We have encircled the entries corresponding to ordinary peaks and thus to prime implicants in the table in section 3.

Proof of Theorem 3: We give an intuitive rather than a formal proof of the theorem: Think of an n-dimensional crocodile or dragon, which is working like a parallel processor and which is able to move one step per time unit in any of the n direction (Patent pending for the RWTH Aachen). In doing so it flattens by its barbs the profile of a derivation table like a smoothing rubber in just the same sense as differentiation works: Look at the crocodile working in x-direction on a given profile:

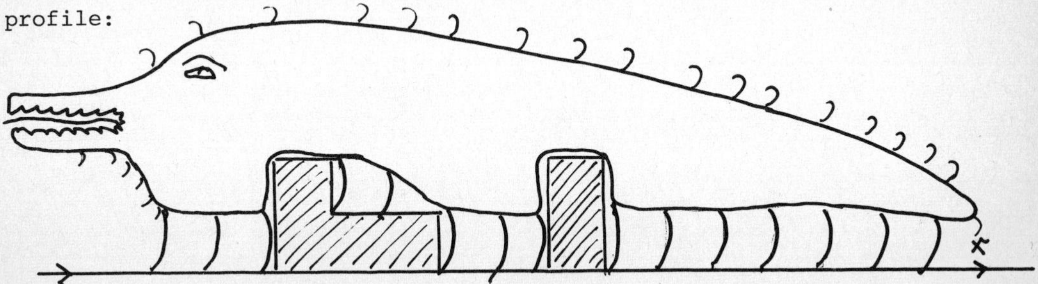

One step later it has produced the following profile:

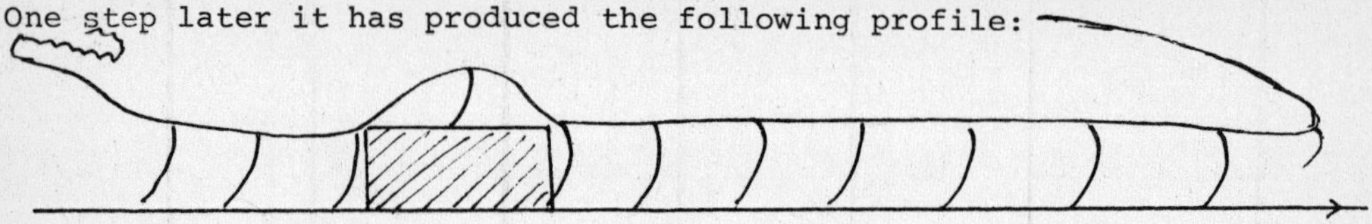

This is the derivation in x-direction, since the crocodile (by construction of its barbed skin) leaves at any point the minimum of the present value and its right neighbour. After two more steps the profile has vanished to zero.

Let us consider the fate of a prime implicant under this procedure. There is one last moment, in which PI survives; this is the peak situation. The height of the PI is the value at the peak; the domain of PI is determined by the moving history of the crocodile, i.e. by the differentiation history of the present table (obviously the order of movement (differentiation) is irrelevant). Thus we can recognize an (ordinary) PI at its northwest corner, which can resist the crocodile for the longest time.

It is possible to treat discrete functions with don't cares in a similar manner; with respect to covering, we have to treat don't care entries as zero-entries, with respect to the search for prime implicants, we should consider those entries as equiped with the highest possible value s-1.

§ 5 The case of improper prime implicants

The sketched proof doesn't work if there is a non-vanishing value of a PI h over a whole direction, i.e. if one or more of the intervals which define the domain D of the block h are improper. In this case we have a constant non-zero line in a marginal table (see the row of 2's in the upper right table of our example). In this case we have to consider a lower dimensional problem by simply neglecting the direction which belongs to the improper interval. In the projected schema we are looking again for ordinary peaks; we reduce dimension again, if it turns out, that a second direction is equiped with an improper interval etc.

In the Boolean case the appearance of an improper interval of a block is the only interesting case: Here the implicants become independent of the corresponding variable, and thus this variable is missing. We have studied in section 2 the frequencies of such events. But with the trend of S to grow larger and larger the case of proper intervals appears to be the normal case. We want to discuss the frequencies for the general case in a forthcoming paper.

§ 6 The complexity of finding all proper prime implicants

The function $f: S^n \to S$ may be given by a table of s^n entries each consisting of $\lceil \log_2 s \rceil$ bits. Taking this as an initial input, we need at any time for the calculation of a new derivation table one earlier table - or even part of it - in the core, other tables may remain in the auxiliary memory. We have the concept of storing or printing the PI's, which result from a table. Thus we only need to keep a table in the memory, if we need it for the calculation of new derivation tables later.

We conceive to have an ideal parallel computer at our disposal. Then we only need two units of time for the establishment of a new table from a former one: One time unit is spent for calcualting the derivation, i.e. for locally determining the minimum of two neighbour-values; the other unit is spent to make the peak test (and finally to bring the resulting PI's into an appropriate buffer). Thus determinig all PI's simply amounts in this situation to an enummeration of all derivation tables according to their d-vectors.

This enummeration can be done by counting in the s-base number system from 0 to $s^n - 1$. It is easy to see, that in counting we never need more than n tables for the future, if we use an appropriate storage-schedule: Counting, i.e. forming the successor of a natural number, is done by augmenting exactly one position of the number (in S-ary representation) by one and making the positions right of it to zero. In order to calculate the successor-table to the d-vector with number value $(S_{n-1}S_{n-2} \ldots S_1S_o)_s$, we later need at most the tables to

$$(S_{n-1}\ 0\ 0 \ldots 0\ 0\ )_S$$
$$(S_{n-1}S_{n-2}\ 0 \ldots 0\ 0)_S$$
$$(S_{n-1}S_{n-2}S_{n-3} \ldots 0\ 0)_S$$
$$\vdots$$
$$(S_{n-1}S_{n-2}S_{n-3} \ldots S_1 0)_S$$
$$(S_{n-1}S_{n-2}S_{n-3} \ldots S_1S_o)_S$$

These tables may be identical, if some S_K are zero. If we have to augment S_K, our schedule is to replace all the tables for numbers of the form

$$(S_{n-1} \ldots S_K\ xx \ldots xx)_S$$

by the table for $(S_{n-1} \ldots S_K{+}1\ 00 \ldots 00)_S$. The table for $(S_{n-1} \ldots S_K\ 0 \ldots 0)$ is the basis for our actual calculation, and we

have to differentiate in the direction given by κ.
Thus the total storage requirement is for n tables of size $s^n\lceil \log_2 s\rceil$ at most, and the total parallel time required is not more than $2s^n$.
If only a serial computer is at our disposal, we have to go over the s^n entries of each table and to compare with the right neighbour in κ-direction. In order to perform the peak test, we have to compare each entry with its 2n neighbours. Assuming, that these comparisons can be done for all of the $\lceil \log_2 s\rceil$ bits simultaneously, we need $\leq (1+2n)s^n$ time units for each table and thus a total time of $(1+2n)s^{2n}$.
As a final remark, we summarize, that for an n-ary discrete function of input size X (using an alphabet with s letters) the storage required to calculate all ordinary PI's is estimated by nX, while the parallel time is linearly bounded by X. For a serial computer we need $O(nX^2)$ of time. In practice, calculation time can often be reduced since the tables belonging to higher derivatives usually have the trend towards zero or towards constant non-vanishing lines.

References

Davio, M., Deschamps, J.P. and Thayse, A., Discrete and switching functions. McGraw-Hill 1978

Friedman, A.D. and Menon, P.R., Theory & design of switching circuits, Pitman 1975

Garey, M.R. and Johnson, D.S., Computers and intractability, Freeman & Co. 1979

Klar, R. Digitale Rechenautomaten (3. Aufl.), de Gruyter 1983

McCluskey, E.J., Minimization of Boolean functions, Bell Syst. Tech. Journ. 35 (1956), 1417-1444

Meo, A.R., On the synthesis of many-variable switching functions. In: Biorci, G. (ed.), Network and switching theory, Academic Press 1968

Mileto, F. and Putzolu, G., Average values of quantities appearing in Boolean Function minimization, IEEE Trans. El. Comp. 13 (1964), 87-92

Mileto, F. and Putzolu, G., Statistical complexity of algorithms for Boolean functions minimization, Journ. ACM 12 (1965), 364-375

Oberschelp, W. and Remlinger, J., Schaltkreistheorie, Schriften zur Informatik und Angewandten Mathematik der RWTH Aachen, to appear 1984

Quine, W.V., The problem of simplyfiying truth functions, Amer. Math. Monthly 59 (1952), 521-531

Savage, J.E., The complexity of computing, Wiley 1976

Spaniol, O., Arithmetik in Rechenanlagen, Teubner 1976

Thayse, A., Boolean calculus of differences, Springer 1981 (Lecture notes in Comp. Science No. 101)

BOUNDS FOR HODES - SPECKER THEOREM

Pavel Pudlák
Mathematical Institute ČSAV
115 67 Praha
Czechoslovakia

Abstract: In [2] Hodes and Specker proved a theorem which implies that certain Boolean functions have nonlinear formula size complexity. I shall prove that the asymptotic bound for the theorem is n.log log n.

§ 0. Introduction

Let f be a Boolean function, i.e. $f:\{0,1\}^n \rightarrow \{0,1\}$ for some positive integer n. The variables of f will be denoted usually by $x_1,\dots,x_n$. Given $1 \leq i_1 < i_2 < \dots < i_r \leq n$, then $f|x_{i_1}\, x_{i_2} \dots x_{i_r}$ denotes the function from $\{0,1\}^r$ into $\{0,1\}$ obtained by substituting 0's for all the variables of f different from $x_{i_1} \dots x_{i_r}$. In § 5 we shall need substitutions containing also 1's; in such a case we add a superscript to the bar. A base is an arbitrary finite set of Boolean functions Ω ; the elements of Ω are called connectives. The (formula size) complexity of f in base Ω is the number $L_\Omega(f)$ equal to the minimum of the total number of occurrences of variables in an expression over Ω equivalent to f , (or ∞ if such an expression does not exist).

The theorem of Hodes and Specker [2] can be expressed as follows:

If Ω is the base of all binary connectives, (or equivalently $\Omega = \{0,1,\wedge,\oplus\}$), then there exists a function $S(r,n)$ such that for every r

$$\lim_{n \rightarrow \infty} S(r,n) \longrightarrow \infty ,$$

and if

(1) $L_\Omega(f) \leq n.\, S(r,n)$, then there exist $1 \leq i_1 < \dots < i_r \leq n$ and a Boolean function $b(x,y)$ of two arguments such that

$$f|x_{i_1} \dots x_{i_r} \equiv b(x_{i_1} \oplus \dots \oplus x_{i_r} ,\ x_{i_1} \vee \dots \vee x_{i_r}).$$

($\oplus$ denotes addition modulo 2).

This form of the statement is convenient for the proof. For the proofs of lower bounds the theorem is used the other way around, namely, if f does not have a restriction of the form above, then it has complexity $n \cdot S(r,n)$, thus nonlinear. The function S constructed by Hodes and Specker grows for fixed r more slowly than $\log^* n$ (which is the inverse function to $2^{2^{\cdot^{\cdot^{2}}}}$ m-times). The aim of this paper is to show that the theorem holds in an arbitrary base Ω with the estimate (1) replaced by

$$L_\Omega(f) \leq \varepsilon_\Omega . n(\log\log n - \log r)$$

for some $\varepsilon_\Omega > 0$. All the logarithms in this paper are of base 2. Further, I shall show that for $\Omega = \{\oplus, \wedge, \vee\}$ and $r \geq 3$ fixed, the bound is asymptotically best; this was done jointly with S. Poljak. Since Hodes-Specker theorem has a direct application to lower bounds on complexity of symmetric Boolean functions, the paper is also a contribution to this still open field.

My method of the proof is based on the Ramsey theorem. This is not surprising since the theorem (in the form above) resembles the Ramsey theorem. The Ramsey theorem has been used by Vilfan [11] for proving a generalization of Hodes-Specker theorem but in a different manner. The idea of my proof is very simple. Let $\alpha(x_1,\ldots,x_n)$ be a formula of small complexity. For every i,j, $1 \leq i < j \leq n$, define the induced formula $\alpha\{x_i,x_j\}$ in a suitable way and so that the number of occurrences of x_i,x_j in the induced formula is the same as in the original one and it does not contain other variables. This concept is almost the same as the one used by Nečiporuk in [8]. Since the complexity of α is small, the number of occurrences of x_i in α is small for many i´s, say for x_i, $i=1,\ldots, m \geq n/2$. If I restrict myself to this set of variables then the number of nonisomorphic induced formulae is small. Now I colour the pairs (i,j) from $\{1,\ldots,m\}$ by "the shape of $\alpha\{x_i,x_j\}$". Using the Ramsey theorem I get a subset $\{i_1,\ldots,i_r\} \subseteq \{1,\ldots,m\}$ such that for every $a<b$, $c<d$, $\alpha\{x_{i_a}, x_{i_b}\}$ is isomorphic to $\alpha\{x_{i_c}, x_{i_d}\}$. The same is true if I use the induced formula $\beta = \alpha\{x_{i_1},\ldots, x_{i_r}\}$, (defined similarly) instead of α. Then it is shown that formulae with this property (it is natural to call them homogeneous formulae) have a very special shape which forces them to be equivalent to $b(x_{i_1} \oplus \ldots \oplus x_{i_2}, x_{i_1} \vee \ldots \vee x_{i_r})$. This is the longest part of the proof since (as it

is often the case in lower bound proofs) one has to consider many cases. The concept of inducing is, of course, chosen so that β is equivalent to the formula obtained from α by substituting 0's for every variable different from $x_{i_1},\dots,x_{i_r}$.

It seams that the method could be used for some other kinds of complexity, (e.g. number of arithmetic operations), therefore in § 1 I prove a theorem about general terms. In § 2 the theorem is applied to Boolean formulae in order to obtain the main result.

§ 1 Structure of homogenous terms

In this paragraph I shall use terms (denoted by α, β, ...) of the language consisting of a set of variables $V = \{x_1,\dots,x_n\}$, (denoted also by $x,y,z,u,v,\dots$) a set of at least binary operations Ω (denoted by a,b,d ,...) and a single constant c. The set of variables which occur in a term α will be denoted by $V(\alpha)$. We shall agree to write $\alpha(u_1,\dots,u_k)$ to indicate that $V(\alpha) \subseteq \{u_1,\dots,u_k\}$.

It is important to visualize terms as rooted trees in which the vertices are the leaves labelled by the elements of V c and junctions labelled by the elements of Ω of the corresponding arity; moreover the leaves are linearily ordered, which induces a linear ordering of the successors of any junction. The tree structure determines an ordering of the vertices, namely, u is below v iff the shortest path from v to the root contains u iff the subterm determined by v is a subterm of the subterm determined by u. The isomorphism of terms will be defined by $\alpha(u_1,\dots,u_k) \cong \beta(v_1,\dots,v_k)$ iff

$$\alpha(u_1,\dots,u_k) = \beta(v_1/u_1,\dots,v_k/u_k),$$

where y/z denotes the substitution of z for every occurrence of y. From this point on I shall use the less precise notation without the substitution sign.

Definition

Let α be a term, $X \subseteq V$. The term induced by X will be denoted by X and defined inductively by

$cX = c$;

$x_iX = x_i$ if $x_i \in X$,
$\quad = c$ otherwise;

$b(\alpha_1,\dots,\alpha_k) = \alpha_iX$ if $\alpha_jX = c$ for all $j \neq i$,
$\quad = b(\alpha_1X,\dots,\alpha_kX)$ otherwise.

(hence $= c$ if $\alpha_j X = c$ for all j).

Less formally the induced term αX can be defined as follows. If $X \cap V(\alpha) = \emptyset$ then $\alpha X = c$. Otherwise
(i) omit from α all the junctions v such that the subformula determined by v is of the form $b(\beta_1,\dots,\beta_k)$, where $V(\beta_i) \cap X \neq \emptyset$ for at most one i;
(ii) omit all the leaves labelled by the elements of $V(\alpha) - X$;
(iii) at the junctions whose arity does not correspond to the number of predecessors, fill the remaining places by c; and preserve the labeling and the orderings. (The operation (iii) is never used if Ω contains only binary symbols.) Whence we have the following lemma.

Lemma 1.1 $X \subseteq Y \Rightarrow X = (\alpha Y)X$. □

If $X = \{x_{i_1},\dots,x_{i_k}\}$, $Y = \{x_{j_1},\dots,x_{j_k}\}$, $i_1 < \dots < i_k$, $j_1 < \dots < j_k$ then I shall write

$$\alpha X \cong \beta Y$$

instead of

$$\alpha X(x_{i_1},\dots,x_{i_k}) \cong \beta Y(x_{j_1},\dots,x_{j_k}).$$

Definition

A term α is called k-homogeneous over Z iff $\forall X,Y \subseteq Z, |X| = |Y| = k$,

$$\alpha X \cong \alpha Y.$$

I shall omit k and/or Z if $k = 2$, $Z = V$.

Lemma 1.2 If $|V| \geqslant 3$ and α is homogeneous then α is 1-homogeneous.
Proof: easily from Lemma 1.1. □

I shall use also the following notation:

$$\alpha[y_1,\dots,y_k]$$

means that every y_i occurs in α exacitly once and α does not contain other variables;

$$\alpha(u,v]$$

means that $V(\alpha) = \{u,v\}$ and v occurs in α exactly once.

If α can be expressed in the form $\varphi(u,\psi]$, where ψ is not a single variable, then u is uniquely determined; if moreover ψ is minimal, then also φ and ψ are uniquely determined.

Examples.

1. $a(b(d(y,z), e(x,x)),x) = \varphi(x,\psi]$ where
   $\varphi(u,v] = a(v,u)$, $\psi = b(d(y,z), e(x,x))$,

or

$\varphi(u,v] = a(b(v,e(u,u)),u)$, $\psi = d(y,z)$.

2. $a(b(x,y), d(x,e(z,x)))$ cannot be decomposed.

Theorem 1.3 (Structure of homogeneous terms.) Let α be a homogeneous term, $V = \{x_1,\dots,x_n\}$, $n \geq 3$, and suppose α does not contain constant subterms other then c. Then at least one of the following conditions holds:

1) $\alpha = c$;
2) $\alpha = b(\beta_1,\dots,\beta_k)$ and $\beta_1,\dots,\beta_k$ are homogeneous;
3) $\alpha = \varphi(x_1, \varphi(x_2,\dots,\varphi(x_n, \delta] \dots]]$ or
   $\alpha = \varphi(x_n, \varphi(x_{n-1}\dots \varphi(x_1, \delta] \dots]]$ and δ is homogeneous;
4) $\alpha = \varphi(x_1, \varphi(x_2,\dots \varphi(x_{n-1}, \psi(x_n)] \dots]]$,
   or the same with the reverse order of variables;
5) $\alpha = \varphi[\delta(x_1),\dots,\delta(x_n)]$ where, for some $b \in \Omega$ and $i<j$, every nontrivial subformula of $\varphi[y_1,\dots,y_n]$ is of the form

$$b(c,\dots,c,\mu,c,\dots,c,\nu,c,\dots,c).$$

   where μ, ν are on the i-th and j-th places.

If moreover α is 3-homogeneous, then at least one of the conditions 1)-4) holds.

An example of a homogeneous term is shown on Fig.1.

Proof:

(i) First I prove the theorem for Ω consisting of binary operations only. The binary operations will be denoted by an infix o, thus different occurrences of o may denote different operations. Let α be a homogeneous term. I omit the trivial case when α is a variable or the constant, and assume $\alpha = \beta \circ \gamma$. Consider the following three main cases (with several subcases):

1. $V(\beta) \cap V(\gamma) = \emptyset$.

1.1. If $V(\beta) = \emptyset$, then $\alpha X = \gamma X$, thus β and γ are homogeneous, which is condition 2) of the theorem. If $V(\gamma) = \emptyset$, then we get the same.

1.2. Let $u \in V(\beta)$, $v \in V(\gamma)$. Then

(1) $\alpha\{u,v\} = \beta\{u\} \circ \gamma\{v\}$.

Let $w \in V$ be an arbitrary element, and let δ be the smallest subterm of α containing all the occurrences of w. If there were an occurrence of z in δ, $z \neq w$, then $\alpha\{w,z\}$ could not be isomorphic to (1). Thus I have proved that

$$\alpha = \varphi[\delta_1(x_1),\dots,\delta_n(x_n)].$$

1-homogeneity of α, (Lemma 1.2), implies

$$\delta_1(t) = \dots = \delta_n(t),$$

and homogeneity implies that all the operations in

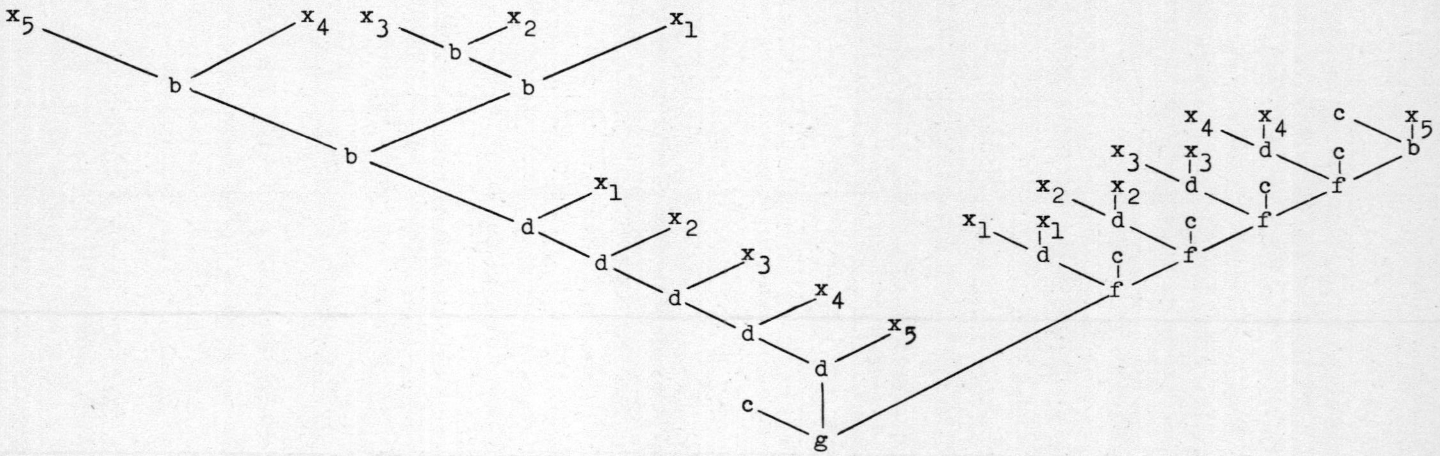

Fig.1. An example of a homogeneous term;

$n = 5$, b,d binary, f,g ternary.

$\varphi[y_1,\dots,y_n]$ are equal, thus we get 5).

2. $V(\beta) = V(\gamma) = V$. Then

$$\alpha X = \beta X \circ \gamma X,$$

thus β, γ are homogeneous - condition 2).

3. $V(\beta) \cap V(\gamma) \neq \emptyset$ and $V(\beta) \subsetneqq V$. (The case $V(\gamma) \subsetneqq V$ is symmetric).

3.1. Claim: $|V(\beta)| = 1$.

If not, then we have $u \in V(\beta)$, $v \in V(\beta) \cap V(\gamma)$, $w \in V(\beta)$, $w \neq v$. Then

$$\alpha\{u,v\} = \beta\{v\} \circ \gamma\{u,v\},$$
$$\alpha\{w,v\} = \beta\{w,v\} \circ \gamma\{w,v\}.$$

(w may not occur in $\gamma\{w,v\}$). This two terms cannot be isomorphic, since $\beta\{v\}$ contains only one variable, while $\beta\{w,v\}$ contains two.

3.2. Claim: $V(\beta) = \{x_1\}$ or $V(\beta) = \{x_n\}$.

If $V(\beta) = \{x_i\}$, $1 < i < n$, then the first occurrence of a variable both in $\alpha\{x_1,x_i\}$ and $\alpha\{x_i x_n\}$ is x_i, thus they would not be isomorphic.

Henceforth assume $V(\beta) = \{x_1\}$ ($=\{x_n\}$ being symmetric). By the assumption of this case $x_1 \in V(\gamma)$, hence $V(\gamma) = V(\alpha) = V$.

3.3 Claim: $\alpha = \varphi_1(x_1, \varphi_2(x_2, \dots \varphi_{n-1}(x_{n-1}, \psi] \dots]]$.

I shall prove by induction for $i = 1,\dots,n-1$,

(2) $\alpha = \varphi_1(x_1,(\varphi_2(x_2,\dots \varphi_i(x_i, \psi_i] \dots]]$.

For $i = 1$ this is just the assumption $V(\beta) = \{x_1\}$.

Suppose (2) for $i < n-1$ and chose ψ_i minimal. Then $\psi_i = \mu \circ \nu$ where $V(\mu)$, $V(\nu)$ are not subsets of $\{x_i\}$.

3.3.1. First I show

$$x_{i+1} \in V(\mu) \cap V(\nu).$$

Suppose not, say $x_{i+1} \in V(\mu) - V(\nu)$. Then $x_j \in V(\nu)$ for some $j \neq i,i+1$. If $i+1 < j$, then

$$\alpha\{x_{i+1},x_j\} = \mu\{x_{i+1},x_j\} \circ \nu\{x_j\} \not\cong$$
$$\alpha\{x_1,x_2\} = \beta\{x_1\} \circ \gamma\{x_1,x_2\},$$

since $x_2 \in V(\gamma)$, - contradiction. If $j < i$, then

$\alpha\{x_j,x_{i+1}\} = \varphi_j(x_j,\mu\{x_j,x_{i+1}\} \circ \nu\{x_j\}] = \varphi'(x_j,\mu\{x_j,x_{i+1}\}]$;

$\alpha\{x_j,x_i\} = \varphi_j(x_j, \varphi_i(x_i, \psi_i\{x_i,x_j\}]]$.

This two terms cannot be isomorphic, since there are more occurrences of u in $\varphi'(u,v]$ than in $\varphi_j(u,v]$, - contradiction.

3.3.2. Now I show

(3) $V(\mu) = \{x_{i+1}\}$.

For $j > i+1$, x_j cannot belong to $V(\mu)$ since

$\alpha\{x_{i+1}, x_j\} = \mu\{x_{i+1}, x_j\} \circ \mu\{x_{i+1}, x_j\} \cong$

$\cong \alpha\{x_1, x_2\} = \beta\{x_1\} \circ \gamma\{x_1, x_2\}$.

If $x_j \in V(\mu)$ for some $j < i+1$, then the following two terms could not be isomorphic

$\alpha\{x_j, x_{i+1}\} = \varphi_j(x_j, \mu\{x_j, x_{i+1}\} \circ \nu\{x_j, x_{i+1}\}]$;

$\alpha\{x_j, x_n\} = \varphi_j(x_j, \mu\{x_j\} \circ \nu\{x_j, x_n\}]$.

Thus, by (3), ψ_i can be expressed in the form

$\psi_i = \varphi_{i+1}(x_{i+1}, \psi_{i+1}]$,

which ends the proof of the claim.

3.4. Claim: If ψ is minimal in 3.3, then

(4) $\varphi_1(u,v = \dots = \varphi_{n-1}(u,v]$.

Let ψ be minimal.If $V(\psi) = \{x_n\}$, then, for $i < n$,

$$\alpha\{x_i, x_n\} = \varphi_i(x_i, \psi\{x_n\}],$$

and the claim follows easily from homogeneity. Otherwise ψ contains at least two variables, hence $\psi = \mu \circ \nu$ for some μ, ν . Using the same argument as in 3.3.1 one can show that $x_n \in V(\mu) \cap V(\nu)$. Thus for $i < n$

(5) $\alpha\{x_i, x_n\} = \varphi_i(x_i, \mu\{x_i, x_n\} \circ \nu\{x_i, x_n\}]$,

whence the claim follows easily.

If $V(\psi) = \{x_n\}$, then α is of the form 4). Therefore it remains to consider the case when $V(\psi)$ is larger. Then $V(\psi) = V$, since α must contain the same number of occurrences of every variable and by 3.3 and 3.4.

3.5. Claim: $\psi = \varphi_n(x_n, \chi]$, for some φ_n and χ. In the proof of 3.4 I have shown (5) (for every $i < n$). This term must be isomorphic to

$\alpha\{x_1, x_2\} = \varphi_1(x_1, \varphi_2(x_2, \psi_2\{x_1, x_2\}]]$.

Moreover we know that $\varphi_1(u,v] = \varphi_2(u,v]$, thus for some term $\int$,

$\varphi_2(x_2, \psi_2\{x_1, x_2\}] = \beta(x_2) \circ \int \cong \mu\{x_i, x_n\} \circ \nu\{x_i, x_n\}$.

Hence $V(\mu) = \{x_n\}$, which proves the claim.

Further I shall assume that χ in 3.5 is minimal.

3.6. Claim: $\varphi(u,v] =_{df} \varphi_1(u,v] = \dots = \varphi_{n-1}(u,v]$ does not contain more occurrences of u then $\varphi_n(u,v]$.

3.6.1. First assume $x_n \in V(\chi)$. Then

(6) $\alpha\{x_1,x_2\} = \varphi(x_1, \varphi(x_2, \chi\{x_1,x_2\}]] \cong \alpha\{x_1,x_n\} = \varphi(x_1, \varphi_n(x_n, \chi\{x_1\}]],$

whence the claim follows. In the rest of the proof of the claim I assume $x_n \in V(\chi)$ and $\chi = \mu \circ \nu$, for some $\mu \circ \nu$.

3.6.2. If $x_j \in V(\mu) \cap V(\nu)$, $j < n$, then $\chi\{x_j,x_n\}$ is the minimal term such that $\alpha\{x_j,x_n\}$ can be expressed as

$$\varphi(x_j, \varphi_n(x_n, \chi\{x_j,x_n\}]].$$

Comparing this decomposition with the decomposition of $\alpha\{x_1,x_2\}$ in (6) the claim follows.

3.6.3. I shall show that $V(\mu) \cap V(\nu) \neq \{x_n\}$. If not, then there are $x_i \in V(\mu) - V(\nu)$, $x_j \in V(\nu) - V(\mu)$, $i,j \neq n$, since χ is minimal. Then

$$\alpha\{x_i,x_n\} = \varphi(x_i, \varphi_n(x_n, \mu\{x_i,x_n\} \circ \nu\{x_n\}]] \not\cong$$
$$\not\cong \alpha\{x_j,x_n\} = \varphi(x_j, \varphi_n(x_n, \mu\{x_n\} \circ \nu\{x_j,x_n\}]],$$

a contradiction.

3.6.4. It remains to consider $V(\mu) \cap V(\nu) = \emptyset$. Suppose it is so and $x_n \in V(\mu)$. (The proof would be similar if $x_n \in V(\nu)$). Choose x_i,x_j as in 3.6.3. If $i < j$, then

(7) $\alpha\{x_i,x_j\} \cong \alpha\{x_i,x_n\},$

hence

$$\varphi(x_j, \mu\{x_j\} \circ \nu\{x_i\}] \cong \varphi_n(x_n, \mu\{x_n\} \circ \nu\{x_i\}].$$

This isomorphism maps $\nu\{x_i\}$ onto itself, hence $\mu\{x_j\} \circ \nu\{x_i\}$ onto $\mu\{x_n\} \circ \nu\{x_i\}$, thus $\varphi(u,v] = \varphi_n(u,v]$, which is even more than I wanted to prove. If $j < i$ then in (7) $\mu\{x_j\}$ must be mapped onto $\nu\{x_i\}$, by the uniqueness of the decomposition, and then $\mu\{x_j\} \circ \nu\{x_i\}$ onto $\mu\{x_n\} \circ \nu\{x_i\}$ which is impossible. Thus 3.6 is proved.

Now I can end Case 3. Since

$\alpha\{x_1\} = \varphi(x_1, \chi\{x_1\}] \cong \alpha\{x_n\} = \varphi_n(x_n, \chi\{x_n\}]$, and by 3.6, either

$$\varphi_n(u,v] = \varphi(u,v],$$

or there is some φ' such that

$$\varphi_n(u,v] = \varphi(u,\varphi'(u,v]].$$

Thus for some term δ,

$$\alpha = \varphi(x_1, \varphi(x_2, \ldots \varphi(x_n, \delta] \ldots]].$$

Since for $i < j$,

$\alpha\{x_i, x_j\} = \varphi(x_i, \varphi(x_j, \delta\{x_i, x_j\}]]$,

δ is homogeneous, i.e., condition 3).

(ii) Now let α be a homogeneous term over a general base Ω . For every operation $b \in \Omega$ of arity $k \geqslant 3$, choose $k-1$ new binary operations $b_1, \ldots, b_{k-1}$. Let α' be the term resulting from α by applying all the possible substitutions of the form

$b(\beta_1, \ldots, \beta_k) \longmapsto b_1(\beta_1, b_2(\beta_2, \ldots b_{k-1}(\beta_{k-1}, \beta_k) \ldots))$.

Claim: α' is homogeneous. This is because similar operations produce $\alpha' X$ from αX, namely, the operations of the form

$b(c, \ldots, c, \alpha_{i_1}, c, \ldots, c, \alpha_{i_2}, c, \ldots \ldots, c, \alpha_{i_j}, c, \ldots, c) \longmapsto$

$\longmapsto b_{i_1}(\alpha_{i_1}, b_{i_2}(\alpha_{i_2}, \ldots b_{i_{j-1}}(\alpha_{i_{j-1}}, \alpha_{i_j}) \ldots))$.

By (i) I know that at least one of the five conditions of the theorem holds for α'. Accordingly I shall consider four cases.

1. $\alpha' = c$, then $\alpha = c$.

$\alpha' = \beta \circ \gamma$, and β, γ are homogeneous, then

$\alpha = b(\beta, \beta_2, \ldots, \beta_k)$, $k \geqslant 2$,

where $V(\beta) = V = V(\gamma) = \bigcup_{i=2}^{k} V(\beta_i)$. Whence

$\alpha X = b(\beta X, \beta_2 X, \ldots, \beta_k X)$,

thus $\beta, \beta_2, \ldots, \beta_k$ are homogeneous - condition 2).

3. If 3) or 4) holds for α' then

$\alpha' = \varphi'(x_1, \varphi'(x_2, \psi_2]]$,

(or with x_n, x_{n-1} instead of x_1, x_2).

3.1. Claim: φ' corresponds to a part of the original term α . Suppose not, then for some operation b of α , $i < j$ and terms $\xi'(x_1), \mu', \nu'$,

(8) $\alpha' = b_i(\xi'(x_1), b_j(\mu', \nu'))$,

or for some term ψ ,

$\alpha' = \psi(x_1, b_i(\xi'(x_1), b_j(\mu', \nu'))$,

where

(9) $b_j(\mu', \nu') = \varphi'(x_2, \psi_2]$.

Without loss of generality I shall consider only (8). Then α is of the following form

$b(\ldots \xi(x_1) \ldots \mu \ldots \nu_1 \ldots \nu_2 \ldots \ldots \nu_m \ldots)$,

where only $\xi, \mu, \nu_1, \ldots, \nu_m$ are different from c , and
$V(\mu) = V(\mu'), \quad \bigcup_{i=1}^{m} V(\nu_i) = V(\nu')$.
By (9), $V(\mu') = \{x_2\}$ or $V(\nu') = \{x_2\}$. In the first case

$$\alpha\{x_1,x_2\} = b(\ldots \xi(x_1)\ldots \mu(x_2) \ldots \nu_1\{x_1,x_2\} \ldots \nu_m\{x_1,x_2\} \ldots),$$

while a term corresponding to $\mu(x_2)$ in $\alpha\{x_1,x_3\}$ is missing - contradiction. The second case is ruled out by a similar argument, thus the claim is proved.

Now the condition 3) or 4) follows easily from the claim. If e.g.
$\alpha' = \varphi'(x_1, \varphi'(x_2,\ldots \varphi'(x_n, \delta']\ldots]]$,
then

$$= \alpha(x_1, \varphi(x_2,\ldots \varphi(x_n, \delta]\ldots]],$$

where φ is the term corresponding to φ'. Homogeneity of δ can be proved in the same way as in (i).

4. $\alpha' = \varphi'[\delta'(x_1),\ldots,\delta'(x_n)]$, and all the operations in $\varphi'[y_1,\ldots,y_n]$ are the same. If this is an operation of Ω then, clearly, $\alpha = \varphi'[\delta_1(x_1),\ldots,\delta_n(x_n)]$. If it is b_i, for some $b \in \Omega$, then different occurrences of b_i in α must correspond to different occurrences of b in α , (since they have the same index i). Thus again $\alpha = \varphi[\delta_1(x_1),\ldots,\delta_n(x_n)]$. The rest is the same as in (i), 1.2.

(iii) Let α be 2- and 3-homogeneous. Suppose 5) holds. Then because of 3-homogeneity also 4) holds. □

Remarks.
1. An easy corollary of the theorem is that if α is 1,2,3-homogeneous, then it is k-homogeneous for <u>every</u> $k \leq n$. It would be more appropriate to reserve the word "homogeneous" for this concept.

2. The theorem can be inverted if the relation of ψ to φ in 4) is specified.

§ 2 The lower bound theorem

In this paragraph I shall apply Theorem 1.3 to Boolean formulae. For sake of convenience I consider only bases which contain nullary connectives (constants) 0,1, unary connectives $\neg(x)$, $\underline{0}(x)$, $\underline{1}(x)$, and do not contain the identity unary connective. The letters will be used in a similar way as in § 1, i.e. $\alpha, \beta, \ldots$, are formulae,

$x, y, z, u, v, \ldots$ variables, $a, b, d, \ldots$ at least unary connectives, c, c_i for $0, 1$, etc. Also I shall assume that the formulae do not contain parts of the form $a_1(a_2(\beta)), (a_1, a_2$ unary connectives), or $(b(c_1, \ldots, c_k)$, (b a k-ary connective and $c_i \in \{0,1\}$).

Definition

Let α be a Boolean formula, X a set of variables. Then αX_B - the formula B-induced by X is the formula produced from α by the following rules:

1) substitute 0 for the variables of X;
2) replace every part $b(c_1, \ldots, c_k)$ by the corresponding constant;
3) replace every part $b(\alpha_1, \ldots \alpha_k)$, where only α_i is a nonconstant, by α_i or $a(\alpha_i)$, where a is the corresponding unary connective;
4) replace every part $a_1(a_2(\beta))$ by β or $a(\beta)$, where a is the corresponding unary connective $(a(x) = a_1(a_2(x)))$.

k-B-homogeneous and B-homogeneous formulae are defined analogically as in § 1.

For a formula α denote by $F(\alpha)$ the term which results from α by omitting the unary connectives of α and replacing 0's and 1's by the symbol c. (Clearly, if α is a formula in the sense of this paragraph then $F(\alpha)$ is a term in the sense of § 1.)

Lemma 2.1

1) $X \subseteq Y \Rightarrow \alpha X_B = (\alpha Y_B) Y_B$;
2) α is B-homogeneous over $Z \Rightarrow \alpha Z_B$ is B homogeneous over Z, (α may contain other variables than those in Z);
3) $F(\alpha X_B) = (F(\alpha))X$;
4) α is B-homogeneous over $Z \Rightarrow F(\alpha)$ is homogeneous over Z;
5) α is B-homogeneous over Z, $|Z| \geq 3 \Rightarrow \alpha$ is 1-B-homogeneous.

Proof:

If $V(\alpha) \cap X \neq \emptyset$, then αX_B can be constructed as follows:

1. construct $(F(\alpha))X$,
2. add the corresponding unary connectives and replace the symbol c by the corresponding constants.

Whence we obtain 3). Then using Lemma 1.1 we get 1). The implications 1) $\Rightarrow$ 2), 1) $\Rightarrow$ 5), 3) $\Rightarrow$ 4) are trivial. □

Lemma 2.2 (Structure of B-homogeneous formulae). Let $\alpha(x_1,\dots,x_n)$ be B-homogeneous over $V = \{x_1,\dots,x_n\}$, $n \geq 3$. Then at least one of the following conditions holds:

1) $\alpha \equiv$ constant;
2) $\alpha = b(\beta_1,\dots,\beta_k)$, $b \in \Omega$, $k \geq 1$, and $\beta_1,\dots,\beta_k$ are B-homogeneous;
3) $\alpha = \varphi(x_1, \varphi(x_2,\dots \varphi(x_{n-1}, \psi(x_n, \delta]]\dots]]$, or with the reverse order of variables, where δ is B-homogeneous, $\varphi(u,v] = \psi(u,v]$ or $\varphi(u,v] = \psi(u, \neg v]$, and $\varphi(u,v]$ is equivalent either to $u \oplus v$ or $u \vee v$ or $\neg u \wedge v$ or v;
4) $\alpha = \varphi(x_1, \varphi(x_2,\dots \varphi(x_{n-1}, \psi(x_n)]\dots]]$, or with the reverse order of variables, where
   (i) either $\varphi(u,v] \equiv u \oplus v$ and ($\psi(u) \equiv u$ or $\psi(u) \equiv \neg u$),
   (ii) or $\varphi(u,v] \equiv u \vee v$ and $\psi(u) \equiv u$,
   (iii) or $\varphi(u,v] \equiv \neg u \wedge v$ and $\psi(u) \equiv \neg u$;
5) $\alpha = \varphi[\delta(x_1),\dots, \delta(x_n)]$, where for some $c \in \{0,1\}$,
   (i) either $\varphi[y_1,\dots,y_n] \equiv c \oplus y_1 \oplus \dots \oplus y_n$,
   (ii) or $\varphi[y_1,\dots,y_n] \equiv c \oplus (y_1 \vee \dots \vee y_n)$ and $\delta(y) \equiv y$,
   (iii) or $\varphi[y_1,\dots,y_n] \equiv c \oplus (y_1 \wedge \dots \wedge y_n)$ and $\delta(y) \equiv \neg y$.

Proof:
Let α be B-homogeneous over V. By Lemma 2.1, $F(\alpha)$ is homogeneous. If $F(\alpha) = c$, then α is constant. Now assume that α is not constant. Then by Theorem 1.3 at least one of the conditions 2)-5) holds for $F(\alpha)$. I shall consider five cases.

1. $\alpha = a(\beta)$, where a is a unary connective. Since α is not constant, $a = \neg$, and $V(\beta) = V$. Let $\emptyset \neq X \subseteq V$, $|X| = 2$, then αX_B is either a variable or a more complicated formula. If αX_B is a variable, or if the main connective of it is at least binary, then $\alpha X_B = \neg \beta X_B$. If $\alpha X_B = b(\gamma)$, $b(x) = \neg(x)$, $\underline{0}(x)$, $\underline{1}(x)$ resp. then $\beta X_B = \gamma$, $\underline{1}(\gamma)$, $\underline{0}(\gamma)$ resp. Thus

$$\alpha X_B \cong \alpha Y_B \Rightarrow \beta X_B \cong \beta Y_B,$$

hence β is homogeneous. This proves condition 2) of the lemma.

Henceforth I shall assume that the main connective of α is at least binary.

2. $F(\alpha) = b(\beta_1,\dots,\beta_k)$, $\beta_1,\dots,\beta_k$ homogeneous. Then $\alpha = b(\gamma_1,\dots,\gamma_k)$ and $V(\gamma_i) = V(\beta_i) = V$ for $i = 1,\dots,k$. Thus
$\alpha X_B = b(\gamma_1 X_B,\dots,\gamma_k X_B)$,

whence $\gamma_1,\dots,\gamma_k$ are B-homogeneous - condition 2).

3. $F(\alpha) = \varphi'(x_1, \varphi'(x_2,\dots, \varphi'(x_n,\delta']]$ where δ' is homogeneous. (I shall not consider the symmetric case where the variables are in the reverse order.) Then α can be decomposed as follows

(1) $\alpha = \varphi_1(x_1, \varphi_2(x_2,\dots \varphi_n(x_{n-1},\delta]\dots]]$.

The decomposition might not be unique because of unary connectives. Therefore I postulate that the unary connective which is possible to include either in φ_i or φ_{i+1}, $i<n$ be included in φ_i. The unary connective between φ_n and δ will be included in φ_n if and only if there is the same connective between φ_1 and φ_2. Now $\varphi_2,\dots,\varphi_n$ have the main connective at least binary; the same is true about φ_1, since this is the assumption above. The formula $\alpha\{x_1\}$, hence any $\alpha\{x_i\}$ too, has the main connective at least binary. This proves that

(2) $\varphi_1(0,v] \equiv \dots \equiv \varphi_{n-1}(0,v] \equiv v$.

Hence, for $i<n$,

$\alpha\{x_i,x_n\} = \varphi_i(x_i, \varphi_n(x_n, \delta]\{x_i,x_n\}]$,

therefore $\varphi_1 = \varphi_2 = \dots = \varphi_{n-1}$. I shall denote these by φ and φ_n by ψ. Since

$F(\varphi(u,v]) = \varphi'(u,v] = F(\psi(u,v])$

and since

$\alpha\{x_1\} \cong \alpha\{x_n\}$,

the only difference between $\varphi(u,v]$ and $\psi(u,v]$ might be the unary connective at v. Recall that the decomposition (1) is chosen so that if there is a unary connective at v in $\psi(u,v]$ then the same is at v in $\varphi(u,v]$. By (2) the connective can be only the negation. Thus

(3) $\varphi(u,v] = \psi(u,v]$ or $\varphi(u,v] = \psi(u,\neg v]$.

Then, by (2),

(4) $\psi(0,v] = v$ or $\psi(0,v] = \neg v$ respectively.

Since $\delta' = F(\delta)$ is homogeneous and α is B-homogeneous, the only difference between δX_B and δY_B, $|X|=|Y|=2$, might be the main unary connective. Using (2), (3) and (4), one can easily compute that the parity of the number of negations before the last junction of δ is always the same. Hence δX_B and δY_B are isomorphic, which proves that δ is B-homogeneous. The rest of condition 3) follows from $\varphi(0,v] \equiv v$.

4. Suppose condition 4) of Theorem 1.3 holds for $F(\alpha)$. Then , using quite a similar argument, one can prove condition 4) of the lemma. The relation between φ and ψ follows easily from B-homogeneity and the assumption that α is nonconstant.

5. Suppose condition 5) of Theorem 3.1 holds for $F(\alpha)$. Then $\alpha = \varphi[\delta_1(x_1),\dots, \delta_n(x_n)]$. This decomposition is not unique again. If I choose $\delta_1,\dots, \delta_n$ so that their main connectives are not unary, then, by 1-B-homogeneity, $\delta_1 = \delta_2 = \dots = \delta_n$. Thus I can drop out the subscripts at δ. Since α is nonconstant, $\delta(x) \equiv x$ or $\equiv \neg x$. The assumption about $F(\alpha)$ implies that there are occurrences of only one connective b of arity $k \geq 2$ in φ, and they are always in a contex

(5) $\dots b(c_1,\dots,c_i, \mu, c_{i+2},\dots,c_j, \nu, c_{j+2},\dots,c_k) \dots .$

Every Boolean formula of two arguments is equivalent to a formula

(6) $$d(e_1(x) \circ e_2(y)),$$

where any of the unary connectives $d, e_1 e_2$ may be missing and $\circ \in \{\oplus, \vee\}$. Choose such a formula for

$$b(c_1,\dots,c_i,x,c_{i+2}\dots c_j,y,c_{j+2},\dots,c_k),$$

replace every occurrence of (5) in φ by this formula, and collect consecutive unary connectives. Denote by ψ the resulting formula. ψ contains only binary and unary connectives. Denote by $\beta = \psi[\delta(x_1),\dots, \delta(x_n)]$.
Since $\psi \equiv \varphi$,
$\beta\{x_i\} = \alpha\{x_i\}$.
A similar argument shows that $\beta\{x_i,x_j\}$, $1 \leq i < j \leq n$ can be constructed from $\alpha\{x_i,x_j\}$ using the same procedure by which I constructed β from α. Thus β is B-homogeneous.

Now I shall consider three cases.

5.1. The binary connectives of ψ are $\oplus$. If there occurs a constant unary connective in ψ, then some, hence by 1-B-homogeneity every, $\beta\{x_i\}$ must be constant. Since the binary connectives are $\oplus$, this is possible only if there is a constant unary connective on every path from a variable to the root. Then ψ, hence also α, is constant - contradiction. If there is no constant unary connective in ψ, then $\psi(y_1,\dots,y_n)$, hence $\varphi(y_1,\dots,y_n)$, is equivalent to $t \oplus y_1 \oplus \dots \oplus y_n$, $t \in \{0,1\}$, i.e. condition 5), (i) of the lemma.

5.2. The binary connectives of ψ are $\vee$ and $\delta(x) \equiv x$. Then $\psi[x_1,\ldots,x_n]$ is B-homogeneous, as it is equivalent to β. Let, for $1 \leq i < j \leq n$,

(7) $\psi\{x_i,x_j\} = f(g_1(x_i) \vee g_2(x_j))$,

where any of the unary connectives f, g_1,g_2 may be missing. If f is not missing, then it is also the main connective of ψ (consider $\psi\{x_1,x_n\}$). It was shown in part 1 of the proof that in such a case after removing the main connective I get a B-homogeneous formula again. Thus without loss of generality I can assume that f is missing.

Claim: The unary connectives may occur only at the variables of ψ.

Suppose not. Take some junction such that there is exactly one unary connective below it. Let i,j be such that the paths from x_i and x_j to the root meet at it for the first time. Then f must occur in (7) - contradiction.

$\underline{1}$ cannot occur in ψ since ψ is not constant. Suppose ψ contains $\neg$. Then by 1-B-homogeneity it must contain at least two occurences of $\neg$. By B-homogeneity it must contain three such occurrences. But then f must occur in (7) again - contradiction. Now we know that only $\underline{0}$ may occur in ψ. But if $\underline{0}$ was at one variable then it must be at every one. This would be a contradiction since ψ is not constant. Thus I have proved that ψ may contain a unary connective only as the main connective. Hence we have 5), (ii) of the lemma.

5.3. The binary connectives of ψ are $\vee$ and $\delta(x) \equiv \neg x$. This is dual to 5.2, hence

$$\psi[y_1,\ldots,y_n] \equiv c \oplus (x_1 \wedge \ldots \wedge x_n), \quad c \in \{0,1\},$$

and 5) (iii) follows.□

<u>Corollary 2.3</u> Let α be B-homogeneous over V, $V(\alpha) \subseteq V = \{x_1,\ldots,x_n\}$ $n \geq 3$. Then, for some Boolean function of two arguments b,

$$\alpha(x_1,\ldots,x_n) \equiv b(x_1 \oplus \ldots \oplus x_n, x_1 \vee \ldots \vee x_n).$$

<u>Proof</u>:

Use induction and Lemma 2.2. For example let

$$\alpha = \varphi(x_1, \varphi(x_2,\ldots\varphi(x_n, \delta]\ldots]],$$

where δ is B-homogeneous and $\varphi(u,v] \equiv \neg u \wedge v$.

Then by the induction assumption

$\delta \equiv b(x_1 \oplus \ldots \oplus x_n, x_1 \vee \ldots \vee x_n)$.

Hence

$\alpha \equiv \neg x_1 \wedge \neg x_2 \wedge \ldots \wedge \neg x_n \wedge b(x_1 \oplus \ldots \oplus x_n, x_1 \vee \ldots \vee x_n) \equiv$
$\equiv (x_1 \vee x_2 \vee \ldots \vee x_n) \wedge b(x_1 \oplus \ldots \oplus x_n, x_1 \vee \ldots \vee x_n)$.

Theorem 2.4 (Main Theorem.) For every base Ω, there exists a positive constant ε_Ω such that, for every $f : \{0,1\}^n \to \{0,1\}$, $r \geq 3$, if

$L_\Omega(f) \leq \varepsilon_\Omega n(\log \log n - \log r)$,

then there exist $1 \leq i_1 < \ldots < i_r \leq n$ and a Boolean function b of two arguments such that

$f|x_{i_1} \ldots x_{i_r} \equiv b(x_{i_1} \oplus \ldots \oplus x_{i_r}, x_{i_1} \vee \ldots \vee x_{i_r})$.

Proof:

1. Let Ω and $r \geq 3$ be given. Let $\beta(x_1,\ldots,x_m)$ be a k-formula, which means that no variable occurs in β more than k-times. Then every B-induced formula βX_B is again a k-formula. There exists a constant $C > 0$ such that the number of k-formulae with two variables is at most $\ell = 2^{C.k}$. Thus ℓ is an upper bound to the number of non-isomorphic formulae βX_B, where $|X| = 2$.

By the Ramsey theorem (see e.g. [7]), if

(1) $m \geq \ell^{r\ell}$

then there exists $Z = \{x_{i_1},\ldots,x_{i_r}\}$, $1 \leq i_1 < \ldots < i_r \leq n$, such that the formulae B-induced by two element subsets of Z are isomorphic, i.e. β is B-homogeneous over Z. By Lemma 2.1, βZ_B is B-homogeneous over Z and, by Corollary 2.3,

$f|x_{i_1} \ldots x_{i_r} \equiv \beta Z_B \equiv b(x_{i_1} \oplus \ldots \oplus x_{i_r}, x_{i_1} \vee \ldots \vee x_{i_r})$.

If

(2) $$k \leq \frac{\log \log m - \log r - \log C}{C+1}$$

then

$C.k + \log k + \log C + \log r \leq (C+1).k + \log C + \log r \leq \log \log m$,

$\Rightarrow 2^{C.k} . C.k.r \leq \log m \Rightarrow \ell^{r\ell} \leq m$.

Hence (2) $\Rightarrow$ (1).

3. Let $f: \{0,1\}^n \to \{0,1\}$, $N = L_\Omega(f)$. Then there exists a formula $\alpha(x_1,\ldots,x_n) \equiv f$ such that the total number of occurrences of va-

riables in α is N. Thus there are $m \geq n/2$ variables which have at most $2N/n$ occurrences each. Let X be the set of these variables. Denote by $\beta = \alpha X_B$. Then β is a k-formula, where $k = \lfloor 2N/n \rfloor$. By 1 and 2, in order to find the required restriction of f, it is enough to have (2), i.e. solving the inequalities,

$$N \leq \frac{1}{2(C+1)} \cdot n \cdot (\log(\log n-1) - \log r - \log C).$$

If $\varepsilon > 0$ is sufficiently small, then the last inequality is implied by

$$N \leq \varepsilon \cdot n \cdot (\log \log n - \log r),$$

for every N and r. □

Remarks.

1. The theorem is true also for $r = 2$, but it is trivial, since, instead of the bound for $L_\Omega(f)$, it is enough to have $n \geq 3$. (Find x_i, x_j, $i \neq j$, such that $f(0^{i-1} 1\, 0^{n-i}) = f(0^{j-1}\, 1\, 0^{n-j})$.) On the other hand $r = 3$ suffices for most applications.

2. For more precise estimates it is better to preserve the constant term $-\log C$ in (2). A direct computation gives for the base Ω of all at most binary connectives the following bound

$$L_\Omega(f) \leq \frac{1}{34} \cdot n \cdot (\log \log n - \log r - 17).$$

3. In the same manner as Hodes and Specker did, I can show that the condition of the theorem can be strengthened to

$$f \mid x_{i_1} \ldots x_{i_r} \equiv b(x_{i_1} \vee \ldots \vee x_{i_r})$$

for some unary Boolean function b , if $\Omega = \{0, 1, \underline{0}, \underline{1}, \neg, \vee, \wedge\}$. In the light of the results of Khrapčenko [3],[4] and Kričevskij [6], see also [10], this is uninteresting however.

§ 3 Some applications

Corollary 3.1 For every base Ω there is a constant $\delta_\Omega > 0$ such that, for every function $f:\{0,1\}^n \longrightarrow \{0,1\}$ and $1 < k < n-3$, if for every vector $\vec{c} \in \{0,1\}^n$

$$\sum c_i = k \Rightarrow f(c) = 0, \text{ (resp. 1)},$$

$$\sum c_i = k + 2 \Rightarrow f(c) = 1, \text{ (resp. 0)},$$

then

$L_{\Omega}(f) \geq \delta_{\Omega} . n . \log \log n .$ □

The proof is almost identical with the proof of Symmetric Function Lower Bound Theorem of [1]. Hence I give only a hint: if $k \leq n/2$, consider a function obtained from f by substituting $k-1$ one's in f; if $k > n/2$, do the same with the dual function. (See also the proof of Theorem 4.2 below.)

A function $f: \{0,1\}^n \rightarrow \{0,1\}$ is called symmetric, if $f(\vec{c})$ depends only on $\sum c_i$. The symmetric threshould functions T_k^n, defined by

$$T_k^n(\vec{c}) = 1 \quad \text{iff} \quad \sum_{i=1}^{n} c_i \geq k,$$

are examples of symmetric functions. Corollary 3.1 can be applied to T_k^n if $1 < k < n-1$, however a more general theorem holds.

Corollary 3.2 For all but sixteen symmetric functions $f: \{0,1\}^n \rightarrow \{0,1\}$,

$L_{\Omega}(f) \geq \delta_{\Omega} . n . \log \log n .$ □

Such a corollary (but with the old bound) was derived by Khrapčenko [5]. The sixteen functions are

$$c_1 \oplus (c_2 \wedge (x_1 \oplus \dots \oplus x_n)) \oplus (c_3 \wedge x_1 \wedge \dots \wedge x_n) \oplus$$
$$\oplus (c_4 \wedge (x_1 \vee \dots \vee x_n)), \; c_1, c_2, c_3, c_4 \in \{0,1\}.$$

I shall show a less direct application of Theorem 2.4. Let $n = m.k$ and let $\underline{C}$ denote a $m \times k$ matrix of 0's and 1's. Define

$g^{m,k}(C) = 1$ iff there are at least two rows with odd numbers of 1', or equivalently,

$$g^{m,k}(A) = T_2^m(c_{11} \oplus \dots \oplus c_{1k}, c_{21} \oplus \dots + \oplus c_{2k}, \dots, c_{m1} \oplus \dots$$
$$\dots \oplus c_{mk}).$$

Proposition 3.3 For any base Ω, $L_{\Omega}(g^{m,k}) \geq \delta_{\Omega} . m . k . \log \log m.$

Proof:

Let $\alpha(x_{11}, \dots, x_{mk}) \equiv g^{m,k}$ be a formula over Ω. Let $1 \leq j \leq k$, and consider the j-th column. Then

$$\alpha\{x_{1j}, x_{2j}, \dots, x_{mj}\} \equiv T_2^m(x_{1j}, \dots, x_{mj}),$$

thus it is of complexity at least

$\delta_\Omega \cdot m \cdot \log \log m.$

Hence there are at least so many occurrences of variables of the j-th column in α . Summing over all the columns I get the bound. □

Notice that I get a nonlinear bound even if k is "large" (say $k = n^{1-\varepsilon}$), when the theorem cannot be used directly. (The same bound can be proved for the function $f^{k,m}$ considered in [1], but they have a better bound for it).

§ 4 Asymptotical optimality of the bound

The following example arose during a discussion with S. Poljak.

I shall use ↑ for exponentiation, i.e. $m \uparrow n$ is m^n; following the usual convention I shall omit brackets in expressions like $m \uparrow (n \uparrow k)$.

Define by induction for $m, k \geq 1$:

$$\alpha_{1,m}(x_1, x_2, \dots, x_{m\uparrow 2}) = \left(\bigvee_{1}^{m} x_i\right) \oplus \left(\bigvee_{m+1}^{2m} x_i\right) \oplus \dots \oplus \left(\bigvee_{(m-1)m+1}^{m\uparrow 2} x_i\right);$$

$$\alpha_{k+1,m}(x_1, x_2, \dots, x_{m\uparrow 2\uparrow (k+1)}) = \alpha_{k,m}(\beta_1, \beta_2 \dots \beta_{m\uparrow 2\uparrow k}),$$

where for $i = 1, \dots, m\uparrow 2\uparrow k$

$$\beta_i = \alpha_{k,m}(x_{(i-1).m\uparrow 2\uparrow k+1}, x_{(i-1).m\uparrow 2\uparrow k+2} \dots x_{i.m\uparrow 2\uparrow k}).$$

Thus every variable x_i, $i \leq m\uparrow 2\uparrow k$ occurs in $\alpha_{k,m}$ exactly once.

Define, for $n = 2\uparrow 2\uparrow r$

$$\alpha_n = \bigwedge_{i=1}^{r} \alpha_{i,2\uparrow 2\uparrow (r-i)} .$$

Hence α_n has n variables, every variable occurs in it $r = \log \log n$ times. Therefore the complexity of α_n is $n.\log \log n$. The reader is recommended to try to visualize α_n as a labeled tree, see Fig. 2.

Lemma 4.1 For every $\vec{c} \in \{0,1\}^n$

$$\sum c_i = 1 \Rightarrow \alpha_n(\vec{c}) = 1,$$

$$\sum c_i = 3 \Rightarrow \alpha_n(\vec{c}) = 0.$$

Proof:

First it is easy to show using induction

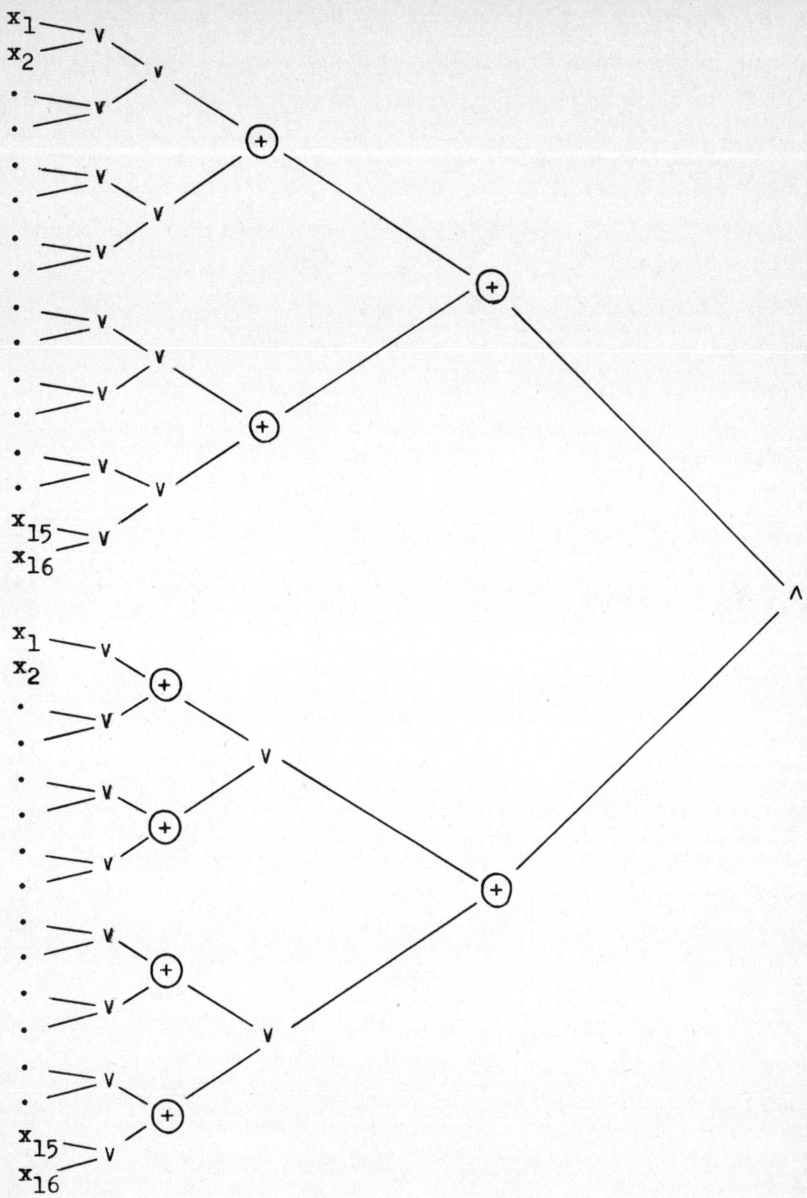

<u>Fig.2.</u> α_{16}

(1) $\sum c_i = 0 \Rightarrow \alpha_{k,m}(\vec{c}) = 0;$

(2) $\sum c_i = 1 \Rightarrow \alpha_{k,m}(\vec{c}) = 1.$

The first implication of the lemma follows from (2). The second one will be proved by induction over $r = \log\log n$. If $r = 1$, then $n = 4$ and

$$\alpha_4 = \alpha_{1,2} = (x_1 \vee x_2) \oplus (x_3 \vee x_4).$$

Thus the implication can be verified easily. Let $r > 1$ and suppose that the implication holds for α_ℓ, $\ell = 2\uparrow 2\uparrow(r-1)$.
Let $\vec{c} \in \{0,1\}^n$ be given, where

$c_{i_1} = c_{i_2} = c_{i_3} = 1,\ i_1 < i_2 < i_3,$

$c_i = 0$ otherwise.

Divide c into ℓ blocks of length ℓ, where the j-th block is

$$\vec{c}^{\,j} = (c_{j\ell+1}, c_{j\ell+2}, \ldots, c_{(j+1)\ell}).$$

Now I shall consider three cases.

1. $c_{i_1}, c_{i_2}, c_{i_3}$ belong to the same block, say the j-th block. By (1), (2), for every $1 < i \leq r$, $m = 2\uparrow 2\uparrow(r-i)$,

$$\alpha_{i,m}(\vec{c}) = \alpha_{i-1,m}(\alpha_{i-1,m}(\vec{c}^{\,1}), \alpha_{i-1,m}(\vec{c}^{\,2}), \ldots \alpha_{i-1,m}(\vec{c}^{\,\ell})) \equiv$$

$$\equiv \alpha_{i-1,m}(0,\ldots,0, \alpha_{i-1,m}(\vec{c}^{\,j}), 0,\ldots,0) \equiv \alpha_{i-1,m}(\vec{c}^{\,j}).$$

Hence

$$\alpha_n(\vec{c}) \equiv \alpha_{1,\ell}(\vec{c}) \wedge \alpha_\ell(\vec{c}^{\,j}),$$

which is equal to 0 by the induction assumption.

2. $c_{i_1}, c_{i_2}, c_{i_3}$ belong to different blocks, say to j_1-th, j_2-th, j_3-th ; $j_1 < j_2 < j_3$. By (1), (2), for every $1 < i \leq r$, $m = 2\uparrow 2\uparrow(r-i)$,

$$\alpha_{i,m}(\vec{c}) \equiv \alpha_{i-1,m}(0,\ldots,0, \alpha_{i-1,m}(\vec{c}^{\,j_1}), 0,\ldots,0, \alpha_{i-1,m}(\vec{c}^{\,j_2}),$$

$$0,\ldots,0, \alpha_{i-1,m}(\vec{c}^{\,j_3}), 0,\ldots,0) \equiv \alpha_{i-1,m}(\vec{d}),$$

where

$d_{j_1} = d_{j_2} = d_{j_3} = 1$; $d_j = 0$ otherwise.

Hence

$\alpha_n(\vec{c}) = \alpha_{1,\ell}(\vec{c}) \wedge \alpha_\ell(\vec{d})$,

which is equal to 0 by the induction hypothesis.

3. Exactly two of c_{i_1}, c_{i_2}, c_{i_3} belong to the same block. Then $\alpha_{1,\ell}(\vec{c}) \equiv 0$, hence $\alpha_n(\vec{c}) \equiv 0$. □

Theorem 4.2 Let $\Omega = \{\oplus, \vee, \wedge\}$, then for every $m \geq 1$, there exists $f:\{0,1\}^m \to \{0,1\}$ such that

$L_\Omega(f) \leq m.\lceil \log\log m \rceil$

and for no $1 \leq i < j < k \leq m$ and no b,

(3) $f|x_ix_jx_k \equiv b(x_i \oplus x_j \oplus x_k, x_i \vee x_j \vee x_k)$.

Proof:

Let m be given, let $r = \lceil \log\log m \rceil$, $n = 2\uparrow 2\uparrow r$, and let α_n be the formula above. Let

$f \equiv \alpha_n\{x_1,\ldots,x_m\}$.

Then the complexity of f is $\leq$ m.r. The righthand side of (3) gives the same value for (1,0,0) and (1,1,1), hence by Lemma 4.1 (3) cannot hold. □

In fact I have proved a little bit more: the bound in Corollary 3.1 is asymptotically best.

§ 5 Related results

1. A generalization of Hodes-Specker theorem to d-ary logic, $d \geq 2$, was considered by Vilfan [11] . Let $D = \{0,1,\ldots, d-1\}$ be the set of logical values. A proper chain $C(x_1,\ldots,x_n,c)$ is, in my notation, every expression

$\varphi(x_1, \varphi(x_2, \ldots \varphi(x_n, c] \ldots]]$,

or with the reverse order of variables, where $c \in D$ and

$\varphi(0, \varphi(x,y]] \equiv \varphi(x,y]$.

(For $d = 2$ the last condition is equivalent to " $\varphi(0,x] \equiv x$ or $\varphi(0,x]$ is constant".) The generalization says, roughly speaking, this: If the complexity of $f:D^n \to D$ is "small", then there exist

$x_{i_1},\dots,x_{i_r}$ such that

$$f \mid x_{i_1} \dots x_{i_r} \equiv b(C_1(x_{i_1},\dots,x_{i_r},c_1),\dots,C_\ell(x_{i_1},\dots,x_{i_r},c_\ell)),$$

where C_i are some proper chains. The actual bound, which is instead of "small", is again worse than $n \cdot \log^* n$. Vilfan derives Hodes-Specker theorem from his generalization only for the base of all at most binary connectives, but this restriction is unessential. Because of the similarity with Theorem 1.3, I believe that Vilfan's theorem can be derived from it with a bound assymptotically equal to $n.\log \log \log n$ for fixed r and d.

3. A related theorem was proved by Fischer, Meyer and Paterson [1]. Their theorem can be expressed as follows: There exists a constant $\eta > 0$ such that if the complexity of $f: \{0,1\}^n \longrightarrow \{0,1\}$ over the base of all binary connectives is

(1) $\eta .n.(\log n - \log r)$,

then there exists a central restriction $f|^A x_{i_1} \dots x_{i_r}$ (i.e. the number of substituted 1's equals to the number of substituted 0's possibly + 1), such that

$$f|^A \; x_{i_1} \dots x_{i_r} \equiv b(x_{i_1} \oplus \dots \oplus x_{i_r}),$$

for some unary Boolean function b. If we replaced (1) by a bound $\mathcal{E} \cdot n (\log \log n - \log r)$, then such a theorem would be a consequence of Theorem 2.4. For comparison consider the symmetric threshould functions. The theorem of Fischer et al. gives better bounds for T^n_k, where $\log n \ll k \ll n - \log n$, for $k \approx \log n$ or $k \approx n - \log n$ both theorems give assymptitically the same bound, and for the other values my theorem is better.

4. Still it is an open problem what is the complexity in a general base of such a simple function as T^n_2. By the results of Kričevskij and others the complexity of T^n_2 is asymptotically $n.\log n$ for the base $\Omega = \{0,1,\neg,\wedge,\vee\}$, see also [10]. This is also the lowest known upper bound for any base. In this paper I have approached to this bound from below. However observe that formula α_n of § 4 can be easily transformed into a formula β_n such that for every $\vec{c}$, if $\sum c_i \leq 4$, then

$$\beta_n(\vec{c}) = T^n_2(\vec{c}),$$

and the complexity of β_n is assymptotically n.log log n. This can be generalized further, namely 4 can be replaced by an arbitrary constant.

References

[1] M.J. Fischer, A.R. Meyer, M.S. Paterson: Ω(n log n) lower bounds on length of Boolean formulas, SIAM J. Comput. Vol. 11, No.3, 1982, 416-427.

[2] L. Hodes, E. Specker: Lengths of formulas and elimination of quantifiers I, in Contributions to Mathematical Logic, H.A. Schmidt, K. Schütte, H.-J.Thiele, eds., North-Holland, (1968), 175-188.

[3] V.M. Khrapčenko: On the complexity of the realization of the linear function in class of π-cirquits, Mat. Zametki 9,1 (1971), 35-40, (Russian).

[4] V.M. Khrapčenko: The complexity of realization of symmetrical functions by formulae, Mat. Zametki 11,1 (1972), 109-120, (Russian), English translation in Math. Notes of the Academy of Sciences of the USSR 11, (1972), 70-76.

[5] V.M. Khrapčenko: Complexity of realization of symmetric Boolean functions on finite basis, Problemy Kibernetiki 31, (1976), 231-234, (Russian).

[6] R.E. Kričevskij: A bound for the formula size complexity of a Boolean function, Diskretnyj Analiz I, (1963), 13-23, also in Dokl. Akad. Nauk SSSR, Vol. 151, No. 4, 803-806, (Russian), English translation in Sov. Phys.-Dokl., Vol. 8, No. 8, (1964), 770-772.

[7] L. Lovász: Combinatorial Problems and Exercises, Akadémiai Kiadó and North-Holland, 1979.

[8] E.I. Nečiporuk: A Boolean function, Dokl. Akad. Nauk SSSR, Vol. 169, No. 4, 765-766, (Russian), English translation in Sov. Math. -Dokl. Vol. 7, No. 4, (1966), 999-1000.

[9] P. Pudlák: Boolean complexity and Ramsey theorems, manuscript.

[10] J.E. Savage: The Complexity of Computing, John Wiley and Sons, (1976).

[11] B. Vilfan: Lower bounds for the size of expressions for certain functions in d-ary logic, Theoretical Comp. Sci. 2, (1976), 249-269.

Ingo Wegener

Proving Lower Bounds on the Monotone Complexity of Boolean Functions

FB-20 Informatik, Johann Wolfgang Goethe-Universität,
6000 Frankfurt a.M., Fed. Rep. of Germany

Abstract

The hardware of computers may be well modelled by Boolean networks computing Boolean functions $f:\{0,1\}^n \to \{0,1\}^m$. Though by counting arguments it is easy to show that nearly all Boolean functions may be computed only by networks of exponential size until now one is not able to prove nonlinear lower bounds for explicitly defined functions. Therefore one has restricted oneself to monotone networks where only $\wedge$-and $\vee$-gates are available. We repeat shortly the known methods for the proof of nonlinear lower bounds for the monotone network complexity of functions of n inputs and outputs. Afterwards we describe the method of using value functions, which was introduced by the author [13] to prove a bound of size $n^2/\log n$, until now the largest bound of this kind. For the Boolean convolution we present a new $\Omega(n^{3/2})$-lower bound proved recently by one of my students (Weiß [14]). There one combines the elimination method and some information flow arguments. Finally we discuss how one may combine the above mentioned methods in order to get perhaps better bounds for the Boolean convolution.

1. Introduction

We consider the computation of Boolean functions $f:\{0,1\}^n \to \{0,1\}^m$ by Boolean networks which model the hardware of computers. A Boolean network is a directed, acyclic graph with n inputs denoted by the variables $x_1,\dots,x_n$. Each other node called gate has indegree 2 where we distinguish the left and right input. The outdegree is not limited. Each gate is denoted by a basic function from $\Omega:=\{f:\{0,1\}^2 \to \{0,1\}\}$. We may now describe the functions computed at the nodes of the network. At the input x_i we compute the projection x_i. At each gate we apply the appropriate basic function to the two functions computed at the inputs. The network computes f if each coordinate function f_i is computed at some gate. The Boolean network complexity of f is the minimum number of gates in a Boolean network computing f.

Since the network complexity of Boolean functions is closely connected to their Turing machine complexity a proof of an exponential lower bound on the complexity of an NP-complete Boolean function is even a proof for NP$\neq$P. By counting arguments it is easy to show that most of the Boolean functions may be computed only by networks of exponential size. Nevertheless one knows a lot astonishingly small upper bounds on the complexity of many Boolean functions. But until now one is not able to prove good lower bounds for explicitly defined Boolean functions. There do not exist nonlinear lower bounds - even for functions of n inputs and n outputs.

Therefore one has restricted oneself to the study of networks where one is not allowed to use all basic functions. The most important model is the monotone one where we may use only $\wedge$-and $\vee$-gates. By these monotone Boolean networks one can compute exactly all monotone Boolean functions. For one-output-functions one nevertheless does not know any nonlinear lower bound. But one knows several nonlinear bounds on the monotone complexity of Boolean functions $f_n:\{0,1\}^n \rightarrow \{0,1\}^n$. Sometimes one has considered important functions and sometimes one has considered functions which have been constructed for the only reason that one was able to prove large lower bounds for them. The second way too can be justified since improving the "world record" for the largest lower bound for explicitly defined Boolean functions was always accompanied by the introduction of new methods.

In Chapter 2 we list the known methods for proving lower bounds on the monotone network complexity. These methods are often easy to explain but hard to apply. In Chapter 3 we shortly state the current world record and especially we introduce the method of using value functions. Afterwards we consider in Chapter 4 the important Boolean convolution. The best known lower bound of size $n^{3/2}$ has been proved recently by one my students, Weiß, by using the elimination method and some information flow arguments. In Chapter 5 we discuss how these methods and the method of using value functions may be combined to improve the lower bound in direction to the supposed bound of size n^2.

We finish this introduction with some definitions and notations. The conjunction is also called product and we suppress the "$\wedge$", i.e. $xy=x\wedge y$. A product of variables is also called monom where the empty monom is the constant 1. Such a monom t is called an implicant of the monotone function g if $t\leq g$, i.e. if for any $a \in \{0,1\}^n$ the fact that $t(a)=1$ implies that $g(a)=1$. An implicant t is even a prime implicant if no proper shortening t' of t is an implicant too. It is a well known fact that each monotone function f may be described as the disjunction of its prime implicants which is called the monotone disjunctive normal form of f. By I(f) resp. PI(f) we denote the set of all implicants resp. prime implicants of f.

2. Methods for the proof of lower bounds

Here we present the methods for the proof of large lower bounds on the monotone complexity of Boolean functions. We omit the new methods which have been developped for the results we describe in the following chapters.

Using graph-theoretical arguments only one can obtain bounds of size n logn (Pippenger/Valiant [8]). Then one tried to consider Boolean sums which are those functions one may compute using $\vee$-gates only. One tried to estimate the use of $\wedge$-gates. Nechiporuk [5], Lamagna/Savage [2] and Tarjan [10] got $\Omega(n^{3/2})$ lower bounds by showing that $\wedge$-gates cannot help for (1,1)-disjoint Boolean sums. In order to apply successfully replacement rules (which will be explained later) Wegener [12] assumed that certain functions are given for free. He could estimate the use of $\wedge$-gates for (1,k)-disjoint Boolean sums. Pippenger [7] and Mehlhorn [3] generalized this result to (h,k)-disjoint Boolean sums yielding $\Omega(n^{5/3})$ lower bounds.

Replacement rules are assertions that networks for f computing s at some gate may be changed in the following way without changing their outputs. We may replace the gate computing s by some network computing some definite s'. If s' is a constant or a variable the given network was not optimal. Thus we obtain by this method results on the structure of optimal networks. The most general replacement rules are given by Mehlhorn/Galil [4].

Another very common method is the elimination method. One tries to fix some variables to constants in order to fulfil two goals. One likes to eliminate a large number of gates, for example since one of its inputs has changed to a constant, and one likes to obtain afterwards a function of the same structure such that one may proceed by induction. Pratt [9], Paterson [6] and Mehlhorn/Galil [4] used the elimination method together with replacement rules in order to compute exactly the monotone complexity of the Boolean matrix product.

In order to apply the replacement rules also in situations where s' is not a constant or a variable Wegener [11],[12], [13], was the first to assume that certain functions and gates are given for free. The functions and gates given for free are chosen in such a way that s' may be computed from s without any new gate. But if one allows that too many functions are given for free it would become easy to compute f. Thus one has to choose very carefully the set of functions which one likes to be given for free. By this method Wegener got the above mentioned results on Boolean sums. In the next chapter we show how these methods have been combined with the method of using value functions to obtain the announced $n^2/\log n$ lower bound. We describe this method a bit more detailed since we need this description for our discussion in Chapter 5.

3. An $\Omega(n^2/\log n)$ lower bound for the generalized matrix product

We investigate the function $f_{MN}^m : \{0,1\}^{mMN} \to \{0,1\}^{M^m}$ which may be called the "direct" product of m matrices of M rows and N columns each. For the output $y_{h_1 \ldots h_m}$ $(1 \leq h_1, \ldots, h_m \leq M)$ we have to decide whether the rows h_1 of matrix $1, \ldots, h_m$ of matrix m altogether have a common 1. Thus

$$y_{h_1 \ldots h_m} := \bigvee_{1 \leq \ell \leq N} x^1_{h_1 \ell} \ldots x^m_{h_m \ell}$$

where the upper index of the variables denotes the matrix to which the variable belongs. We always assume $m \geq 2$ and $M \geq 2$. For m=2 the function f_{MN}^2 computes the ordinary Boolean matrix product of the first matrix and the transposed of the second matrix.

Each output of f_{MN}^m obviously consists of N prime implicants, altogether we have NM^m prime implicants. It is easy to show that f_{MN}^m may be computed with less than $3NM^m$ gates. On the other hand we believe that we need at least one $\wedge$-gate for the computation of each prime implicant, that means we claim the lower bound NM^m for the monotone complexity of f_{MN}^m, denoted by $C_{\wedge,\vee}(f_{MN}^m)$. Indeed we were able to prove a lower bound of $\frac{1}{2} NM^m$ [13].

Theorem 1: $\frac{1}{2} NM^m \leq C_{\wedge,\vee}(f_{MN}^m) \leq 3NM^m$.

Corollary: If $m := \lfloor \log n \rfloor$, $M := 2$, $N := \lfloor n/2\log n \rfloor$, f_{MN}^m has at most n variables and n outputs. Its monotone complexity is of size $n^2/\log n$.

The proof of Theorem 1 uses most of the methods explained in Chapter 2 and beyond that for the first time value functions.

Since we like to prove the existence of one $\wedge$-gate for each prime implicant we may assume that all $\vee$-gates and all monoms of less than m variables are given for free. This enables us to make use of the replacement rules. We may change each monotone network for f_{MN}^m without increasing its cost to another monotone network for f^m_{MN} with the following additional property. Each input and each output of an $\wedge$-gate has at most one prime implicant or one proper shortening of a prime implicant of type ℓ. By a prime implicant of type ℓ we mean a prime implicant $x^1_{h_1\ell}\ldots x^m_{h_m\ell}$. Thus all prime implicants of the same type are computed for the first time at different $\wedge$-gates. We have proved the existence of M^m different $\wedge$-gates where the prime implicants of type N become constructed.

In [11] we used the elimination method. We proved the existence of some k such that the following holds. If we fix the last column of the kth matrix to 0 and all other last columns to 1, then we compute $f^m_{M,N-1}$ (this is always correct), and we may eliminate at least $\frac{2}{m} M^m$ gates (proved by the pigeon-hole-principle). Proceeding by induction we obtain the lower bound $\frac{2}{m} NM^m$ yielding only an $n^2/\log^2 n$ lower bound. By the elimination method we cannot eliminate all the M^m gates whose existence has been proved.

Indeed we may create at $\wedge$-gates prime implicants of different types. Taking this fact into account we have to admit that we cannot associate with each gate a definite prime implicant. Gates may be useful for several prime implicants. Here we like to mention that all methods explained until now have local character only. But it seems to be obvious that one needs global arguments to get strong results on the complexity of hard Boolean functions. The method of value functions tries to estimate the value of each gate for each prime implicant. The value function $v(G,t)$ measures the value of the $\wedge$-gate G for the prime implicant t of the function f computed by the network. Since each gate should be counted only once a value function must fulfil for each G

$$\sum_{t \in PI(f)} v(G,t) \leq 1.$$

For an optimal network $\sum_G \sum_{t \in PI(f)} v(G,t)$ is a lower bound on the monotone complexity of f. On the other hand we believe that we need for each prime implicant a certain value to compute it. Thus we like to prove that for some function b on $PI(f)$, which should be as large as possible, we can prove for all $t \in PI(f)$

$$\sum_G v(G,t) \geq b(t).$$

That means that $\sum_{t \in PI(f)} b(t)$ is a lower bound on the monotone complexity of f. In order to define a good value function one needs good intuition.

For the generalized matrix product we have $b \equiv \frac{1}{2}$ and thus

$$\sum_{t \in PI(f^m_{MN})} b(t) = \frac{1}{2} NM^m .$$

We define the value function v as the sum of v_ℓ and v_r, two value functions referring to the left and right input of the gate. For $v_\ell(G,\cdot)$ (v_r is defined in the same way) we count the number of prime implicants of f^m_{MN} which are computed at the output of G but not at the left input of G. If this number is q_ℓ all these prime implicants get value $(2q_\ell)^{-1}$ while all other prime implicants get value 0.
Thus by definition

$$\sum_{t \in PI(f^m_{MN})} v_\ell(G,t) \leq \frac{1}{2} \text{ and } \sum_{t \in PI(f^m_{MN})} v(G,t) \leq 1.$$

By our result on the structure of the reconstructed networks we distribute at each gate positive value only to one prime implicant of a definite type. If q_ℓ and q_r are large, each prime implicant gets only a small value. Why can we hope that we are able to prove , that each prime implicant gathers on the whole the value $\frac{1}{2}$? If q_ℓ and q_r are large we do not compute only the desired prime implicants but by the law of distributivity also some "ugly" terms, which are monoms consisting of variables of at least two types. In order to get rid of these ugly terms the prime implicants have to gather value elsewhere. The proof that these heuristic ideas are correct should not be given here in detail (see [13]), since we here have to learn only how this method works. We believe that this method will turn out to be useful in many situations because of its global character.

4. An $\Omega(n^{3/2})$ lower bound for the Boolean convolution

Definition 1: The Boolean convolution is the Boolean function $f^{c,n}:\{0,1\}^{2n} \to \{0,1\}^{2n-1}$ where

$$f^{c,n}_k(x_1,\dots,x_n,y_1,\dots,y_n) := \bigvee_{i+j=k} x_i y_j \quad \text{for } 2 \leq k \leq 2n.$$

By graph-theoretical arguments Pippenger/Valiant [8] proved an $\Omega(n \log n)$ lower bound for the Boolean convolution.
Blum [1] could prove an $\Omega(n^{4/3})$ lower bound by a clever and sophisticated application of the elimination method. Recently one of my students (Weiß [14]) proved the now largest lower bound for the Boolean convolution, which is of size $n^{3/2}$.
Here we like to mention that one believes that the monotone disjunctive normal form is optimal for the Boolean convolution, i.e. $C_{\wedge,\vee}(f^{c,n}) \overset{!}{=} 2n^2-2n+1$.

Weiß's proof is a combination of the elimination method and information flow arguments. We like to fix the variables $x_1,\dots,x_n$ one after another to 0. In order to proceed by induction we have to consider also those functions which are produced from the Boolean convolution by fixing some x-variables to 0. Thus we define the set of semidisjoint bilinear functions.

<u>Definition 2</u>: A Boolean function f on some x-variables and some y-variables is called to be semidisjoint bilinear iff

(i) all prime implicants consist of exactly one x-and one y-variable, and

(ii) no two outputs have a common prime implicant, and

(iii) each output has for each variable at most one prime implicant.

Obviously the Boolean convolution and its subfunctions, where some variables are set to 0, are semidisjoint bilinear.

<u>Theorem 2</u>: If $f:\{0,1\}^{n+n'} \to \{0,1\}^N$ is semidisjoint bilinear on $\{x_1,\dots,x_n,y_1,\dots,y_{n'}\}$ and if r_i is the number of outputs with a prime implicant containing x_i , then

$$C_{\wedge,\vee}(f) \geq \sum_{1\leq i\leq n} r_i^{1/2}.$$

Since for the Boolean convolution $r_1=\dots=r_n=n$ we get immediately the following corollary

<u>Corollary</u>: $C_{\wedge,\vee}(f^{c,n}) \geq n^{3/2}$.

<u>Proof of Theorem 2</u>: It is sufficient to prove

$$C_{\wedge,\vee}(f) \geq C_{\wedge,\vee}(f_{|x_1=0}) + r_1^{1/2} ,$$

since $f_{|x_1=0}$ is again semidisjoint bilinear and since $r_2,\dots,r_n$ are the same for f and $f_{|x_1=0}$.

Let s_1 be the number of outputs with exactly one prime implicant, where this prime implicant contains x_1. By fixing x_1 to 0 we may eliminate the s_1 $\wedge$-gates where these outputs are computed. Besides these s_1 $\wedge$-gates we prove that we eliminate at least $(r_1-s_1)^{1/2}$ $\vee$-gates. Altogether the number of eliminated gates is at least

$$s_1 + (r_1-s_1)^{1/2} \geq s_1^{1/2} + (r_1-s_1)^{1/2} \geq r_1^{1/2}.$$

We consider the following set G* of $\vee$-gates for elimination. An $\vee$-gate G belongs to G* iff it is the first $\vee$-gate on a path from the input x_1 to an output depending on x_1, where for some $i \neq 1$ the monom $x_i y_j$ for some j is an implicant of the function computed at G. These are the first $\vee$-gates where something useful besides the terms containing x_1 is computed. We show that G^* is not too small and that we may eliminate all gates of G^* by fixing x_1 to 0.

Claim 1: Each path from x_1 to an output f_k, where f_k depends on x_1 and has at least two prime implicants, contains an element of G^*.

Proof: Let $G_1,\ldots,G_m$ be such a path and let us assume that all these gates do not belong to G^*. Since $f_{k|x_1=0}$ is not constant, this path has to contain $\vee$-gates. Let G_ℓ be an $\vee$-gate on this path. Since all gates on the path do not belong to G^* by definition $x_i y_j$ $(i\neq 1)$ is not an implicant of $\mathrm{res}(G_\ell)$, the function computed at G_ℓ. Thus all prime implicants of $\mathrm{res}(G_\ell)$ either contain x_1 or at least two x-variables or at least two y-variables. Fixing $x_1=0$ $\mathrm{res}(G_\ell)$ is changed to a function where no prime implicant is a prime implicant or a shortening of a prime implicant of some output. By the first replacement rule of Mehlhorn/Galil [4] the function can be replaced by 0. This is also valid for the last $\vee$-gate on this path and $f_{k|x_1=0}$ would be equal to 0 in contradiction to our assumptions. Thus one of the $\vee$-gates on this path, say G_ℓ , is an element of G^*.

By the proof of Claim 1 we can conclude that fixing $x_1=0$ replaces the input of the path and $\mathrm{res}(G_1),\ldots,\mathrm{res}(G_{\ell-1})$ to 0 either directly or by an application of the replacement rule. Thus one input of G_ℓ is replaced by 0 and we have proved

Claim 2: After fixing $x_1=0$ we may eliminate all gates of G^*.

At last we like to prove

Claim 3: $m:=|G^*| \geq (r_1-s_1)^{1/2}$

Proof: Let us assume $G^*=\{G_1,\ldots,G_m\}$. By Claim 1 all paths from x_1 to outputs, which depend on x_1 and have at least two prime implicants, contain an element of G^*. Thus G^* is a bottleneck where all information concerning x_1 has to flow through. We show that this bottleneck cannot be too tight. We look for variables such that fixing these variables but not x_1 to 1 replaces $\mathrm{res}(G_1),\ldots,\mathrm{res}(G_m)$ by 1. Afterwards no information can flow from x_1 to the outputs having more than one prime implicant. But in order to make these outputs independent from x_1 the number of variables fixed to 1 must be large.

Now we come to the details. By definition of G^* there exist $i_1,\ldots,i_m\neq 1$ and $j_1,\ldots,j_m$ such that $x_{i_\ell} y_{j_\ell} \in I(\mathrm{res}(G_\ell))$. We fix $x_{i_1}=\ldots=x_{i_m}=y_{j_1}=\ldots=y_{j_m}=1$. Obviously $\mathrm{res}(G_1),\ldots,\mathrm{res}(G_m)$ are replaced by 1 and no output having more than one prime implicant can depend on x_1. We consider those r_1-s_1 outputs depending on x_1 and having at least two prime implicants. Before we have fixed the variables each of these outputs, for example f_k, had a prime implicant x_1y_p for some p. The output is now changed to f'_k . Since f'_k is independent of x_1 some shortening of x_1y_p not containing x_1 is prime implicant of f'_k . Thus $y_p \in PI(f'_k)$ or $f'_k=1$.

Since x_1y_p was by Definition 2 the only prime implicant of f_k containing y_p it is impossible that $y_p \in PI(f'_k)$. Thus $f'_k=1$ and both variables of some prime implicant of f_k have been fixed to 1. We have fixed at most m x-variables and at most m y-variables to 1. We can form at most m^2 different pairs out of these variables. Since by Definition 2 different outputs have only different prime implicants at most m^2 outputs are replaced by the constant 1. But on the other hand we proved that all the considered r_1-s_1 outputs are replaced by 1. Thus we can conclude that

$m^2 \geq r_1-s_1$ which implies that $m \geq (r_1-s_1)^{1/2}$.

5. How to improve the bound for the Boolean convolution

We have seen that the proof of Chapter 4 combines two well known concepts namely the elimination method and replacement rules with information flow arguments. The bound we get for the Boolean convolution is of size $n^{3/2}$ while we still believe that the monotone complexity of this function is indeed of size n^2. Furthermore by considering those networks we believe to be optimal we see that their bottleneck indeed has size n-1. If we could prove that this must be the case we would obtain the desired quadratic lower bound. That means we should improve Claim 3. But it is sufficient to fix $n^{1/2}$ x-variables and $n^{1/2}$ y-variables in order to replace all outputs depending on x_1 and having at least two prime implicants by 1. We may choose for $0 \leq \ell \leq n^{1/2}-1$ the variables $x_{\ell+2}$ and $y_{\ell n^{1/2}+1}$.

But this argument is rather crude. We have only distinguished functions depending on x_1 and constant functions. Especially we did not consider the whole structure of semidisjoint bilinear functions or only of the Boolean convolution and its subfunctions.

We know something about the structure of the functions computed at the bottleneck and all about the output functions. The question is whether the outputs are computable from the functions at the bottleneck and the variables $x_2,\ldots,x_n$ and $y_1,\ldots,y_n$. We may change the functions at the bottleneck in the following way. Let b* be such a bottleneck function. At least one input depends on x_1 but b* must not depend on x_1. If b* does not depend on x_1 we may replace b* by 0 since we may compute b* from the other variables. In the other case we may assume by the first replacement rule of Mehlhorn/Galil [4] that all prime implicants containing x_1 have besides x_1 at most one y-variable. This part of b* is called b_1 and may be written as $b_1=x_1 \wedge b_1'$ where b_1' is a (perhaps empty) Boolean sum of y-variables.

Furthermore we know that $x_i y_j \in I(b)$ for some $i \neq 1$ and j. We claim that we may replace b* by $b:=b_1 \vee b_2$ where $b_2=x_i y_j$. This claim has to be explained carefully. We do not claim that after replacing the given bottleneck function b* by b the network still computes the Boolean convolution. But we claim that after this replacement the Boolean convolution is computable from the set of bottleneck functions and all variables without x_1. This claim is easy to show. The change of the prime implicants containing x_1 is justified by the replacement rule cited above. Let b_3 be the function consisting of all prime implicants of b* not containing x_1. b_3 may be computed because we only excluded x_1. Now it is easy to see that by the cited replacement rule b* may be replaced by $b \vee b_3$. All prime implicants of $b \vee b_3$ are prime implicants of b* while the prime implicants of b* are either prime implicants of $b \vee b_3$ or they are superfluous. Thus replacing b* by $b \vee b_3$ we again compute the Boolean convolution.

In order to determine whether we may compute with a given bottleneck size the Boolean convolution the best method seems to be to apply the method of value functions. Until now we got only examples that a bottleneck of size $n^{1/2}$ is not large enough. As yet we do not know how much the $n^{3/2}$ lower bound can be improved. In the best case we would obtain a new "world record" namely a quadratic bound. In any case we learn about the structure of networks for the Boolean convolution and we learn how to combine the most important methods for the proof of lower bounds in network complexity theory.

6. References

[01] N.Blum: An $\Omega(n^{4/3})$ lower bound on the monotone network complexity of the n-th degree convolution, 22-th Symp. on Foundations of Comp. Science (1981) 101-108

[02] E.A.Lamagna/J.E.Savage: Combinational complexity of some monotone functions, Proc. 15th SWAT Conference, New Orleans (1974) 140-144

[03] K.Mehlhorn: Some remarks on Boolean sums, Acta Informatica 12 (1979), 371-375

[04] K.Mehlhorn/Z.Galil: Monotone switching circuits and Boolean matrix product, Computing 16 (1976) 99-111

[05] E.I.Nechiporuk: On a Boolean matrix, Systems Res.Theory 21 (1971) 236-239

[06] M.S.Paterson: Complexity of monotone networks for Boolean matrix product, Theoret. Comput. Sci. 1 (1975) 13-20

[07] N.Pippenger: On another Boolean matrix, IBM Research Report 6914 (1977)

[08] N.Pippenger/L.G.Valiant: Shifting graphs and their applications, J.ACM 23 (1976) 423-432

[09] V.R.Pratt: The power of negative thinking in multiplying Boolean matrices, SIAM J.Comput. 4 (1975) 326-330

[10] R.E.Tarjan: Complexity of monotone networks for computing conjunctions, Ann. Discrete Math. 2 (1978) 121-133

[11] I.Wegener: Switching functions whose monotone complexity is nearly quadratic, Theoret. Comput. Sci. 9 (1979) 83-97

[12] I.Wegener: A new lower bound on the monotone network complexity of Boolean sums, Acta Informat. 13 (1980) 109-114

[13] I.Wegener: Boolean functions whose monotone complexity is of size $n^2/\log n$, Theoret. Comput. Sci. 21 (1982) 213-224

[14] J.Weiß: Eine $n^{3/2}$-untere Schranke für die Boolesche Konvolution Diplomarbeit, Univ. Bielefeld (1982)